Marine and Freshwater Botany

AF843123

Biology of Polar Benthic Algae

Edited by Christian Wiencke

DE GRUYTER

Editor
Prof. Christian Wiencke
Department Seaweed Biology
Section Functional Ecology
Alfred Wegener Institute
for Polar and Marine Research
Am Handelshafen 12
27570 Bremerhaven
Germany
christian.wiencke@awi.de

This book has 36 figures and 22 tables.

Front cover image
The cover shows the endemic Antarctic brown macroalga *Cystosphaera jacquinotii* (Montagne) Skottsberg. *Cystosphaera jacquinotii* is the only member of the Fucales in the Antarctic. Its distinctive, elegant fronds reach a length of 3 m and have been found down to 50 m depth. The photo was taken in Potter Cove, King George Island, Antarctica (© Alfred Wegener Institute, Bremerhaven, Germany).

ISBN 978-3-11-022970-7
eISBN 978-3-11-022971-4

Library of Congress Cataloging-in-Publication Data
Biology of polar benthic algae / edited by Christian Wiencke.
p. cm. – (Marine and freshwater botany)
Includes bibliographical references.
ISBN 978-3-11-022970-7
1. Marine algae–Polar regions. 2. Benthic plants–Polar regions. I. Wiencke, Christian.
QK579.B56 2010
579.8′17760911–dc22 2010037549

Bibliographic information published by the Deutsche Nationalbibliothek
The Deutsche Nationalbibliothek lists this publication in the Deutsche Nationalbibliografie; detailed bibliographic data are available in the Internet at http://dnb.d-nb.de.

© 2011 by Walter de Gruyter GmbH & Co. KG, Berlin/New York

Printing and Binding: Hubert & Co. GmbH & Co. KG, Göttingen
Printed in Germany
www.degruyter.com

The citation of registered names, trade names, trade marks, etc. in this work does not imply, even in the absence of a specific statement, that such names are exempt from laws and regulations protecting trade marks etc. and therefore free for general use.

To my wife Christine
and our children Antje, Friederike and Jan-Christian

Preface

This book on the *Biology of Polar Benthic Algae* is the outcome of investigations over about three decades in the Arctic and in the Antarctic carried out by scientists from all over the world. The main aim of the book was to synthesise the present state of knowledge and to develop perspectives as basis for future research. A preliminary version has been published already as a Special Issue of Botanica Marina (Vol. 52, Issue 6; 2009), edited by myself and Margaret N. Clayton (Monash University, Victoria, Australia).

I would like to express my sincere thanks to my wife Christine for her ever-lasting encouragement, inspiration and patience throughout the years. Thanks go also to my collaborators, students and all other authors of this book for their contribution. Last not least I would like to thank my institution, the Alfred Wegener Institute for Polar and Marine Research, Bremerhaven, for the strong support I received and the publisher for the opportunity to publish this book.

Bremerhaven, October 2010 *Christian Wiencke*

List of contributing authors

Adil Al-Handal
Department of Marine Ecology
University of Gothenburg
P.O. Box 461
405 30 Göteborg, Sweden
e-mail: adil.yousif@marecol.gu.se

Charles D. Amsler
Department of Biology
University of Alabama at Birmingham
Birmingham
AL 35294-1170, USA
e-mail: amsler@uab.edu

Margaret O. Amsler
Department of Biology
University of Alabama at Birmingham
Birmingham
AL 35294-1170, USA
e-mail: mamsler@uab.edu

Bill J. Baker
Department of Chemistry
University of South Florida
Tampa
Florida 33620, USA
e-mail: bjbaker@cas.usf.edu

Inka Bartsch
Department Seaweed Biology
Section Functional Ecology
Alfred Wegener Institute for
Polar and Marine Research
Am Handelshafen 12
27570 Bremerhaven, Germany
e-mail: inka.bartsch@awi.de

Susanne Becker
Department of Marine Botany
University of Bremen
Leobener Strasse, NW2
28359 Bremen, Germany
e-mail: sbecker@ipoe.uni-kiel.de

Kai Bischof
Department of Marine Botany
University of Bremen
Leobener Strasse, NW2
28359 Bremen, Germany
e-mail: kbischof@uni-bremen.de

Gabriela Laura Campana
Departamento de Biología Costera
Instituto Antártico Argentino
Cerrito 1248 (1010AAZ)
Buenos Aires, Argentina
e-mail: gcampana@dna.gov.ar
and
CONICET
Rivadavia 1917 (1033AAJ)
Buenos Aires, Argentina
and
Departamento de Ciencias Básicas (PIEA)
Universidad Nacional de Luján
Rutas 5 y 7 (6700), Luján
Buenos Aires, Argentina

Margaret N. Clayton
School of Biological Sciences
Monash University
Victoria 3800, Australia
e-mail: margaret.clayton@ monash.edu

Ken Dunton
Marine Science Institute
University of Texas at Austin
750 Channel View Drive
Port Aransas
TX 78373, USA
e-mail: ken.dunton@mail.utexas.edu

Suzanne Fredericq
Department of Biology
University of Louisiana at Lafayette
Lafayette
LA 70504-2451, USA
e-mail: slf9209@louisiana.edu

Jana Fredersdorf
Institute for Polar Ecology
University of Kiel
Wischhofstrasse 1–3
24148 Kiel, Germany
and
Institute of Marine Botany
University of Bremen
Leobener Strasse
28359 Bremen, Germany
e-mail: jafre@gmx.de

Anna Fricke
Department Seaweed Biology
Section Functional Ecology
Alfred Wegener Institute
for Polar and Marine Research
Am Handelshafen 12
27570 Bremerhaven, Germany
and
Leibniz Center for Tropical Marine Ecology
Fahrenheitstr. 6
28359 Bremen, Germany
e-mail: anna.fricke@zmt-bremen.de
and
Bremen International Graduate School
for Marine Sciences "Global Change
in the Marine Realm" (GLOMAR)
Bremen International Graduate School
for Marine Sciences
Leobener Str.
28359 Bremen, Germany

Ronnie N. Glud
Southern Danish University
Institute of Biology &
Nordic Center for Earth Evolution
(NordCEE)
e-mail: rnglud@biology.sdu.dk
and
Scottish Association for Marine Science
(SAMS)
Scottish Marine Institute
Oban
PA37 1QA, UK
and
Greenland Climate Research Centre (c/o
Greenland Institute of Natural Resources)
BOX 570
3900 Nuuk, Greenland

Iván Gómez
Instituto de Biología Marina
Universidad Austral de Chile
Casilla 567
Valdivia, Chile
email: igomezo@uach.cl

Dieter Hanelt
Biozentrum Klein Flottbek
University of Hamburg
Ohnhorststr. 18
22609 Hamburg, Germany
e-mail:
Dieter.Hanelt@botanik.uni-hamburg.de

Max H. Hommersand
Department of Biology
University of North Carolina
Chapel Hill
NC 27599-3280, USA
e-mail: Hommersand@bio.unc.edu

Pirjo Huovinen
Instituto de Biología Marina
Universidad Austral de Chile
Casilla 567
Valdivia, Chile
e-mail: pirjo.huovinen@docentes.uach.cl
and
Centro i-mar
Universidad de Los Lagos
Casilla 557
Puerto Montt, Chile

Katrin Iken
School of Fisheries and Ocean Sciences
University of Alaska Fairbanks
P.O. Box 757220
Fairbanks, AK 99775-7220, USA
e-mail: iken@ims.uaf.edu

Ulf Karsten
Institute of Biological Sciences
Applied Ecology
University of Rostock
Albert-Einstein-Str. 3
18051 Rostock, Germany
e-mail: ulf.karsten@uni-rostock.de

Michael Kühl
Marine Biological Laboratory
Department of Biology
University of Copenhagen
Strandpromenaden 5
3000 Helsingør, Denmark
e-mail: mkuhl@bio.ku.dk

Thomas Laepple
Section Paleoclimate Dynamics
Alfred Wegener Institute
for Polar and Marine Research
Bussestrasse 24
27570 Bremerhaven, Germany
e-mail: thomas.laepple@awi.de

Peter Leopold
Institute of Biological Sciences
Applied Ecology
University of Rostock
Albert-Einstein-Str. 3
18051 Rostock, Germany
e-mail: peter.leopold@gmx.de

James B. McClintock
Department of Biology
University of Alabama at Birmingham
Birmingham
AL 35294-1170, USA
e-mail: mcclinto@uab.edu

Richard L. Moe
University and Jepson Herbaria
University of California
Berkeley
CA 94720-2465, USA
e-mail: rlmoe@berkeley.edu

Markus Molis
Department of Seaweed Biology
Section of Functional Ecology
Biologische Anstalt Helgoland
Alfred Wegener Institute
for Polar and Marine Research
Marine Station
Kurpromenade 201
27498 Helgoland, Germany
e-mail: markus.molis@awi.de

Ruth Müller
Department Seaweed Biology
Section Functional Ecology
Alfred Wegener Institute for
Polar and Marine Research
Am Handelshafen 12
27570 Bremerhaven, Germany
and
Section Adaptation and Climate
LOEWE Biodiversity and Climate Research
Centre
Georg-Voigt-Straße 16
60325 Frankfurt/Main, Germany
e-mail: RuthMueller@bio.uni-frankfurt.de

María Liliana Quartino
Departamento de Ciencias del MarInstituto
Antártico Argentino
Dirección Nacional del Antártico
Cerrito 1248, C1010AAZ
Buenos Aires, Argentina
e-mail: lquartino@dna.gov.ar
and
Museo Argentino de Ciencias Naturales
"B. Rivadavia"
Av. A. Gallardo 470 (C1405DJR)
Buenos Aires, Argentina

Ralf Rautenberger
Institute for Polar Ecology
Christian Albrechts University of Kiel
Wischhofstr. 1–3, Building 12
24148 Kiel, Germany

and
Department of Botany
University of Otago
PO Box 56
Dunedin, New Zealand
e-mail: ralf.rautenberger@botany.otago.ac.nz

Michael Y. Roleda
Institute for Polar Ecology
University of Kiel
Wischhofstrasse 1–3
24148 Kiel, Germany
and
Department of Botany
University of Otago
Dunedin 9054
New Zealand
e-mail: michael.roleda@botany.otago.ac.nz

Søren Rysgaard
Greenland Climate Research Centre (c/o Greenland Institute of Natural Resources)
Box 570
3900 Nuuk, Greenland
e-mail: rysgaard@natur.gl

Rhena Schumann
Institute of Biological Sciences
Applied Ecology
University of Rostock
Albert-Einstein-Str. 3
18051 Rostock, Germany
e-mail: rhena.schumann@uni-rostock.de

Franciska S. Steinhoff
Institute of Marine Botany
University of Bremen
Leobener Strasse
28359 Bremen, Germany
e-mail: franciska.steinhoff@gu.se

Bettina Walter
Department of Marine Botany
University of Bremen
Leobener Strasse, NW2
28359 Bremen, Germany
e-mail: bettiwalter@web.de

Christian Wiencke
Department Seaweed Biology
Section Functional Ecology
Alfred Wegener Institute
for Polar and Marine Research
Am Handelshafen 12
27570 Bremerhaven, Germany
e-mail: christian.wiencke@awi.de

Jana Woelfel
Institute of Biological Sciences
Applied Ecology
University of Rostock
Albert-Einstein-Str. 3
18051 Rostock, Germany
e-mail: jana.woelfel@uni-rostock.de

Angela Wulff
Department of Marine Ecology
University of Gothenburg
P.O. Box 461
405 30 Göteborg, Sweden
e-mail: angela.wulff@marecol.gu.se

Katharina Zacher
Department Seaweed Biology
Section Functional Ecology
Alfred Wegener Institute for Polar and Marine Research
Am Handelshafen 12
27570 Bremerhaven, Germany
e-mail: Katharina.Zacher@awi.de

Contents

1. Introduction: Biology of polar benthic algae
Christian Wiencke and Margaret N. Clayton 1

Environment, biogeography and biodiversity

2. The abiotic environment of polar marine benthic algae
Katharina Zacher, Ralf Rautenberger, Dieter Hanelt, Angela Wulff and Christian Wiencke 9
3. Biodiversity, biogeography and zonation of marine benthic micro- and macroalgae in the Arctic and Antarctic
Angela Wulff, Katrin Iken, María Liliana Quartino, Adil Al-Handal, Christian Wiencke and Margaret N. Clayton 23
4. Notes on the systematics and biogeographical relationships of Antarctic and sub-Antarctic Rhodophyta with descriptions of four new genera and five new species
Max H. Hommersand, Richard L. Moe, Charles D. Amsler and Suzanne Fredericq 53

Chemical ecology

5. Defenses of polar macroalgae against herbivores and biofoulers
Charles D. Amsler, Katrin Iken, James B. McClintock and Bill J. Baker 101
6. Field studies on deterrent properties of phlorotannins in Antarctic brown algae
Katrin Iken, Charles D. Amsler, Margaret O. Amsler, James B. McClintock and Bill J. Baker 121

Primary production and ecophysiology

7. Benthic microalgal production in the Arctic: applied methods and status of the current database
Ronnie N. Glud, Jana Woelfel, Ulf Karsten, Michael Kühl and Søren Rysgaard 141
8. Microphytobenthic biomass along gradients of physical conditions in Arctic Kongsfjorden, Svalbard
Jana Woelfel, Rhena Schumann, Peter Leopold, Christian Wiencke and Ulf Karsten 163
9. Phenology and seasonal physiological performance of polar seaweeds
Christian Wiencke, Iván Gómez and Ken Dunton 181
10. Light and temperature demands of marine benthic microalgae and seaweeds in polar regions
Iván Gómez, Angela Wulff, Michael Y. Roleda, Pirjo Huovinen, Ulf Karsten, María Liliana Quartino, Ken Dunton and Christian Wiencke 195
11. Freezing tolerance and photosynthetic performance of polar seaweeds at low temperatures
Susanne Becker, Bettina Walter and Kai Bischof 221

Polar benthic algae in a changing world

12. Impact of oceanic warming on the distribution of seaweeds in polar and cold-temperate waters
Ruth Müller, Inka Bartsch, Thomas Laepple and Christian Wiencke 237

13. Physiological responses of polar benthic algae to ultraviolet radiation
Ulf Karsten, Angela Wulff, Michael Y. Roleda, Ruth Müller, Franciska S. Steinhoff, Jana Fredersdorf and Christian Wiencke 271

14. Drivers of colonization and succession in polar benthic macro- and microalgal communities
Gabriela Laura Campana, Katharina Zacher, Anna Fricke, Markus Molis, Angela Wulff, Maria Liliana Quartino and Christian Wiencke 299

15. Conclusion and outlook: Future perspectives on the investigation of polar benthic algae
Christian Wiencke and Margaret N. Clayton 321

Acknowledgements 325

Index 329

1 Introduction
Biology of polar benthic algae

Christian Wiencke and Margaret N. Clayton

Motivation

The main motivation for this book and the special issue of *Botanica Marina* is to document the considerable progress that has been achieved during the last few decades in the biology of polar benthic marine algae. Following significant research outputs in the 1970s and early 1990s, the first review of the ecophysiology of polar phytoplankton, ice algae and seaweeds was published in 1995 by Kirst and Wiencke (1995). A subsequent review (Wiencke 1996) was centred on the investigation of Antarctic seaweeds. A synopsis on Antarctic seaweeds appeared in 2002 (Wiencke and Clayton 2002). More recently, Wiencke et al. (2007) summarised our knowledge of the life strategy, ecophysiology and ecology of seaweeds from the Arctic and Southern Oceans.

This short list indicates that many new results were obtained on the biology of marine benthic seaweeds from polar waters during those years. In contrast, benthic microalgae from polar regions received little attention until recently, and their biology has not yet been reviewed. Moreover, all reviews published so far have been centered around basic research, and no attempt has been made to include the impacts of global climate changes on marine benthic algae, which are frequently regarded as the most important primary producers in Arctic and Antarctic coastal waters. Hence, the aim of this book is to synthesise the current state of knowledge and to present an outlook on the biology of marine benthic micro- and macroalgae from both polar regions.

Following a summary of the environmental conditions in both polar regions, the focus of this book will be on biodiversity and biogeography of seaweeds and marine benthic microalgae. The recent progress in chemical ecology will be emphasised in a separate section before primary production and ecophysiology of polar benthic algae are discussed. The final section is devoted to the responses of polar benthic algae in a changing world. Reviews and original research articles are combined to summarise recent research and to give a brief overview of current investigations. It will become obvious, however, that many gaps still exist and these are addressed in most of the chapters and in an outlook at the end of this book.

Environment, biodiversity and biogeography

Reviews in this section deal with the state of knowledge of the abiotic environment inhabited by benthic algae (Chapter 2; Zacher et al. 2009), their biodiversity, biogeography and zonation (Chapter 3; Wulff et al. 2009) and include an original contribution by Hommersand et al. (2009; Chapter 4) describing several new taxa of Antarctic and sub-Antarctic red algae together with a much-needed appraisal of the systematics and biogeography of this highly diverse and ancient group.

The environment of polar marine benthic algae is characterised by a strong seasonality of the light conditions and low temperatures. Nutrient levels in the Southern Ocean are high throughout the year, whereas nutrients are depleted in Arctic waters during summer. These and other abiotic factors strongly affecting polar algae are addressed in the review by Zacher et al. (2009; Chapter 2). This chapter also includes a summary of the presently occurring and expected changes in the environmental conditions due to global climate changes.

The review by Wulff et al. (2009; Chapter 3) makes it clear that the systematics and biodiversity of the benthic microalgae of polar regions are poorly known and that further research is required before any conclusions can be drawn or any meaningful study of their biogeography is possible. In contrast, the more conspicuous and accessible polar macroalgae have been studied more intensively. The systematics of Arctic and Antarctic brown and green macroalgae and the commoner red algae are comparatively well known and it is well established that both macroalgal biodiversity and endemism are greater in Antarctica than in the Arctic and that this is explained by the contrasting biogeographical histories of the two regions. Experimental studies have shown that the local distribution or zonation of polar micro- and macroalgae is influenced by their responses to a range of environmental factors including their ability to resist and survive grazing.

Chapter 4 by Hommersand et al. (2009) contains a timely update on the knowledge of the red algal flora of Antarctica based on recent collections made by Moe, Fredericq, Amsler and co-workers and includes descriptions of four new genera and five new species. Phylogenetic analyses of *rbcL* sequences are used as a basis for discussing the hypothetical biogeographic relationships of Antarctic red algae. The authors conclude that the composition of the Antarctic red algal flora can be attributed to vicariance resulting from plate tectonics and sea floor spreading and to long distance dispersal. They argue that climatic change over geological time and changes in ocean currents also influenced the composition of the present-day Antarctic algal flora.

Chemical ecology

This section comprises two chapters by Amsler et al. (2009; Chapter 5) and Iken et al. (2009; Chapter 6) dealing with recent research on the biotic interactions of polar macroalgae with herbivores and with biofouling organisms. Amsler et al. (2009; Chapter 5) argue in their review that chemical defences of Antarctic macroalgae, in particular against important amphipod mesoherbivores, appear to be of greater significance than physical defense mechanisms, in contrast to Arctic species whose physical defences may provide greater protection against herbivory. In addition, phlorotannins from some species of Antarctic macroalgae are shown to have antifouling activity against diatoms. Iken et al. (2009; Chapter 6) report directly on field studies examining the effect of phlorotannins from species of brown algae as deterrents against three different herbivores, a diatom and bacteria. They found that in general deterrent properties were specific to the algal species and also to the grazers, diatoms and bacteria tested.

Primary production and ecophysiology

This section comprises five chapters, three reviews and two original contributions. Two (Chapter 7, 8; Glud et al. 2009, Woelfel et al. 2009) deal with benthic microalgae, two (Chapter 9, 11; Becker et al. 2009, Wiencke et al. 2009) with macroalgae and one with both (Chapter 10; Gómez et al. 2009). The review by Glud et al. (2009; Chapter 7) makes it clear that the database on primary production of polar benthic microalgae is limited, as it is for biodiversity and biogeography of this algal group (Wulff et al. 2009; Chapter 3). The available data demonstrate, however, that Arctic benthic microalgae contribute significantly to productivity in coastal areas down to 30 m depth and exceed that in the pelagic zone. For the coastal Arctic ecosystem as a whole, the authors estimate a benthic microalgal contribution of between 1.1 and 1.6×10^7 t C year^{-1}. This enormous figure is based on a robust comparative analysis of all available datasets taking into account the various applied methods.

The first data on microphytobenthic biomass in Kongsfjorden (Spitsbergen), a glacial fjord in the Arctic, are presented in the chapter by Woelfel et al. (2009; Chapter 8). On soft bottom substrata, chl-*a* varied depending on current/wave exposure and geographic location. The highest values were obtained close to the glaciers. Biomass doubled or tripled in shallow water between mid and late summer, but was comparatively stable during the season at greater depths. The data underline the importance of microalgae for marine benthic productivity in coastal waters of the Arctic, as also pointed out by Glud et al. (2009; Chapter 7).

The phenology and physiology of polar seaweeds is strongly attuned to the extreme seasonal changes in daylength. This fact is revealed very clearly in the review by Wiencke et al. (2009; Chapter 9). Polar season anticipators grow and reproduce predominantly during winter and early spring. This strategy makes necessary a strong interrelation between seasonal photosynthesis and the formation and remobilisation of storage carbohydrates. Indeed, photosynthetic rates in almost all species studied are highest in spring when the light conditions are good. In kelps and kelp-like species, the photoassimilates formed are stored in summer as reserve compounds, and in winter and early spring they are translocated to the meristem to fuel growth.

The light requirements for photosynthesis and growth of both polar macro- and microalgae are very low, allowing the algae to grow down to considerable depths (Chapter 10; Gómez et al. 2009). In seaweeds, the stages characterised by most efficient light use are microscopic, i.e., their spores and microscopic gametophytes. Similarly, benthic diatoms are strongly shade-adapted. In this context, it is not surprising that benthic diatoms and few-celled microscopic stages of seaweeds are able to withstand long periods of darkness. Antarctic seaweeds and – as far as we know – benthic microalgae from the Antarctic are adapted to low temperatures to a greater extent than species from the Arctic. These characteristics are based on special physiological and metabolic properties that allow benthic marine algae from both polar regions to make considerable contributions to coastal primary productivity, as pointed out above for benthic microalgae.

A special focus in the investigation of polar seaweeds is seen in studies on the intrinsic freezing tolerance and photosynthetic performance of benthic algae at low temperatures. A pilot study by Becker et al. (2009; Chapter 11) of two eulittoral/upper sublittoral seaweeds from the Arctic and Antarctic shows that photosynthesis of Arctic

Fucus distichus L. is remarkably insensitive to freezing. On the other hand, low water temperatures in combination with high irradiance severely affected photosynthesis of *Palmaria decipiens* (Reinsch) R.W. Ricker from Antarctica, prompting the need for more intensive studies on the interactive effects of various factors.

Polar benthic algae in a changing world

Global warming due to the greenhouse effect and enhanced UV radiation due to stratospheric ozone depletion are major threats to marine life, especially in polar regions. The original contribution by Müller et al. (2009; Chapter 12) focuses on the effects of global warming on the geographic distribution of cold-water key structural seaweeds. For the first time, modelled changes of seawater temperatures in the 21st century have been used as basis for the prediction of changes in seaweed distribution. Fourteen species were studied and the result was that under the moderate scenario of temperature change, North Atlantic polar to cold-temperate seaweeds will extend their distribution into the high Arctic, but will retreat along the coasts of the northeastern Atlantic. In contrast, the southern hemisphere species studied will presumably not significantly change their geographic distribution. Several cold-temperate regions were identified where seaweed composition and abundance will certainly alter under the scenario tested. These changes will provoke substantial and cascading effects for biodiversity and functioning of seaweed-dominated ecosystems.

The review by Karsten et al. (2009; Chapter 13) summarises our knowledge of the effect of UV radiation on the physiology and metabolism of polar marine benthic micro- and macroalgae. The cellular targets of UV radiation are various biomolecules, in particular the DNA and proteins, which directly absorb UV radiation or are indirectly affected by various UV-induced photochemical reactions, e.g., the formation of reactive oxygen radicals. In this way, photosynthesis and other cellular processes can be strongly inhibited. On the other hand, there is a considerable potential for acclimation to UV radiation through various repair and protective mechanisms. DNA damage can be repaired enzymatically, and oxidative stress is counteracted by enzymatic defence systems and scavenging by antioxidants. UV-absorbing substances such as mycosporine-like amino acids and brown algal phlorotannins may prevent damage. If damaging effects prevail, cell fine structure is damaged and growth, as well as reproduction, can be constrained, which can lead to severe ecological consequences.

The ecological effects of UV radiation are an important element in the review by Campana et al. (2009; Chapter 14). Ambient UV radiation shapes the composition of polar seaweed communities during early succession and can exert persistent effects on later stages. In contrast to macroalgal propagules and germination stages, the UV-resistance of benthic microalgal communities is exceptionally high, which has positive implications for seaweed recruitment by providing UV-free space. In addition to UV radiation, other abiotic and biotic factors driving the structure of polar marine benthic algal communities are summarised. Primary succession starts with a rapid colonisation by diatoms and ephemeral macroalgae, whereas colonisation by perennial seaweeds is slower and highly seasonal. Global warming will presumably affect algal succession as increasing temperatures will affect the vitality of individual species and will promote the invasion of cold-temperate species, as pointed out also by Müller et al. (2009; Chapter 11). Retreating glaciers provide on the one hand open space for colonisation, but change nutrient supply, reduce salinity, increase sedimentation and thereby decrease light availability. Altogether, this will affect algal succession and presumably lead to changes in structure and function of polar benthic communities.

References

Amsler, C.D., K. Iken, J.B. McClintock and B.J. Baker. 2009. Defenses of polar macroalgae against herbivores and biofoulers. *Bot. Mar. 52*: 535–545.

Becker, S., B. Walter and K. Bischof. 2009. Freezing tolerance and photosynthetic performance of polar seaweeds at low temperatures. *Bot. Mar. 52*: 609–616.

Campana, G.L., K. Zacher, A. Fricke, M. Molis, A. Wulff, M.L. Quartino and C. Wiencke. 2009. Drivers of colonization and succession in polar benthic macro- and microalgal communities. *Bot. Mar. 52*: 655–667.

Glud, R.N., J. Woelfel, U. Karsten, M. Kühl and S. Rysgaard. 2009. Benthic microalgal production in the Arctic: applied methods and status of the current database. *Bot. Mar. 52*: 559–571.

Gómez, I., A. Wulff, M.Y. Roleda, P. Huovinen, U. Karsten, M.L. Quartino, K. Dunton and C. Wiencke. 2009. Light and temperature demands of marine benthic microalgae and seaweeds in polar regions. *Bot. Mar. 52*: 593–608.

Hommersand, M.H., R.L. Moe, C.D. Amsler and S. Fredericq. 2009. Notes on the systematics and biogeographical relationships of Antarctic and sub-Antarctic Rhodophyta with descriptions of four new genera and five new species. *Bot. Mar. 52*: 509–534.

Iken, K., C.D. Amsler, M.O. Amsler, J.B. MacClintock and B.J. Baker. 2009. Field studies on deterrent properties of phlorotannins in Antarctic brown algae. *Bot. Mar. 52*: 547–557.

Karsten, U., A. Wulff, M.Y. Roleda, R. Müller, F.S. Steinhoff, J. Fredersdorf and C. Wiencke. 2009. Physiological responses of polar benthic algae to ultraviolet radiation. *Bot. Mar. 52*: 639–654.

Kirst, G.O. and C. Wiencke. 1995. Ecophysiology of polar algae. *J. Phycol. 31*: 181–199.

Müller, R., T. Laepple, I. Bartsch and C. Wiencke. 2009. Impact of oceanic warming on the distribution of seaweeds in polar and cold-temperate waters. *Bot. Mar. 52*: 617–638.

Wiencke, C. 1996. Recent advances in the investigation of Antarctic macroalgae. *Polar Biol. 16*: 231–240.

Wiencke, C. and M.N. Clayton. 2002. Antarctic seaweeds. *In*: (J.W. Wägele, ed) *Synopses of the Antarctic benthos 9*. A.R.G. Gantner, Ruggell, Liechtenstein. pp. 239.

Wiencke, C., M.N. Clayton, I. Gómez, K. Iken, U.H. Lüder, C.D. Amsler, U. Karsten, D. Hanelt, K. Bischof and K. Dunton. 2007. Life strategy, ecophysiology and ecology of seaweeds in polar waters. *Rev. Environ. Sci. Biotechnol. 6*: 95–126.

Wiencke, C., I. Gómez and K. Dunton. 2009. Phenology and seasonal physiological performance of polar seaweeds. *Bot. Mar. 52*: 585–592.

Woelfel, J., R. Schumann, P. Leopold, C. Wiencke and U. Karsten. 2009. Microphytobenthic biomass along gradients of physical conditions in Arctic Kongsfjorden, Svalbard. *Bot. Mar. 52*: 573–583.

Wulff, A., K. Iken, M.L. Quartino, A. Al-Handal, C. Wiencke and M.N. Clayton. 2009. Biodiversity, biogeography and zonation of marine benthic micro- and macroalgae in the Arctic and Antarctic. *Bot. Mar. 52*: 491–507.

Zacher, K., R. Rautenberger, D. Hanelt, A. Wulff and C. Wiencke. 2009. The abiotic environment of polar benthic algae. *Bot. Mar. 52*: 483–490.

Environment, biogeography and biodiversity

2 The abiotic environment of polar marine benthic algae

Katharina Zacher, Ralf Rautenberger, Dieter Hanelt, Angela Wulff and Christian Wiencke

Introduction

In the polar regions, both light conditions and temperatures are exceptionally low for a large part of the year. Another important feature of polar regions is the strong seasonality, especially of the light regime, distinguishing the Arctic and the Antarctic from all other regions of the world. The present review gives an account of the environmental factors to which benthic algae are exposed in polar regions, explains differences between the Arctic and Antarctic and summarizes our present knowledge of global and anthropogenic changes on polar benthic ecosystems.

Differences between polar regions

While many environmental features are similar in the two polar regions, the Arctic and the Antarctic Oceans differ considerably in their origins, their cold-water history and their present environmental conditions. The Arctic Ocean is surrounded by continental land masses with a vast coastline that is continuously connected to the temperate coasts of America and Eurasia. In contrast, Antarctica is an ice-covered continent completely surrounded by the Southern Ocean without any land bridge to temperate regions since the late Mesozoic (Lüning 1990). A very important feature of the Southern Ocean amplifying this separation is the Antarctic Circumpolar Current (ACC), which flows in a clockwise direction around the continent driven by westerly winds (Lüning 1990). The ACC is thought to have developed 25 million years ago, thermally isolating the waters around the Antarctic continent. The Antarctic polar front (or Antarctic Convergence) delimits the Southern Ocean sharply to the north (Orsi et al. 1995). In contrast, the southern limit of the Arctic Ocean is poorly defined as the Arctic polar front is discontinuous and subject to strong variations (Clarke 1990). There is a considerable inflow of warm North Atlantic waters through the Fram Strait, where the West Spitsbergen Current mixes with Arctic waters.

Another major feature differentiating the two polar regions is their different cold-water history. The water temperatures in the Southern Ocean have been low for 14 million years, following the first glaciation of East Antarctica (Crame 1993). By comparison, water temperatures in the Arctic decreased much later. Glaciation and a winter ice cover did not develop in the Arctic before 2 million years ago, at the beginning of the Pleistocene (Thiede 1986, Clarke 1990, Zachos et al. 2001). The Arctic Ocean is covered by multi-year pack ice, whereas the Southern Ocean pack ice is thinner and most of it melts during summer. Due to these differences in oceanographic and geological features, the benthic algal communities between both polar regions have developed differently (Chapter 3, 10; Gómez et al. 2009, Wulff et al. 2009). The Antarctic marine seaweed flora is characterized by a high degree of endemism (33% in Antarctic seaweeds), whereas in the Arctic, only few endemic species have been found so far (Chapter 3; Wulff et al. 2009). The kelp forests formed by the Laminariales, present in temperate and Arctic regions, are completely absent in Antarctica. The same ecological niche is occupied in Antarctic regions by endemic Desmarestiales (Wiencke and Clayton 2002). Endemism of polar benthic microalgae remains to be studied (Chapter 3; Wulff et al. 2009).

Seasonal changes of light conditions

At the northern and southern distribution limits of benthic algae in the Arctic and the Antarctic at 80° N and 77° S, respectively, the annual solar radiation is 30–50% lower than in temperate to tropical regions, and the polar night lasts for about 4 months (Lüning 1990). At lower latitudes, e.g., at the northern limit of the Antarctic region, around the South Shetland Islands, daylengths vary between 5 h in winter and 20 h in summer (Wiencke 1990). This seasonal variation of light conditions is extreme, with large associated variations in the coverage of sea ice and snow. The strongest seasonal variation in sea ice conditions is found in the Southern Ocean. Here, it increases from ≈4 million km^2 in summer to ≈20 million km^2 in winter (Thomas and Dieckmann 2002). In the Arctic, sea ice cover varies from ≈4–5 million km^2 in late summer to ≈15 million km^2 in late winter (Drobot et al. 2008). If the ice is covered by snow, irradiance directly below can be diminished to <2% of the surface value. In northeastern Greenland, the sea is covered by ice for 9–10 months out of the year, the water temperature at 20 m depth is 0 °C year around and there are months of continuous darkness during winter (Rysgaard et al. 1998). Hence, benthic algae can be exposed for long periods of the year to darkness or very low light conditions (see also other examples from the Canadian Arctic, the Antarctic Peninsula region and East Antarctica: Chapman and Lindley 1980, Dunton 1990, Miller and Pearse 1991, Drew and Hastings 1992, Klöser et al. 1993).

These conditions suddenly change in concert with the sea ice break up. At this time, the dim light conditions come to an end as surface solar radiation may reach high values and the water is very clear. On sunny days, up to 1400 μmol photons $m^{-2}\,s^{-1}$ photosynthetically active radiation (PAR, 400–700 nm), 19 W m^{-2} UV-A radiation (UVAR, 315–400 nm) and 1.09 W m^{-2} UV-B radiation (UVBR, 280–315 nm) have been measured in the air above Kongsfjorden, a glacial fjord off Arctic Spitsbergen (Bischof et al. 1998a, Wiencke et al. 2006). On King George Island (South Shetland Islands, Antarctica), levels of 1700 μmol photons $m^{-2}\,s^{-1}$ PAR, 44 W m^{-2} UVAR and 2.3Wm^{-2} UVBR have been measured (Quartino et al. 2005, Zacher et al. 2007a). On Signy Island (South Orkney Islands, Antarctica), the 1% depth of surface values for PAR was at 29 m after ice break-up, indicating high water transparency (Brouwer 1996). These relatively high light conditions prevail for about 2 months. In mid summer and autumn, light transmittance decreases due to the formation of phytoplankton blooms and/or the discharge of turbid meltwater (Drew and Hastings 1992, Gómez et al. 1997, Hanelt et al. 2001, Richter et al. 2008).

The changing light climate during the year is the main factor influencing the development of benthic macro- and microalgae. Polar algae are extremely low-light adapted but must be able to tolerate high light after ice break-up (Chapter 10; Gómez et al. 2009). Only recently were the seasonal changes in the photosynthetic characteristics related to adjustments of the photosynthetic apparatus by altering the content and composition of photochemically active and accessory pigments (Brey 2009).

Many polar seaweeds, like the endemic Arctic *Laminaria solidungula* J. Agardh or the Antarctic Desmarestiales grow or even reproduce in late winter/spring. This type of seasonal development is controlled by photoperiodisms and/or circannual rhythms synchronized by daylengths (Chapter 9; Wiencke et al. 2009), further showing the significance of the changing light climate in polar regions. Moreover, the seasonal physiological performance of polar algae is strongly linked to the life-strategies of individual species.

Environmental conditions in the eulittoral and supralittoral

Organisms inhabiting the supralittoral and eulittoral zones are periodically subjected to strong changes in abiotic environmental conditions due to tidal water movements. On King George Island (Antarctica), amplitudes of the semi-diurnal tides range between 1.20 (mean neap tide) and 1.48 m (mean spring tide) (Schöne et al. 1998). In the Arctic, tides are diurnal and their heights vary between 0.20 and 1.70 m in Kongsfjorden, and can reach up to 4 m in southern Greenland (Ito and Kudoh 1997, Svendsen et al. 2002). The seawater temperature varies only slightly between −1.8 °C and +2.2 °C off the Antarctic Peninsula region (Drew and Hastings 1992, Klöser et al. 1993, Kang et al. 2002). In Kongsfjorden – although located at 79 °N – temperature increases up to 6.5 °C, as determined by Brey (2009) in the summer (June) of 2004. In tide pools, however, organisms may be exposed to water temperatures up to 14 °C, which may occur within a few hours on sunny days in summer in Antarctica (Klöser and Arntz 1994). In the supralittoral, temperatures can change even more dramatically, for example the thallus temperature of a lichen growing on exposed coastal rocks on King George Island reached +27.4 °C in summer due to heating-up by the sun and decreased down to −27.3 °C in winter (Schroeter et al. 1995).

During low tide, intertidal algae are exposed to extreme changes in abiotic conditions. Benthic algae may regularly dry out for several hours, increasing the risk of cellular damage (Lüning 1990). When air temperatures fall significantly below 0 °C, as is frequently the case in polar regions, benthic algae may freeze. However, frozen algae have an even higher degree of desiccation tolerance than non-frozen algae (Kang et al. 2002). Species from this zone are able to tolerate freezing, as shown for *Fucus distichus* Linnaeus from Spitsbergen (Chapter 11; Becker et al. 2009). Only algae in tidal pools remain directly protected from dessication and freezing.

In summer, when rising air temperatures lead to thawing of snow and glaciers, both the supralittoral and the eulittoral zones are flushed by fresh meltwater. Consequently, the salinity within the upper few meters of the water column is reduced from 34.5 psu to 31 psu or even 27 psu in tide pools, as measured in Potter Cove (King George Island) (Klöser and Arntz 1994), and to 23 psu in Arctic waters of Kongsfjorden (Hanelt et al. 2001, Svendsen et al. 2002). Hence, these algal communities are often exposed to changing osmotic conditions of the seawater, requiring a high physiological acclimation potential, as demonstrated in the euryhaline macroalga *Fucus distichus* from the upper sublittoral in Kongsfjorden (Karsten 2007). The species-specific acclimation potentials resulting from the vertical salinity gradient may be one factor influencing macroalgal vertical zonation patterns (Karsten 2007). During calm weather conditions, there is a stable stratification of the water column in sheltered regions in the inlets (Klöser et al. 1993, Hanelt et al. 2001, Kang et al. 2002). A halocline (pycnocline) between 1 and 5 m depth has been recorded under such conditions (Abele et al. 1999): surface water with lower salinity is not mixed in this case with water of higher salinity from greater depths.

Gradients of PAR and UVR in the water column represent major factors influencing macroalgal depth distributions. Algae from upper littoral zones are more tolerant of high irradiances of both PAR and UVR than those from greater depths (Bischof et al. 1998b, Hanelt 1998), which stems from highly active and efficient photoprotective mechanisms (Chapter 13; Franklin et al. 2003, Bischof et al. 2006, Karsten et al. 2009). However, Antarctic benthic diatoms from both the inter- and subtidal seem to be able to acclimatize to comparably high UV radiation (Chapter 13; Zacher et al. 2007a, Wulff et al. 2008, Karsten et al. 2009). Furthermore, during low tides, benthic algae are air-exposed and can be exposed to very high PAR and UVR. However, it is not yet known whether a vertical zonation due to the amount of light is also the case for microphytobenthos.

Ice represents another key physical factor influencing algal zonation along the coastal depth profile in polar and subpolar regions as a consequence of mechanical damage (Barnes 1999, Gutt 2001). Abrading effects of ice scour by drifting sea ice and icebergs may either prevent coverage of the entire intertidal zone with macroalgae or restricts algal vegetation to sheltered locations and microniches such as crevices in the rocks and rock pools (Chapter 3; Gutt 2001, Wulff et al. 2009). As a result of this, only an annual or pseudoperennial algal flora can develop in the upper eulittoral zones due to its higher regeneration capacity and growth rates (Chapter 3, 14; Post and Larkum 1993, Clayton 1994, Gutt 2001, Wiencke et al. 2004, Campana et al. 2009, Wulff et al. 2009). Therefore, in both polar regions, benthic algae almost exclusively colonize in the sublittoral at depths >5 or 10 m (Chapter 3; Wulff et al. 2009). In winter when the freezing sea surface (fast ice) meets the hard substratum, an ice foot is formed that encases the rock surface and the inhabiting biota (Barnes 1999). Due to tidal movements, its thickness increases. The ice foot may further raft sessile benthic algae away when it disintegrates in summer (Gutt 2001). Annual, less susceptible macroalgae such as *Enteromorpha*, *Ulva*, *Ulothrix* and *Urospora* remain attached to the rock or re-colonize denuded hard substratum quickly, within <15 days (Gutt 2001 and references therein). Benthic microalgae can be completely removed by ice-scouring (Teixido et al. 2007) and, furthermore, frazil ice can lift the microalgae and incorporate the benthic microflora into the ice-algal community (Garrison et al. 1989).

Levels of nutrients (nitrate and phosphate) at the surface of coastal waters are characteristically high throughout the year along the Antarctic Peninsula. The concentration of nitrate (NO_3^-) can range between 14 and 33 μM, with even higher peaks (Schloss et al. 2002, Peters et al. 2005). Average phosphate (PO_4^{3-}) concentrations also vary between 2.0 and 3.2 μM (Schloss et al. 2002). Nutrient concentrations may decrease in February as a consequence of dilution with meltwater (Brandini and Rebello 1994), as well as a consequence of phytoplankton blooms in November/December (Kang et al. 2002). However, reduced concentrations of nitrate and phosphate are generally not a limiting factor for macroalgal growth throughout the year in Antarctica, as demonstrated by high tissue nitrogen levels as well as low C : N ratios (Drew and Hastings 1992, Weykam et al. 1996, Peters et al. 2005). Productivity of pelagic phytoplankton is limited by the nutrient iron despite sufficient levels of other available nutrients. In comparison, Arctic nitrate and phosphate concentrations are relatively high in winter (nitrate: ~10 μM and phosphate ~1 μM), yet they are depleted during spring and summer (nitrate and phosphate: <0.5 and 0.2 μM, respectively, Chapman and Lindley 1980, Aguilera et al. 2002). Nitrate concentrations in the uppermost water zone are especially low due to phytoplankton blooms (Sejr et al. 2007).

Environmental conditions in the sublittoral

As previously explained, benthic algae in the sublittoral zone are exposed in winter to long periods of darkness (or extremely low light conditions) for periods up to ten months. For example, on Arctic Spitsbergen, the polar night ends on the 14th of February but the break-up of sea ice usually occurs months later (Svendsen et al. 2002). Extremely low maximal photon flux densities of PAR (6.5 μmol photons m^{-2} s^{-1}), irradiances of UVAR (0.42 W m^{-2}) and zero UVBR have been measured in June directly below a sea ice cover of 1 m thickness and a snow layer of 0.30 m (Hanelt et al. 2001). On snow-free ice, PAR decreases down to about 8.5% of the value determined in air. On ice covered with snow, this value is reduced to 2.4% (Hanelt et al. 2001). After the sea ice break-up, both PAR and UVR are able to penetrate deeply into the water column due to the clear water conditions, corresponding to Type 1 (Jerlov 1976) in Kongsfjorden (Hanelt et al. 2001, Svendsen et al. 2002). PAR suddenly increases up to nearly 600 μmol photons m^{-2} s^{-1} in the upper sublittoral (4 m depth, Bischof et al. 2002, Brey 2009) and to more than 300 μmol photons m^{-2} s^{-1} at 8–9 m depth (McMinn et al. 2004). Simultaneously, biologically relevant effective UVBR is able to penetrate down to the considerable depth of 10 m (Bischof et al. 1998a). However, as a consequence of increasing water turbidity caused by inflow of meltwater carrying sediments and (chromophoric) dissolved organic matter (DOM) and mixing of the water column in summer and autumn, water transparency is then significantly reduced (Hanelt et al. 2001). In the Antarctic, the water column is usually less turbid than in the Arctic (Hoyer et al. 2003). Seasonal changes of optical properties of the coastal waters are similar to those in Arctic waters despite a lower relative turbidity. The transparency of Antarctic coastal waters in Potter Cove changes slowly from spring to autumn. For example, the 10% depth of PAR relative to surface values decreases continuously from ~12 m in November and 8 m in December to 2 m in January/February in Potter Cove (Klöser et al. 1993, Richter et al. 2008). The 1% depth of surface radiation of UVBR can reach maximum penetration depths of 16 m (Zacher et al. 2007a). The vertical attenuation coefficient (K_d), a parameter representing light transmittance in the water column, increases dramatically from 0.19 m^{-1} (measured before melting) to 0.74 m^{-1} in Kongsfjorden in summer (Hanelt et al. 2001, Brey 2009). This water turbidity is particularly persistent in half-open bays and fjords, such as Kongsfjorden, until September due to slower water exchange (Svendsen et al. 2002). Sites that remain relatively unaffected by meltwater inflow due to the presence of continuous currents of clear oceanic waters are generally characterized by lower K_d values (e.g., 0.2–0.4 m^{-1} at Peñón de Pesca, King George Island) compared to meltwater-influenced sites (e.g., 0.5–0.8 m^{-1} at Peñón Uno, King George Island; Richter et al. 2008). In order to estimate marine benthic primary production more precisely, continuous recordings of the underwater light regime *in situ* are urgently needed (see also chapter 9; Glud et al. 2009).

Water temperature and salinity regimes change only slightly below water depths of 15 m. The water temperature ranges between −1.2 °C and +1.0 °C from spring to autumn in Maxwell Bay on King George Island (Klöser et al. 1993). Salinity remains stable around 34 psu (Klöser et al. 1993). In Kongsfjorden, temperature in deep waters (20 m) rose slowly from 0.1 °C at the beginning of June to 4.2 °C in the middle of August (Hanelt et al. 2001). Correspondingly, salinity decreased from 34.4 to 27.7 psu by mixing with melt water. In the sublittoral (40 m) of Tyrolerfjord – Young Sound in the Arctic (Greenland), water temperature varied minimally between −0.85 °C and −1.74 °C, and salinity ranged slightly between 31.4 and 32.6 psu (Sejr et al. 2007). The annual sedimentation rate at the same location was 114 g m^{-2} $year^{-1}$ and 1.4 g m^{-2} $year^{-1}$ for total matter and carbon, respectively, with the highest rates measured in July during ice-free conditions (Sejr et al. 2007). In Kongsfjorden, sedimentation rates are as high as 20 kg m^{-2} $year^{-1}$ directly in front of the glacier due to ice calving events. Towards the outer fjord, rates decrease significantly down to 0.2 kg m^{-2} $year^{-1}$ (Svendsen et al. 2002).

Anchor ice, which is formed between 0 and 30 m depths in Antarctica, has a great impact on benthic biota (Dayton 1989). It develops on the seafloor and encloses benthic algae, as described for the Ross Sea (Miller and Pearse 1991). Ice scouring by floating icebergs, calving glaciers or broken parts of fast ice grounding and scraping on the seabed results in patchy removal of algal benthic communities in depths >15 m in boreal and austral summer and autumn (Barnes 1999, Gutt 2001). Thus, the physical factor of ice scouring represents a major influence shaping macroalgal flora (and fauna) in shallow waters (Chapter 3, 14; Campana et al. 2009, Wulff et al. 2009). Exposed sublittoral zones are highly affected by mechanical disturbances, leading to a poor biodiversity and species number (Gutt 2001). Dropped stones released from thawing ice reflect further agents of disturbances to the benthos but, on the other hand, they also provide new substrata for colonization by benthic flora and associated fauna. There is still, however, little information available on the impact of ice scour on benthic algal communities.

In addition to the abiotic factors discussed above, substratum characteristics are additionally important for the distribution of benthic algae. In contrast to hard substratum, which is densely colonized by benthic macroalgae, soft bottom is mostly devoid of seaweeds (Chapter 3; Klöser et al. 1996, Wulff et al. 2009). However, the sublittoral soft bottom of the inner part of Potter Cove (King George Island) is dominated by *Himantothallus grandifolius* (A. *et* E.S. Gepp) Zinova, which inhabits boulders and stones of moraine deposits down to 9 m (Klöser et al. 1996). As steep slopes with promontories and inlets are largely sheltered from ice scouring, e.g., by icebergs in summer, dense belts of ice-sensitive seaweeds such as *Desmarestia anceps* Montagne grow in water depths down to 25 m (Klöser et al. 1996). In contrast to macroalgae, benthic microalgae are not limited to hard substratum. They inhabit ice, soft bottoms and live epiphytically (Chapter 3; Wulff et al. 2009). However, grain size is an important factor for benthic diatoms; the epipelic community seems to be dependent on more sheltered conditions and silt accumulations, while the episammic community occurs preferentially in the presence of larger sand grains within the sediment (e.g., de Jonge 1985, Sabbe 1997).

A steep light gradient in the subtidal zone can occur within dense benthic algal communities, due to canopy forming large brown macroalgae or kelps (Hanelt et al. 2004). Irradiances of PAR and UVR are strongly attenuated by absorption and scattering processes by the canopy algae. Consequently, understory algae are exposed to lower irradiances as well as to a changed spectral composition of transmitted light: the green and far red regions of the electromagnetic spectrum are dominant in the subcanopy. Two examples of this are the red macroalgae *Phycodrys austrogeorgica* Skottsberg and *Delesseria lancifolia* (J.D. Hooker) J. Agardh, both living in the shade of the canopy of *D. anceps* and *H. grandifolius* in the middle sublittoral in Antarctica (Klöser et al. 1996). They are both characterized by high sensitivity to UVR, which is absent in the understory (Bischof et al. 1998b). Steep light gradients also occur in the microscale range within microbial mats and in the sediment habitat where light attenuation coefficients decrease with increasing particle size (Kühl et al. 1994).

Signs of global change and other anthropogenic influences

Climate change in polar regions is expected to be among the largest and most rapid of any region on Earth. In both polar regions, there is strong evidence of the ongoing impacts of climate change (Anisimov et al. 2007). In particular, the Atlantic sector of the Arctic and the Antarctic Peninsula will suffer from major physical, ecological, sociological, and economic impacts (Chapter 12; Anisimov et al. 2007, Müller et al. 2009).

The Antarctic atmosphere is clean, dry and relatively isolated from the rest of the atmosphere by the surrounding Southern Ocean and by the polar vortex. The Arctic, on the other hand, is surrounded by landmasses and industrialized countries, and is strongly affected by seasonal atmospheric transport and anthropogenic emissions (Simpson et al. 2007). Ozone measurements spanning the last 20 years show clear evidence of stratospheric ozonedepletion (Randel et al. 1999). This is largely due to emissions of industrially produced halocarbons (particular chlorofluorocarbons; WMO 2007). There are also natural sources of these substances, e.g., the release of volatile organohalogenic compounds from macroalgae (Laturnus 2001). Although the release of ozone-depleting substances is currently declining, no clear answer can be given as to whether ozone levels will ever recover to pre-1980s values (Weatherhead and Andersen 2006). Furthermore, Northern Hemisphere continents are likely to continue to warm rapidly during winter, and Arctic ozone recovery may therefore be substantially delayed. Depletion of stratospheric ozone over polar regions has caused an increase in UVBR and may potentially affect macroalgal distribution patterns (Chapter 13; Bischof et al. 2006, Wiencke et al. 2006, Karsten et al. 2009). Polar benthic microalgae in general and macroalgae from upper littoral zones seem largely unaffected by relatively high UVR irradiances (Zacher et al. 2007a, Rautenberger and Bischof 2008, Wulff et al. 2008), whereas subtidal macroalgae are strongly affected by higher levels of UVR (Wiencke et al. 2006, Roleda et al. 2007). UVBR is therefore regarded as one of the most important factors in controlling depth zonation of macroalgae due to effects from the cellular (DNA and protein damage) to the ecosystem level (change in community structure, diversity; Chapter 13, 14; Bischof et al. 2006, Zacher et al. 2007b, Campana et al. 2009, Karsten et al. 2009).

The western Antarctic Peninsula has faced a recent rapid regional warming with a mean temperature increase of 3.4 K per century (Vaughan et al. 2003). The glacial systems on the Antarctic Peninsula show direct responses to the climatic changes recorded over recent decades in this region; retreat of ice fronts, increased melt water production and massive retreat of glaciers have been documented since 1956 (e.g., Braun and Gossmann 2002). In the Arctic as well, a warming trend has been observed over recent decades (Turner et al. 2007). In the Arctic Ocean, total freshwater inflow increased by ≈7% from 1936 to 1999, with potential for further increases with Arctic warming (Bluhm and Gradinger 2008). Increased freshwater discharge (resulting from climatic change) acts on biological communities, nutrient supply, turbidity, and the inflow of low salinity water (Pivovarov et al. 2003).

Judging from satellite data, sea ice has declined by ≈17% at the end of an Arctic summer over the last 25 years (Turner et al. 2007). Light attenuation by sea ice might protect primary producers from UVR, but could also limit primary production, and consequently many ecosystem processes are closely tied to the few months of open water (Horner and Schrader 1982, Rysgaard et al. 1998, 1999, Glud et al. 2007). An increase in the duration of the ice-free period due to global warming may further increase primary productivity in the open water. For example, the thinning of sea ice and the increase in the open-water period is expected to enhance primary productivity in the Young Sound area off Greenland due to a ≈50% increase in light availability (Rysgaard and Glud 2007). Furthermore, an increased nutrient supply by higher river run-off rates may support an increase in primary productivity of opportunistic species, whereas a decrease in salinity may counteract this effect.

An effect of the increase of human visitor frequency may have contributed to the introduction of invasive species, especially in combination with global climate change and the associated increase in temperature of the water column. The Southern Ocean around Antarctica was once thought to be free of invasive species due to its extreme environment and isolation (Tavares and de Melo 2004). In 2004, however, the first invasive animal species, the spider crab *Hyas araneus* Linnaeus, was recorded off the Antarctic Peninsula (Tavares and de Melo 2004). Antarctic waters are characterized by a high level of endemism and are therefore probably one of the ecosystems most sensitive to invasion (Barnes and Fraser 2003). Clayton et al. (1997) discussed the possibility that some of the cosmopolitan macroalgal species found in Antarctica, were perhaps introduced via earlier shipping activities. For example, in 1965 the green alga *Ulva intestinalis* Linnaeus did not occur on Half Moon Island, but was found 30 years later by Clayton et al. (1997). An increased water temperature observed off the Antarctic Peninsula (Meredith and King 2005, Schloss et al. 2008) might stimulate invasion of temperate species as the distribution limits of temperate algae are likely to change with fluctuations in seawater temperatures (Chapter 12; Müller et al. 2009). Many polar species are very vulnerable to climate change because they are narrowly adapted to the environment they inhabit and it is postulated that loss of a keystone species might have cascading effects on the entire polar ecosystem (Anisimov et al. 2007 and references therein).

Without doubt, the global climate is changing, especially in some polar regions (the Western Antarctica Peninsula and the Atlantic sector of the Arctic). Yet little is known about the general community structure and trophic interactions of these systems, in particular with regard to microalgae (Chapter 3; Wulff et al. 2009) and even less as to how the changing climate affects these delicate ecosystems. Because the macroalgal communities in the Arctic and Antarctic are quite different, comparative studies are crucial in determining whether both systems will react in the same manner or differently in response to environmental changes. Experiments with natural light climates and field experiments manipulating abiotic parameters at the community level are particularly lacking. Furthermore, long-term monitoring projects including the long winter months are crucial to get data that could be used for climate models. Much research still has to be undertaken in these extreme ecosystems in order to get a better picture of the whole systems and how ongoing climate change is affecting and will affect polar regions in the future.

References

Abele, D., G.A. Ferreyra and I.R. Schloss. 1999. H_2O_2 accumulation from photochemical production and atmospheric wet deposition in Antarctic coastal and off-shore waters of Potter Cove, King George Island, South Shetland Islands. *Antarct. Sci. 11*: 131–139.

Aguilera, J., K. Bischof, U. Karsten, D. Hanelt and C. Wiencke. 2002. Seasonal variation in ecological patterns in macroalgae from an Arctic fjord. II. Pigment accumulation and biochemical defence systems against high light stress. *Mar. Biol. 140*: 1087–1095.

Anisimov, O.A., D.G. Vaughan, T.V. Callaghan, C. Furgal, H. Marchant, T.D. Prowse, H. Vilhjálmsson and J.E. Walsh. 2007. Polar regions (Arctic and Antarctic). *In*: (M.L. Parry, O.F. Canziani, J.P. Palutikof, P.J. van der Linden and C.E. Hanson, eds) *Climate change 2007: impacts, adaptation and vulnerability. Contribution of working Group II to the fourth assessment report of the intergovernmental panel on climate change.* Cambridge University Press, Cambridge. pp. 653–685.

Barnes, D.K.A. 1999. The influence of ice on polar nearshore benthos. *J. Mar. Biol. Ass. UK 79*: 401–407.

Barnes, D.K.A. and K.P.P. Fraser. 2003. Rafting by five phyla on man-made flotsam in the Southern Ocean. *Mar. Ecol. Prog. Ser. 262*: 289–291.

Becker, S., B. Walter and K. Bischof. 2009. Freezing tolerance and photosynthetic performance of polar seaweeds at low temperatures. *Bot. Mar. 52*: 609–616.

Bischof, K., D. Hanelt, H. Tüg, U. Karsten, P. Brouwer and C. Wiencke. 1998a. Acclimation of brown algal photosynthesis to ultraviolet radiation in Arctic coastal waters (Spitsbergen, Norway). *Polar Biol. 20*: 388–395.

Bischof, K., D. Hanelt and C. Wiencke. 1998b. UV-radiation can affect depth-zonation of Antarctic macroalgae. *Mar. Biol. 131*: 597–605.

Bischof, K., D. Hanelt, J. Aguilera, U. Karsten, B. Vögele, T. Sawall and C. Wiencke. 2002. Seasonal variation in ecophysiological patterns in macroalgae from an Arctic fjord. I. Sensitivity of photosynthesis to ultraviolet radiation. *Mar. Biol. 140*: 1097–1106.

Bischof, K., I. Gómez, M. Molis, D. Hanelt, U. Karsten, U. Lüder, M.Y. Roleda, K. Zacher and C. Wiencke. 2006. Ultraviolet radiation shapes seaweed communities. *Rev. Environ. Sci. Biotechnol. 5*: 141–166.

Bluhm, B.A. and R. Gradinger. 2008. Regional variability in food availability for Arctic marine mammals. *Ecol. Appl. 18*: S77–S96.

Brandini, F. and J. Rebello. 1994. Wind field effect on hydrography and chlorophyll dynamics in the coastal pelagial of Admiralty Bay, King George Island, Antarctica. *Antarct. Sci. 6*: 433–442.

Braun, M. and H. Gossmann. 2002. Glacial changes in the areas of Admiralty Bay and Potter Cove, King George Island, Maritime Antarctica. *In*: (L. Beyer and M. Bölter, eds) *Geoecology of Antarctic ice-free coastal landscapes*. Ecological Studies 154. Springer-Verlag, Heidelberg. pp. 75–90.

Brey, L. 2009. *Acclimation of kelp photosynthesis to seasonal changes in the underwater radiation regime of an Arctic fjord system*. Dissertation, University of Bremen. http://elib.suub.uni-bremen.de/diss/docs/00011524.pdf.

Brouwer, P.E.M. 1996. *In situ* photosynthesis and estimated annual production of the red macroalga *Myriogramme mangini* in relation to underwater irradiance at Signy Island (Antarctica). *Antarct. Sci. 8*: 245–252.

Campana, G., K. Zacher, A. Fricke, M. Molis, A. Wulff and C. Wiencke. 2009. Drivers of colonization and succession in polar benthic macro- and microalgal communities. *Bot. Mar. 52:* 655–667.

Chapman, A.R.O. and J.E. Lindley. 1980. Seasonal growth of *Laminaria solidungula* in the Canadian High Arctic in relation to irradiance and dissolved nutrient concentration. *Mar. Biol. 57*: 1–5.

Clarke, D.L. 1990. Arctic Ocean ice cover. Geologic history and climatic significance. *In*: (A. Grantz, L. Johnson and J.L. Sweeney, eds) *The Arctic Ocean region*. The Geological Society of North America. Boulder, Colorado. pp. 53–62.

Clayton, M. 1994. Evolution of the Antarctic marine benthic algal flora. *J. Phycol. 30*: 897–904.

Clayton, M., C. Wiencke and H. Klöser. 1997. New records of temperate and sub-Antarctic marine benthic macroalgae from Antarctica. *Polar Biol. 17*: 141–149.

Crame, J.A. 1993. Latitudinal range fluctuations in the marine realm through geological time. *Trends Ecol. Evol. 8*: 162–166.

Dayton, P. 1989. Interdecadal variation in an Antarctic sponge and its predators from oceanographic climate shifts. *Science 243*: 1484–1486.

de Jonge, V. 1985. The occurrence of "epipsammic" diatom populations: a result of interaction between physical sorting of sediment and certain properties of diatom species. *Estuar. Coast. Shelf. Sci. 21*: 607–622.

Drew, E.A. and R.M. Hastings. 1992. A year-round ecophysiological study of *Himantothallus grandifolius* (Desmarestiales, Phaeophyta) at Signy Island, Antarctica. *Phycologia 31*: 262–277.

Drobot, S., J. Stroeve, J. Maslanik, W. Emery, C. Fowler and J. Kay. 2008. Evolution of the 2007–2008 Arctic sea ice cover and prospects for a new record in 2008. *Geophys. Res. Lett. 35*: L19501.

Dunton, K.H. 1990. Growth and production in *Laminaria solidungula*: relation to continuous underwater light levels in the Alaskan High Arctic. *Mar. Biol. 106*: 297–304.

Franklin, L.A., C.B. Osmond and A.W.D. Larkum. 2003. Photoinhibition, UV-B and algal photosynthesis. *In*: (A.W.D. Larkum, S.E. Douglas and J.A. Raven, eds) *Photosynthesis in algae*. Kluwer Academic Publishers, Dordrecht. pp. 351–384.

Garrison, D.L., A.R. Close and E. Reimnitz. 1989. Algae concentrated by frazil ice-evidence from laboratory experiments and field-measurements. *Antarct. Sci. 1*: 313–316.

Glud, R.N., S. Rysgaard, M. Kühl and J.W. Hansen. 2007. The sea ice in Young Sound: implications for carbon cycling. *In*: (S. Rysgaard and R.N. Glud, eds) *Carbon cycling in Arctic marine ecosystems: case study Young Sound*. Meddelelser Grønland Biosci. 58. The Commission for Scientific Research in Greenland, Copenhagen. pp. 62–85.

Glud, R.N., J. Woelfel, U. Karsten, M. Kühl and S. Rysgaard. 2009. Benthic microalgal production in the Arctic: applied methods and status of the current database. *Bot. Mar. 52*: 559–571.

Gómez, I., G. Weykam, H. Klöser and C. Wiencke. 1997. Photosynthetic light requirements, metabolic carbon balance and zonation of sublittoral macroalgae from King George Island (Antarctica). *Mar. Ecol. Prog. Ser. 148*: 281–293.

Gómez, I., A. Wulff, M.Y. Roleda, P. Huovinen, U. Karsten, L.M. Quartino and C. Wiencke. 2009. Light and temperature demands of marine benthic microalgae and seaweeds in polar regions. *Bot. Mar. 52*: 593–608.

Gutt, J. 2001. On the direct impact of ice on marine benthic communities, a review. *Polar Biol. 24*: 553–564.

Hanelt, D. 1998. Capability of dynamic photoinhibition in Arctic macroalgae is related to their depth distribution. *Mar. Biol. 131*: 361–369.

Hanelt, D., H. Tüg, K. Bischof, C. Groß, H. Lippert, T. Sawall and C. Wiencke. 2001. Light regime in an Arctic fjord: a study related to stratospheric ozone depletion as a basis for determination of UV effects on algal growth. *Mar. Biol. 138*: 649–658.

Hanelt, D., K. Bischof and C. Wiencke. 2004. The radiation, temperature and salinity regime in Kongsfjorden. *Ber. Polarforsch. Meeresforsch. 492*: 14–25.

Horner, R.A. and G.C. Schrader. 1982. Relative contribution of ice algae, phytoplankton, and benthic microalgae to primary production in nearshore regions of the Beaufort Sea. *Arctic 35*: 485–503.

Hoyer, K., U. Karsten and C. Wiencke. 2003. Inventory of UV-absorbing mycosporine-like amino acids in polar macroalgae and factors controlling their content. *In*: (A.H.L. Huiskes, W.W. C. Gieskes, J. Rozema, R.M.L. Schorno, S.M. van der Vies and W.J. Wolff, eds) *Antarctic biology in the global context*. Backhuys, Leiden. pp. 56–62.

Ito, H. and S. Kudoh. 1997. Characteristics of water in Kongsfjorden, Svalbard. Proc. NIPR Symp. *Polar Meteorol. Glaciol. 11*: 211–232.

Jerlov, N. 1976. *Marine optics*. 2nd edition. Elsevier, Amsterdam. pp. 231.

Kang, J.-S., S.-H. Kang, J. Lee and S. Lee. 2002. Seasonal variation of microalgal assemblages at a fixed station in King George Island, Antarctica, 1996. *Mar. Ecol. Prog. Ser. 229*: 19–32.

Karsten, U. 2007. Salinity tolerance of Arctic kelps from Spitsbergen. *Phycol. Res. 55*: 257–262.

Karsten, U., A. Wulff, M. Roleda, R. Müller, F.S. Steinhoff, J. Fredersdorf and C. Wiencke. 2009. Physiological responses of polar benthic algae to ultraviolet radiation. *Bot. Mar. 52*: 639–654.

Klöser, H., G. Ferreyra, I. Schloss, G. Mercuri, F. Laturnus and A. Curtosi. 1993. Seasonal variation of algal growth conditions in sheltered Antarctic bays: the example of Potter Cove (King George Island, South Shetlands). *J. Mar. Syst. 4*: 289–301.

Klöser, H. and W. Arntz. 1994. Research on Antarctic shallow coastal and littoral systems. *Polarforschung 64*: 27–41.

Klöser, H., M.L. Quartino and C. Wiencke. 1996. Distribution of macroalgae and macroalgal communities in gradients of physical conditions in Potter Cove, King George Island, Antarctica. *Hydrobiologia 333*: 1–17.

Kühl, K., C. Lassen and B.B Jørgensen. 1994. Light penetration and light intensity in sandy marine sediments measured with irradiance and scalar irradiance fiber-optic microprobes. *Mar. Ecol. Prog. Ser. 105*: 139–148.

Laturnus, F. 2001. Marine macroalgae in polar regions as natural sources for volatile organohalogens. *Environ. Sci. Pollut. Res. 8*: 103–108.

Lüning, K. 1990. *Seaweeds. Their environment, biogeography, and ecophysiology*. John Wiley & Sons, New York. pp. 527.

McMinn, A., J. Runcie and M. Riddle. 2004. Effect of seasonal sea ice breakout on the photosynthesis of benthic diatom mats at Casey, Antarctica. *J. Phycol. 40*: 62–69.

Meredith, M.P. and J.C. King. 2005. Rapid climate change in the ocean west of the Antarctic Peninsula during the second half of the 20th century. *Geophys. Res. Lett. 32*: L19604.

Miller, A.K. and J.S. Pearse. 1991. Ecological studies of seaweeds in McMurdo Sound, Antarctica. *Amer. Zool. 31*: 35–48.

Müller, R., T. Laepple, I. Bartsch and C. Wiencke. 2009. Impact of oceanic warming on the distribution of seaweeds in polar and cold-temperate waters. *Bot. Mar. 52*: 617–638.

Orsi, A.H., T. Whitworth III and D.N. Worth Jr. 1995. On the meridional extent and fronts of the Antarctic Circumpolar Current. *Deep Sea Res. I 42(5)*: 641–673.

Peters, K.J., C.D. Amsler, M.O. Amsler, J.B. McClintock, R.B. Dunbar and B.J. Baker. 2005. A comparative analysis of the nutritional and elemental composition of macroalgae from the western Antarctic Peninsula. *Phycologia 44*: 453–463.

Pivovarov, S., R. Schlitzer and A. Novikhin. 2003. River runoff influence on the water mass formation in the Kara Sea. *In*: (R. Stein, K. Fahl, D.K. Fütterer, E.M. Galimov and O.V. Stepanets,

eds) *Siberian river run-off in the Kara Sea: characterisation, quantification, variability, and environmental significance*. Elsevier, Amsterdam. pp. 9–26.
Post, A. and A. Larkum. 1993. UV-absorbing pigments, photosynthesis and UV exposure in Antarctica: comparison of terrestrial and marine algae. *Aquat. Bot. 45*: 231–243.
Quartino, M., H. Zaixso and A. Boraso de Zaixso. 2005. Biological and environmental characterization of marine macroalgal assemblages in Potter Cove, South Shetland Islands, Antarctica. *Bot. Mar. 48*: 187–197.
Randel, W.J., R.S. Stolarski, D.M. Cunnold, J.A. Logan, M.J. Newchurch and J.M. Zawodny. 1999. Trends in the vertical distribution of ozone. *Science 285*: 1689–1692.
Rautenberger, R. and K. Bischof. 2008. UV-susceptibility of photosynthesis of adult sporophytes of four brown Antarctic macroalgae (Phaeophyceae). *Ber. Polarforsch. Meeresforsch. 571*: 263–269.
Richter, A., S. Wuttke and K. Zacher. 2008. Two years of *in situ* UV measurements at Dallmann Laboratory/Jubany Station. *Ber. Polarforsch. Meeresforsch. 571:* 12–19.
Roleda, M.Y., C. Wiencke, D. Hanelt and K. Bischof. 2007. Sensitivity of the early life stages of macroalgae from the Northern Hemisphere to ultraviolet radiation. *Photochem. Photobiol. 83*: 851–862.
Rysgaard, S., B. Thamdrup, N. Risgaard-Petersen, H. Fossing, P. Berg, P.B. Christensen and T. Dalsgaard. 1998. Seasonal carbon and nitrogen mineralization in a high-Arctic coastal marine sediment, Young Sound, Northeast Greenland. *Mar. Ecol. Prog. Ser. 175*: 261–276.
Rysgaard, S., T.G. Nielsen and B. Hansen. 1999. Seasonal variation in nutrients, pelagic primary production and grazing in a high-Arctic coastal marine ecosystem, Young Sound, Northeast Greenland. *Mar. Ecol. Prog. Ser. 179*: 13–25.
Rysgaard, S. and R.N. Glud. 2007. Carbon cycling and climate change: predictions for a high Arctic marine ecosystem (Young Sound, NE Greenland) *In*: (S. Rysgaard and R.N. Glud, eds) *Carbon cycling in Arctic marine ecosystems: case study Young Sound*. Meddelelser Grønland Biosci. 58. The Commission for Scientific Research in Greenland, Copenhagen. pp. 206–214.
Sabbe, K. 1997. *Systematics and ecology of intertidal benthic diatoms of the Westernschelde estuary (The Netherlands)*. Doctoral thesis, University of Gent.
Schloss, I., G. Ferreyra and D. Ruiz-Pino. 2002. Phytoplankton biomass in Antarctic shelf zones: a conceptual model based on Potter Cove, King George Island. *J. Mar. Syst. 36*: 129–143.
Schloss, I., G. Ferreyra, O. González, A. Atencio, V.L. Fuentes, G. Tosonotto, G. Mercuri, R. Sahade, M. Tatián and D. Abele. 2008. Long-term hydrographic conditions and climate trends in Potter Cove. *Ber. Polarforsch. Meeresforsch. 571*: 382–389.
Schöne, T., M. Pohl, A. Zakrajsek and H. Schenke. 1998. Tide gauge measurements – a contribution for the long-term monitoring of the sea level. *Ber. Polarforsch. Meeresforsch. 299*: 12–14.
Schroeter, B., M. Olech, L. Kappen and W. Heitland. 1995. Ecophysiological investigations of *Usnea antarctica* in the Maritime Antarctic. I. Annual microclimatic conditions and potential primary production. *Antarct. Sci. 7*: 251–260.
Sejr, M.K., S. Rysgaard, D. Mikkelsen, M. Frederiksen, E.R. Frandsen, K. Lennert and T. Juul-Pedersen. 2007. Zackenberg Basic: the Marine Basic programme. *In*: (A.B. Klitgaard and M. Rasch, eds) *Zackenberg ecological research operations. 13th Annual Report, 2007.* Danish Polar Center, Danish Agency for Science, Technology and Innovation, Ministry of Science, Technology and Innovation, Copenhagen. pp. 57–69.
Simpson, W.R., R. Glasow von, K. Riedel, P. Anderson, P. Ariya, J. Bottenheim, J. Burrows, L. Carpenter, U. Frieß, M.E. Goodsite, D. Heard, M. Hutterli, H.-W. Jacobi, L. Kaleschke, B. Neff, J. Plane, U. Platt, A. Richter, H. Roscoe, R. Sander, P. Shepson, J. Sodeau, A. Steffen, T. Wagner and E. Wolff. 2007. Halogens and their role in polar boundary-layer ozone depletion. *Atmos. Chem. Phys. Discuss. 7*: 4285–4403.
Svendsen, H., A. Beszczynska-Moeller, J. Hagen, B. Lefauconnier, V. Tverberg, S. Gerland, J. Ørbaek, K. Bischof, C. Papucci, M. Zajaczkowski, R. Attolini, O. Bruland, C. Wiencke,

J. Winther and W. Dallmann. 2002. The physical environment of Kongsfjorden Krossfjorden, an Arctic fjord system in Svalbard. *Polar Res. 21*: 133–166.

Tavares, M. and G.A.S. de Melo. 2004. Discovery of the first known benthic invasive species in the Southern Ocean: the North Atlantic spider crab *Hyas araneus* found in the Antarctic Peninsula. *Antarct. Sci. 16*: 129–131.

Teixido, N., J. Garrabou, J. Gutt and W.E. Arntz. 2007. Iceberg disturbance and successional spatial patterns: the case of the shelf Antarctic benthic communities. *Ecosystems 10*: 142–157.

Thiede, J. 1986. *Zur geologischen Geschichte der Polarmeere und ihrer Wassermassen.* Geologisch-Paläontologisches Institut der Universität Kiel, Kiel. pp. 312.

Thomas, D.N. and G.S. Dieckmann. 2002. Antarctic sea ice: a habitat for extremophiles. *Science 295*: 641–644.

Turner, J., J.E. Overland and J.E. Walsh. 2007. An Arctic and Antarctic perspective on recent climate change. *Int. J. Climatol. 27*: 277–293.

Vaughan, D.G., G.J. Marshall, W.M. Connolley, C. Parkinson, R. Mulvaney, D.A. Hodgson, J. C. King, C.J. Pudsey and J. Turner. 2003. Recent rapid regional climate warming on the Antarctic Peninsula. *Clim. Change 60*: 243–274.

Weatherhead, E.C. and S.B. Andersen. 2006. The search for signs of recovery of the ozone layer. *Nature 441*: 39–45.

Weykam, G., I. Gómez, C. Wiencke, K. Iken and H. Klöser. 1996. Photosynthetic characteristics and C:N ratios of macroalgae from King George Island (Antarctica). *J. Exp. Mar. Biol. Ecol. 204*: 1–22.

Wiencke, C. 1990. Seasonality of brown macroalgae from Antarctica – a long-term culture study under fluctuating Antarctic daylengths. *Polar Biol. 10*: 589–600.

Wiencke, C. and M. Clayton. 2002. Antarctic seaweeds. *In*: (J.W. Wägele, ed) *Synopsis of the Antarctic benthos.* Volume 9. A.R.G. Gantner Verlag KG, Ruggell/Lichtenstein. pp. 239.

Wiencke, C., B. Vögele, N.A. Kovaltchouk and H. Hop. 2004. Species composition and zonation of marine benthic macroalgae at Hansneset in Kongsfjorden. *Ber. Polarforsch. Meeresforsch. 492*: 55–62.

Wiencke, C., M.Y. Roleda, A. Gruber, M.N. Clayton and K. Bischof. 2006. Susceptibility of zoospores to UVR determines upper depth distribution limit of Arctic kelps: evidence through field experiments. *J. Ecol. 94*: 455–463.

Wiencke, C., I. Gómez and K. Dunton. 2009. Phenology and seasonal physiological performance of polar seaweeds. *Bot. Mar. 52*: 585–592.

WMO (World Meteorological Organization). 2007. *Scientific assessment of ozone depletion: 2006.* Global ozone research and monitoring project. Report 50. Geneva. pp. 572.

Wulff, A., K. Zacher, D. Hanelt, A. Al-Handal and C. Wiencke. 2008. UV radiation – a threat to Antarctic benthic marine diatoms? *Antarct. Sci. 20*: 13–20.

Wulff, A., K. Iken, M.L. Quartino, A. Al-Handal, C. Wiencke and M.N. Clayton. 2009. Biodiversity, biogeography and zonation of benthic micro- and macroalgae in the Arctic and Antarctic. *Bot. Mar. 52*: 491–507.

Zacher, K., D. Hanelt, C. Wiencke and A. Wulff. 2007a. Grazing and UV radiation effects on an Antarctic intertidal microalgal assemblage: a long-term field study. *Polar Biol. 30*: 1203–1212.

Zacher, K., A. Wulff, M. Molis, D. Hanelt and C. Wiencke. 2007b. Ultraviolet radiation and consumer effects on a field-grown intertidal macroalgal assemblage in Antarctica. *Global Change Biol. 13*: 1201–1215.

Zachos, J., M. Pagani, L. Sloan, E. Thomas and K. Billups. 2001. Trends, rhythms, and aberrations in global climate 65 Ma to present. *Science 292*: 686–693.

3 Biodiversity, biogeography and zonation of marine benthic micro- and macroalgae in the Arctic and Antarctic

Angela Wulff, Katrin Iken, María Liliana Quartino, Adil Al-Handal, Christian Wiencke and Margaret N. Clayton

Introduction

Marine benthic macroalgae (seaweeds) and microalgae were first recorded from Arctic and Antarctic regions in the first half of the 19th century. Most early studies were taxonomic and biogeographical, and the majority concentrated on the most conspicuous forms, the seaweeds. Papenfuss (1964) published a comprehensive catalogue of the early studies of Antarctic and sub-Antarctic benthic marine algae. The first major studies on the Arctic benthic seaweeds included the works of Kjellman (1883), Rosenvinge (1898) and Lund (1959a,b) and various other publications detailed by Wiencke et al. (2007). The first group of polar microalgae to attract taxonomic attention was the diatoms. Knowledge of other microalgal taxa followed, but in general, polar benthic microalgae are still much less well known than the macroalgae. Since the mid 20th century, knowledge and understanding of seaweed diversity, ecology and ecophysiology has advanced greatly, particularly following the introduction of diving. However, as this chapter indicates, many gaps still remain in our knowledge of the benthic macroalgae and microalgae inhabiting these remote and relatively inaccessible regions of the planet.

The scientific authorities for binomials of microalgae and macroalgae are presented in Tables 3.1 and 3.2, respectively.

Composition and zonation of macroalgae in the Arctic

Macroalgae are common elements of nearshore hard substratum communities in both polar regions, but there are distinct differences in the composition of marine macroalgae in the Arctic and Antarctic. Arctic macroalgae are of Atlantic and Pacific origin, with only a few cosmopolitan or endemic species, and many species with pan-Arctic distributions (Lüning 1990, Wilce 1990). Contrary to earlier suggestions that Arctic macroalgae are largely of Atlantic origin (Wilce 1990, Dunton 1992), recent evidence based on molecular data for Arctic endemic species and the Pacific and Atlantic distribution of their sister species suggests that Arctic macroalgae are to a large extent of Pacific origin (Lindstrom 2001, Adey et al. 2008). In the European Arctic, macroalgal composition and zonation patterns are most well-known from Svalbard. Up to 70 algal species are reported from various fjord regions in Svalbard, with species numbers differing depending on the depth range investigated (Florczyk and Latala 1989, Weslawski et al. 1993, Hop et al. 2002). In Kongsfjorden, macroalgae occur in distinct

Table 3.1 Marine benthic diatom species observed in Arctic sea ice and as microphytobenthos or periphyton in the Arctic or Antarctica.

	Arctic ice algae	Arctic/Antarctic microphytobenthos/Periphyton	Comment
Actinocyclus curvatulus Janisch	von Quillfeldt 2000	Al-Handal and Wulff 2008a	
Amphora coffeaeformis (C. Agardh) Kützing	Okolodkov 1993	Al-Handal and Wulff 2008b*	*cf. *coffeaeformis*
Amphora laevis var. *laevissima* (Gregory) Cleve	Poulin and Cardinal 1983, Okolodkov 1993, von Quillfeldt 2000, von Quillfeldt et al. 2003	von Quillfeldt et al. 2003	
Berkeleya rutilans (Trentepohl *ex* Roth) Grunow	von Quillfeldt 1997, 2000	Klöser 1998, von Quillfeldt et al. 2003	Confirmed
Biremis ambigua (Cleve) D.G. Mann	von Quillfeldt 2000, von Quillfeldt et al. 2003	von Quillfeldt et al. 2003	
Cocconeis scutellum Ehrenberg	Okolodkov 1993, von Quillfeldt 2000	von Quillfeldt et al. 2003, Al-Handal and Wulff 2008a, b	
Cylindrotheca closterium (Ehrenberg) Reimann *et* Lewin	Horner and Schrader 1982, Okolodkov 1993, von Quillfeldt 1997, 2000, von Quillfeldt et al. 2003	Horner and Schrader 1982, Klöser 1998, von Quillfeldt et al. 2003, Wulff et al. 2008a, b	Confirmed
Entomoneis gigantea Grunow	Poulin and Cardinal 1983*, Okolodkov 1993*, von Quillfeldt 1997*, Cremer 1999*	Klöser 1998, Al-Handal and Wulff 2008b	var. *septentrionalis*, confirmed *different varieties, confirmed
Entomoneis kjellmanii (Cleve) Poulin *et* Cardinal	Horner and Schrader 1982, Poulin and Cardinal 1983, Okolodkov 1993, von Quillfeldt 1997*, Cremer 1999, von Quillfeldt 2000*, von Quillfeldt et al. 2003*	Al-Handal and Wulff 2008a, Wulff et al. 2008b	
Entomoneis paludosa (W. Smith) Reimer	Horner and Schrader 1982*, Poulin and Cardinal 1983*, Okolodkov 1993*, von Quillfeldt 1997*, Cremer 1999**, von Quillfeldt 2000*	Al-Handal and Wulff 2008a	*var. *hyperborea*, **var. *borealis*
Fragilaria striatula Lyngbye	von Quillfeldt 1997, 2000	Klöser 1998, Blazewicz-Paszkowycz and Ligowski 2002, Karsten et al. 2006, Al-Handal and Wulff 2008b, Campana et al. 2008	Confirmed

Gyrosigma fasciola (Ehrenberg) Griffith *et* Henfrey	Horner and Schrader 1982, von Quillfeldt 2000	Zacher et al. 2007, Al-Handal and Wulff 2008a, Wulff et al. 2008b, c	Confirmed
Gyrosigma tenuissimum (W. Smith) Griffith *et* Henfrey	Okolodkov 1993*, von Quillfeldt 1997*, 2000*, von Quillfeldt et al. 2003	von Quillfeldt et al. 2003	*var. *hyperborea*
Licmophora gracilis (Ehrenberg) Grunow	von Quillfeldt 2000	Zacher et al. 2007, Al-Handal and Wulff 2008a,b	Confirmed
Navicula cancellata Donkin	Horner and Schrader 1982	Horner and Schrader 1982	
Navicula directa (W. Smith) Ralfs	Horner and Schrader 1982, Poulin and Cardinal 1982*, Poulin 1990a, Okolodkov 1993*, von Quillfeldt 1997, Cremer 1999, von Quillfeldt 2000, von Quillfeldt et al. 2003	Horner and Schrader 1982, Klöser 1998, Blazewicz-Paszkowycz and Ligowski 2002, von Quillfeldt et al. 2003, Zacher et al. 2007, Al-Handal and Wulff 2008a, b, Wulff et al. 2008b, c	*var. *directa*, confirmed
Navicula glaciei Van Heurck	Poulin and Cardinal 1982, von Quillfeldt 1997, 2000, von Quillfeldt et al. 2003	Blazewicz-Paszkowycz and Ligowski 2002	
Navicula spicula (Hickie) Cleve	Horner and Schrader 1982	Horner and Schrader 1982	
Navicula transitans Cleve	Horner and Schrader 1982, Poulin and Cardinal 1982*, Cremer 1999, von Quillfeldt et al. 2003	von Quillfeldt et al. 2003	*var. *transitans*
Navicula transitans var. *derasa* (Grunow) Cleve	Horner and Schrader 1982	Horner and Schrader 1982	
Nitzschia angularis W. Smith	Horner and Schrader 1982, Poulin and Cardinal 1983, Cremer 1999	Horner and Schrader 1982	
Nitzschia cylindrus (Grunow) Hasle	Horner and Schrader 1982	Horner and Schrader 1982	
Nitzschia distans W. Gregory	Poulin and Cardinal 1983, Okolodkov 1993*, von Quillfeldt 1997*	Zacher et al. 2007	*var. *erratica*, confirmed
Nitzschia hybrida Grunow	Poulin and Cardinal 1983, Poulin 1990a, Okolodkov 1993, von Quillfeldt 2000, von Quillfeldt et al. 2003	Klöser 1998, von Quillfeldt et al. 2003, Zacher et al. 2007, Al-Handal and Wulff 2008a, b	Confirmed
Pinnularia quadratarea (A. Schmidt) Cleve	Horner and Schrader 1982, Okolodkov 1993*, von Quillfeldt 1997, Cremer 1999, von Quillfeldt 2000, von Quillfeldt et al. 2003	Klöser 1998, Blazewicz-Paszkowycz and Ligowski 2002, von Quillfeldt et al. 2003, Al-Handal and Wulff 2008a, b, Wulff et al. 2008b, c	*Several varieties, confirmed

Table continued on next page.

Table 3.1 (Continued)

	Arctic ice algae	Arctic/Antarctic microphytobenthos/Periphyton	Comment
Odontella aurita (Lyngbye) C.A. Agardh	Okolodkov 1993, von Quillfeldt 1997	Wulff et al. 2008c*	*var. *aurita* and *obtusa*, confirmed
Paralia sulcata (Ehrenberg) Cleve	von Quillfeldt 1997, Cremer 1999	von Quillfeldt et al. 2003	
Plagiotropis longa (Cleve) Kuntze	von Quillfeldt 2000, von Quillfeldt et al. 2003	von Quillfeldt et al. 2003	
Pseudogomphonema kamtschaticum (Grunow) Medlin	von Quillfeldt 1997, Cremer 1999	Klöser 1998, Blazewicz-Paszkowycz and Ligowski 2002, Zacher et al. 2007, Al-Handal and Wulff 2008a, b	Confirmed
Rhabdonema arcuatum (Lyngbye) Kützing	von Quillfeldt 2000	Blazewicz-Paszkowycz and Ligowski 2002, Al-Handal and Wulff 2008b	
Tabularia tabulata (C. Agardh) Snoeijs	von Quillfeldt 2000	Al-Handal and Wulff 2008b	
Tabularia investiens (W. Smith) Williams *et* Round	von Quillfeldt 2000*	Klöser 1998	Confirmed
Trachyneis aspera (Ehrenberg) Cleve	von Quillfeldt 2000	Klöser 1998, Blazewicz-Paszkowycz and Ligowski 2002, Al-Handal and Wulff 2008a, b, Wulff et al. 2008a, b*	*var. *aspera*, confirmed
Triceratium arcticum (Brightwell) Cleve	von Quillfeldt 2000	Klöser 1998, Blazewicz-Paszkowycz and Ligowski 2002, Al-Handal and Wulff 2008b	Confirmed

'Confirmed' indicates the species has been reported as living/growing in the benthos. In some cases, varieties have not been reported. Taking into account the work by Cremer 1999, all samples are from surface sediment, but the original habitat has not been confirmed. The group designated "Arctic-marine-epibenthic" has been included in sea ice algae but could also be sediment-living, as indicated by Cremer 1999: "[...] these surface sediment assemblages are originally composed of a mixture of accumulated [...] diatom valves and, to a certain degree, still photosynthetically active benthic diatoms, especially at shallower water depths [...]."
For explanation of asterisks (*, **; column 2), please see comment column.

vertical zonation mainly between the intertidal down to 20 m depth, but occasionally, some foliose red algae can grow as deep as 60 m (Hop et al. 2002, Wiencke et al. 2004, C. Wiencke unpublished data). About 20 species can be found intertidally in various locations in Svalbard (Florczyk and Latala 1989, Weslawski et al. 1993). Where intertidal algae occur, the assemblages are dominated by the brown algae *Fucus evanescens*, *Pylaiella littoralis*, and sometimes *Elachista fucicola*, and the green algae *Ulothrix* spp. and *Urospora penicilliformis* (Florczyk and Latala 1989, Hansen and Haugen 1989). In exposed areas, the phaeophyte *Chordaria flagelliformis* can also be abundant (Hop et al. 2002). Species richness increases in the upper sublittoral (<2 m), where, aside from the intertidal species, the green algae *Acrosiphonia* spp. and *Spongomorpha* spp. and the red alga *Dumontia contorta* (cited as *D. incrassata*) also become abundant (Florczyk and Latala 1989, Hop et al. 2002). Between 5 and 15 m, the benthic community becomes dominated by a perennial or annual kelp canopy of *Saccorhiza dermatodea, Alaria esculenta, Laminaria digitata* and *Saccharina latissima* (formerly *Laminaria saccharina*). *Laminaria solidungula* is a shade-adapted undergrowth and deep-water species (Florczyk and Latala 1989, Weslawski et al. 1993, Hop et al. 2002, Wiencke et al. 2004, Carlsen et al. 2007). The latter species is reported from Kongsfjorden but not from the more southerly Hornsund. In addition, the phaeophytes *Desmarestia aculeata* and *D. viridis* can be very abundant locally (e.g., Beuchel and Gulliksen 2008). The understory and deeper assemblages are mainly composed of red algae including *Devaleraea ramentacea, Palmaria palmata, Phycodrys rubens* and *Ptilota gunneri* (Hop et al. 2002, Wiencke et al. 2004). In addition, encrusting coralline red algae abundantly cover open rock surface in the sublittoral zone.

The algal flora in Svalbard is generally composed of species also found in other regions of the North Atlantic and very few species with a strong distributional centre in the Arctic (e.g., *Laminaria solidungula, Devaleraea ramentacea*). The southern archipelago may also be the northern distribution limit for some boreal species, for example the kelp *Laminaria hyperborea* (Chapter 12, Müller et al. 2009). This coincides with the species composition and distribution in the sub-Arctic Bear Island, which overall, is very similar to that described for Svalbard, but where these boreal species are more prominent and the Arctic-endemic species do not occur (Weslawski et al. 1997).

A flora similar to that within the Svalbard archipelago extends eastwards along the Russian Arctic coast. Overall, the macroalgal flora along the Russian Arctic coast contains an impoverished assemblage of the cold-temperate flora reported from Svalbard. Dominant sublittoral species in the White Sea region of the Murman coast are similar to those in Svalbard, with kelps like *Saccharina latissima, Laminaria digitata, Alaria esculenta* and *Saccorhiza dermatodea* dominating the subtidal community (Tzetlin et al. 1997, Mikhaylova 1999) and red algae such as *Phycodrys rubens* as well as *Odonthalia dentata, Ahnfeltia plicata, Palmaria palmata*, and *Devaleraea ramentacea* forming much of the understory (Schoschina 1996, Makarov et al. 1999). At shallower depths (upper sublittoral), the phaeophytes *Fucus vesiculosus*, *F. distichus*, and *Saccharina latissima,* and the chlorophytes *Cladophora rupestris* and *Chaetomorpha melagonium* are common (Dobretsov 1999, Makarov et al. 1999), although the rockweed *Ascophyllum nodosum* is also reported (Vershinin et al. 2006). The kelp *Laminaria hyperborea*, previously assumed to have its eastern distribution limit on the northeastern-most coast of Norway, does occur off the outer Murman coast in exposed locations (Schoschina 1997).

Table 3.2 List of polar seaweeds species mentioned in the text.

Acrosiphonia arcta (Dillwyn) Gain
Acrosiphonia sonderi (Kützing) Kornmann
Adenocystis utricularis (Bory de Saint-Vincent) Skottsberg
Agarum clathratum Dumortier
Ahnfeltia plicata (Hudson) Fries
Alaria esculenta (Linnaeus) Greville
Ascophyllum nodosum (Linnaeus) Le Jolis
Ascoseira mirabilis Skottsberg
Ballia callitricha (C. Agardh) Kützing
Bangiadulcis atropurpurea (Roth) W.A. Nelson
Chaetomorpha linum (O.F. Müller) Kützing
Chaetomorpha melagonium (F. Weber *et* Mohr) Kützing
Chordaria flagelliformis (O.F. Müller) C. Agardh
Chukchia endophytica (S. Lund) R.T. Wilce, P.M. Pedersen *et* S. Sekida
Chukchia pedicellata R.T. Wilce, P.M. Pedersen *et* S. Sekida
Cladophora rupestris (Linnaeus) Kützing
Clathromorphum obtectulum (Foslie) Adey
Coccotylus truncatus (Pallas) M.J. Wynne *et* J.N. Heine
Curdiea racovitzae Hariot
Cystosphaera jacquinotii (Montagne) Skottsberg
Desmarestia aculeata (Linnaeus) J.V. Lamouroux
Desmarestia anceps Montagne
Desmarestia antarctica R.L. Moe *et* P.C. Silva
Desmarestia chordalis J.D. Hooker *et* Harvey
Desmarestia menziesii J. Agardh
Desmarestia viridis (O.F. Müller) J.V. Lamouroux
Devaleraea ramentacea (Linnaeus) Guiry
Dictyosiphon foeniculaceus (Hudson) Greville
Dilsea integra (Kjellman) Rosenvinge
Dumontia contorta (S.G. Gmelin) Ruprecht
Elachista fucicola (Velley) J.E. Areschoug
Enteromorpha bulbosa (Suhr) Montagne
Fucus distichus Linnaeus
Fucus evanescens C. Agardh
Fucus vesiculosus Linnaeus
Geminocarpus geminatus (J.D. Hooker *et* Harvey) Skottsberg
Georgiella confluens (Reinsch) Kylin
Gigartina skottsbergii Setchell *et* N.L. Gardner
Gymnogongrus turquetii Hariot
Himantothallus grandifolius (A. Gepp *et* E.S. Gepp) Zinova
Iridaea cordata (Turner) Bory de Saint-Vincent
Lambia antarctica (Skottsberg) Delépine
Laminaria digitata (Hudson) J.V. Lamouroux
Laminaria hyperborea (Gunnerus) Foslie
Laminaria solidungula J. Agardh
Leptophytum coulmanicum (Foslie) Adey
Monostroma hariotii Gain
Neuroglossum ligulatum (Reinsch) Skottsberg
Odonthalia dentata (Linnaeus) Lyngbye
Pachymenia orbicularis (Zanardini) Setchell *et* N.L. Gardner
Palmaria decipiens (Reinsch) R.W. Ricker
Palmaria palmata (Linnaeus) Kuntze

Table 3.2 (Continued)

Pantoneura baerii (Ruprecht) Kylin
Pantoneura plocamioides Kylin
Petrocelis polygyna (Kjellman) F. Schmitz
Phaeurus antarcticus Skottsberg
Phycodrys rubens (Linnaeus) Batters
Phyllophora antarctica A. Gepp *et* E.S. Gepp
Phymatolithon foecundum (Kjellman) L. Düwel *et* S. Wegeberg
Phymatolithon tenue (Rosenvinge) L. Düwel *et* S. Wegeberg
Picconiella plumosa (Kylin) De Toni fil.
Platysiphon verticillatus Wilce
Plocamium cartilagineum (Linnaeus) P.S. Dixon
Polysiphonia stricta (Dillwyn) Greville
Porphyra endiviifolia (A. Gepp *et* E.S. Gepp) Y.M. Chamberlain
Prasiola crispa (Lightfoot) Kützing
Ptilota gunneri P.C. Silva, Maggs *et* L.M. Irvine
Punctaria glacialis Rosenvinge
Pylaiella littoralis (Linnaeus) Kjellman
Rhodomela confervoides (Hudson) P.C. Silva
Rhodomela tenuissima (Ruprecht) Kjellman
Saccharina latissima (Linnaeus) C.E. Lane, C. Mayes, Druehl *et* G.W. Saunders
Saccorhiza dermatodea (Bachelot de la Pylaie) J.E. Areschoug
Scagelia americana (Harvey) Athanasiadis
Sphacelaria arctica Harvey
Sphacelaria plumosa Lyngbye
Stictyosiphon tortilis (Ruprecht) Reinke
Trematocarpus antarcticus (Hariot) Fredericq *et* R. Moe
Turnerella pennyi (Harvey) F. Schmitz
Ulothrix flacca (Dillwyn) Thuret
Urococcus foslieanus Hansgirg
Urospora penicilliformis (Roth) J.E. Areschoug

Distribution records of macroalgae around Greenland are sparse. The assemblages are similar, though reduced, to those described for Svalbard (Lee 1973). In eastern Greenland (Young Sound) *Saccharina latissima*, *Fucus evanescens* and *Desmarestia aculeata* are the dominant seaweeds (Chapter 7; Kühl et al. 2001, Borum et al. 2002, Glud et al. 2009). Rocks are often covered with encrusting coralline red algae (*Phymatolithon tenue* and *P. foecundum*), with increasing cover at greater depth (Roberts et al. 2002). In western Greenland (Disko Island), a rich macroalgal flora exists, within which, in addition to the brown algae mentioned above, green algae are also common, including *Acrosiphonia arcta*, *A. sonderi* and *Urospora penicilliformis*, as well as the brown algae *Pylaiella littoralis* and *Chordaria flagelliformis*, and some red algae (e.g., *Devaleraea ramentacea*, *Polysiphonia stricta*, *Bangiadulcis atropurpurea*) (Bischoff and Wiencke 1993).

The seaweed flora of the Canadian High Arctic (mainly the Baffin Bay area) is similar to that found in Svalbard, but with a large portion of algae of clear Pacific affinity, and a total record of about 55 species (Cross et al. 1987). The zonation patterns consist of an intertidal belt of *Fucus evanescens* and *Pylaiella littoralis* with a number of ephemeral green algae such as *Urococcus foslieanus*, *Ulothrix flacca*, *Acrosiphonia arcta* and *Prasiola crispa* (Ellis and Wilce 1961). The lower intertidal zone is dominated in biomass by *Fucus distichus*, *Stictyosiphon tortilis*, *Dictyosiphon foeniculaceus*, and *Pylaiella littoralis* with an understory of *Sphacelaria arctica*, *S. plumosa*, *Chaetomorpha linum* and *C. melagonium* (Cross et al. 1987). The sublittoral kelp zone contains *Saccharina latissima*, *Laminaria solidungula* and also the red algae *Dilsea integra*, *Devaleraea ramentacea*, *Rhodomela confervoides*, other brown algae such as *Punctaria glacialis*, *Desmarestia* sp. and *Chorda* sp. and green algae of the genus *Chaetomorpha* (Ellis and Wilce 1961, Cross et al. 1987). At some locations, *Alaria* sp. is reported (Ellis and Wilce 1961) and at greater depth, the kelp *Agarum clathratum* may become dominant (Cross et al. 1987).

West of the Canadian Arctic Archipelago, on the Alaskan Beaufort Sea shelf, macroalgae are restricted to small areas where hard substrata protected from ice scour are available (Mohr et al. 1957, Dunton et al. 1982). Most information on Beaufort Shelf seaweeds comes from the Boulder Patch in Stefansson Sound. The algal composition had been considered to be mostly of Atlantic origin (Dunton 1992), but recently, Lindstrom (2001) and Adey et al. (2008) challenged this view and proposed both Atlantic and Pacific affinities. There are no intertidal algae. Kelp beds develop in about 5–10 m depth with *Laminaria solidungula*, *Saccharina latissima* and occasionally *Alaria esculenta* (Dunton and Schonberg 2000). The understory is a dense cover of the red algae *Phycodrys rubens* and *Coccotylus truncatus*. Other common understory red algae include *Odonthalia dentata*, *Rhodomela confervoides*, *Ahnfeltia plicata* and *Dilsea integra* (Dunton and Schonberg 2000). Exposed rock is often encrusted with coralline red algae (*Lithothamnium* spp., Dunton et al. 1982, Konar and Iken 2005). Recent investigations increased the total number of species in the Boulder Patch from about 10 (Dunton and Schonberg 2000) to 15. New records in the Boulder Patch include the green alga *Chaetomorpha melagonium*, the red algae *Rhodomela tenuissima* and *Scagelia* cf. *americana* and the brown algae *Sphacelaria plumosa* and *S. arctica* (K. Iken, unpublished data). Similar macroalgae also occur in other isolated locations along the Alaskan Beaufort Sea, such as Demarcation Point at the Alaska-Yukon border and Camden Bay (K. Dunton and K. Iken, respectively, unpublished data), and, supposedly, southwest of Barrow in the Chukchi Sea (Mohr et al. 1957).

Composition and zonation of macroalgae in the Antarctic

The rocky shores along the Western Antarctic Peninsula are covered by extensive subtidal macroalgal assemblages, often providing 70% or more of bottom coverage and with standing biomass stocks that rival those in temperate kelp forests (Amsler et al. 1995, Brouwer et al. 1995, Klöser et al. 1996, Quartino et al. 2001). Since the early Antarctic explorations in this region that included botanical observations, the number of recorded macroalgal species has increased from 96 species (Skottsberg 1964) to about 120–130 (Wiencke and Clayton 2002), most of them restricted to the Western Antarctic along the Peninsula. A further increase in species number is expected. Only recently Hommersand et al. (2009; Chapter 4) described four new genera and five new red algal species. The Antarctic marine flora is characterized by a relatively high proportion of endemism (Wiencke and Clayton 2002; see also below).

Distinct depth zonation patterns have been described from macroalgal assemblages along the Antarctic Peninsula, mainly from the region between King George Island and Anvers Island. Intertidal communities are rare, but can establish in sheltered locations.

Where they occur, they mostly consist of the red alga *Porphyra endiviifolia* in protected, shaded rock crevices, the green algae *Prasiola crispa* in the supralittoral, and *Urospora penicilliformis*, *Ulothrix* spp., *Enteromorpha bulbosa* and *Acrosiphonia arcta* covering moist rock surfaces or tide pools (DeLaca and Lipps 1976, Westermeier et al. 1992, Chung et al. 1994, Wiencke and Clayton 2002, Amsler et al. 2005, Wiencke et al. 2007). Occasionally, the brown alga *Adenocystis utricularis* and the red algae *Iridaea cordata* and *Curdiea racovitzae* can form intertidal stands (DeLaca and Lipps 1976, Heywood and Whitaker 1984, Castilla and Rozbaczylo 1985, Chung et al. 1994). Similarly, the upper 3–5 m of the subtidal zone are often devoid of fleshy macroalgae because of severe ice abrasion. However, rocks are abundantly covered with encrusting coralline red algae (e.g., *Lithothamnion* sp., Heywood and Whitaker 1984, Wiencke and Clayton 2002).

Overall, the deeper macroalgal assemblages are dominated by canopies of large brown algae, which are perennial, non-acidic members of the order Desmarestiales that ecologically replace the Laminariales occurring in temperate waters and the Arctic (Clayton 1994). *Desmarestia anceps*, *D. menziesii* and *Himantothallus grandifolius* are the main species with the highest macroalgal biomass along the Antarctic Peninsula (Amsler et al. 1995, Brouwer 1996, Quartino et al. 2001, 2005). *D. menziesii* typically occurs just below the ice abrasion zone and is dominant to about 10 m depth. At greater depths it is replaced by *D. anceps*, which dominates to about 15–20 m, where it is replaced by *H. grandifolius*. The latter species is common to depths of at least 35–40 m (DeLaca and Lipps 1976, Lamb and Zimmermann 1977, Zielinski 1990, Klöser et al. 1994, Amsler et al. 1995, Brouwer et al. 1995, Quartino et al. 2001, Quartino and Boraso de Zaixso 2008). Occasionally, the annual *D. antarctica* can be locally important in the upper *D. menziesii* zone (Amsler et al. 1995). Similarly, the large phaeophyte *Ascoseira mirabilis*, the sole member within the Antarctic-endemic, mono-specific order Ascoseirales, may be common within the *Desmarestia* belt; and *Cystosphaera jacquinotii* can co-dominate with *H. grandifolius* below 25 m, usually in more wave-exposed areas (Zielinski 1990, Chung et al. 1994, Klöser et al. 1994, 1996).

These brown algal stands are accompanied by diverse understory assemblages. In the upper *Desmarestia menziesii* zone, the green algae *Monostroma hariotii* and *Enteromorpha bulbosa*, the brown algae *Adenocystis utricularis* and *Phaeurus antarcticus*, and a variety of red algae such as *Iridaea cordata*, *Palmaria decipiens* and *Curdiea racovitzae* are common (DeLaca and Lipps 1976, Zielinski 1981, Heywood and Whitaker 1984, Quartino et al. 2001). At intermediate depths, the *Desmarestia* belt is interspersed with *Gigartina skottsbergii*, *Trematocarpus antarcticus*, *Neuroglossum ligulatum*, *Gymnogongrus turquetii*, *Plocamium cartilagineum*, *Pantoneura plocamioides*, *Georgiella confluens* (Klöser et al. 1994, Quartino et al. 2005). At greater depth among *Himantothallus grandifolius* individuals, the red algae *Georgiella confluens*, *Ballia callitricha*, *Plocamium cartilagineum,* and *Picconiella plumosa* compose the common understory (Quartino et al. 2001, 2005). Differences in species composition within the depth zones are mainly related to substratum type and exposure (Richardson 1979, Klöser et al. 1994). The high diversity at intermediate depths can be explained by patches of different successional stages caused by physical disturbance such as iceberg scouring (Quartino et al. 2005).

The main depth distribution patterns are similar along the Antarctic Peninsula but species richness and biomass decline with increasing latitude (DeLaca and Lipps

1976, Moe and DeLaca 1976). For example, *Desmarestia anceps, D. antarctica,* and *Cystosphaera jacquinotii* disappear south of 66° S. Recent investigations report sparse macroalgal assemblages at Adelaide Island (67° S, 68° W), with only a few red (*Trematocarpus antarcticus*, *Hildenbrandia* sp., encrusting corallines) and brown (*Desmarestia antarctica*) algae present (Bowden 2005). This low algal species richness at Adelaide Island is similar to patterns described from other Antarctic locations at similar latitudes (e.g., Zaneveld 1966, Dhargalkar et al. 1988, Kirkwood and Burton 1988, Johnston et al. 2007), and has been attributed to the decrease of light availability at these higher latitudes (Moe and DeLaca 1976). In the Lützow-Holm Bay (Eastern Antarctic, 69° S, 39° E), macroalgae are sparse and only a small population of *Desmarestia chordalis* has been reported (Takeuchi and Watanabe 2002). In the East Antarctic Vestfold Hills (68° S, 77° E), nine species of fleshy macroalgae have been reported, although this may not be a complete collection of the area (Dhargalkar et al. 1988). Among these, *Desmarestia menziesii* and *Himantothallus grandifolius* are common brown algae and *Phyllophora antarctica*, *Iridaea cordata* and *Palmaria decipiens* are common red algae, with *Phyllophora antarctica* being dominant. In other regions of Ellis Fjord (Vestfold Hills), *P. antarctica* is the only recorded macroalgal species (Kirkwood and Burton 1988). Farther east, along the Windmill Island coast (66° S, 110° E), *Himantothallus grandifolius*, *Desmarestia* sp., *Palmaria decipiens* and encrusting coralline red algae are the dominant species together with an undefined "algal turf understory" (Johnston et al. 2007).

The southern-most location of open water where macroalgae occur is in the Ross Sea. There is a distinct latitudinal gradient in macroalgal diversity along the Victoria Land coast in the Ross Sea, with 17 algal species reported from Terra Nova Bay (74° S, 165° E), with four green algal, nine red algal and four brown algal species (Cormaci et al. 1992). Most abundant in this region are the encrusting coralline red alga *Clathromorphum obtectulum*, and the fleshy red algae *Iridaea cordata*, *Phyllophora antarctica*, *Plocamium cartilagineum*, as well as the green algae *Urospora penicilliformis*, *Monostroma hariotii* and *Prasiola crispa*. The large brown algae *Himantothallus grandifolius*, *Desmarestia anceps* and *D. menziesii*, not found in Terra Nova Bay, have a patchy distribution in the northern Ross Sea and seem to be restricted to regions north of Cape Hallett (72° S, 170° E; Smith et al. 2007). Only three red algal species occur regularly further south in the Ross Sea, in McMurdo Sound: *Iridaea cordata*, *Phyllophora antarctica* and the coralline crust *Leptophytum coulmanicum* (Miller and Pearse 1991, Cormaci et al. 1996, 1998). The presence of the green alga *Monostroma hariotii* has not been confirmed since early investigations, and a few other collections remained unidentified (Zaneveld 1966, Miller and Pearse 1991). Among the regularly occurring red algae, *Iridaea cordata* occupies shallower depths of 3–5 m, while *Phyllophora antarctica* and *Leptophytum coulmanicum* are abundant at depths below 10 m (Miller and Pearse 1991). *Phyllophora antarctica* often occurs detached and is transported to greater depth, but is viable in its unattached state (Miller and Pearse 1991, Amsler et al. 1998, Norkko et al. 2004).

Composition and distribution of microalgae in the Arctic

While information about the biodiversity of Antarctic marine benthic microalgae is scarce, it is virtually absent for the Arctic region. The majority of studies on Arctic benthic microalgae have focused on freshwater habitats and ice algae (Poulin 1990a and references therein, Douglas and Smol 1995, Michelutti et al. 2003). In particular, ice algae in the Canadian High Arctic have been well studied (Poulin 1990a, b, Poulin 1991, Lim et al. 2008 and references therein). However, it is worth mentioning that Arctic benthic diatoms and ice algae were collected during the Vega expedition as early as 1883 (Cleve 1883). In surface sediment samples from the Laptev Sea, Cremer (1999) observed several typical benthic diatoms although they were never separated into dead vs. living cells. Also, Polyakova (1989, and references therein) gives long species lists of benthic diatoms from Arctic shallow seas sediments, but their true habitat is impossible to establish from the information given. Due to concern about climate change, the marine ecosystem of Arctic Kongsfjorden (Svalbard, Norway) has also received considerable attention lately (Hop et al. 2002). However, in this extensive review (op. cit.), benthic microalgae were not included due to the fact that they "remain a largely unknown part of the autotrophic system" (Hop et al. 2002). This gap has been partly filled by a study of Woelfel et al. (2009; chapter 8) on microphytobenthic biomass at several coastal sites of Kongsfjorden. Around Greenland, Arctic microphytobenthos has been in focus recently due to its contribution to shallow bottom (<30 m) primary productivity (e.g., Glud et al. 2002, Glud 2008). In these studies (op. cit.), the composition of the microphytobenthos, dominated by pennate diatoms, was described only superficially (in total, three genera were mentioned: *Navicula*, *Nitzschia* and *Pinnularia*). Several pennate genera are found both in the ice and on soft bottom sediments, for example *Navicula*, *Nitzschia*, *Fragilaria*, *Gyrosigma* and *Pleurosigma* (Horner 1990, Poulin 1990a, Al-Handal and Wulff 2008a, b). This is not surprising since it has been suggested that the origin of sea ice algae might be benthic diatoms. The underlying transfer mechanism is likely bottom and water column frazil ice movement, which lifts benthic diatoms (and sediment) to the surface ice (Reimnitz et al. 1987, Garrison et al. 1989). However, when the ice melts, the sinking ice algae do not seem to be incorporated into the microphytobenthos (Matheke and Horner 1974, Horner and Schrader 1982). When comparing published species lists of polar ice algae with the very species-limited lists from soft bottom and epiphytic habitats, it is apparent that at least some genera seem to thrive in both habitats (Table 3.1). Hoppenrath and Okolodkov (2000) reported that the benthic dinoflagellate *Amphidinium glabrum*, a sediment-living species, is also present in Arctic sea ice. The genus *Amphidinium* also grows frequently on soft bottom sediment in sub-polar or temperate regions (Wulff et al. 1999). Clearly, there is a substantial knowledge gap for the distribution of polar benthic microalgae in general, and particularly for living cells.

Composition and distribution of microalgae in Antarctica

Interest in Antarctic microalgae may have started with the study of Ehrenberg (1843) who received samples collected during the *Erebus* and *Terror* expedition (1839–43). He introduced the first diatom record from that remote region. A number of expeditions followed, including the *Challenger* (1872–76), *Antarctica* (1895), *Belgica* (1897–99), *Southern Cross* (1899–1909) and *Scotia* (1902–4). These and other expeditions yielded major works on the benthic diatoms such as those of Van Heurck (1909), West and West (1911), Fritsch (1912), Mangin (1915), Peragallo (1921) and Heiden and Kolbe (1928).

The majority of investigations on benthic Antarctic microalgae, both marine and freshwater, focused on diatoms, and to a lesser extent on dinoflagellates and silicoflagellates. The Antarctic microphytobenthos is dominated by diatoms, which have been investigated both ecologically and taxonomically (e.g., Manguin 1957, 1960, Hustedt 1958, Krebs 1983, Krebs et al. 1987, Leventer 1992, Tanimura 1992, Moisan and Fryxell 1993, Riaux-Gobin 1994, Andreoli et al. 1995, Ahn et al. 1997, Riaux-Gobin et al. 2000, Sabbe et al. 2001, Ligowski 2002, Cremer et al. 2003, Longhi et al. 2003, Zacher et al. 2007, Al-Handal and Wulff 2008a,b, Wulff et al. 2008a–c). Further, based on their good preservation potential in sediments and their sensitivity to environmental changes, diatoms have been widely used as biological markers for reconstructing paleolimnological and paleooceanographical Antarctic environments (Spaulding and McNight 1999, Gersonde and Zielinski 2000, Cremer et al. 2003). However, despite the large number of investigations made on marine Antarctic diatoms, it is still difficult to draw a complete picture of their biogeographic distribution and relative abundance.

Little is known about the biodiversity and distribution of the non-diatom benthic microalgae (Cyanobacteria excluded) in the marine environments around Antarctica, and most studies are from Antarctic freshwater habitats. There is a wide and variable system of freshwater habitats on the Antarctic continent and its off-lying islands, ranging from small streams to large lakes. Here also, diatoms dominate but chlorophytes form an important component of the microphytobenthos. Fumanti and Cavacini (2005) counted more than 300 taxa of freshwater microalgae of which diatoms and chlorophytes constituted 97%.

The biodiversity and distribution of the benthic microalgae is controlled by the unique and extreme physical and chemical characteristics of Antarctic habitats. Most important among these are the presence of pack ice and its seasonal cycle, and the variability of the light regime, which has long light and dark periods. Sea ice is the major habitat for microalgae; the ice flora creates a potential inoculum for the spring algal flora at the marine ice-edge zone (Palmisano and Garrison 1993). Another important component of the benthic microalgae comprises epiphytic diatoms; they are important because of their attachment to a living substratum and the possibility of biological interactions occurring between epiphytes and their hosts, as well as their considerable contribution to the primary production of the coastal marine ecosystem. Epiphytic diatoms have not been widely investigated in Antarctic marine habitats. However, there have been three investigations: one by Thomas and Jiang (1986) at Davis Station who listed nine diatoms growing on 17 different macrophytes, Ligowski (2002) who found 83 epiphytes, and Al-Handal and Wulff (2008b) who reported 50 diatom species thriving on one chlorophyte, seven phaeophytes and 11 rhodophytes.

Endemism and biogeographical relationships

Although various features of both polar biogeographic regions are similar today, there are strong differences between them, especially with respect to their genesis and their cold-water histories (Chapter 2; Zacher et al. 2009a). The Arctic Ocean is a "Mediterranean" ocean, whereas the Southern Ocean forms a ring ocean around the Antarctic continent. The coasts of the Arctic Ocean are continuously connected to the temperate coasts of America and Eurasia, whereas the Southern Ocean has had no land bridge to temperate regions since the late Mesozoic when it became further separated from the neighbouring continents to the north by the Antarctic Circumpolar Current (Hempel 1987, Lüning 1990). The sea ice cover in the Arctic did not develop before 2 Ma ago (Clarke 1990), whereas Antarctica was already glaciated 14 Ma ago (Clarke 1990, Crame 1993, Zachos et al. 2001). So both polar regions have very different cold-water histories. These differences in their genesis and cold-water histories are the main reasons for the different degree of endemism between the regions and are strongly reflected in the biogeographical relationships. A first summary of the biogeographic relationships based on molecular data is given by Hommersand et al. (2009; Chapter 4).

So far, very few endemic seaweed species have been found in the Arctic, e.g., the phaeophytes *Laminaria solidungula, Punctaria glacialis, Platysiphon verticillatus Chukclia* and the rhodophytes *Petrocelis polygyna, Neodilsea integra, pedicellata, C. endophytica*, *Turnerella pennyi*, and *Pantoneura baerii* (Lüning 1990, Wilce 1990, 2009). The geographic distribution of most species with their centres of distribution in the Arctic extends well into the temperate region, e.g., the phaeophyte *Laminaria solidungula* and the rhodophytes *Devaleraea ramentacea, Turnerella pennyi, Neodilsea integra* and *Pantoneura baerii*. In contrast to the Arctic, there is a high degree of endemism in Antarctica. Overall, about a third of the 120 species recorded so far are endemic to the Antarctic region (Wiencke and Clayton 2002). The endemism is highest in the Heterokontophyta (Phaeophyceae and Chrysophyceae; 44%) and the Rhodophyta (32%) and is lowest in the Chlorophyta (18%). One order, the kelp-like brown algal order Ascoseirales, and several genera are endemic to the Antarctic. The endemic genera include the phaeophytes *Himantothallus, Cystosphaera* and *Phaeurus*, the rhodophytes *Gainia Notophycus* and *Antarcticothamnion*, and the chlorophytes *Lambia* and *Lola* (Wiencke and Clayton 2002). Well-known Antarctic endemic species are the phaeophytes *Himantothallus grandifolius*, *Cystosphaera jacquinotii*, the rhodophyte *Porphyra endiviifolia* and the chlorophyte *Lambia antarctica*.

The Southern Ocean is the only region in the world devoid of kelps, brown algae of the order Laminariales (Moe and Silva 1977). This order is ecologically replaced by the Desmarestiales, which is considered to have originated in the Antarctic region and subsequently radiated into the Northern Hemisphere according to phylogenetic, morphological and ecophysiological studies (Peters et al. 1997). It has been hypothesized that over 90% of the seaweeds in the Arctic originated from Atlantic populations (Dunton 1992), but newest examinations of molecular data show a strong biogeographic link of Arctic macroalgae to the North Pacific (Adey et al. 2008). The Atlantic biogeographic connection is particularly obvious in regions with strong influx of Atlantic waters, e.g., in Svalbard (Hop et al. 2002, Svendsen et al. 2002). There are, however, also affinities to the Pacific and the Indo-Pacific region (Wilce 1990, Lindstrom 2001) as indicated by the high proportion of macroalgae of Pacific origin e.g., in the Canadian Arctic (Cross et al. 1987). Based on phylogenetic trees of Arctic endemic red algae and their sister species, a prevalence of Pacific origin for many Arctic species was established (Adey et al. 2008). Another example of an invader from the Pacific may be *Acrosiphonia arcta*. Populations of this species from the Arctic and the North Atlantic seem to originate from the Pacific according to molecular studies (van Oppen et al. 1994). Although genetic evidence of Pacific origin is less strong in Laminariales because of their high dispersal and speciation rates (Adey et al.

2008), kelps in general are thought to have evolved in the North Pacific and radiated into the Atlantic Ocean via the trans-Arctic route (Lüning and tom Dieck 1989, Lane et al. 2006).

In the Southern Ocean, the second largest group of seaweed species after the endemic Antarctic species includes those occurring both in the Antarctic region and on sub-Antarctic islands and Tierra del Fuego such as the rhodophyte *Iridaea cordata*, the phaeophyte *Geminocarpus geminatus* and the chlorophyte *Enteromorpha bulbosa* (Wiencke and Clayton 2002, Hayward and Morley 2005). Some species even occur in South Africa, New Zealand and/or Australia, e.g., the rhodophytes *Pachymenia orbicularis* and *Ballia callitricha* or the phaeophyte *Adenocystis utricularis* (Womersley 1990, Stegenga et al. 1997).

A few seaweed species have a disjunct amphiequatorial distribution and occur both in the Arctic and Antarctic, e.g., the chlorophyte *Acrosiphonia arcta* and the phaeophyte *Desmarestia viridis/confervoides*. According to molecular studies, the biogeographic disjunctions of these species are recent and probably date back to the maximum of the Würm/Wisconsin glaciation 18,000 years ago (van Oppen et al. 1993). The units of dispersal were almost certainly the microscopic stages in the life cycle, which are more resistant to high temperatures, allowing a passage through the tropics during times of lowered water temperatures (Peters and Breeman 1992, Wiencke et al. 1994, Bischoff and Wiencke 1995).

For polar benthic microalgae, the extent of endemism remains to be researched. The diversity of the Antarctic benthic microalgae (as well as sea ice and plankton taxa) so far reported contains low numbers of endemic species (Broady 1996, Jones 1996). This, however, may be attributed to the wide use of the European literature for identification (Sabbe et al. 2003). Furthermore, “the cosmopolitan view” (Finlay 2002, 2004) claiming that global microbiota consist of a small number of cosmopolitan species has recently been challenged (Logares 2006). In contrast to “the cosmopolitan view”, molecular data point at a microbial diversification with a higher level of endemism than previously thought (Logares 2006). So far, the majority of the cosmopolitan taxa are those belonging to the Heterokontophyta (mostly diatoms and dinoflagellates), with very few taxa that are Antarctic endemics. Several species of microalgae have been recently described from Antarctica and have not been observed elsewhere, indicating a considerable level of endemism (Vincent 2000). There are far too few studies on marine Arctic benthic microalgae to be able to point in any direction. The increasing number of studies on Arctic marine microphytobenthos might lead to conclusions similar to those for Antarctica, that is, there might be a considerable level of endemism (Vincent 2000). Besides traditional taxonomic studies, modern molecular techniques such as rRNA gene sequencing are required to provide insight into the genetic diversity and distribution of both the Antarctic and Arctic benthic microalgae.

Structuring factors for polar marine algal assemblages

Polar marine micro- and macroalgal assemblages are structured in ways similar to temperate and tropical assemblages, i.e., by a variety of environmental and biological factors. A basic requirement for polar macroalgae, as for most macroalgae in other geographic regions, is the occurrence of hard substrata (excluding the few forms that are able to anchor in sand by their rhizoids). In the Arctic, coastal areas are often characterized by soft-silty substratum because of glacial or river deposits (Lee 1973, Svendsen et al. 2002). This is particularly obvious in the Beaufort Sea where macroalgae are restricted to isolated boulder deposits (Dunton et al. 1982), although other Arctic regions have abundant rocky littoral areas. In the Antarctic, hard substratum is abundant along the Peninsula but many other regions of the shallow coastline are permanently covered by glaciers or shelf ice. Different Antarctic algal species have varied occurrences on different substrata from bedrock to pebbles, but, overall, they are more abundant on rock-covered areas than sandy substrata (Richardson 1979). Among the dominant large brown algae, *Himantothallus grandifolius* is mostly attached to small pebbles while *Desmarestia* spp. are restricted to larger boulders and bedrock, and some species are able to take hold in soft sediments, like the green alga *Lambia antarctica* (Neushul 1965, Delépine et al. 1966, DeLaca and Lipps 1976, Richardson 1979, Zielinski 1990, Brouwer et al. 1995, Klöser et al. 1996).

Polar benthic microalgae inhabit soft bottom sediments, rocks or live as epiphytes on other micro- or macroalgae. The dominating group of microalgae comprises pennate diatoms, but dinoflagellates, cyanobacteria, cryptomonads, chrysophytes, euglenoids and unicellular green algae are also present. Whereas epipelic diatoms have the ability to migrate within the sediment, the epipsammic diatoms are firmly attached, e.g., to sandgrains (Round 1971). Thus, epipelic diatoms can actively seek more optimal light and nutrient conditions compared to the epipsammon (Harper 1977). Epipelic diatoms seem to depend on more sheltered conditions and silt accumulation, while the epipsammic diatoms prefer the presence of larger sand grains within the sediment (Sabbe 1997). Tidal currents are a stress factor causing regular physical disturbance of the sediment, and the microalgae are subjected to frequent events of suspension and settling. Resuspended benthic diatoms are thought to be a substantial carbon source for benthic macrofauna in both polar regions (Gilbert 1991a,b, Ahn et al. 1994, Glud et al. 2002).

The influence of seasonal or permanent ice obviously is of great importance to benthic organisms in polar regions (Chapter 2; Zacher et al. 2009a). Intertidal regions in both polar systems are often ice-covered in winter, and are also scoured intensively during other times of the year. However, the severity of ice scour in the intertidal depends much on local conditions such as exposure, currents, vicinity to glaciers etc. (Lee 1973). While ice impact often prevents the establishment of permanent intertidal macroalgal assemblages, there are many regions in the Arctic where perennial species like *Fucus* spp. survive and dominate the intertidal zone (Ellis and Wilce 1961, Cross et al. 1987, Florczyk and Latala 1989). Ephemeral species such as *Chordaria flagelliformis*, *Pylaiella littoralis*, and *Acrosiphonia* spp., can also successfully inhabit the intertidal because they survive ice-covered winters as microscopic stages or rhizoidal cushions (Hop et al. 2002). In the Antarctic, intertidal macroalgal assemblages are less common than in the Arctic, possibly due to more severe ice effects (Neushul 1965, Moe and DeLaca 1976). Some intertidal species are restricted to protected rock crevices (*Porphyra endiviifolia*) or to ephemeral assemblages (*Enteromorpha bulbosa*, *Urospora penicilliformis*) in sheltered locations (DeLaca and Lipps 1976, Westermeier et al. 1992, Amsler et al. 2005). Occasionally, more developed assemblages occur, including species like *Adenocystis utricularis* and *Monostroma hariotii* (Heywood and Whitaker 1984). At high latitudes in the Ross Sea, crusts of the red alga *Iridaea cordata* can survive heavy ice scour and anchor ice in shallow depths (Miller and Pearse 1991).

Although both benthic and planktonic microalgae frequently inhabit sea ice, ice is also in itself a structuring factor. Benthic microalgae can be completely removed by ice-scouring (Teixido et al. 2007), and bottom and water column frazil ice can lift the microalgae (mostly pennate diatoms) and sediment to the ice (Reimnitz et al. 1987, Garrison et al. 1989). However, when the ice melts, the sinking microalgae do not seem to be incorporated into the microphytobenthos, possibly due to the very low salinity meltwater lens immediately under the ice, transport of the ice algae by coastal currents away from the bottom (Matheke and Horner 1974, Horner and Schrader 1982), or effective removal by pelagic and benthic suspension and surface deposit feeders (e.g., McMahon et al. 2006, Søreide et al. 2006).

The availability of nutrients can restrict the occurrence or affect the assemblage composition of macroalgae in the Arctic, where inorganic nutrient concentrations can be seasonally depleted (Lee 1973, Chapman and Lindley 1980, Dunton and Schell 1986). In response, some species like the Arctic endemic kelp *Laminaria solidungula* grow mostly in winter and early spring when inorganic nitrogen concentrations are high (Chapter 9; Wiencke et al. 2009). In contrast, the Southern Ocean is a nutrient-replete system where nutrients rarely become limiting for macroalgae (Chapter 2; Gutkowski and Maleszewski 1989, Drew and Hastings 1992, Zacher et al. 2009a) and Antarctic macroalgae do not have strongly evolved storage mechanisms for internal nitrogen (Korb and Gerard 2000).

Micronutrients are seldom limiting for benthic microalgae and the nitrogen concentrations are often orders of magnitude higher in sediment pore water compared to the overlying water. However, nutrient limitation has been observed in microbenthic communities (e.g., Nilsson et al. 1991) and addition of micronutrients has been shown to mitigate negative effects of potential stress factors such as UV radiation (Wulff et al. 2000). Furthermore, periods of high photosynthetic activity can cause inorganic carbon to become limiting even when concentrations in the overlying water or in the deeper sediment layers are high (cf. Admiraal 1984). A strategy to cope with nutrient limitation is the uptake of organic compounds such as free amino acids (Nilsson and Sundbäck 1996 and references therein), which is of particular importance to survival during the dark polar winter (cf. Wulff et al. 2008c).

Light can determine both the lower and upper distribution limits of micro- and macroalgae (Chapter 10; Gómez et al. 1997, 2009). Sufficient light is required to reach the compensation point for photosynthesis, but too much light can induce photoinhibition, and UV radiation in shallow waters can have many harmful effects at the cellular and molecular levels (Bischof et al. 2006). Light for polar micro- and macroalgae can be seasonally limiting in winter, and light availability at other times of the year may also be limited due to the low solar altitude angle and high sediment loads in the water (Chapter 2; Dunton 1990, Gómez et al. 1997, Zacher et al. 2009a). In the Arctic, coralline red algae (*Phymatolithon* spp.) are able to grow down to 50 m depth (Roberts et al. 2002). The kelp *Saccharina latissima* in Greenland is adapted to low light by high pigment concentrations and photosynthetic efficiency while simultaneously having low light compensation and saturation points (Borum et al. 2002). In the Alaskan Beaufort Sea, however, *Saccharina latissima* is less able to compensate for the extremely low light availability in that region than *Laminaria solidungula* with its extreme low-light adaption, which may contribute to the success of the latter species in that region (Dunton 1990, Dunton and Dayton 1995). In the Antarctic, macroalgae also show distinct depth zonation patterns according to their photosynthetic abilities, where ambient PAR levels are sufficient to slightly exceed reported saturation and compensation points of photosynthesis (Wiencke et al. 1993, Klöser et al. 1994, Weykam et al. 1996). Some of the large brown algae like *Desmarestia anceps* and *Himantothallus grandifolius* require annual light doses only half of what is commonly needed by temperate kelp species (Wiencke 1990a, b, Runcie and Riddle 2006). Adaptation to low light may also give some species a competitive advantage over others, as suggested for *Himantothallus grandifolius* over *Desmarestia anceps* at greater depths (Klöser et al. 1994).

Despite this fact, polar micro- and macroalgae also have the ability to dynamically protect photosynthesis from excessive radiation through photoinhibition, which also structures their depth distribution (Hanelt 1998, Kühl et al. 2001). In the Arctic, faster recovery of photosynthesis after photoinhibition was measured in shallower species, indicating their adaptation to the variable light levels in shallow waters. For example, intertidal *Fucus distichus* does not exhibit photoinhibition at high light levels (Aguilera et al. 1999). Other species, like Antarctic encrusting coralline red algae are highly sensitive to increased light levels (Irving et al. 2005).

The effects of UV radiation on polar macroalgae have received much recent attention due to increasing UV levels in polar regions (Chapter 13; Bischof et al. 2006, Karsten et al. 2009). Intertidal and shallow-water species are less UV-inhibited in their photosynthesis, experience less DNA damage and exhibit faster recovery and repair mechanisms than deeper species (Hanelt et al. 1997, Wiencke et al. 2000, Karsten et al. 2001). UV effects are also more pronounced in Arctic species, e.g., in the kelp *Laminaria solidungula*, than in other, cold-temperate species in the Arctic (Michler et al. 2002). Generally, the UV sensitivity of reproductive stages is correlated to the depth distribution of the adults; hence survival of reproductive stages is an important factor for adult zonation patterns (Wiencke et al. 2000, 2006, Roleda et al. 2007, 2009, 2010, Zacher et al. 2009b). The differential sensitivity of the algal propagules influences recruitment and hence succession (Chapter 14; Zacher et al. 2007, Campana et al. 2009). In Antarctica, benthic diatoms have been shown to be quite resistant to UV radiation both in the field (Zacher et al. 2007, Campana et al. 2008) and in the laboratory (Wulff et al. 2008a–c). See also the review by Karsten et al. (2009; Chapter 13).

Although little studied, salinity can be a factor driving the occurrence and zonation patterns of polar macroalgae. Changing salinities are mainly an issue in intertidal and upper sublittoral zones (Chapter 2; Zacher et al. 2009a). In the Arctic, eulittoral species are more tolerant to salinity variation than sublittoral species, and salinity tolerance is suggested to be one of the driving factors for this depth distribution (Karsten 2007). Similarly, Antarctic supratidal species like the green alga *Prasiola crispa* are adapted to extreme salinity fluctuations (Jacob et al. 1991, Karsten et al. 1991). For benthic diatom species, Williams (1964) indicated several decades ago that many estuarine benthic diatom species occur over wide ranges of salinity. Admiraal (1977) and Admiraal and Peletier (1980) confirmed the conclusion by showing that marine benthic diatoms were photosynthetically active when subjected to changes between 4 and 60 PSU. Ice algae are also known to tolerate a wide range of salinities (Grant and Horner 1976).

Apart from these abiotic factors, polar micro- and macroalgal distribution patterns are also influenced by grazing pressure from herbivores. While much of the macroalgal production in both polar regions enters the detrital food web before it is consumed (Dunton and Schell 1987, Dunton 2001), macroalgae are susceptible to grazing (Iken et al. 1997, Iken 1999, Wessels et al. 2006). In the Arctic, chitons [(*Tonicella marmorea* (O. Fabricius), amphipods (*Gammarellus homari* J.C. Fabricius) and sea urchins (*Strongylocentrotus droebachiensis* O.F. Müller)] are important macroalgal grazers (Cross et al. 1987, Hop et al. 2002, Wessels et al. 2006). As is typical for non-polar macroalgal-grazer interactions, grazer preferences are species-specific. Arctic urchins preferentially feed on kelps over other algal species. This can exert strong top-down grazing pressure on the kelp canopy in regions where urchins are abundant and create urchin barrens, which is usually the case in more exposed areas with high salinity as compared to low-salinity inner fjord regions (Sivertsen 1997, Wessels et al. 2006). In the Antarctic, invertebrate mesograzers include limpets, gastropods, polychaetes and amphipods (Iken 1999, Huang et al. 2006), but their structuring effect on the macroalgal assemblage is mostly uncertain. Limpets can have a strong structuring effect on macroalgal assemblage composition in the Antarctic intertidal zone where grazing reduces overall algal biomass but increased diversity (Zacher et al. 2007). The lack of filamentous epiphytes and the prominence of filamentous endophytes led to the suggestion that mesoherbivory has a strong structuring effect on the community in the Antarctic (Peters 2003). This hypothesis has recently been substantiated by Amsler et al. (Chapter 5; 2009a,b). Fishes (especially *Notothenia coriiceps*) are selective feeders on macroalgae, which has the potential to be a structuring effect on the macroalgal assemblages (Klöser et al. 1994, Iken et al. 1997). In response, macroalgae are able to produce secondary metabolites that deter grazers (Amsler et al. 2005).

In a 15-week intertidal field study in Antarctica, grazing significantly reduced benthic diatom biomass and altered the species composition (Chapter 14; Zacher et al. 2007, Campana et al. 2009). The grazers (dominated by the gastropod *Nacella concinna*) were ineffective when the diatom biomass was low, and firmly attached diatoms such as *Cocconeis* were shown to be protected from grazing by amphipods (Zacher et al. 2007). Also in the subtidal zone, grazers significantly reduced diatom biomass and changed the species composition (Chapter 14; Campana et al. 2008, 2009). In this study, the grazers comprised amphipods (mostly *Gondogeneia antarctica*) and gastropods (*Nacella concinna* and *Laevilacunaria antarctica*). The largest impact was observed for overstory colony-forming species like *Fragilaria stiatula* while the small individuals of *Navicula* cf. *perminuta* were largely unaffected by the grazers. The gut content of the bottom-dwelling crustacean *Nototanais antarcticus* inhabiting Admiralty Bay, King George Island was dominated (97%) by benthic diatoms (Blazewicz-Paszkowycz and Ligowski 2002). Of particular interest is that at 16–65 m depth, benthic (epiphytic) diatoms still constituted the major part of the gut contents. Hydroids also feed on benthic diatoms: the genus *Cocconeis* was found to be the major food source for *Silicularia rosea* Meyen growing at 3 m depth (Gili et al. 1996). Also in the Arctic, benthic diatoms serve as high-quality carbon source for macrofauna such as the bivalves *Hiatella arctica* Linnaeus and *Mya truncata* Linnaeus (Sejr and Christensen 2007). In addition to the macrofaunal grazers, meiofauna (<500 μm) can be an important structuring factor for benthic microalgae in both polar regions (Rysgaard et al. 2000, de Skowronski and Corbisier 2002).

Concluding remarks

The urgent need for more studies of all aspects of the biodiversity and ecology of polar algae has never been more apparent than at present. General acceptance of the evidence that Earth's climate is changing and that the impacts are already evident, especially in polar regions, is accompanied by an increased imperative to describe and document the diversity and ecology of organisms before the changes are too great. The remoteness and inaccessibility of much of Antarctica and the Arctic means that the species of algae and their ecology are much less well known than in other regions. The biodiversity and ecology of algae are keys to understanding their evolutionary histories, together with other aspects of their biology, such as knowledge of their physiology, including how they respond to a changing environment. This knowledge will improve our ability to predict the potential impact of climate change and allow science to influence appropriate management responses at an international level.

References

Adey, W.H., S.C. Lindstrom, M.H. Hommersand and K.M. Müller. 2008. The biogeographic origin of Arctic endemic seaweeds: a thermogeographic view. *J. Phycol.* 44: 1384–1394.

Admiraal, W. 1977. Salinity tolerance of benthic estuarine diatoms as tested with a rapid polarographic measurement of photosynthesis. *Mar. Biol. 39*: 11–19.

Admiraal, W. 1984. The ecology of estuarine sediment-inhabiting diatoms. *Prog. Phycol. Res. 3*: 269–322.

Admiraal, W. and H. Peletier. 1980. Distribution of diatom species on an estuarine mudflat and experimental analysis of the selective effect of stress. *J. Exp. Mar. Biol. Ecol. 46*: 157–175.

Aguilera, J., U. Karsten, H. Lippert, B. Vögele, E. Phillip, D. Hanelt and C. Wiencke. 1999. Effects of solar radiation on growth, photosynthesis and respiration of marine macroalgae from the Arctic. *Mar. Ecol. Prog. Ser. 191*: 109–119.

Ahn, I.Y., H. Chung, J.S. Kang and S.H. Kang. 1994. Preliminary studies on the ecology of neritic marine diatoms in Maxwell Bay, King George Island, Antarctica. *Korean J. Phycol. 9*: 37–45.

Ahn, I.Y., H. Chung, J.S. Kang and S.H. Kang. 1997. Diatom composition and biomass variability in nearshore waters of Maxwell Bay, Antarctica, during the 1992/1993 austral summer. *Polar Biol. 17*: 123–130.

Al-Handal, A.Y. and A. Wulff. 2008a. Marine benthic diatoms from Potter Cove, King George Island, Antarctica. *Bot. Mar. 51*: 51–68.

Al-Handal, A.Y. and A. Wulff. 2008b. Marine epiphytic diatoms of the shallow littoral zone in Potter Cove, King George Island, Antarctica. *Bot. Mar. 51*: 411–435.

Amsler, C.D., R.J. Rowley, D.R. Laur, L.B. Quentin and R.M. Ross. 1995. Vertical distribution of Antarctic peninsular macroalgae: cover, biomass and species composition. *Phycologia 34*: 424–430.

Amsler, C.D., J.B. McClintock and B.J. Baker. 1998. Chemical defense against herbivory in the Antarctic marine macroalgae *Iridaea cordata* and *Phyllophora antarctica* (Rhodophyceae). *J. Phycol. 34*: 53–59.

Amsler, C.D., K. Iken, J.B. McClintock, M.O. Amsler, K.J. Peters, J.M. Hubbard, F.B. Furrow and B.J. Baker. 2005. Comprehensive evaluation of the palatability and chemical defenses of subtidal macroalgae from the Antarctic Pensinsula. *Mar. Ecol. Prog. Ser. 294*: 141–159.

Amsler, C.D., M.O. Amsler, J.B. McClintock and B.J. Baker. 2009a. Filamentous algal endophytes in macrophytic Antarctic algae: prevalence in hosts and palatability to mesoherbivores. *Phycologia 48*: 324–334.

Amsler, C.D., K. Iken, J.B. McClintock, and B.J. Baker. 2009b. Defenses of polar macroalgae against herbivores and biofoulers. *Bot. Mar. 52*: 535–545.

Andreoli, C., C. Tolomio, I. Moro, M. Radice, E. Moschin and S. Bellato. 1995. Diatoms and dinoflagellates in Terra Nova Bay (Ross Sea-Antarctica) during austral summer 1990. *Polar Biol. 15*: 465–475.

Beuchel, F. and B. Gulliksen. 2008. Temporal patterns of benthic community development in an Arctic fjord (Kongsfjorden, Svalbard): results of a 24-year manipulation study. *Polar Biol. 31*: 913–924.

Bischoff, B. and C. Wiencke. 1993. Temperature requirements for growth and survival of macroalgae from Disko Island (Greenland). *Helgol. Meeresunters. 47*: 167–191.

Bischoff, B. and C. Wiencke. 1995. Temperature ecotypes and biogeography of Acrosiphoniales (Chlorophyta) with Arctic-Antarctic disjunct and Arctic/cold-temperate distributions. *Eur. J. Phycol. 30*: 19–27.

Bischof, K., I. Gómez, M. Molis, D. Hanelt, U. Karsten, U. Lüder, M.Y. Roleda, K. Zacher and C. Wiencke. 2006. Ultraviolet radiation shapes seaweed communities. *Rev. Environ. Sci. Biotechnol. 5*: 141–166.

Blazewicz-Paszkowycz, M. and R. Ligowski. 2002. Diatoms as food source indicator for some Antarctic Cumacea and Tanaidacea (Crustacea). *Antarct. Sci. 14*: 11–15.

Borum, J., M.F. Pedersen, D. Krause-Jensen, P.B. Christensen and K. Nielsen. 2002. Biomass, photosynthesis and growth of *Laminaria saccharina* in a high-arctic fjord, NE Greenland. *Mar. Biol. 141*: 11–19.

Bowden, D.A. 2005. Quantitative characterization of shallow marine benthic assemblages at Ryder Bay, Adelaide Island, Antarctica. *Mar. Biol. 146*: 1235–1249.

Broady, P.A. 1996. Diversity, distribution and dispersal of Antarctic terrestrial algae. *Biodiv. Conserv. 5*: 1307–1335.

Brouwer, P.E.M. 1996. Decomposition *in situ* of the sublittoral Antarctic macroalga *Desmarestia anceps* Montagne. *Polar Biol. 16*: 129–137.

Brouwer, P.E.M., E.F.M. Geilen, N.J.M. Gremmen and F. Van Lent. 1995. Biomass, cover and zonation pattern of sublittoral macroalgae at Signy Island, South Orkney Islands, Antarctica. *Bot. Mar. 38*: 259–270.

Campana, G., M.L. Quartino, A. Yousif and A. Wulff. 2008. Impacts of UV radiation and grazing on the structure of a subtidal benthic diatom assemblage in Antarctica. *Ber. Polarforsch. Meeresforsch. 571*: 302–310.

Campana, G., K. Zacher, A. Fricke, M. Molis, A. Wulff, L. Quartino and C. Wiencke. 2009. Drivers of colonization and succession in polar benthic macro- and microalgal communities. *Bot. Mar. 52*: 655–667.

Carlsen, B.P., G. Johnsen, J. Berge and P. Kuklinski. 2007. Biodiversity patterns of macrofauna on different lamina parts of *Laminaria digitata* and *Saccharina latissima* collected during spring and summer 2004 in Kongsfjorden, Svalbard. *Polar Biol. 30*: 939–943.

Castilla, J.C. and N. Rozbaczylo. 1985. Rocky intertidal assemblages and predation on the gastropod *Nacella* (*Patinigera*) *concinna* at Robert Island, South Shetlands, Antarctica. *INACH Serie Cientifica 32*: 65–74.

Chapman, A.R.O. and J.E. Lindley. 1980. Seasonal growth of *Laminaria solidungula* in the Canadian High Arctic in relation to irradiance and dissolved nutrient concentrations. *Mar. Biol. 57*: 1–5.

Chung, H., Y.S. Oh, I.K. Lee and D-Y. Kim. 1994. Macroalgal vegetation of Maxwell Bay in King George Island, Antarctica. *Kor. J. Phycol. 9*: 47–58.

Clarke, D.L. 1990. Arctic Ocean ice cover; geologic history and climatic significance. *In*: (A. Grantz, L. Johnson and J.L. Sweeney, eds) *The Arctic Ocean region*. Geol. Soc. America, Boulder, Colorado. pp. 53–62.

Clayton, M.N. 1994. Evolution of the Antarctic Marine Benthic Algal Flora. *J. Phycol. 30*: 897–904.

Cleve, P.T. 1883. Diatoms collected during the expedition of the Vega. *Vega – Exped. Vetenskap. Iakttag.* 3: 455–517.

Cormaci, M., G. Furnari and B. Scammacca. 1992. The benthic algal flora of Terra Nova Bay (Ross Sea, Antarctica). *Bot. Mar. 35*: 541–552.

Cormaci, M., G. Furnari, B. Scammacca and G. Alongi. 1996. Summer biomass of a population of *Iridaea cordata* (Gigartinaceae, Rhodophyta) from Antarctica. *Hydrobiologia 326/ 327*: 267–272.

Cormaci, M., G. Furnari, B. Scammacca, G. Alongi and M. Catra. 1998. Summer biomass of a population of *Phyllophora antarctica* (Phyllophoraceae, Rhodophyta) from Antarctica. *Hydrobiologia 362*: 85–91.

Crame, J.A. 1993. Latitudinal range fluctuations in the marine realm through geological time. *Trends Ecol. Evol. 8*: 162–266.

Cremer, H. 1999. Distribution patterns of diatom surface sediment assemblages in the Laptev Sea (Arctic Ocean). *Mar. Micropaleontol. 38*: 39–67.

Cremer, H., D. Roberts, A. McMinn, D. Gore and M. Melles. 2003. The Holocene diatom flora of marine bays in Windmill Islands, East Antarctica. *Bot. Mar. 46*: 82–106.

Cross, W.E., R.T. Wilce and M.F. Fabijan. 1987. Effects of experimental releases of oil and dispersed oil on Arctic nearshore macrobenthos. III. Macroalgae. *Arctic 40*: 211–219.

de Skowronski, R.S. and T.N. Corbisier. 2002. Meiofauna distribution in Martel Inlet, King George Island (Antarctica): sediment features versus food availability. *Polar Biol. 25*: 126–134.

DeLaca, T.E. and J.H. Lipps. 1976. Shallow-water marine associations, Antarctic Peninsula. *Antarctic J. US. 11*: 12–20.

Delépine, R., J.M. Lamb and M.H. Zimmermann. 1966. Preliminary report on the vegetation of the Antarctic peninsula. *Proc. Int. Seaweed Symp. 5*: 107–116.

Dhargalkar, V., H. Burton and J. Kirkwood. 1988. Animal associations with the dominant species of shallow water macrophytes along the coastline of the Vestfold Hills, Antarctica. *Hydrobiologia 165*: 141–150.

Dobretsov, S.V. 1999. Effects of macroalgae and biofilm on settlement of blue mussel (*Mytilus edulis* L.) larvae. *Biofouling 14*: 153–165.

Douglas, M.S.V. and J.P. Smol. 1995. Perpiphytic diatom assemblages from high arctic ponds. *J. Phycol. 31*: 60–69.

Drew, E.A. and R.M. Hastings. 1992. A year-round ecophysiological study of *Himantothallus grandifolius* (Desmarestiales, Phaeophyta) at Signy Island, Antarctica. *Phycologia 31*: 262–277.

Dunton, K.H. 1990. Growth and production in *Laminaria solidungula*: relation to continuous underwater light levels in the Alaskan High Arctic. *Mar. Biol. 106*: 297–304.

Dunton, K.H. 1992. Arctic biogeography: The paradox of the marine benthic fauna and flora. *Trends Ecol. Evol. 7*: 183–189.

Dunton, K.H. 2001. $\delta^{15}N$ and $\delta^{13}C$ measurements of Antarctic Peninsula fauna: trophic relationships and assimilation of benthic seaweeds. *Am. Zool. 41*: 99–112.

Dunton, K.H. and P.K. Dayton. 1995. The biology of high latitude kelp. *In*: (H.R. Skjoldal, K.E. Erikstad and H.P. Leinaas, eds). *Ecology of Fjords and coastal waters*. Elsevier Science. pp. 499–507.

Dunton, K.H. and D.M. Schell. 1987. Dependence of consumers on macroalgal (*Laminaria solidungula*) carbon in an arctic kelp community: $\delta^{13}C$ evidence. *Mar. Biol. 93*: 615–625.

Dunton, K.H. and S. Schonberg. 2000. The benthic faunal assemblage of the Boulder Patch kelp community. *In:* (J.C. Truett and S.R. Johnson, eds) *The natural history of an arctic oil field*. Academic Press, London. pp. 371–398.

Dunton, K.H., E. Reimnitz and S. Schonberg. 1982. An Arctic kelp community in the Alaskan Beaufort Sea. *Arctic 35*: 465–484.

Ehrenberg, C.G. 1843. Verbreitung und Einfluss des mikroskopischen Lebens in Süd- und Nord-America. *Abh. Königl. Akad. Wiss. Berlin. 1*: 291–445.

Ellis, D.V. and R.T. Wilce. 1961. Arctic and subarctic examples of intertidal zonation. *Arctic 14*: 224–235.

Finlay, B.J. 2002. Global dispersal of free-living microbial eukaryote species. *Science 296*: 1061–1063.

Finlay, B.J. 2004. Protist taxonomy: an ecological perspective. *Phil. Trans. R. Soc. Lond. B 359*: 599–610.

Florczyk, I. and A. Latala. 1989. The phytobenthos of the Hornsund fjord, SW Spitzbergen. *Polar Res. 7*: 29–41.

Fritsch, F.E. 1912. National Antarctic 'Discovery' Expedition 1902–1904. Freshwater algae 6. British Museum Natural History, London. pp. 1–60.

Fumanti, B. and P. Cavacini. 2005. *ANTADATA: a database on biogeography of non-diatom algae in continental Antarctica*. CD-ROM. Casa Editrice Universita La Sapienza, Rome.

Garrison, D.L., A.R. Close and E. Reimnitz. 1989. Algae concentrated by frazil ice-evidence from laboratory experiments and field-measurements. *Antarct. Sci. 1*: 313–316.

Gersonde, R. and U. Zielinski. 2000. The reconstruction of late Quaternary Antarctic sea-ice distribution – the use of diatoms as a proxy for sea-ice. *Paleogeogr. Paleoclimat. Paleoecol. 162*: 263–286.

Gilbert, N.S. 1991a. Primary production by benthic microalgae in nearshore marine sediments of Signy Island, Antarctica. *Polar Biol. 11*: 339–346.

Gilbert, N.S. 1991b. Microphytobenthic seasonality in near-shore marine sediments at Signy Island, South Orkney Islands, Antarctica. *Estuar. Coast Shelf Sci. 33*: 89–104.

Gili, J.M., V. Alva, F. Pages, H. Klöser and W.E. Arntz. 1996. Benthic diatoms as the major food source in the sub-Antarctic marine hydroid *Silicularia rosea*. *Polar Biol. 16*: 507–512.

Glud, R.N. 2008. Oxygen dynamics of marine sediments. *Mar. Biol. Res. 4:* 243–289.

Glud, R.N., M. Kühl, F. Wenzhofer and S. Rysgaard. 2002. Benthic diatoms of a high Arctic fjord (Young Sound, NE Greenland): importance for ecosystem primary production. *Mar. Ecol. Prog. Ser. 238*: 15–29.

Glud, R., J. Woelfel, U. Karsten, M. Kühl and S. Rysgaard. 2009. Benthic microalgal production in the Arctic: applied methods and status of the current database. *Bot. Mar. 52*: 559–571.

Gómez, I., G. Weykam, H. Klöser and C. Wiencke. 1997. Photosynthetic light requirements, daily carbon balance and zonation of sublittoral macroalgae from King George Island (Antarctica). *Mar. Ecol. Prog. Ser. 148*: 281–293.

Gómez, I., A. Wulff, K. Dunton, U. Karsten, M. Roleda and C. Wiencke. 2009. Light and temperature demands of marine benthic microalgae and seaweeds in polar regions. *Bot. Mar. 52*: 593–608.

Grant, W.S. and R. Horner. 1976. Growth responses to salinity variation in four arctic sea ice diatoms. *J. Phycol. 12*: 180–185.

Gutkowski, R. and S. Maleszewski. 1989. Seasonal changes of the photosynthetic capacity of the Antarctic macroalga *Adenocystis utricularis* (Bory) Skottsberg. *Polar Biol. 10*: 145–148.

Hanelt, D. 1998. Capability of dynamic photoinhibition in Arctic macroalgae is related to their depth distribution. *Mar. Biol. 131*: 361–369.

Hanelt, D., B. Melchersmann, C. Wiencke and W. Nultsch. 1997. Effects of high light stress on photosynthesis of polar macroalgae in relation to depth distribution. *Mar. Ecol. Prog. Ser. 149*: 255–266.

Hansen, J.R. and I. Haugen. 1989. Some observations of intertidal communities on Spitsbergen (79° N), Norwegian Arctic. *Polar Res. 7*: 23–27.

Harper, M.A. 1977. Movements. *In:* (D. Werner, ed). *The biology of diatoms*. Blackwell, Oxford. pp. 224–249.

Hayward, B.W. and M.S. Morley. 2005. Zonation and biogeography of the intertidal biota of subantarctic Campbell and Auckland Islands, New Zealand. *Records Auckland Museum 42*: 7–33.

Heiden, H. and R.W. Kolbe. 1928. Die marinen Diatomeen der Deutschen Südpolar-Expedition 1901–1903. *In*: (E. von Drygalski, ed) *Deutsche Südpolar-Expedition 1901–1903*, VIII. Band Botanik. De Gruyter, Berlin. pp. 447–715.

Hempel, G. 1987. Die Polarmeere – ein biologischer Vergleich. *Polarforschung 57*: 173–189.

Heywood, R.B. and T.M. Whitaker. 1984. The Antarctic marine flora. *In*: (R.M. Laws, ed) *Antarctic ecology*. Academic Press, London. pp. 373–219.

Hommersand, M.H., R.L. Moe, C.D. Amsler and S. Fredericq. 2009. Notes on the systematics and biogeographical relationships of Antarctic and sub-Antarctic Rhodophyta with descriptions of for new genera and five new species. *Bot. Mar. 52*: 509–534.

Hop, H., T. Pearson, E.N. Hegseth, K.M. Kovacs, C. Wiencke, S. Kwasniewski, K. Eiane, F. Mehlum, B. Gulliksen, M. Wlodarska-Kowalczuk, C. Lydersen, J.M. Weslawski, S. Cochrane, G.W. Gabrielsen, R.J.G. Leakey, O.J. Lønne, M. Zajaczkowski, S. Falk-Petersen, M. Kendall, S.A. Wängberg, K. Bischof, A.Y. Voronkov, N.A. Kovaltchouk, J. Wiktor, M. Poltermann, G. di Prisco, C. Papucci and S. Gerland. 2002. The marine ecosystem of Kongsfjorden, Svalbard. *Polar Res. 21*: 167–208.

Hoppenrath, M. and Y.B. Okolodkov. 2000. *Amphidinium glabrum* sp. nov. (Dinophyceae) from the North German Wadden Sea and European Arctic sea ice: morphology, distribution and ecology. *Eur. J. Phycol. 35*: 61–67.

Horner, R. 1990. Ice-associated ecosystems. *In*: (L. Medlin and J. Priddle, eds) *Polar marine diatoms*. British Antarctic Survey, Cambridge. pp. 9–14.

Horner, R. and G.C. Schrader. 1982. Relative contributions of ice algae, phytoplankton, and benthic microalgae to primary production in nearshore regions of the Beaufort Sea. *Arctic 35*: 485–503.

Huang, Y.M., J.B. McClintock, C.D. Amsler, K.J. Peters and B.J. Baker. 2006. Feeding rates of common Antarctic gammarid amphipods on ecologically important sympatric macroalgae. *J. Exp. Mar. Biol. Ecol. 329*: 55–65.

Hustedt, F. 1958. Diatomeen aus der Antarktis und dem Südatlantik. *Deutsche Antarktische Expedition 1938/3 (2)*: 103–191.

Iken, K. 1999. Feeding ecology of the Antarctic herbivorous gastropod *Laevilacunaria antarctica* Martens. *J. Exp. Mar. Biol. Ecol. 236*: 133–148.

Iken, K., E.R. Barrera-Oro, M.L. Quartino, R.J. Casaux and T. Brey. 1997. Grazing in the Antarctic fish *Notothenia coriiceps*: evidence for selective feeding on macroalgae. *Antarctic Sci. 9*: 386–391.

Irving, A.D., S.D. Connell, E.L. Johnston, A.J. Pile and B.M. Gillanders. 2005. The response of encrusting coralline algae to canopy loss: an independent test of predictions on an Antarctic coast. *Mar. Biol. 147*: 1075–1083.

Jacob, A., G.O. Kirst, C. Wiencke and H. Lehmann. 1991. Physiological responses of the Antarctic green alga *Prasiola crispa* ssp. *antarctica* to salinity stress. *J. Plant Physiol. 139*: 57–62.

Johnston, E., S. Connell, A. Irving, A. Pile and B. Gillanders. 2007. Antarctic patterns of shallow subtidal habitat and inhabitants in Wilke's Land. *Polar Biol. 30*: 781–788.

Jones, V.J. 1996. The diversity, distribution and ecology of diatoms from Antarctic inland waters. *Biodivers. Conserv. 5*: 1433–1449.

Karsten, U. 2007. Salinity tolerance of Arctic kelps from Spits-bergen. *Phycol. Res. 55*: 257–262.

Karsten, U., C. Wiencke and G. Kirst. 1991. The effect of salinity changes upon physiology of eulittoral green macroalgae from Antarctica and Southern Chile. I. Cell viability, growth, photosynthesis and dark respiration. *J. Plant Physiol. 138*: 667–673.

Karsten, U., K. Bischof and C. Wiencke. 2001. Photosynthetic performance of Arctic macroalgae after transplantation from deep to shallow waters. *Oecologia 127*: 11–20.

Karsten, U., R. Schumann, S. Rothe, I. Jung and L. Medlin. 2006. Temperature and light requirements for growth of two diatom species (Bacillariophyceae) isolated from an Arctic macroalga. *Polar Biol. 29*: 476–486.

Karsten, U., A. Wulff, M. Roleda, R. Müller, F.S. Steinhoff, J. Fredersdorf and C. Wiencke. 2009. Physiological responses of polar benthic algae to ultraviolet radiation. *Bot. Mar. 52*: 639–654.

Kirkwood, J. and H. Burton. 1988. Macrobenthic assemblages in Ellis Fjord, Vestfold Hills, Antarctica. *Mar. Biol. 97*: 445–457.

Kjellman, F.R. 1883. The algae of the Arctic Sea: a survey of the species, together with an exposition of the general characters and the development of the flora. *Kungl. Svenska Vetenskaps-Akademiens Handlingar 20 (5)*: 1–351, 31 plates.

Klöser, H. 1998. Habitats and distribution patterns of benthic diatoms in Potter Cove (King George Island, Antarctic) and its vicinity. *Ber. Polarforsch. Meeresforsch. 299*: 95–105.

Klöser, H., G. Mercuri, F. Laturnus, M.L. Quartino and C. Wiencke. 1994. On the competitive balance of macroalgae at Potter Cove (King George Island, South Shetlands). *Polar Biol. 14*: 11–16.

Klöser, H., M.L. Quartino and C. Wiencke. 1996. Distribution of macroalgae and macroalgal communities in gradients of physical conditions in Potter Cove, King George Island, Antarctica. *Hydrobiologia 333*: 1–17.

Konar, B. and K. Iken. 2005. Competitive dominance among sessile marine organisms in a high Arctic boulder community. *Polar Biol. 29*: 61–64.

Korb, R.E. and V.A. Gerard. 2000. Effects of concurrent low temperature and low nitrogen supply on polar and temperate seaweeds. *Mar. Ecol. Prog. Ser. 198*: 73–82.

Krebs, W.N. 1983. Ecology of neritic marine diatoms, Arthur Harbor, Antarctica. *Micropaleontology 29*: 267–297.

Krebs, W.N., J.H. Lipps and L.H. Burckle. 1987. Ice diatom floras, Arthur Harbor, Antarctica. *Polar Biol. 7*: 163–171.

Kühl, M., R.N. Glud, J. Borum, R. Roberts and S. Rysgaard. 2001. Photosynthetic performance of surface-associated algae below sea ice as measured with a pulse-amplitude-modulated (PAM) fluorometer and O_2 microsensors. *Mar. Ecol. Prog. Ser. 223*: 1–14.

Lamb, I.M. and M.H. Zimmermann. 1977. Benthic marine algae of the Antarctic Peninsula. *Antarctic Research Series 23*: 130–229.

Lane C.E., C. Mayes, L.D. Druehl and G.W. Saunders. 2006. A multigene molecular investigation of the kelp (Laminariales, Phaeophyceae) supports substantial taxonomic reorganization. *J. Phycol. 42*: 492–512.

Lee, R.K.S. 1973. General ecology of the Canadian Arctic benthic marine algae. *Arctic 26*: 32–43.

Leventer, A. 1992. Modern distribution of diatoms in sediments from the George V Coast, Antarctica. *Mar. Micropaleontol. 19*: 315–332.

Ligowski, R. 2002. Diatoms in different habitats of Admiralty Bay, Antarctica. *In:* (J. John, ed.), *Proceedings of the 15th International Diatom Symposium, Perth, Australia.* A.R.G. Gantner Verlag, Ruggell, Liechtenstein. pp. 173–191.

Lim, D.S.S., J.P. Smol and M.S.V. Douglas. 2008. Recent environmental changes on Banks Island (NWT, Canadian Arctic) quantified using fossil diatom assemblages. *J. Paleolimnol. 40*: 385–398.

Lindstrom, S.C. 2001. The Bering Strait connection: dispersal and speciation in boreal macroalgae. *J. Biogeogr. 28*: 243–251.

Logares, R.E. 2006. Does the global microbiota consist of a few cosmopolitan species. *Ecologia Austral. 16:* 85–90.

Longhi, M.L., I.R. Schloss and C. Wiencke. 2003. Effect of irradiance and temperature on photosynthesis and growth of two antarctic benthic diatoms, *Gyrosigma subsalinum* and *Odontella litigiosa*. *Bot. Mar. 46*: 276–284.

Lund, S. 1959a. The marine algae from East Greenland. I. Taxonomical part. *Meddr. Grønland 156(1)*: 1–274.

Lund, S. 1959b. The marine algae from East Greenland. II. Geographic distribution. *Meddr. Grønland 156(2)*: 1–70.

Lüning, K. 1990. *Seaweeds. Their environment, biogeography and ecophysiology*. John Wiley & Sons Inc., New York. pp. 527.

Lüning, K. and I. tom Dieck. 1989. The distribution and evolution of the Laminariales: North Pacific–Atlantic relationships. *In*: (D. Garbary and G.R. South, eds.) *Evolutionary biogeography of the North Atlantic*. Springer-Verlag, Berlin. pp. 187–204.

Makarov, V.N., M.V. Makarov and E.V. Schoschina. 1999. Seasonal dynamics of growth in the Barents Sea seaweeds: endogenous and exogenous regulation. *Bot. Mar. 42*: 43–49.

Mangin, L. 1915. Phytoplancton de l'Antarctique. *Documents scientifiques de la Deuxième Expédition Antarctique Francaise 1908–1910, commandée par le Dr Jean Charcot, Sciences Naturelles*. Masson, Paris. pp. 1–95.

Manguin, E. 1957. Premier inventaire des diatomées de la Terre Adélie Antarctique. Espèces nouvelles. *Rev. Algol. 3*: 111–134.

Manguin, E. 1960. Les diatomées de la Terre Adélie Antarctique. Campagne du "Commandant Charcot" 1949–1950. *Ann. Sci. Nat. Bot. 12*: 223–263.

Matheke, G.E.M. and R. Horner. 1974. Primary productivity of the benthic microalgae in the Chukchi Sea near Barrow, Alaska. *J. Fish. Res. Board Canada 31*: 1779–1786.

McMahon, K.W., W.G. Ambrose Jr., B.J. Johnson, M.Y. Sun, G.R. Lopez, L.M. Clough and M.L. Carroll. 2006. Benthic community response to ice algae and phytoplankton in Ny Ålesund, Svalbard. *Mar. Ecol. Prog. Ser. 310*: 1–14.

Michelutti, N., A.J. Holtham, M.S.V. Douglas and J.P. Smol. 2003. Periphytic diatom assemblages from ultra-oligotrophic and UV transparent lakes and ponds on Victoria Island and comparisons with other diatom surveys in the Canadian Arctic. *J. Phycol. 39*: 465–480.

Michler, T., J. Aguilera, D. Hanelt, K. Bischof and C. Wiencke. 2002. Long-term effects of ultraviolet radiation on growth and photosynthetic performance of polar and cold-temperate macroalgae. *Mar. Biol. 140*: 1117–1127.

Mikhaylova, T.A. 1999. The initial stages of experimental forming of *Laminaria* communities in the White Sea. *Botanicheskii Zhurnal 84*: 56–66.

Miller, K.A. and J.S. Pearse. 1991. Ecological studies of seaweeds in McMurdo Sound, Antarctica. *Am. Zool. 31*: 35–48.

Moe, R.L. and T.E. DeLaca. 1976. Occurrence of macroscopic algae along the Antarctic Peninsula. *Antarctic J. U.S. 11*: 20–24.

Moe, R.L. and P.C. Silva. 1977. Antarctic marine flora: uniquely devoid of kelps. *Science 196*: 1206–1208.

Mohr, J.L., N.J. Wilimovsky and E.Y. Dawson. 1957. An Arctic Alaskan kelp bed. *Arctic 10*: 45–52.

Moisan, T.A. and G.A. Fryxell. 1993. The distribution of Antarctic diatoms in the Weddell Sea during austral winter. *Bot. Mar. 36*: 489–497.

Müller, R., T. Läpple, I. Bartsch and C. Wiencke. 2009. Impact of oceanic warming on the distribution of seaweeds in polar and cold-temperate waters. *Bot. Mar. 52*: 617–638.

Neushul, M. 1965. Diving observations of subtidal Antarctic marine vegetation. *Bot. Mar. 8*: 234–243.

Nilsson, C. and K. Sundbäck. 1996. Amino acid uptake in natural microphytobenthic assemblages studied by microautoradiography. *Hydrobiologia 332*: 119–129.

Nilsson, P., B. Jönsson, I. Lindström Swanberg and K. Sundbäck. 1991. Response of a marine shallow-water sediment system to an increased load of inorganic nutrients. *Mar. Ecol. Prog. Ser. 71*: 275–290.

Norkko, A., S.F. Thrush, V.J. Cummings, G.A. Funnell, A.M. Schwarz, N.L. Andrew and I. Hawes. 2004. Ecological role of *Phyllophora antarctica* drift accumulations in coastal soft-sediment communities of McMurdo Sound, Antarctica. *Polar Biol. 27*: 482–494.

Okolodkov, Y.B. 1993. A checklist of algal species found in the East Siberian Sea in May 1987. *Polar Biol. 13*: 7–11.

Palmisano, A.C. and D.L. Garrison. 1993. Microorganisms in Antarctic sea-ice. In: (E. Friedmann, ed.) *Antarctic microbiology*. Wiley-Liss, New York. pp. 167–218.

Papenfuss, G.F. 1964. Catalogue and bibliography of Antarctic and Subantarctic benthic marine algae. *Am. Geophys. Union, Antarctic Res. Ser. 1*: 1–76.

Peragallo, M. 1921. *Deuxième Expédition Antarctique Française, 1908–1910, commandée par le Dr. Jean Charcot. Botanique. Diatomées d'eau douce et diatomées d'eau salée.* Masson, Paris. pp. 1–98.

Peters, A.F. 2003. Molecular identification, taxonomy and distribution of brown algal endophytes, with emphasis on species from Antarctica. *Proc. Int. Seaweed Symp. 17*: 293–302.

Peters, A.F. and A.M. Breeman. 1992. Temperature responses of disjunct temperate brown algae indicate long-distance dispersal of microthalli across the tropics. *J. Phycol. 28*: 428–438.

Peters, A.F., M.J.H. van Oppen, C. Wiencke, W.T. Stam and J.L. Olsen. 1997. Phylogeny and historical ecology of the Desmarestiaceae (Phaeophyceae) support a Southern Hemisphere origin. *J. Phycol. 33*: 294–309.

Polyakova, E.I. 1989. Diatoms in arctic shallow seas sediment. *In*: (Y. Herman, ed) *The Arctic Seas. Climatology, oceanography, geology, and biology.* Van Nostrand Reinhold, New York. pp. 481–496.

Poulin, M. 1990a. Ice diatoms: The Arctic. *In:* (L. Medlin and J. Priddle, eds) *Polar marine diatoms*. British Antarctic Survey, Cambridge. pp. 15–18.

Poulin, M. 1990b. Sea ice diatoms (Bacillariophyceae) of the Canadian Arctic. 1. The genus *Stenoneis*. *J. Phycol. 26*: 156–167.

Poulin, M. 1991. Sea ice diatoms (Bacillariophyceae) of the Canadian Arctic. 2. A taxonomic, morphological and geographical study of *Gyrosigma concilians*. *Nord. J. Bot. 10*: 681–688.

Poulin, M. and A. Cardinal. 1982. Sea ice diatoms from Manitounuk Sound, southeastern Hudson Bay (Quebec, Canada). II. *Naviculaceae*, genus *Navicula*. *Can. J. Bot. 60*: 2825–2845.

Poulin, M. and A. Cardinal. 1983. Sea ice diatoms from Manitounuk Sound, south-eastern Hudson Bay (Quebec, Canada). III. Cymbellaceae, Entomoneidaceae, Gomphonemataceae, and Nitzschiaceae. *Can. J. Bot.* 61: 107–118.

Quartino, M.L. and A.L. Boraso de Zaixso. 2008. Summer macroalgal biomass in Potter Cove, South Shetland, Antarctica: its production and flux to the ecosystem. *Polar Biol. 31*: 281–294.

Quartino, M.L., H. Klöser, I.R. Schloss and C. Wiencke. 2001. Biomass and associations of benthic marine macroalgae from the inner Potter Cove (King George Island, Antarctica) related to depth and substrate. *Polar Biol. 24*: 349–355.

Quartino, M.L., A.L. Boraso de Zaixso and H. Zaixso. 2005. Biological and environmental characterization of marine macroalgal assemblages in Potter Cove, South Shetland Islands, Antarctica. *Bot. Mar. 48*: 187–197.

Reimnitz, E., E.W. Kempema and P.W. Barnes. 1987. Anchor ice, seabed freezing, and sediment dynamics in shallow Arctic Seas. *J. Geophys. Res. C 92*: 14671–14678.

Riaux-Gobin, C. 1994. A check list of the *Cocconeis* species (Bacillariophyceae) in Antarctic and Subantarctic areas with special focus on Kerguelen Islands. *Cryptogam. Algol. 15*: 135–146.

Riaux-Gobin, C., P. Tréguer, M. Poulin and G. Vétion. 2000. Nutrients, algal biomass and communities in land-fast ice and seawater of Adélie Land, Antarctica. *Antarct. Sci. 12*: 160–171.

Richardson, M.G. 1979. The distribution of Antarctic marine macroalgae related to depth and substrate. *British Antarctic Survey Bulletin 49*: 1–13.

Roberts, R.D., M. Kühl, R.N. Glud and S. Rysgaard. 2002. Primary production of crustose coralline red algae in a high Arctic fjord. *J. Phycol. 38*: 273–283.

Roleda, M.Y., U.H. Lüder and C. Wiencke. 2010. UV-susceptibility of zoospores of the brown macroalga *Laminaria digitata* from Spitsbergen. Polar Biol. 33: 577–588.

Roleda, M.Y., C. Wiencke, D. Hanelt and K. Bischof. 2007. Sensitivity of early life stages of macroalgae from the northern hemisphere to ultraviolet radiation. *Photochem. Photobiol. 83:* 851–862.

Roleda, M.Y., G. Campana, C. Wiencke, D. Hanelt, M.L. Quartino and A. Wulff. 2009. Sensitivity of Antarctic *Urospora penicilliformis* (Ulotrichales, Chlorophyta) to ultraviolet radiation is life stage dependent. *J. Phycol. 45*: 600–609.

Rosenvinge, L.K. 1898. Deuxième mémoire sur les algues marines du Groenland. *Meddr. Grønland 20*: 1–125.

Round, F.E. 1971. Benthic marine diatoms. *Oceanogr. Mar. Biol. Annu. Rev. 9*: 83–139.

Runcie, J.W. and M.J. Riddle. 2006. Photosynthesis of marine macroalgae in ice-covered and ice-free environments in East Antarctica. *Eur. J. Phycol. 41*: 223–233.

Rysgaard, S., P.B. Christensen, M.V. Sørensen, P. Funch and P. Berg. 2000. Marine meiofauna, carbon and nitrogen mineralization in sandy and soft-sediments of Disko Bay, West Greenland. *Aquat. Microb. Ecol. 21*: 59–71.

Sabbe, K. 1997. *Systematics and ecology of intertidal benthic diatoms of the Westernschelde estuary* (The Netherlands). Doctoral thesis. Dept. Biology, Section Protistology and Aquatic Ecology. University of Gent, Belgium. pp. 347.

Sabbe, K., K. Vanhoutte, R.L. Lowe, E.A. Bergey, B.J.F. Biggs, S. Francoeur, D. Hodgson and W. Vyverman. 2001. Six new *Actinella* (Bacillariophyceae) species from Papua New Guinea, Australia and New Zealand: further evidence for widespread diatom endemism in the Australasia region. *Eur. J. Phycol. 36*: 321–340.

Sabbe, K., E. Vereleyn, D.A. Hodgson, K. Vanhoutte and W. Vyverman. 2003. Benthic diatom flora of freshwater and saline lakes in the Larsemann Hills and Rauer Islands, East Antarctica. *Antarct. Sci. 15*: 227–248.

Schoschina, E.V. 1996. Seasonal and age dynamics of growth and reproduction of *Phycodrys rubens* (Rhodophyta) in the Barents and White Seas. *Aquat. Bot. 55*: 13–30.

Schoschina, E.V. 1997. On *Laminaria hyperborea* (Laminariales, Phaeophyceae) on the Murman coast of the Barents Sea. *Sarsia 82*: 371–373.

Sejr, M.K. and P.B. Christensen. 2007. Growth, production and carbon demand of macrofauna in Young Sound, with special emphasis on the bivalves *Hiatella arctica* and *Mya truncata*. *In:* (S. Rysgaard and R. Glud, eds.). *Carbon cycling in Arctic marine ecosystems: case study Young Sound.* Copenhagen: Danish Polar Center.-Meddelelser om Grønland. *Bioscience 58:* 121–136.

Sivertsen, K. 1997. Geographic and environmental factors affecting the distribution of kelp beds and barren grounds and changes in biota associated with kelp reduction at sites along the Norwegian coast. *Can. J. Fish. Aquat. Sci. 54*: 2871–2887.

Skottsberg, C.J.F. 1964. Antarctic phycology. *In*: (Carrick, R. M., Holdgate and J. Prévost, eds) *Biologie Antarctique.* 1st SCAR Symposium. Premier Symposium. Hermann, Paris. pp. 147–154.

Smith, W.O. Jr., D.G. Ainley and R. Cattaneo-Vietti. 2007. Trophic interactions within the Ross Sea continental shelf ecosystem. *Phil. Trans. R. Soc. B 362*: 95–111.

Søreide, J.E., H. Hop, M.L. Carroll, S. Falk-Petersen and E.N. Hegseth. 2006. Seasonal food web structures and sympagic-pelagic coupling in the European Arctic revealed by stable isotopes and a two-source food web model. *Prog. Oceanogr. 71*: 59–87.

Spaulding, S.A. and D.M. McKnight. 1999. Diatoms as indicators of environmental changes in Antarctic freshwaters. *In*: (Stoermer, E.F. and J.P. Smol, eds) *The diatoms, applications for environmental and earth sciences*. Cambridge University Press, Cambridge. pp. 263–345.

Stegenga, H., J. Bolton and R.J. Anderson. 1997. Seaweeds of the south African West Coast. *Contributions Bolus Herbarium 18*: 1–655.

Svendsen, H., A. Beszczynska-Møller, J.O. Hagen, B. Lefauconnier, V. Tverberg, S. Gerland, K. Bischof, C. Papucci, J.B. Ørbæk, M. Zajaczkowski, R. Azzolini, O. Bruland, C. Wiencke, J.G. Winther and W. Dallmann. 2002. The Physical Environment of Kongsfjorden-Krossfjorden, an Arctic fjord system in Svalbard. *Polar Res. 21*: 133–166.

Takeuchi, I. and K. Watanabe. 2002. Mobile epiphytic invertebrates inhabiting the brown macroalga, *Desmarestia chordalis* under the coastal fast ice of Lützow-Holm Bay, East Antarctica. *Polar Biol. 25*: 624–628.

Tanimura, Y. 1992. Distribution of diatom species in the surface sediments of Lützow-Holm Bay, Antarctica. *In*: (K. Ishizaki and T. Satio, eds) *Centenary of Japanese micropaleontology*. Terra Scientific Publishing Company, Tokyo. pp. 399–411.

Teixido, N., J. Garrabou, J. Gutt and W.E. Arntz. 2007. Iceberg disturbance and successional spatial patterns: the case of the shelf Antarctic benthic communities. *Ecosystems 10*: 142–157.

Thomas, D.P. and J. Jiang. 1986. Epiphytic diatoms of the inshore marine area near Davis Station. *Hydrobiologia 140*: 193–198.

Tzetlin, A.B., V.O. Mokievsky, A.N. Melnikov, M.V. Saphonov, T.G. Simdyanov and I.E. Ivanov. 1997. Fauna associated with detached kelp in different types of subtidal habitats of the White Sea. *Hydrobiologia 355*: 91–100.

Van Heurck, H. 1909. Diatomées. *In: Expèdition Antarctique Belge. Resultats du voyage due S. A. Belgica en 1897, 1898 1899 sous le commandement de A. de Gerlach de Gomery.* Rapports scientifiques, botaniques. J.-E. Buschmann, Anvers. pp. 129.

van Oppen, M.J.H., J.L. Olsen, W.T. Stam, C. van den Hoek and C. Wiencke. 1993. Arctic-Antarctic disjunctions in the benthic seaweeds *Acrosiphonia arcta* (Chlorophyta) and *Desmarestia viridis/willii* (Phaeophyta) are of recent origin. *Mar. Biol. 115*: 381–386.

van Oppen, M.J.H., O.E. Diekmann, C. Wiencke, W.T. Stam and J.L. Olsen. 1994. Tracking dispersal routes: phylogeography of Arctic-Antarctic disjunct seaweed *Acrosiphonia arcta* (Chlorophyta). *J. Phycol. 30*: 67–80.

Vershinin, A., A. Moruchkov, S.L. Morton, T.A. Leighfield, M.A. Quilliam and J.S. Ramsdell. 2006. Phytoplankton composition of the Kandalaksha Gulf, Russian White Sea: *Dinophysis* and lipophilic toxins in the blue mussel (*Mytilus edulis*). *Harmful Algae 5*: 558–564.

Vincent, W.F. 2000. Evolutionary origins of Antarctic microbiota: invasion, selection and endemism. *Antarct. Sci. 12:* 274–385.

von Quillfeldt, C.H. 1997. Distribution of diatoms in the Northeast Water Polynya, Greenland. *J. Mar. Syst. 10*: 211–240.

von Quillfeldt, C.H. 2000. Common diatom species in Arctic spring blooms: Their distribution and abundance. *Bot. Mar. 43*: 499–516.

von Quillfeldt, C.H., W.G. Ambrose and L.M. Clough. 2003. High number of diatom species in first-year ice from the Chukchi Sea. *Polar Biol. 26*: 806–818.

Weslawski, J.M., J. Wiktor, M. Zajaczkowski and S. Swerpel. 1993. Intertidal zone of Svalbard: 1. Microorganism distribution and biomass. *Polar Biol. 13*: 73–79.

Weslawski, J.M., M. Zajaczkowski, J. Wiktor, M. Szymelfenig. 1997. Intertidal zone of Svalbard: 3. Littoral of a subarctic, oceanic island: Bjornoya. *Polar Biol. 18*: 45–52.

Wessels, H., W. Hagen, M. Molis, C. Wiencke and U. Karsten. 2006. Intra- and interspecific differences in palatability of Arctic macroalgae from Kongsfjorden (Spitzbergen) for two benthic sympatric invertebrates. *J. Exp. Mar. Biol. Ecol. 329*: 20–33.

West, W. and G.S. West. 1911. Freshwater algae. *In*: (J. Murray, ed.) *Biology. Vol. 1. Reports on the scientific investigations. British Antarctic Expedition 1907–1909*. Heinemann, London. pp. 263–298.

Westermeier, R., I. Gómez, P.J. Rivera and D.G. Müller. 1992. Macroalgas marinas antarticas: distribucion, abundancia y necromasa en isla Rey Jorge, Shetland del Sur, Antartica. *Ser. Cient. INACH 42*: 21–34.

Weykam, G., I. Gómez, C. Wiencke, K. Iken, H. Klöser. 1996. Photosynthetic characteristics and C:N ratios of macroalgae from King George Island (Antarctica). *J. Exp. Mar. Biol. Ecol. 204*: 1–22.

Wiencke, C. 1990a. Seasonality of brown macroalgae from Antarctica – a long-term culture study under fluctuating Antarctic daylengths. *Polar Biol. 10*: 589–600.

Wiencke, C. 1990b. Seasonality of red and green macroalgae from Antarctica – a long-term culture study under fluctuating Antarctic daylengths. *Polar Biol. 10*: 601–607.

Wiencke, C. and M.N. Clayton. 2002. Antarctic seaweeds. *In*: (J.W. Wägele and J. Sieg, eds) *Synopses of the Antarctic benthos 9*. A.R.G. Gantner Verlag KG, Ruggell, Liechtenstein. pp. 239.

Wiencke, C., J. Rahmel, U. Karsten, G. Weykam and G.O. Kirst. 1993. Photosynthesis of marine macroalgae from Antarctica: light and temperature requirements. *Botanica Acta 106*: 78–87.

Wiencke, C., I. Bartsch, B. Bischoff, A.F. Peters and A.M. Breeman. 1994. Temperature requirements and biogeography of Antarctic, Arctic and amphiequatorial seaweeds. *Bot. Mar. 37*: 247–259.

Wiencke, C., I. Gómez, H. Pakker, A. Flores-Moya, M. Altamirano, D. Hanelt, K. Bischof and F.L. Figueroa. 2000. Impact of UV radiation on viability, photosynthetic characteristics and DNA of brown algal zoospores: implications for depth zonation. *Mar. Ecol. Prog. Ser. 197*: 217–229.

Wiencke, C., B. Vögele, N.A. Kovaltchouk and H. Hop. 2004. Species composition and zonation of marine benthic macroalgae at Hansneset in Kongsfjorden, Svalbard. *Ber. Polarforsch. Meeresforsch. 492*: 55–62.

Wiencke, C., M.Y. Roleda, A. Gruber, M.N. Clayton and K. Bischof. 2006. Susceptibility of zoospores to UV radiation determines upper depth distribution limit of Arctic kelps. Evidence through field experiments. *J. Ecol. 94*: 455–463.

Wiencke, C., M.N. Clayton, I. Gómez, K. Iken, U.H. Lüder, C.D. Amsler, U. Karsten, D. Hanelt, K. Bischof and K. Dunton. 2007. Life strategy, ecophysiology and ecology of seaweeds in polar waters. *Rev. Environ. Sci. Biotechnol. 6*: 95–126.

Wiencke, C., I. Gómez and K. Dunton. 2009. Phenology and seasonal physiological performance of polar seaweeds. *Bot. Mar. 52*: 585–592.

Wilce, R.T. 1990. Role of the Arctic Ocean as a bridge between Atlantic and Pacific Oceans: fact and hypothesis. *In*: (D.J. Garbary and G.R. South, eds) *Evolutionary biogeography of the marine algae of the North Atlantic*. Springer Verlag, Berlin. pp. 323–348.

Wilce, R.T., P.M. Pedersen and S. Sekida. 2009. *Chukchia pedicellata* and *C. endophytica* nov. comb., arctic endemic brownalgae (Phaeophyceae). *J. Phycol. 45*: 272–286.

Williams, R.B. 1964. Division rates of salt marsh diatoms diatoms in relation to salinity and cell size. *Ecology 45*: 877–880.

Woelfel, J., R. Schumann, P. Leopold, C. Wiencke and U. Karsten. 2009. Microphytobenthic biomass along gradients of physical conditions in Arctic Kongsfjorden, Svalbard. *Bot. Mar. 52*: 573–583.

Womersley, H.B.S. 1990. Biogeography of Australasian marine macroalgae. *In*: (M.N. Clayton and R.J. King, eds) *Biology of marine plants*. Longman Cheshire, Melbourne. pp. 367–381.

Wulff, A., C. Nilsson, K. Sundbäck, S.A. Wängberg and S. Odmark. 1999. UV radiation effects on microbenthos: a four-month field experiment. *Aquat. Microb. Ecol. 19*: 269–278.

Wulff, A., S.A. Wängberg, K. Sundbäck, C. Nilsson and G.J.C. Underwood. 2000. Effects of UVB radiation on a marine microphytobenthic community growing on a sand-substratum under different nutrient conditions. *Limnol. Oceanogr. 45*: 1144–1152.

Wulff, A., M.Y. Roleda, K. Zacher and C. Wiencke. 2008a. UV radiation effects on pigments, photosynthetic efficiency and DNA of an Antarctic marine benthic diatom community. *Aquat. Biol. 3*: 167–177.

Wulff, A., K. Zacher, D. Hanelt, A. Al-Handal and C. Wiencke. 2008b. UV radiation – a threat to Antarctic benthic marine diatoms? *Antarct. Sci. 20*: 13–20.

Wulff, A., M.Y. Roleda, K. Zacher and C. Wiencke. 2008c. Exposure to sudden light burst after prolonged darkness: a case study on benthic diatoms in Antarctica. *Diatom Res. 23*: 519–532.

Zacher, K., D. Hanelt, C. Wiencke and A. Wulff. 2007. Grazing and UV radiation effects on an Antarctic intertidal microalgal assemblage: a long-term field study. *Polar Biol. 30*: 1203–1212.

Zacher, K., R. Rautenberger, D. Hanelt, A. Wulff and C.Wiencke. 2009a. The abiotic environment of polar benthic algae. *Bot. Mar. 52*: 483–490.

Zacher, K., M.Y. Roleda, A. Wulff, D. Hanelt and C. Wiencke. 2009b. Responses of Antarctic *Iridaea cordata* (Rhodophyta) tetraspores exposed to ultraviolet radiation. *Phycol. Res. 57*: 186–193.

Zachos, J., M. Pagani, L. Sloan, E. Thomas and K. Billups. 2001. Trends, rhythms, and aberrations in global climate 65 Ma to present. *Science 292*: 686–693.

Zaneveld, J.S. 1966. Vertical zonation of Antarctic and subantarctic benthic marine algae. *Antarctic J. US. 1*: 211–213.

Zielinski, K. 1981. Benthic macroalgae of Admirality Bay (King George Island, South Shetland Islands) and circulation of algal matter between water and the shore. *Polish Polar Res. 2*: 71–94.

Zielinski, K. 1990. Bottom macroalgae of the Admiralty Bay (King George Island, South Shetlands, Antarctica). *Polish Polar Res. 11*: 95–131.

4 Notes on the systematics and biogeographical relationships of Antarctic and sub-Antarctic Rhodophyta with descriptions of four new genera and five new species

Max H. Hommersand, Richard L. Moe, Charles D. Amsler and Suzanne Fredericq

Introduction

Antarctica, the South Shetland, South Orkney, and South Georgia islands form a biogeographic province that includes some distantly related genera and species clusters found in temperate regions in New Zealand and Tasmania, South Africa and Namibia, and southern South America, the Falkland Islands and the islands of the West Wind Drift, such as the Marion and Prince Edward islands, the Crozet and Kerguelen islands and the Macquarie, Campbell and Auckland islands. Antarctic marine habitats are by no means uniform: the most extreme "high Antarctic" conditions are reached in McMurdo Sound in East Antarctica at a latitude of about 78° S. Most of the Antarctic coastline lies south of the Antarctic Circle, and because of the relative scarcity of stable substrata in the euphotic zone, it provides only discontinuous habitats for marine macroalgae. Conditions are less extreme along the Antarctic Peninsula (much of which lies north of the Antarctic Circle) and in the islands of the Scotia Arc that lead to South Georgia (54.5° S). Generally speaking, the Antarctic Province is defined by the boundary of the Antarctic Polar Front (formerly Antarctic Convergence), which incorporates such poorly known regions as the Heard and McDonald islands, and the Balleny and Scott islands (Chapter 2; Zacher et al. 2009), and may have extended further north than its present boundary during the Pleistocene. In general, the diversity of marine algae increases from high latitudes to lower latitudes (Chapter 3; Wulff et al. 2009). Wiencke and Clayton (2002) consolidated our knowledge of the Antarctic Rhodophyta in a recent volume on the Antarctic seaweeds. Even so, many outstanding taxonomic problems remain unresolved. In the present chapter, we have attempted to update information about the red algae of the Antarctic Peninsula and South Shetland Islands and to provide descriptions of several species that have been either overlooked or misidentified previously. In addition, we have made a first attempt at placing the Antarctic Rhodophyta in a broader biogeographic context.

Specimens used in this study were collected along the Antarctic Peninsula by R. Moe (1973–1975, 1989) and by C. Amsler and co-workers (2005–2008) and from King George Island by S. Fredericq (1994). Descriptions of the new taxa are based on examinations of material preserved in formalin as well as pressed material. Drawings were made with the aid of camera lucida. Scans of herbarium specimens were made using an Epson (Epson America, Inc., Long Beach, USA) XL10000 scanner at 400 dpi. These are available at (http://ucjeps.berkeley.edu/images/antarctic_algae/). Material is deposited in UC or NCU. Herbarium abbreviations follow Thiers (2010, continuously updated at http://sweetgum.nybg.org/ih/).

Part 1: Taxonomic list of the Antarctic Rhodophyta

The following list of Antarctic species is drawn mainly from Wiencke and Clayton (2002). Classes, orders and families have been arranged according to the recent classification by Schneider and Wynne (2007). All new information comes from collections taken from the Antarctic Peninsula or the South Shetland Islands. Current names, basionyms and type localities are given for each species, including the authority and date of publication as it appears in Guiry and Guiry (2009). Only those references that appear in parentheses are cited in the bibliography. No attempt has been made to clarify the names of species we have not seen. Representative specimens available in the Herbarium of the University of California, Berkeley (UC) are cited. In several entries we enclose generic or species names of binomials in double quotation marks. This is intended to convey that we are using a name that we know (or suspect) to be incorrect. When we have corrected the original spelling of a name, we have added the original spelling in double quotation marks. Most of the entries in the taxonomic list include comments or original taxonomic or ecological observations.

Class Compsopogonophyceae
Order Erythropeltidales
Family Erythrotrichiaceae
***Erythrotrichia carnea* (Dillwyn) J. Agardh 1883**

Basionym *Conferva carnea* Dillwyn 1807.

Type locality Near Loughor (Llwchwr), Glamorgan, Wales.

Note The worldwide distribution of this species has not been critically investigated.

Class Bangiophyceae
Order Bangiales
Family Bangiaceae
***Bangia atropurpurea* (Roth) C. Agardh 1824**

Basionym *Conferva atropurpurea* Roth 1806.

Neotype locality Ysselmeer, Netherlands.

Note The order Bangiales has been shown by Nelson et al. (2006) to be a genetically diverse assemblage in the Southern Hemisphere with *Bangia atropurpurea* restricted to freshwater environments.

***Porphyra endiviifolia* ("*endiviifolium*") (A. Gepp *et* E.S. Gepp) Y.M. Chamberlain 1963**

Basionym *Monostroma endiviifolium* ("*endiviaefolium*") A. Gepp *et* E.S. Gepp 1905.

Type locality South Orkney Islands.

Representative specimens Lysted Island, Melchior Archipelago, Antarctic Peninsula, Siple, 6 March 1941, UC677389; Kazak Island, Vestfold Hills, Davis, East Antarctica, intertidal, Ricker 2561, 8 December 1981, UC1497964; Arthur Harbor, Anvers Island, Antarctic Peninsula, high intertidal, Moe, 12 January 1988, UC1557625.

Note See Wiencke and Clayton (1998). This species is usually found in the high intertidal, particularly in cracks and crevices.

***Porphyra plocamiestris* R.W. Ricker 1987**

Type locality Goat Bay, Macquarie Island.

Representative specimens Arthur Harbor, Anvers Island, Antarctic Peninsula, −6 m (on *Gigartina*), Moe 344c, 9 March 1974, UC1571406; Arthur Harbor, −8 m (on *Georgiella*), Moe 941a, 15 January 1975, UC1571407; Arthur Harbor, −8 m (on *Georgiella*), Aumack, 8 January 2008, Aumack in NCU.

Note The plant that passes under this name in Antarctica is collected most often on *Georgiella confluens*, but has also been found on other macroalgae including *Delesseria salicifolia*, *Plumariopsis peninsularis*, *Plocamium*, *Rhodokrambe*, and *Desmarestia* spp. It can also occasionally be observed growing on rocks.

Class Florideophyceae
Subclass Hildenbrandiophycidae
Order Hildenbrandiales
Family Hildenbrandiaceae
***Hildenbrandia lecannellieri* Hariot 1887**

Syntype localities Orange Bay, Fuegia, Strait of Magellan and Tierra del Fuego.

Representative specimens Arthur Harbor, Anvers Island, Antarctic Peninsula, −10 m, Moe 8, 23 December 1973, UC1571379; Arthur Harbor, intertidal, Moe 396, 11 April 1974, UC1571378; Arthur Harbor, intertidal, Moe 536, 9 July 1974, UC1571811.

Note In the littoral zone of Fuegia and the Falkland Islands, this species is prevalent and conspicuous as a thick, rugose crust. Algae identifiable with certainty as *Hildenbrandia* in West Antarctica are thin, smooth crusts that grow sublittorally. West Antarctic records are probably based on an undescribed species of *Hildenbrandia* or on the basal crusts of any of several species.

Subclass Nemaliophycidae
Order Acrochaetiales
Family Acrochaetiaceae
***Rhodochorton purpureum* (Lightfoot) Rosenvinge 1900**

Basionym *Byssus purpurea* Lightfoot 1777.

Type locality Abbey at Y-Columb-Kill, Scotland (destroyed).

Representative specimens Arthur Harbor, Anvers Island, Antarctic Peninsula, high intertidal, Moe, 20 January 1988, UC1557586.

Note Maggs et al. (2007) investigated the relative contributions of vicariance and dispersal in determining the distributions of *Rhodochorton purpureum,* which is circumpolar in both the Northern and Southern Hemispheres. They sequenced the nuclear marker, ribosomal ITS2, and plastid marker, *tuf*A-*rps*7-*rps*12-*rpl*31, and found three cryptic species in the North Pacific, which they interpreted as the imprints of vicariant diversification, probably in the Miocene. These were accompanied by multiple crossings of different lineages through the Bering Strait and Arctic Ocean to the Atlantic Ocean at various times. The shallow divergence between Arctic and Antarctic isolates indicated recent transequatorial passage and confirmed the bipolar distribution of the species.

Order Colaconematales
Family Colaconemataceae
***Colaconema membranaceum* (Magnus) Woelkerling 1973**

Basionym *Callithamnion membranaceum* Magnus 1875.

Type locality Korsör, Denmark.

Representative specimens Arthur Harbor, Anvers Island, Antarctic Peninsula, −28 m in bryozoan, Moe 489, 10 June 1974, UC1571372.

Note In Antarctica, this species is found in anemones, bryozoans, hydroids, and pycnogonids.

Order Palmariales
Family Meiodiscaceae Claydon *et* Saunders 2010
***Meiodiscus concrescens* (K.M. Drew)**
P.W. Gabrielson in Gabrielson et al. 2000

Basionym *Rhodochorton concrescens* K.M. Drew 1928.

Type locality Carmel, California.

Representative specimens Arthur Harbor, Anvers Island, Antarctic Peninsula, −3 m on *Ballia*, Moe 617, 20 August 1974, UC1571439.

Note Southern Hemisphere representatives of this species should be compared with plants from the type locality, even though they seem to be identical morphologically. In California, *Meiodiscus concrescens* grows on sessile animals; in Antarctica and South Africa it has been found only as an epiphyte on *Ballia callitricha.*

Family Palmariaceae
***Palmaria decipiens* (Reinsch) R.W. Ricker 1987**

Basionym *Rhodymenia decipiens* Reinsch 1888.

Type locality South Georgia.

Representative specimens Arthur Harbor, Anvers Island, Antarctic Peninsula, −2 m, Moe 1021, 4 February 1975, UC1571510; New Rock, Deception Island, South Shetland Islands, mid intertidal, Moe 122, 11 January 1974, UC1571509; Arthur Harbor, −5 m, Amsler, 10 January 2008, Amsler in NCU.

Note An ITS sequence for *Palmaria decipiens* from King George Island (Lindstrom et al. 1996) places this species close to *P. callophylloides* from Alaska, suggesting a Northern Hemisphere origin and bipolar distribution for *Palmaria*. The numerous records of this and other species from localities elsewhere in the Southern Hemisphere (See Ricker 1987, Page et al. 2002) require reinvestigation. *P. decipiens* is generally restricted to the shallow subtidal (<6 m) and intertidal in relatively low wave energy areas and where it occurs in shallow water it can be very common. It can be present but is rare in deeper water.

Order Balliales
Family Balliaceae
***Ballia callitricha* (C. Agardh) Kützing 1843**

Basionym *Sphacelaria callitricha* C. Agardh 1824.

Type locality Falkland Islands.

Representative specimens Livingston Island, South Shetland Islands, DeLaca, January 1974, UC1497938; Arthur Harbor, Anvers Island, Antarctic Peninsula, shallow subtidal, Moe, 17 January 1988, UC1571437; Arthur Harbor, –18 m, Moe, January 1988, UC1571438.

Note This species is said to be widely distributed in Antarctica and the south temperate zone. As elsewhere, in Antarctica it bears the encrusting coralline, *Synarthrophyton patena*, the filamentous red alga, *Meiodiscus concrescens*, and the ascomycete, *Spathulospora*.

***Ballia sertularioides* (Suhr) Papenfuss 1940**

Basionym *Callithamnion sertularioides* Suhr 1840.

Type locality Table Bay, South Africa.

Note The Ross Sea record (Cormaci et al. 1992, Page et al. 2002) requires reinvestigation. It represents a striking range extension from temperate South Africa and Argentina.

Subclass Corallinophycidae
Order Corallinales
Family Hapalidiaceae
Subfamily Melobesioideae
***Clathromorphum obtectulum* (Foslie) W.H. Adey 1970**

Basionym *Lithothamnion kerguelenum* f. *obtectulum* ("*obtectula*") Foslie 1898.

Type locality Kerguelen.

Taxonomic synonyms *Lithophyllum discoideum* f. *aequabile* ("*aequabilis*") Foslie 1905, type locality: South Orkney Islands [=*Lithophyllum aequabile* (Foslie) Foslie 1906, *Pseudolithophyllum aequabile* (Foslie) Adey 1970, *Antarcticophyllum aequabile* (Foslie) Mendoza 1976; *Lithophyllum discoideum* f. *wandelicum* ("*wandelica*") Foslie 1906, type locality: Antarctic Peninsula].

Representative specimens Arthur Harbor, Anvers Island, Antarctic Peninsula, −10 m, Moe 10, 20 December 1973, UC1571389; Arthur Harbor, −9 m, Moe 534, 9 July 1974, UC1469731; Joubin Islands, Anvers Island, Antarctic Peninsula, −20 m, Moe 949, 16 January 1975, UC1571375.

Note This species is very common in intertidal pools and in the subtidal along the Antarctic Peninsula. It often grows on the shells of limpets. The crust reaches 1.3 cm in thickness (Mendoza and Cabioch 1985) and has a relatively smooth surface. Conceptacles are embedded in the thallus, so it can be difficult to determine that the tetrasporangial conceptacles are multiporate.

Clathromorphum lemoineanum M.L. Mendoza *et J. Cabioch 1985*

Type locality Booth Wandel Island, Antarctic Peninsula.

Misapplied names (*fide* Mendoza and Cabioch, 1985): *Lithophyllum subantarcticum* sensu Lemoine 1913, *Antarcticophyllum subantarcticum* sensu Mendoza 1979.

Note This species occurs along the Antarctic Peninsula much less frequently than *Clathromorphum obtectulum*. The crust is thinner and cell and conceptacle dimensions are smaller. Ecological differences need to be investigated. Woelkerling and Lamy (1998) asserted that *C. lemoineanum* is an illegitimate name because the type specimen of *Lithophyllum aequabile* was cited in its protologue. We interpret the citation as part of an itemization of specimens examined, not as the inclusion of a specimen in the circumscription of the new species, and therefore regard the name as legitimate.

Lithothamnion granuliferum Foslie 1905

Type locality Observatory Island, Tierra del Fuego.

Representative specimens Arthur Harbor, Anvers Island, Antarctic Peninsula, −6 m, Moe 302, 24 February 1974, UC1571366; Arthur Harbor, −10 m, Moe 466, 16 May 1974, UC1571376; Arthur Harbor, shallow subtidal (with *Hydrolithon subantarcticum*), Moe 870, 11 December 1974, UC1571369.

Note It is possible that Antarctic specimens identified as this species are referable to *Phymatolithon foecundum*, which seems to be widespread in the Antarctic and is morphologically similar.

Phymatolithon foecundum (Kjellman) Düwel *et* Wegeberg 1996

Basionym *Lithothamnion foecundum* Kjellman 1883.

Type locality Kara Sea, Arctic Ocean.

Taxonomic synonym *Lithothamnion coulmanicum* Foslie 1905, type locality: Cape Wadsworth, Coulman Island, Ross Sea [=*Leptophytum coulmanicum* (Foslie) W.H. Adey 1970].

Misapplied names (*fide* Alongi et al. 2002): *Clathromorphum lemoineanum* sensu Cormaci et al. (1992); *Leptophytum laeve sensu* Zaneveld and Sanford (1980); *Phymatolithon lenormandii sensu* Zaneveld and Sanford (1980).

Note Alongi et al. (2002) concluded that only one species of Corallinaceae is present in the Ross Sea. This conclusion is supported by observations by one of us in McMurdo Sound (R. Moe, personal observations in 1982 and 1983). The name assigned to the species is uncertain, however, and depends on the typification of *Leptophytum laeve*; see Adey et al. (2001) whose reinstatement of *Leptophytum* Adey 1966 based on *L. laeve* (Foslie) Adey requires a re-evaluation of the nomenclatural situation.

***Synarthrophyton patena* (J.D. Hooker *et* Harvey) R.A. Townsend 1979**

Basionym *Melobesia patena* J.D. Hooker *et* Harvey 1849.

Type locality Flat Point, New Zealand.

Taxonomic synonym *Melobesia verrucata* var. *antarctica* J.D. Hooker *et* Harvey, type locality: Southern Ocean [=*Lithothamnion antarcticum* (J.D. Hooker *et* Harvey) Heydrich 1901].

Representative specimens Arthur Harbor, Anvers Island, Antarctic Peninsula, −3 m on *Ballia*, Moe 617, 20 August 1974, UC1571439; Arthur Harbor, −22 m on *Phyllophora ahnfeltioides*, Moe 741, 24 October 1974, UC1571399; Lavebrua Island, Deception Island, South Shetland Islands, −13 m on *Phyllophora ahnfeltioides*, Moe 969, 1 January 1975, UC1469676.

Note As elsewhere, this species grows epiphytically on *Ballia callitricha*. It [or perhaps *S. schmitzii* (type locality: Fuegia)] also occurs commonly on *Phyllophora ahnfeltioides* and on the lower axes of *Myriogramme manginii*.

Family Corallinaceae
Subfamily Mastophoroideae
***Hydrolithon subantarcticum* (Foslie) M.L Mendoza *et* Cabioch 1986**

Basionym *Lithophyllum decipiens* f. *subantarcticum* Foslie 1906.

Type locality Fuegia.

Taxonomic synonym *Lithothamnion manginii* Lemoine *et* Rosenvinge 1913, type locality: Antarctic Peninsula.

Representative specimens Arthur Harbor, Anvers Island, Antarctic Peninsula, shallow subtidal, Moe 870, 11 December 1974, UC1571369.

Note This species seems to be of irregular occurrence in West Antarctica; it is small and thin and often overgrown by other crusts.

Subclass Ahnfeltiophycidae
Order Ahnfeltiales
Family Ahnfeltiaceae
***Ahnfeltia plicata* (Hudson) E.M. Fries 1835**

Basionym *Fucus plicatus* Hudson 1762.

Type locality England.

Note We have not examined any specimens on which Antarctic records are based. The record in Kylin and Skottsberg (1919) is based on Skottsberg's memory of a station near Deception Island from which the material was lost. Skottsberg (1941) listed the species from that station with a query. Reinsch (1890) recorded the species from South Georgia, but it has not been recorded from there again. Ricker (1987) illustrated non-reproductive specimens from Macquarie Island (which we consider sub-Antarctic).

Subclass Rhodymeniophycidae
Order Peyssonneliales
Family Peyssonneliaceae
***Peyssonnelia harveyana* P.L. Crouan *et* H.M. Crouan *ex* J. Agardh 1851**

Type locality Rade de Brest, Brittany, France.

Note R. Moe has examined the specimens (in PC) on which this record is based (Hariot 1907). They appear to represent the basal crust of *Gymnogongrus turquetii* Hariot, an alga that is common in the habitat from which the specimens were collected.

Order Bonnemaisoniales
Family Bonnemaisoniaceae
***Delisea pulchra* (Greville) Montagne 1844**

Basionym *Bowiesia pulchra* Greville 1830.

Type locality Possibly Macquarie Island, Australia.

Representative specimens Arthur Harbor, Anvers Island, Antarctic Peninsula, −20 m, Moe 53, 2 January 1974, UC1571545; Arthur Harbor, −10 m, Moe 939, 15 January 1975, UC1571546; Joubin Islands, Anvers Island, Antarctic Peninsula, −21 m, Moe 947, 16 January 1975, UC1557764.

Note Unpublished *rbc*L sequence analyses by M. Volovsek and M. Hommersand establish that *Delisea pulchra* from Antarctica is distinct from *Delisea* species that have been sequenced from Tasmania, New Zealand and South Africa. Hooker (1847) reported *D. pulchra* as being common in Kerguelen and noted that the original collection by Mr. Fraser (lectotype in the Greville herbarium at E) may have come from Macquarie Island, where it is uncommon according to Ricker (1987). Antarctic samples need to be compared to specimens from these localities. They can be moderately common in some locations, often at depths between 8 and 20 m. Fresh specimens produce a strong odor of violets.

***Ptilonia magellanica* (Montagne) J. Agardh 1852**

Basionym *Thamnophora magellanica* Montagne 1842.

Type locality Fuegia.

Representative specimens Arthur Harbor, Anvers Island, Antarctic Peninsula, −25 m, Moe 665, 7 October 1974, UC1571364; Arthur Harbor, −35 m, Moe 1087, 26 February 1975, UC1557584.

Note This species has been collected three times from Arthur Harbor, Anvers Island (R. Moe in UC). The Hariot (1907) record from the Antarctic Peninsula (in PC) is representative of *Delisea*.

Order Plocamiales
Family Plocamiaceae
***Plocamium "cartilagineum"* (Linnaeus) P.S. Dixon 1967**

Basionym *Fucus cartilagineus* Linnaeus 1753.

Type locality Doubtful.

Representative specimens Arthur Harbor, Anvers Island, Antarctic Peninsula, –24 m, Moe 566, 8 August 1974, UC1557630; English Strait, Greenwich Island, South Shetland Islands, –8 m, Moe 1018, 26 January 1975, UC1557605; Horseshoe Island, Marguerite Bay, Antarctic Peninsula, –13 m, Moe 1037, 12 February 1975, UC1469661.

Note Unpublished *rbc*L sequence analyses by S. Fredericq show that *P.* "*cartilagineum*" from Antarctica is distinct from all other species of *Plocamium* investigated so far. Its nearest neighbor is either a similar species from New Zealand or one from Chile. The Antarctic species can be very abundant in moderately protected waters, particularly below 8 m depth.

***Plocamium hookeri* Harvey in J.D. Hooker *et* Harvey 1845**

Type locality Kerguelen Island.

Note Antarctic Peninsula records (e.g., Moe and DeLaca 1976) probably represent specimens of *P.* "*cartilagineum*" with larger than average branchlets, and require reinvestigation.

***Plocamium secundatum* (Kützing) Kützing 1866**

Basionym *Plocamium coccineum* var. *secundatum* Kützing 1849.

Type locality Fuegia.

Note Antarctic Peninsula records (e.g., Moe and DeLaca 1976) probably represent specimens of *P.* "*cartilagineum*" with a slightly altered branching pattern, and require reinvestigation.

Family Sarcodiaceae
***Sarcodia montagneana* (J.D. Hooker *et* Harvey) J. Agardh 1852**

Basionym *Rhodymenia montagneana* J.D. Hooker *et* Harvey 1845.

Type locality Bay of Islands, New Zealand.

Note Many records of this species from the Antarctic Peninsula (e.g., Kylin and Skottsberg 1919, Moe and DeLaca 1976) are referable to *Trematocarpus antarcticus*.

Additional studies are needed to determine whether any species of *Sarcodia* are present in Antarctica.

Trematocarpus antarcticus **(Hariot) Fredericq *et* R.L. Moe comb. nov.**

Basionym *Kallymenia* ("*Callymenia*") *antarctica* Hariot 1907 (Hariot 1907, p. 7).

Type locality Booth Wandel Island, Antarctic Peninsula.

Representative specimens Arthur Harbor, Anvers Island, Antarctic Peninsula, −27 m, DeLaca, January 1974, UC1497879; Henkes Islands, Marguerite Bay, Antarctic Peninsula, −15 m, Moe 1041, 15 February 1975, UC1571362; Argentine Islands, Antarctic Peninsula, −22 m, Moe 1072, 19 February 1975, UC1557592.

Note This large, flattened species is distinct from any other species of *Trematocarpus*. It occupies a basal position in unpublished *rbc*L analyses by S. Fredericq and P.W. Gabrielson and is distantly related to *T. acicularis* from New Zealand. It occurs over a wide depth range from 5 m to deeper than 30 m, but is more commonly found in deep waters. This species was referred to as *Sarcodia montagneana* (Hooker *et* Harvey) J. Agardh in Amsler et al. (1995).

Order Gracilariales
Family Gracilariaceae
Curdiea racovitzae **Hariot 1900**

Type locality Antarctic Peninsula.

Representative specimens Hope Bay, Antarctic Peninsula, −13 m, Moe 215, 24 January 1974, UC1557573; Arthur Harbor, Anvers Island, Antarctic Peninsula, −7 m, Moe 730, 24 October 1974, UC1557583; Henkes Islands, Marguerite Bay, Antarctic Peninsula, −15 m, Moe 1039, 15 February 1975, UC1571365.

Note *rbc*L analyses (Gurgel and Fredericq 2004) show that *Curdiea racovitzae* is only distantly related to *Curdiea* species from New Zealand and may represent a separate genus. It was recently reported from the Ross Sea (Page et al. 2002). The species occurs across a broad range of both depth and wave-exposure.

Order Halymeniales
Family Halymeniaceae
"Pachymenia orbicularis" **(Zanardini) Setchell *et* N.L. Gardner 1934**

Basionym *Gigartina orbicularis* Zanardini 1874.

Type locality Port Phillip, Victoria, Australia.

Representative specimens Arthur Harbor, Anvers Island, Antarctic Peninsula, −10 m, Moe 527, 26 June 1974, Moe in UC; Arthur Harbor, −8 m, Moe 542, 9 July 1974, Moe in UC; Arthur Harbor, shallow subtidal, Moe, 17 January 1988, Moe in UC.

Note The plant recorded under *"Pachymenia orbicularis"* in Antarctica is unrelated to the Australian species. It is separated by long base pair distances from other *Pachymenia* species (Hommersand and Fredericq 2003), and probably should be placed in a new genus. Its nearest relative appears to be *"Pachymenia"* sp. from Macquarie Island (Ricker 1987). The two plants share a color dimorphism with male plants being reddish brown and female plants dark purple, drying to nearly black. A similar feature is reported for "*Pachymenia saxicola*" Taylor 1945 from the Galapagos Islands. On the Antarctic Peninsula the species occurs commonly from intertidal pools to depths of 22 m. Reports of this species in Amsler et al. (2005a, b) and Peters et al. (2005) were incorrect and actually represent *Varimenia macropustulosa* R.L. Moe.

Order Rhodymeniales
Family Rhodymeniaceae
***Rhodymenia coccocarpa* (Montagne) M.J. Wynne 2007**

Basionym *Phyllophora coccocarpa* Montagne 1852.

Type locality Strait of Magellan.

Taxonomic synonyms (see Wynne 2007) *Rhodymenia subantarctica* R.W. Ricker 1987 [= substitute name for *Rhodymenia cuneifolia* (J.D. Hooker *et* Harvey) W.R. Taylor 1939; basionym: *Phyllophora cuneifolia* J.D. Hooker *et* Harvey 1845; type locality: St. Salvador Bay, East Falkland Island].

Note We have not examined any of the Antarctic records of this species.

Family Fryeellaceae (see Le Gall et al. 2008)
***Hymenocladiopsis prolifera* (Reinsch) M.J. Wynne 2004**

Basionym *Gracilaria prolifera* Reinsch 1888.

Type locality South Georgia.

Taxonomic synonym *Hymenocladiopsis crustigena* R.L. Moe 1986, type locality: Arthur Harbor, Anvers Island, Antarctic Peninsula.

Representative specimens Elephant Island, South Shetland Islands, between −6 m and −12 m, 15 January 1974 (Temnikow leg.), UC1557653; Arthur Harbor, Anvers Island, Antarctic Peninsula, −18 m, Moe 331, 25 March 1974, UC1557652; Henkes Islands, Marguerite Bay, Antarctic Peninsula, between −10 m and −16 m, Moe 1040, 15 February 1975, UC1571393.

Note It seems likely that Antarctic plants are conspecific with South Georgia plants; new collections from South Georgia are needed to be certain.

***Hymenocladiopsis* sp.**

Representative specimens Cape Bellue, Antarctic Peninsula, −22 m, Moe 166, 5 February 1974, Moe in UC; Arthur Harbor, Anvers Island, Antarctic Peninsula, between −6 m and −33 m, Amsler, 11 May 2008, Amsler in NCU; Argentine Islands, Antarctic Peninsula, −24 m, Moe 1069, 19 February 1975, Moe in UC.

Note There is at least one additional undescribed species of *Hymenocladiopsis* in Antarctica based on unpublished *rbc*L sequence analyses by P.W. Gabrielson; the morphological variation encountered in this genus is such that there may be other species as well. The plant illustrated under *Gymnogongrus turquetii* in Wiencke and Clayton (2002) seems to be misidentified; it falls within the range of variation of specimens of *Hymenocladiopsis* that we have examined.

Order Ceramiales
Family Ceramiaceae Tribe Pterothamnieae
***Pterothamnion antarcticum* (Kylin) R.L. Moe *et* P.C. Silva 1980**

Basionym *Antithamnion antarcticum* Kylin in Kylin and Skottsberg 1919.

Type locality Astrolabe Island, Antarctic Peninsula.

Representative specimens Arthur Harbor, Anvers Island, Antarctic Peninsula, DeLaca, December–March 1973/1974, UC1497844.

Note This species has seldom been collected and is known primarily from the studies of Moe and Silva (1980).

Pterothamnion simile (J.D. Hooker *et* Harvey) Nägeli 1862

Basionym *Callithamnion simile* J.D. Hooker *et* Harvey 1845.

Type locality Kerguelen.

Note Specimens from Enderby Land, East Antarctica were described and illustrated by Skottsberg. Plants from Macquarie Island are well illustrated by Ricker (1987).

Tribe Ceramieae
***Ceramium involutum* Kützing 1849**

Type locality (Not specified) Cape Horn and Falkland Islands.

Note The presence of this species in Antarctica requires confirmation; Skottsberg's report (Skottsberg 1953) is based on a single fragment in the drift at the Mac Robertson Coast.

Tribe Callithamnieae
***Georgiella confluens* (Reinsch) Kylin 1956**

Basionym *Ptilota confluens* Reinsch 1888.

Type locality South Georgia.

Representative specimens Joubin Islands, Anvers Island, Antarctic Peninsula, Moe 173 (DeLaca leg.), 2 February 1974, UC1571431; Arthur Harbor, Anvers Island, Antarctic Peninsula, –9 m, Moe 444, 1 May 1974, UC1571430; Argentine Islands, Antarctic Peninsula, 22 m, Moe 1073, 19 February 1975, UC1571432.

Note Originally described from South Georgia, this species is also widespread in Antarctica (Moe and Silva 1983, Page et al. 2002). *Georgiella confluens* was identified as a member of the tribe Callithamnieae by Hommersand et al. (2005).

***Falklandiella harveyi* (J.D. Hooker) Kylin 1956**

Basionym *Ptilota harveyi* J.D. Hooker 1845.

Lectotype locality Hermite Island, Cape Horn.

Note The placement of *Ptilota harveyi* in *Dasyptilon* by Papenfuss was rejected by Silva (1993). For the recognition of *Falklandiella* as a member of the tribe Callithamnieae, see Hommersand et al. (2005). Records of *Falklandiella* from Antarctica are probably misidentifications of *Georgiella confluens* or *Plumariopsis peninsularis* (Moe and Silva 1983).

Tribe incertae sedis
***Plumariopsis peninsularis* R.L. Moe *et* P.C. Silva 1983**

Type locality Arthur Harbor, Anvers Island, Antarctic Peninsula.

Representative specimens Arthur Harbor, Anvers Island, Antarctic Peninsula, −21 m, Moe 278, 18 January 1974, UC1475008; Hurd Peninsula, False Bay, Livingston Island, South Shetland Islands, −20 m, Moe 188, 28 January 1974, UC1557620; Arthur Harbor, −25 m, Moe 669, 7 October 1974, UC1557621.

Note Widely distributed in the Antarctic Peninsula, this species is sister to *Dasyptilon pellucidum* (Harvey) G. Feldmann 1950 from New Zealand based on unpublished *rbc*L analyses by D.W. Freshwater.

***Plumariopsis eatonii* (Dickie) De Toni 1903**

Basionym *Ptilota eatonii* Dickie 1876.

Type locality Kerguelen Island.

Note We have not encountered this species in the Antarctic continent or in the South Shetland Islands.

Tribe Antarcticothamnieae
***Antarcticothamnion polysporum* R.L. Moe *et* P.C. Silva 1979**

Type locality Elephant Island, South Shetland Islands.

Representative specimens New Rock, Deception Island, South Shetland Islands, Moe 104, 11 January 1974, UC1571457; Elephant Island, South Shetland Islands, Moe 139, 15 January 1974, UC1571456; Kazak Island, Vestfold Hills, Davis, East Antarctica, −14 m, Ricker 3017, 15 January 1982, UC1497976.

Note Known from the South Shetland Islands and the Antarctic Peninsula and recently reported from the Ross Sea (Page et al. 2002), this species has been collected infrequently.

Tribe Griffithsieae
***Griffithsia antarctica* J.D. Hooker *et* Harvey in J.D. Hooker 1847**

Type locality (Not specified) Fuegia and the Falkland Islands.

Note The Etcheverry (1983) record is from the Antarctic Peninsula, and the Lucas (1919) record is from East Antarctica.

Family Delesseriaceae
Subfamily Delesserioideae
Tribe Botryocarpeae [Hemineureae in Lin et al. (2001)]
***Pantoneura plocamioides* Kylin in Kylin *et* Skottsberg 1919**

Type locality South Georgia.

Representative specimens Arthur Harbor, Anvers Island, Antarctic Peninsula, −20 m, Moe 56, 2 January 1974, UC1469698; Arthur Harbor, −20 m, Moe 43, 30 December 1973 (Zumwalt leg.), UC1571526; English Strait, Greenwich Island, South Shetland Islands, −8 m, Moe 1012, 26 January 1975, UC1469697.

Note Unpublished *rbc*L sequence data by S.-M. Lin establish that this genus belongs in the tribe Botryocarpeae. It is unrelated to the Arctic species: *Pantoneura fabriciana* (Lyngbye) Wynne 1997 [formerly *P. baerii* (Ruprecht) Kylin 1924], a species closely related to *Membranoptera alata* (Hudson) Stackhouse 1809. Along the Antarctic Peninsula, the species is locally abundant at depths exceeding 20 m.

***Rhodokrambe laingioides* R.L. Moe gen. *et* sp. nov. (Figures 4.1 and 4.2)**

Vegetative structure Thallus composed of one to several blades borne on short, branched or unbranched stipe (Figure 4.1). Blades salmon-pink when living, brick-red to dark purplish red when dry, coarse, 20–50 cm tall and up to 20 cm broad, with an ovate to lanceolate outline (Figure 4.1). Blades with percurrent midribs and unbranched, opposite lateral veins. Both midribs and veins flattened and often inconspicuous in young plants. Branching, which is sparse, taking place by proliferation from the stipe and lobing from the blade margin, with a lateral vein of the main blade continuing as the midrib of a branch. In young plants the margins of blades smooth and entire, but in old plants becoming irregularly torn. Apical cells dividing transversely, embedded in a shallow notch in a rounded apex. Cells in the primary cell row (derivatives of the apical cell) not undergoing intercalary division; some cells in secondary rows dividing intercalarily (Figure 4.2). Blades, except at the apex, polystromatic, to 400 μm thick, consisting of a central layer with one or more cortical layers on either side. Central layer consisting of coplanar primary cells that are derivatives of the primary axial cell row. These expanding enormously, reaching 0.5 mm long and bearing numerous secondarily derived rhizoidal cells. Central layer in veins compact, with rhizoids interweaving densely among large cells; between veins central layer loose. Cortex composed of small isodiametric cells. Some of the cells of the primary layer near the margin larger than surrounding cells and staining more intensely.

Reproductive structure Procarps produced along the axial cell row of small bladelets that proliferate between the veins from the upper and lower surfaces of blades. These fertile bladelets, borne in clusters on short, branched petioles, are polystromatic and about 2 mm long. Several procarps, each with one carpogonial branch and two sterile

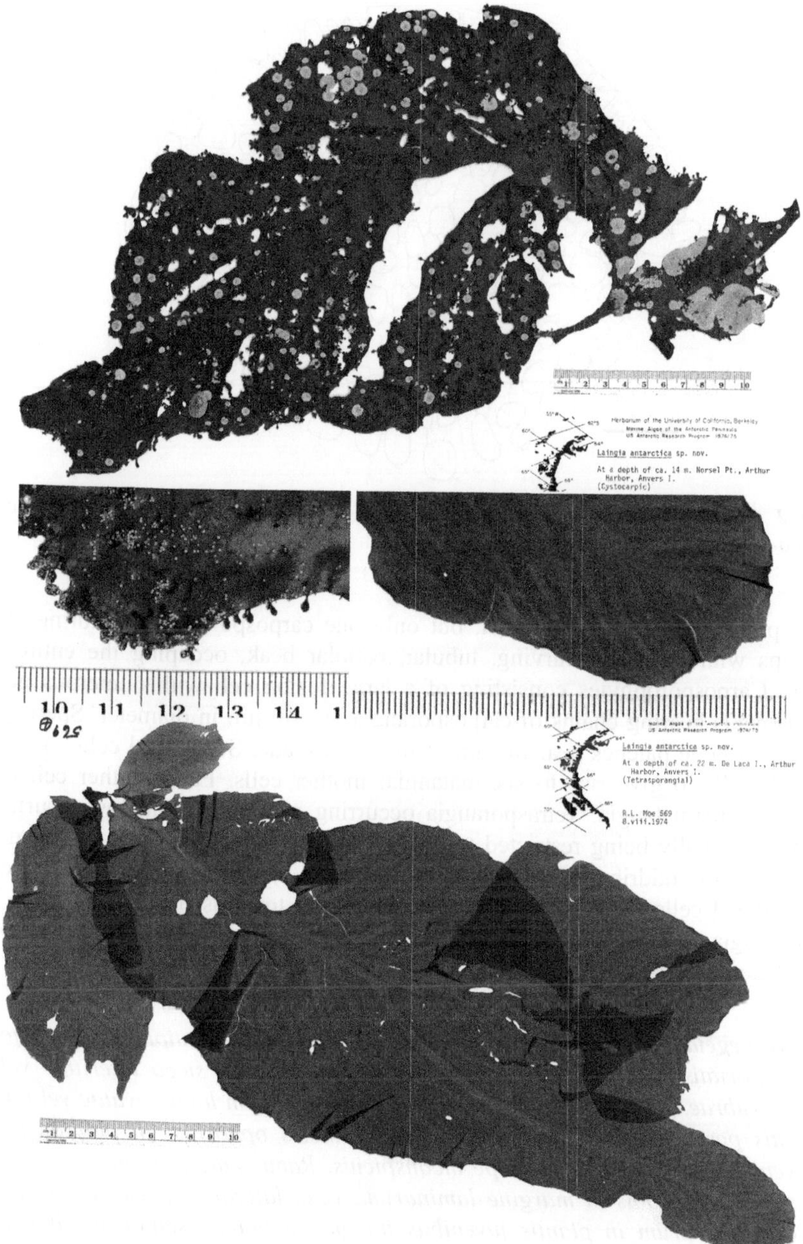

Figure 4.1 *Rhodokrambe laingioides*: digitized scans of herbarium specimen.
(Upper) Type, UC1935488; scale: 10 cm. Cystocarpic blade. (Center left) Detail showing clusters of cystocarps at margin; scale in mm. (Center right) Detail showing tetrasporangia distributed between veins; scale in mm. (Lower) Tetrasporangial blade showing marginal branching; Moe 569 in UC; scale: 10 cm.
Scan reproduced with kind permission from the University Herbarium, University of California at Berkeley (UC) allowing us to publish scans of these specimens.

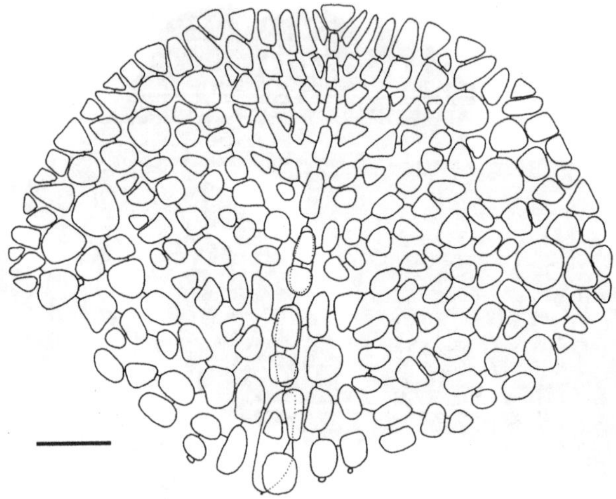

Figure 4.2 *Rhodokrambe laingioides*: diagram of the apex of young plant. Intercalary divisions in second-order cell rows are evident; scale = 50 μm.

groups, produced by each branchlet, but only one carposporophyte maturing. Mature cystocarps with upwardly curving, tubular, ostiolar beak, occuping the entire fertile bladelet. Carposporophytes consisting of a large fusion cell with numerous gonimoblast filaments bearing chains of carposporangia 30–40 μm in diameter. Spermatangia occurring in large patches near margin of the main blade. Superficial cells dividing by anticlinal walls to give rise to spermatangial mother cells. Each mother cell bearing one or two spermatangia. Tetrasporangia occurring over upper and lower surfaces of main blade, usually being restricted to the area between the veins, but sometimes distributed over the midrib and veins as well (Figure 4.1). Tetrasporangial primordia cut off by cortical cells below the surface and enlarging towards the center of the blade. Tetrasporangia spherical, 90–120 μm in diameter at maturity, completely embedded in the blade.

Structura vegetativa *Thallus e laminis singulis vel aliquot in stipite brevi ramoso vel eramoso portatis constans. Laminae in vivo salmoneae, in sicco lateritiae vel fusce purpureo-rubrae, grossae, 20–50 cm altae, usque ad 20 cm latae, ovatae vel lanceolatae, costis percurrentibus, venis lateralibus eramosis oppositis, in plantis juvenibus costis venisque complanatis et saepe inconspicuis. Rami sparsi, orientes e prolificationibus stipitis et e lobis in margine laminarum, vena laterali extensa in costam rami. Margines laminarum in plantis juvenibus integri, in plantis senioribus irregulariter rupti. Cellulae apicales transverse dividentes, ad apicem rotundatum inclusae in incisura non-profunda. Cellulae in serie primaria (e cellula apicali ortae) non-subeuntes divisiones intercalares; cellulae aliquot in seriebus secundariis subeuntes divisiones intercalares. Laminae praeter apicem polystromaticae 400 μm crassae, e strato centrale et stratis corticalibus singulis magisve in quoque facie constantes. Stratum centrale constans e cellulis primariis – progeniebus coplanaribus seriei primarii cellularum axialium, valde dilatatis et usque ad 0.5 mm longis – et multis cellulis secundariis* *rhizoidalibus, in venis compactum rhizoideis inter cellulas magnas dense implexis, inter venas laxum. Cortex e cellulis parvis isodiametricis constans. Aliquot cellulae strati primarii prope marginem cellulis ambientibus majores et labe magis coloratae.*

Structura reproductiva *Procarpia e laminulis parvis prodientia secus seriem cellularum axialem. Illae laminulae fertiles prolificatae inter venas e superficiebus ambobus lamellarum, fasciculatae in petiolis brevibus ramosis, polystromaticae, circa 2 mm longae. Procarpia uno ramo carpogoniali et duobus gregibus sterilibus, aliquot per laminulam sed unum tantum carposporophytum maturum parientia. Cystocarpia matura totam laminulam fertilem occupantia, rostro tubulari ostiolato sursum curvato. Carposporophyta constans e cellula fusionis magna multis filamentis gonimoblasti ferentibus carposporangia catenata 30–40 µm in diametro. Spermatangia in areis magnis prope marginem laminae primariae aggregata. Cellulae superficiales divisionibus anticlinalibus cellulas maternas spermatangiorum efferentes, quaque cellula materna spermatangia singula duave ferenti. Tetrasporangia per superficiem superiorem inferioremque laminae primariae disposita, plerumque inter venas sed interdum atque supra costis venisque. Primordia tetrasporangiorium cellulis corticalibus infra superficie abscissa, centrum laminae versus aucta. Tetrasporangia sphaerica, 90–120 µm in diametro in maturitatem, omnino lamina inclusa.*

Holotype UC1935498: Moe 228; 2 March 1974; south side Norsel Point, Arthur Harbor, Anvers Island, Antarctic Peninsula (64°46′ S, 64°06′ W); –14 m; cystocarpic.

Etymology The greek word 'krambe' refers to the kale-like appearance of this plant in the field and its resemblance to the genus *Laingia*.

Distribution Antarctic Peninsula, South Shetland Islands, over broad depth range, but never in abundance. Most often found in shallow (<8 m) depths in moderately protected to strongly wave-exposed waters, but also found at depths >20 m.

Representative specimens Arthur Harbor, Anvers Island, Antarctic Peninsula, –14 m, Moe 228, 2 March 1974, UC1935498; Arthur Harbor, –22 m, Moe 569, 8 August 1974, Moe in UC; Joubin Islands, Anvers Island, Antarctic Peninsula, –20 m, Moe 945, 16 January 1975, Moe in UC.

Note An *rbc*L sequence published for a plant from Antarctica bearing the name *Pseudophycodrys pulcherrima* Baardseth in Lin et al. (2001) is referable to this taxon. It has been informally referred to as 'Kale' by C.D. Amsler and his associates because the thallus shape of fresh material often resembles the garden vegetable. While it bears a resemblance to *Pseudophycodrys phyllophora* (J. Agardh) Skottsberg from southern Chile and the Falkland Islands, the cystocarps are borne on surface proliferations, much as in *Laingia hookeri* (Lyall) Kylin (Wagner 1954). The evidence from unpublished *rbc*L sequence analyses by S.-M. Lin suggests that the new species is most closely related to *Pantoneura plocamioides*.

Microrhinus carnosus **(Reinsch) Skottsberg 1923**

Basionym *Delesseria carnosa* Reinsch 1888.

Type locality South Georgia.

Representative specimens Cumberland Bay, South Georgia, Skottsberg, 25 April 1902, UC205727; Byers Peninsula, Livingston Island, South Shetland Islands, in drift, Moe 89, January 1974, UC1557694.

Note This species is rare in Antarctica, having been found only in the South Shetland Islands. Skottsberg (1923) argued that *Microrhinus* is most closely related to *Pantoneura*. We have not sequenced *Microrhinus* and cannot confirm the speculation.

Tribe Apoglosseae
"Delesseria" lancifolia J. Agardh 1872

Type locality Hermite Island, Cape Horn (intertidal and dredged in 5–6 fathoms).

Representative specimens Arthur Harbor, Anvers Island, Antarctic Peninsula, −6 m, Moe 925, 30 December 1975, UC1469706; Joubin Islands, Anvers Island, Antarctic Peninsula, Moe 956, 16 January 1975, UC1557733; Arthur Harbor −10 m, Moe 1088, 1 March 1975, UC1469707.

Note This species belongs in *Paraglossum* J. Agardh (1898, pp. 213–218) and is one of two possible lectotype species (see Lin et al. 2001). A cursory examination of type material from Hermite Island suggests that Antarctic records identified as "*Delesseria lancifolia*" may represent a new species of *Paraglossum*. It is most common in shallower (<8 m) depths in moderately to heavily wave-exposed areas, but occasionally occurs deeper, to 30 m or greater.

"Delesseria" salicifolia **Reinsch 1888**

Type locality South Georgia.

Taxonomic synonym "*Delesseria*" *stephanocarpa* (A. Gepp *et* E.S. Gepp) Skottsberg 1941.

Basionym *Hydrolapathum stephanocarpum* A. Gepp *et* E.S. Gepp 1905, type locality: South Orkney Islands.

Representative specimens Arthur Harbor, Anvers Island, Antarctic Peninsula, −13 m, Moe 657, 5 October 1974, UC1557758; Arthur Harbor, Anvers Island, Antarctic Peninsula, −22 m, Moe 739, 24 October 1974, UC1557759; Joubin Islands, Anvers Island, Antarctic Peninsula, −20 m, Moe 948, 16 January 1975, UC1557733.

Note This species belongs in *Paraglossum* based on its morphology and on unpublished *rbc*L analyses of Antarctic specimens by S.-M. Lin. It is locally common at depths exceeding 25 m and is extremely variable in form depending on age.

Subfamily Phycodryoideae
Tribe Phycodryeae

Phycodrys antarctica **(Skottsberg) Skottsberg 1923**

Basionym *Delesseria antarctica* Skottsberg in Kylin and Skottsberg 1919.

Type locality Antarctic Peninsula.

Representative specimens Cape Bellue, Antarctic Peninsula, deeper than −22 m, Moe 167b, 5 February 1974, UC1557683; Arthur Harbor, Anvers Island, Antarctic Peninsula, −60 m, Moe 959, 16 January 1975, UC1557684; Admiralty Bay, King George Island, South Shetland Islands, −35 m, Zielinski, January 1983, UC1501415.

Note This is a morphologically variable species that occurs only in deep water along the Antarctic Peninsula, but at all depth ranges in high latitudes. It has been recorded widely throughout Antarctica (Wiencke and Clayton 2002) including the Ross Sea (Page et al. 2002).

***Phycodrys austrogeorgica* Skottsberg 1923**

Type locality South Georgia.

Representative specimens Arthur Harbor, Anvers Island, Antarctic Peninsula, −22 m, Moe 264, 4 March 1974, UC1469670; Cape Rey, Antarctic Peninsula, −15 m, Moe 1054, 15 February 1975 (Laine leg.), UC1557680; Arthur Harbor, −10 m, Moe, 18 January 1988, UC1469674.

Note Antarctic specimens, like those from South Georgia, are distinguished by proliferous branching from both the margins and the midribs. The species is sister to but clearly separated from *Phycodrys antarctica* in unpublished *rbc*L sequence analyses by S.-M. Lin. It is usually collected at depths <12 m, but occurs as deep as 22 m.

***Phycodrys quercifolia* (Bory) Skottsberg 1922**

Basionym *Delesseria quercifolia* Bory 1826.

Type locality Falkland Islands.

Note Plants from Antarctica that resemble *Phycodrys quercifolia* as traditionally figured from South America and the Falkland Islands have a different *rbc*L sequence and are not closely related, nor are they related to species bearing that name from New Zealand (Lin and Nelson 2010). The Antarctic specimens so identified may all be forms of *Phycodrys antarctica*.

Tribe Myriogrammeae
***Myriogramme livida* (J.D. Hooker and Harvey) Kylin 1924**

Basionym *Nitophyllum lividum* J.D. Hooker *et* Harvey 1845.

Type locality Falkland Islands.

Note This species has been reported from the Antarctic Peninsula and the South Shetland Islands, but we have never collected it, nor have we examined the reported specimens.

***Myriogramme manginii* ("*mangini*") (Gain) Skottsberg 1953**

Basionym *Nitophyllum manginii* ("*mangini*") Gain 1911.

Type locality Antarctic Peninsula.

Representative specimens Arthur Harbor, Anvers Island, Antarctic Peninsula, −22 m, Moe 262, 4 March 1974, UC1557700; Arthur Harbor, −10 m, Moe 564, 8 August 1974, UC1557701; Henkes Islands, Marguerite Bay, Antarctic Peninsula, −15 m, Moe 1038, 15 February 1975, UC1571492.

Note *Myriogramme manginii* clearly belongs in the tribe Myriogrammeae based on *rbc*L analyses (Lin et al. 2001). The Antarctic species is distant from *M. livida,* the type species of *Myriogramme,* in *rbc*L analyses. The blade is tristromatic compared to the typically monostromatic condition found in blades of most species of *Myriogramme*. It may constitute a separate genus. See additional comments under *Neuroglossum delesseriae*. *M. manginii* can be very abundant in wave-exposed areas, particularly at depths shallower than 10 m.

***Neuroglossum delesseriae* (Reinsch) M.J. Wynne 1997**

Basionym *Choreocolax delesseriae* Reinsch 1890.

Type locality South Georgia.

Taxonomic synonym *Neuroglossum ligulatum* Skottsberg in Kylin and Skottsberg 1919; basionym: *Delesseria ligulata* Reinsch 1888 (not *D. ligulata* S.F. Gray 1821); type locality: South Georgia.

Note *Choreocolax delesseriae* was based on proliferations (originally thought to represent a parasite) from the thallus of *Delesseria ligulata* Reinsch. *C. delesseriae* and *D. ligulata* are thus taxonomic synonyms. Although Reinsch published *D. ligulata* before *C. delesseriae*, the former name is a later homonym of *D. ligulata* S.F. Gray; accordingly, *C. delesseriae* has nomenclatural priority (see Wynne 1997). The type species of *Neuroglossum*, *N. binderianum* Kützing 1843 belongs in the Schizoserideae (Lin et al. 2001). Specimens from Antarctica that resemble *Delesseria ligulata* Reinsch from South Georgia all have *rbc*L sequences that are identical to those of plants identifiable as *Myriogramme manginii* and may represent a form or stage in the development of that species. Quartino et al. (2005) place *Myriogramme manginii* from the South Shetland Islands under *Neuroglossum delesseriae*. A new combination under *Myriogramme* or a new genus name will be required should plants from South Georgia and Antarctica prove to represent the same species.

Tribe Cryptopleureae
***Myriogramme smithii* (J.D. Hooker *et* Harvey) Kylin**

Basionym *Nitophyllum smithii* J.D. Hooker *et* Harvey 1845.

Type locality Falkland Islands.

Representative specimens Arthur Harbor, Anvers Island, Antarctic Peninsula, −22 m, Moe 510, 20 April 1974, Moe in UC; Arthur Harbor, −22 m, Moe 857, 11 December 1974, (with gall), Moe in UC; Henkes Islands, Marguerite Bay, Antarctic Peninsula, −15 m, Moe 1045b, 15 February 1975, Moe in UC.

Note Antarctic material referred to *Myriogramme smithii* is a new species of *Hymenena* based on morphology and *rbc*L sequence analysis (S.-M. Lin and M. Hommersand, unpublished data). It falls in the clade that contains *Hymenena venosa* (Linnaeus) Krauss 1846 from South Africa, the type species of *Hymenena* Greville 1830 (see Wagner 1954). In Antarctica, "*M. smithii*" is usually restricted to water exceeding 25 m and can be locally abundant.

Family Rhodomelaceae
Tribe Polysiphoneae
***Polysiphonia abscissa* J.D. Hooker *et* Harvey 1845**

Type locality Fuegia.

Note Antarctic Peninsula records are based on fragments obtained in deep-water dredge (Gain 1912) and require reinvestigation; East Antarctic records are based on a specimen from Cape Dennison figured by Skottsberg (1953).

Tribe incertae sedis
***Picconiella plumosa* (Kylin) J. De Toni 1936**

Basionym *Pteronia plumosa* Kylin in Kylin and Skottsberg 1919.

Representative specimens Joubin Islands, Anvers Island, Antarctic Peninsula, −35 m, Moe 160, 8 February 1974, UC1469673; English Strait, Greenwich Island, South Shetland Islands, −8 m, Moe 1019, 26 January 1975, UC1571415; Arthur Harbor, Anvers Island, Antarctic Peninsula, between −12 and −15 m, Amsler, 27 March 2007, Amsler in NCU.

Type locality (Not indicated), Antarctic Peninsula and South Georgia.

Note The phylogenetic position and taxonomic placement of *Picconiella* have not been investigated. The species is locally abundant at depths exceeding 20 m.

***Antarctocolax lambii* Skottsberg 1953**

Type locality Antarctic Peninsula.

Representative specimens Arthur Harbor, Anvers Island, Antarctic Peninsula, −30 m, Moe 461, 7 May 1974, UC1557754; Arthur Harbor, −23 m, Moe 839a, 4 December 1974, UC1557799.

Note This species is parasitic on *Picconiella plumosa*, and is relatively common.

Order Gigartinales
Family Gainiaceae
***Gainia mollis* R.L. Moe 1985**

Type locality Henkes Islands, northern Marguerite Bay, Antarctic Peninsula.

Representative specimens Henkes Islands, Marguerite Bay, Antarctic Peninsula, −15 m, Moe 1044, 15 February 1975, UC1475017; Kazak Island, Vestfold Hills, Davis, East Antarctica, −14 m, Ricker 3012, 15 January 1982, UC1497986.

Family Kallymeniaceae
***Nereoginkgo adiantifolia* Kylin in Kylin *et* Skottsberg 1919**

Type locality Syntype localities South Georgia, Antarctic Peninsula.

Representative specimens New Rock, Deception Island, South Shetland Islands, −27 m, Moe 108, 11 January 1974, UC1571512; Arthur Harbor, Anvers Island, Antarctic Peninsula, −22 m, Moe 1089, 1 March 1974, UC1571514; Arthur Harbor, between −27 and −37 m, Amsler, 24 March 2007, Amsler in NCU.

Note This species is usually restricted to waters >25 m depth. It occurs at a variety of sites but never in abundance.

***Nereoginkgo populifolia* R.L. Moe sp. nov. (Figure 4.3, upper right; Figure 4.4, upper right; Figure 4.5, upper right)**

Vegetative structure Thallus consisting of several (number indeterminate) blades borne on a prostrate, branched or unbranched stipe. Each blade composed of stipitate portion (to 2.5 cm long) and orbiculate, basally attenuate lamina (to 10 cm in diameter). Lower part of stipitate portion beset with small spines, some of these developing into blades. Stipe irregularly sympodial in development. Holdfast initially discoid, becoming hapteroid. Blades brownish-red, delicate, but not soft or slippery, 100–200 μm thick, composed of thin cortex and loose medulla. Outer cells of cortex anticlinally elongated (5–7 μm × 7–12 μm), interior cortical cells spherical (5–12 μm in diameter). Medullary cells mostly thin-walled, elongated in plane of blade (50 μm × 150–200 μm), surrounded loosely by filaments of small, subspherical pigmented cells. Greatly elongated or stretched cells (presumably the primary cells of medulla) sparingly present, stellate cells absent. Medulla at margin comprising single layer of spherical cells. Medulla of stipe composed of elongate cells densely packed together and united by numerous secondary pit-connections.

Reproductive structure Spermatangia (1 μm × 3 μm) borne on small bladelets at base of blade on plants that also bear cystocarps. Carpogonial branch systems produced at boundary of cortex and medulla, each consisting of a supporting cell bearing a 3-celled carpogonial branch and 1–3 subsidiary cells, initially in a dense cluster of small, darkly-staining cells, with carpogonial branch and trichogyne evident before subsidiary cells are cut off. All cells of system except carpogonium becoming multinucleate and lobing into medulla, with the system at maturity being 50 μm in diameter. Trichogyne coiled beneath cortex with tip extending to surface. Auxiliary cell systems consisting of spherical auxiliary cell with 2–4 subsidiary cells produced near carpogonial branch systems. Supporting cell, subsidiary cells, and first cell of carpogonial branch coalescing to form fusion cell. Fusion cells rotund (to 60 μm in diameter). Elongate cells cut off from fusion cells developing into filaments (presumptive connecting filaments). Cystocarps developing over entire blade, to 700 μm in diameter. Carposporangia spherical, uninucleate (15 μm in diameter), separated into indistinct groups by sterile filaments, released by breakdown of cystocarp wall. Tetrasporangia scattered over upper and lower surface of blade (20 μm × 30 μm), spores irregularly arranged in sporangium, sometimes appearing nearly cruciate, sometimes nearly zonate.

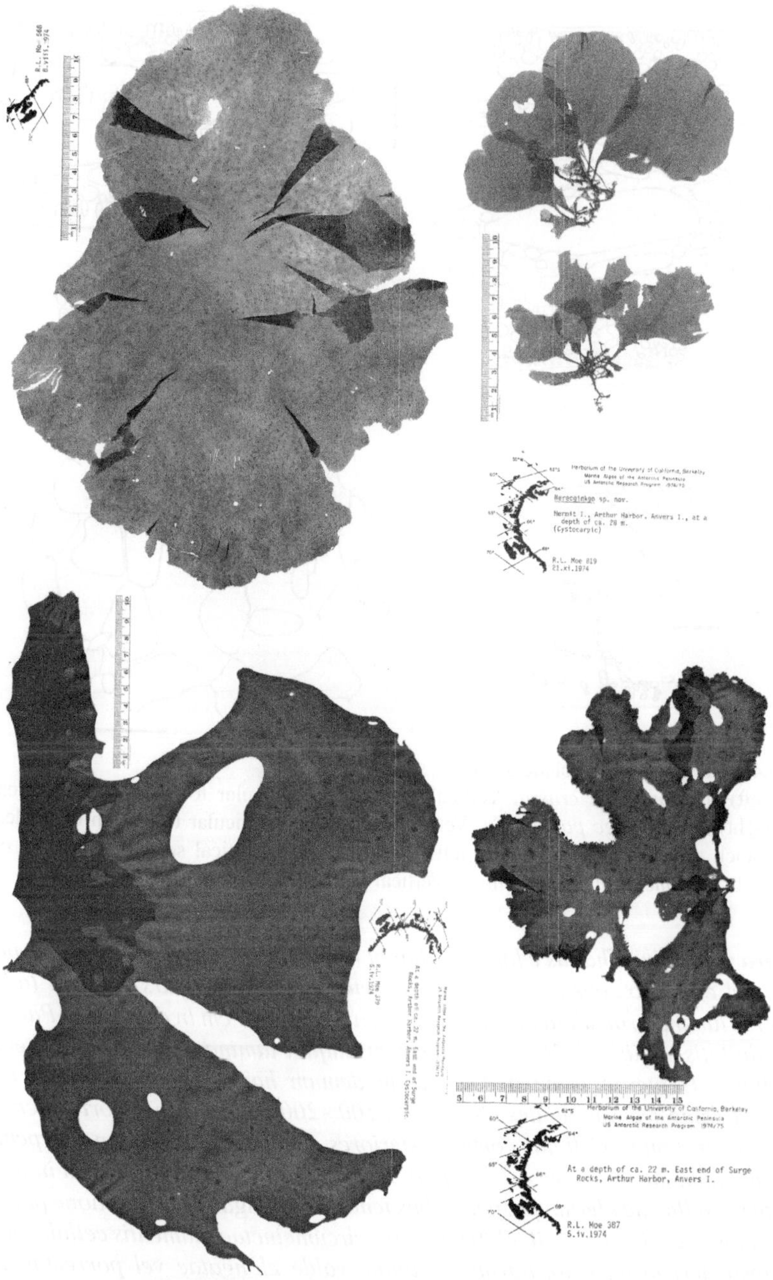

Figure 4.3 New taxa of Kallymeniaceae (all scales = 10 cm).
(Upper left) *Leniea lubrica*. Type, UC1935500. (Upper right) *Nereoginkgo populifolia*. Type, UC1935497. (Lower left) *Varimenia macropustulosa*. UC1935499. (Lower right) *Austropugetia crassa*. Type, UC1935496.
Scan reproduced with kind permission from the University Herbarium, University of California at Berkeley (UC) allowing us to publish scans of these specimens.

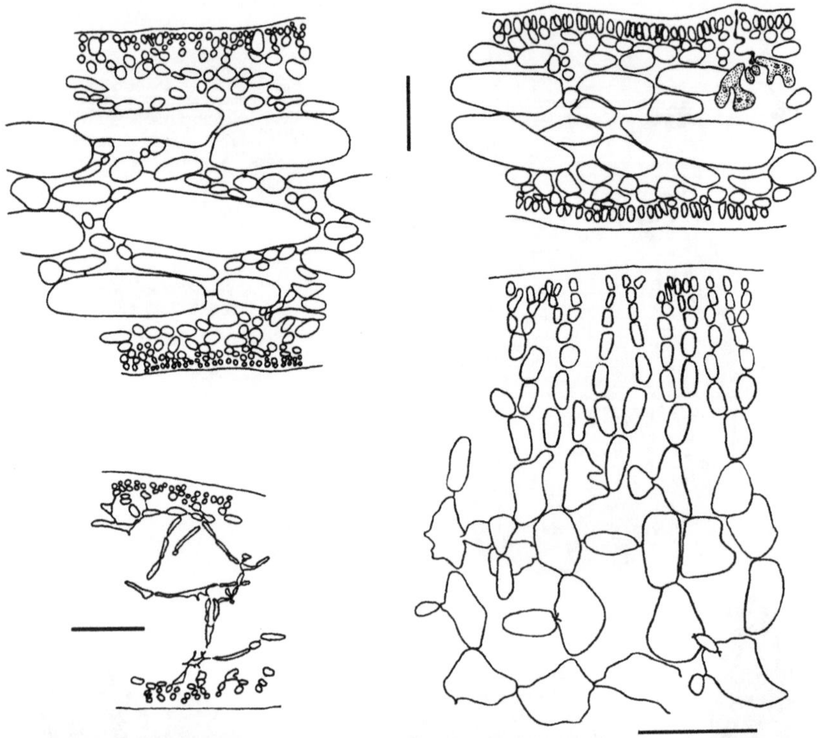

Figure 4.4 Kallymeniaceae blade sections; scales = 50 μm.
(Upper left) *Austropugetia crassa*. Vertical section perpendicular to margin of first-year plant. (Upper right) *Nereoginkgo populifolia*. Vertical section perpendicular to margin of blade. Carpogonial branch system stippled. (Lower left) *Leniea lubrica*. Vertical section in middle of blade. (Lower right) *Varimenia macropustulosa*. Vertical section of cortex.

Structura vegetativa *Thallus e laminis aliquot (numerorum variorum) in stipite prostrato ramoso vel eramoso constans. Lamina in partes duas divisa: pars stipitata (ad 2.5 cm longam) et lamina orbiculata basin versus attenuata (ad 10 cm in diametro). Pars inferior partis stipitatis spinis parvis munita, quarum aliquot laminas fiunt. Stipes irregulariter sympodialis. Hapteron primum discoideum demum hapteroideum. Laminae brunneo-rubrae delicatae sed non-molles lubricaeve, 100–200 μm crassae, e cortice tenui et medulla laxa constantes. Cellulae corticis exteriores elongatae in directione perpendiculari ad planum laminae (5–7 μm × 7–12 μm), interiores sphaericae (5–12 μm in diametro). Cellulae medullariae plerumque parietibus tenuibus, elongatae in directione parallela ad planum laminae (50 μm × 150–200 μm), laxe circumcinctae filamentis cellularum parvarum subsphaericarum coloratarum. Cellulae valde elongatae vel porrectae (cellulae primariae medullae) sparsae, cellulae stellatae nullae. Medulla ad marginem uniseriata cellulis sphaericis, in stipite e cellulis elongatis arcte contiguis et conjunctionibus foveatis secundariis multis unitis constans.*

Structura reproductiva *Spermatangia (1 μm × 3 μm) in laminulis parvis ad basin laminae portata, in plantis etiam cystocarpias ferentibus. Fasciculi carpogoniales ad*

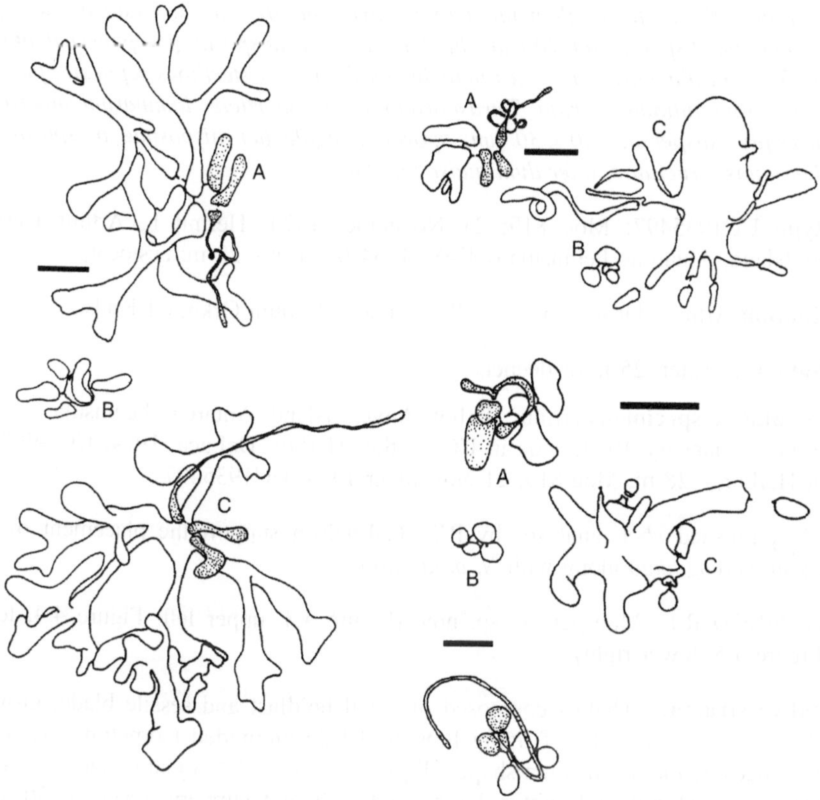

Figure 4.5 Kallymeniaceae: carpogonial branch systems from squash preparations; scales = 20 μm. (Left) *Varimenia macropustulosa*: A, carpogonial branch system with a supporting cell bearing a three-celled carpogonial branch and three subsidiary cells; B, auxiliary cell system; C, fusion cell. Note that carpogonial branch has not been incorporated.
(Upper right) *Nereoginkgo populifolia*: A, carpogonial branch system with a supporting cell bearing a three-celled carpogonial branch and one subsidiary cell. B, auxiliary cell system; C: fusion cell emitting connecting filaments.
(Middle right) *Austropugetia crassa*: A, carpogonial branch system with a supporting cell bearing a three-celled carpogonial branch and one subsidiary cell; B, auxiliary cell system; C, fusion cell emitting connecting filaments.
(Lower right) *Leniea lubrica*: carpogonial branch system with a supporting cell bearing a three-celled carpogonial branch and one subsidiary cell.

limitem inter corticem et medullam prodientia, e cellula sustinenti ramum carpogonialem 3-cellulum ferenti et 1–3 cellulis subsidiariis constantia, primum in grege denso cellularum parvarum fusce labe coloratarum, ramo carpogoniali et trichogyne manifestis ante abscissionem cellularum subsidiarium; omnes cellulae fasciculi praeter carpogonium multinucleatae factae et lobos in medullam emittentes; fasciculo maturo 50 μm in diametro. Fasciculi cellularum auxiliarium e cellula auxiliari sphaerica et cellulis subsidiariis constantes, prope fasciculos carpogoniales prodientes. Cellula sustinens, cellulae subsidiariae, et cellula prima rami carpogonialis coalescentes in cellulam conjunctionis rotundam usque ad 60 μm in diametro. Cellulae elongatae e cellulis conjunctionis abscissae in filamenta (conjunctiva) maturescentes. Cystocarpia super totam laminam dispersa, ad 700 μm in diametro. Carposporangia sphaerica uninucleata 15 μm in diametro, in greges indistinctos filamentis sterilibus separata, parietis cystocarpii dissolutione liberata. Tetrasporangia per superficies laminarum superiores inferioresque dispersa, 20 × 30 μm; sporae irregulariter dispositae in sporangio, interdum quasi cruciatae, inter-dum quasi zonatae.

Holotype UC1935497: Moe 819; 21 November 1974; Hermit I., Arthur Harbor, Anvers Island, Antarctic Peninsula (64°48′ S, 64°02′ W); −28 m; cystocarpic.

Distribution Arthur Harbor, Antarctic Peninsula and South Orkney Islands.

Habitat Clear water, 25 m or deeper.

Representative specimens Arthur Harbor, Anvers Island, Antarctic Peninsula, −22 m, Moe 144, 18 January 1974, Moe in UC; Arthur Harbor, DeLaca, 1974, UC1497839; Arthur Harbor, −28 m, Moe 819, 21 November 1974, UC1935497.

Note Unpublished *rbc*L analyses by P.W. Gabrielson support the placement of this species in *Nereoginkgo* along with *N. adiantifolia*.

Leniea lubrica R.L. Moe gen. *et* sp. nov. (Figure 4.3, upper left; Figure 4.4, lower left; Figure 4.5, lower right)

Vegetative structure Thallus composed of small holdfast and sessile blade, growing on other algae or on shells of living limpets. Blade undivided (sometimes variously torn or lobed), without a distinct shape, elliptical or lanceolate to subcircular, dark red (living) or pinkish-red to bright red (pressed), soft and slippery, reaching 30 cm in length, 200 μm thick. Blade cortex 1–3 layers of cells, the outermost ellipsoidal, 5 μm × 8 μm, the innermost subspherical, 10 μm in diameter. Medulla consisting of sparse filaments in gelatinous matrix. Cells linear, 2–4 μm × 10–60 μm, or stellate with central body 15 μm in diameter and processes 2–6 μm in diameter and reaching 500 μm.

Reproductive structure Spermatangia, carpogonia, tetrasporangia formed on separate thalli. Spermatangia borne over both blade surfaces, 2 × 2.5 μm. Carpogonial branches produced abundantly over both blade surfaces. Supporting cells borne on elongate cells at the boundary of cortex and medulla, spherical, 10–12 μm in diameter, each bearing one (rarely 2) carpogonial branch and two (1–3) subsidiary cells. Carpogonial branches with two spherical cells (~8 μm, not enlarging or lobing during development) and smaller, pyramidal carpogonium; trichogyne curving inwardly before ascending through cortex. Auxiliary cell groups similar to carpogonial branch groups. Post-fertilization development unclear. Cystocarps scattered in clusters everywhere on blade, up to 6 in same stage of development per cluster. Mature cystocarps 300–500 μm in diameter. Carposporangia in several indistinct groups separated by sparse sterile filaments. Pericarp tenuous, filamentous, lacking ostiole, breaking down to release carpospores. Tetrasporangia embedded in cortex, 15 μm × 20 μm, borne on an inner cortical cell, with the pit-connection perpendicular to the long axis, cruciately divided. Blades deliquescing rapidly in formalin.

Structura vegetativa *Thallus ex haptero parvo et lamina sessili constans, in algis aliis vel in conchis Nacellarum vivurum crescens. Lamina indivisa (sed interdum varie laciniata lobatave), sine forma propria, elliptica lanceolata vel subcircularis, in vivo fusce rubra, in sicco roseo-rubra vel vivide rubra, mollis lubricaque, 30 cm in longitudine et 200 μm in crassitudine attingens. Cortex laminarum e 1–3 stratis cellularum constans, extimarum ellipsoidearum 5 × 8 μm, intimarum subsphaericarum 10 μm in diametro. Medulla e filamentis sparsis in matrice gelatinosa constans, cellulis linearibus 2–4 μm × 10–60 μm vel stellatis corpore centrali 15 μm in diametro et processibus 2–5 μm in diametro 500 μm in longitudine attingentibus.*

Structura reproductiva *Spermatangia et carpogonia et tetrasporangia in thallis separatis facta. Spermatangia per superficies utrosque laminarum portata, 2 × 2.5 μm. Rami carpogoniales per superficies utrosque laminarum abundanter prodientes. Cellulae sustinentes in cellulis elongatis ad limen inter corticem medullamque portatae, sphaericae 10–12 μm in diametro, unum (raro 2) ramum carpogonialem et duos (1–3) cellulas subsidiarias ferentes. Rami carpogoniales duabus cellulis sphaericis (~8 μm, maturatione nec auctis nec lobatis) et carpogonio parviore pyramidali; trichogyne incurva dein per corticem ascendens. Fasciculi cellularum auxiliarium fasciculis carpogonialibus similes. Cystocarpia in gregibus per totam laminam dispersis, usque ad 6 in eodem statu maturationis in quoque grege; cystocarpia matura 300–500 μm in diametro. Carposporangia in greges aliquot indistinctos filamentis sterilibus sparsis separata. Pericarpium tenue filamentosum sine ostiolo, dissolvens et carposporas liberans. Tetrasporangia in cortice inclusa 15 × 20 μm, in cellula interiore corticis portata, conjunctione foveata ad axem longitudinalem perpendiculari, cruciatim divisa. Laminae in formalino prompte deliquescentes.*

Holotype UC1935500: Moe 568; 8 August 1974; DeLaca Island, Arthur Harbor, Anvers Island, Antarctic Peninsula (64°47′ S, 64°07′ W); −21 m; cystocarpic.

Type locality Arthur Harbor, Anvers Island.

Etymology The generic name honors Pieter Lenie, captain of the RIV Hero in the 1970s and 1980s.

Distribution Known only from the central Antarctic Peninsula (Arthur Harbor, Anvers Island, and Argentine Islands).

Habitat Found both in inshore and offshore water.

Representative specimens Arthur Harbor, Anvers Island, Antarctic Peninsula, −21 m, Moe 568, 8 August 1974, UC1935500; Arthur Harbor, −22 m, Moe 927, 30 December 1974, Moe in UC; Argentine Islands, Antarctic Peninsula, −18 m, Moe 1078, 19 February 1975, Moe in UC.

Note P.W. Gabrielson's unpublished *rbc*L sequences of this species place it in a group in the Kallymeniaceae that contains *Glaphyrymenia pustulosa* and *Polycoelia laciniata* from Australia and, perhaps, New Zealand, but separated by long base pair distances.

***Austropugetia crassa* R.L. Moe gen. *et* sp. nov. (Figure 4.3, lower right; Figure 4.4, middle right; Figure 4.5, upper left)**

Vegetative structure Thallus consisting of a single blade (or several) borne on a short, cylindrical stipe arising from a discoid holdfast. Plants perennial, the appearance varying greatly with age; young blades papery and bright red, usually with dark red spots scattered over surface, rotundate to strap-shaped, undivided or lobed, 300 μm–450 μm thick; blade margins entire or ruffled, becoming proliferous. Old blades cartilaginous and reddish-black, irregularly tattered and pertuse, with secondary blades (resembling young plants) proliferating from margins and edges of holes, to 750 μm thick. Maximum dimension 30 cm. Blade cortex up to 45 μm thick, comprising several layers of cells (3–6 μm in diameter) in anticlinal files. Medulla comprising irregularly arranged ellipsoid cells (150 μm × 300 μm [–800 μm]) surrounded by chains of small, pigmented cells (originating from medulla and inner cortex and becoming very abundant in old plants). Medulla at margin of young blades comprising one layer of spherical cells (30 μm in diameter). Stellate cells absent, but large medullary cells often compressed and stretched so as to appear filiform.

Reproductive structure Spermatangia (2 μm × 3 μm) and carpogonia borne on separate plants on small bladelets proliferating from margins of blades or holes in blades. Carpogonial branch systems produced from cells of inner cortex, each consisting of a supporting cell bearing one (rarely 2) 3-celled carpogonial branch and one (0–2) subsidiary cell. Supporting cell, subsidiary cell, and basal cell of the carpogonial branch elongating and lobing slightly (especially the supporting cell). All cells of system except carpogonium fusing and cutting off small cells (presumptive connecting filaments). Presumptive auxiliary cell systems composed of 3 or 4 spherical, darkly-staining cells near carpogonial branches. Cystocarps protruding from one or both sides of fertile bladelet, 450 μm in diameter at maturity, with distinct ostiole passing through pericarp surmounted by excrescences. Carposporangia (10–15 μm in diameter) in indistinct groups separated by sterile filaments. Tetrasporangia embedded in cortex of main blade (sometimes in proliferous bladelets), 12–15 μm × 25 μm, regularly cruciately divided.

Structura vegetativa *Thallus e lamina singula (vel aliquot) portata in stipite brevi cylindrico ex haptero discoideo oriente constans. Plantae perennes, aetate aspectu valde mutabiles. Laminae juvenes chartaceae, vivide rubrae plerumque maculis atrorubentibus per superficiem dispersis, rotundatae vel ligulatae, indivisae vel lobatae, 300–450 μm crassae, marginibus integris vel undulatis, demum proliferae. Laminae veteres cartiligineae rubronigrae, irregulariter laceratae pertusaeque, laminis secundariis (plantas juvenes simulantibus) e marginibus laminarum et foraminum orientibus, usque ad 750 μm crassae, dimensione maxima 30 cm. Cortex laminae usque ad 45 μm crassa, e stratis aliquot cellularum 3–6 μm in diametro in seriebus anticlinalibus dispositarum constans. Medulla constans e cellulis ellipsoideis (150 × 300 [–800] μm) irregulariter dispositis, catenis cellularum parvarum pigmentosarum (e medulla et cortice interiori orientium et in plantis veteribus abundantissimarum) circumcinctis. Medulla ad marginem laminarum juvenum uniseriata cellulis sphaericis 30 μm in diametro. Cellulae stellatae nullae sed cellulae medullariae magnae saepe compressae prolongataeque ut filiformes videantur.*

Structura reproductiva *Spermatangia 2 × 3 µm et carpogonia in plantis separatis portatis in laminulis parvis e marginibus laminarum et foraminum orientibus. Fasciculi carpogoniales e cellulis corticis interioris prodientes, quoque e cellula sustinenti unum (raro 2) ramum carpogonialem 3-cellularem ferenti et una (0–2) cellula subsidiaria composito. Cellula sustinens, subsidiaria, et basalis rami carpogonialis elongatae et (praesertim cellula sustinens) parum lobatae. Omnes cellulae fasciculi praeter carpogonium coalescentes et cellulas parvas (filamenta conjunctiva) abscindentes. Fasciculi cellularum auxiliarium e cellulis 3–4 sphaericis fusce labe coloratis prope ramos carpogoniales constantes. Cystocarpia e superficie una vel utraque laminulae fertilis, 450 µm in diametro in maturitate, ostiolo distincto pericarpium penetranti ad apicem excrescentiis ornato. Carposporangia 10–15 µm in diametro in greges indistinctos filamentis sterilibus separata. Tetrasporangia in cortice laminae principalis (interdum in laminulis proliferatis) inclusa, 12–15 µm × 25 µm, regulatim cruciatim divisa.*

Holotype UC1935496: Moe 387; April 5 1974; south side Surge Rocks, Arthur Harbor, Anvers Island, Antarctic Peninsula (64°47.5′ S, 64°05′ W); −22 m; cystocarpic.

Distribution Antarctic Peninsula, South Shetland Islands, South Orkney Islands (Gepp and Gepp, 1905 in BM).

Etymology "Southern Pugetia" – a reference to the morphological similarity of the carposporophyte to that of *Pugetia*.

Habitat Usually offshore water 20 m or deeper.

Representative specimens Arthur Harbor, Anvers Island, Antarctic Peninsula, −22 m, Moe 231, 2 March 1974, Moe in UC; Arthur Harbor, −22 m, Moe 387, 5 April 1974, UC1935496; Arthur Harbor, shallow subtidal, Moe 877, 14.xii.1974, Moe in UC.

Note Although being similar in carposporophyte development to *Pugetia*, this unusual species is separated from "*Pugetia*" *delicatissima* Norris 1957 from New Zealand, and from *Pugetia fragilissima* Kylin 1925 from Washington, USA by long branch lengths when *rbc*L sequences were compared by P.W. Gabrielson (unpublished data). Fresh material often has a slight pungent odor. This species was referred to as "undescribed species 1" in Amsler et al. (2005a) and Peters et al. (2005).

Callophyllis tenera J. Agardh 1849

Type locality South Shetland Islands (supposedly collected by Archibald Menzies, but this is not possible since the South Shetland Islands were not known when Menzies was in the Royal Navy).

Note Several species of "*Callophyllis*" from Antarctica that were sequenced by P.W. Gabrielson form a clade with *Nereoginkgo* and may belong in this genus.

Callophyllis atrosanguinea (J.D. Hooker *et* Harvey) Hariot 1887

Basionym *Rhodymenia variegata* var. *atrosanguinea* J.D. Hooker *et* Harvey in Hooker 1847.

Type locality Kerguelen.

Representative specimens Arthur Harbor, Anvers Island, Antarctic Peninsula, −22 m, Moe 146, 18 January 1974, Moe in UC; Arthur Harbor, −25 m, Moe 677, October 7 1974, Moe in UC; Arthur Harbor, −23 m, Moe 839a, 4 December 1974, Moe in UC.

Note Plants from Antarctica that resemble *Callophyllis atrosanguinea* from Kerguelen need to be critically compared with authentic material and an evaluation of the generic assignment is necessary.

***Callophyllis variegata* (Bory) Kützing 1843**

Basionym *Halymenia variegata* Bory in Duperrey 1828.

Syntype localities Concepción, Chile; New Guinea.

Note Although this species has been reported from South Georgia and Macquarie Island, we think it unlikely that Bory's plant occurs in Antarctica; a Gepp and Gepp (1905) record (in BM) is referable to *Austropugetia crassa*, but we have not examined other material.

***Varimenia macropustulosa* R.L. Moe gen. *et* sp. nov. (Figure 4.3, lower left; Figure 4.4, lower right; Figure 4.5, lower right)**

Vegetative structure Thallus composed of small, discoid holdfast and sessile or short-stipitate blade. Blade bright red, rubbery and soft, but not lubricous, extremely variable in shape: undivided, lacerate, perforated, lobed, or divided into orbiculate, obtuse or acute segments, reaching 1 m in breadth and 300−1300 μm thick. Cortex comprising 4−6 layers of pigmented, spherical or anticlinally elongated cells, ca. 4 μm at surface, 10 μm in diameter internally. Subcortical cells spherical or periclinally elongate, interconnected by numerous secondary pit-connections, 20 μm in diameter outwardly, 100 μm inwardly. Medulla comprising dense assemblage of filiform cells, irregularly elongated and stretched cells, and short chains of thin-walled, spherical cells. Regularly stellate cells and cells with refractive contents absent. Juvenile blades (~5 cm) with tapering margin and medulla at edge comprising a single layer of filaments; mature blades with truncate margins and medulla polystromatic to edge.

Reproductive structure Spermatangia, carpogonia, tetrasporangia formed on separate thalli. Male plants similar in size to tetrasporangial and carpogonial plants, but less robust and with less dense medulla. Spermatangia borne over both blade surfaces, restricted to marginal sori 1−4 cm broad that are light in color. Carpogonial branch systems borne by subcortical cells; each carpogonial group consisting of seven cells, a supporting cell bearing a 3-celled carpogonial branch and 3 subsidiary cells. All cells of system initially spherical (8−10 μm in diameter), elongating (except for carpogonium) away from pit-connections into outer medulla, becoming filled with dense, refractive substance with the supporting cell and subsidiary cells beginning to branch distally into antler-like lobes. Presumptive auxiliary cell systems consisting of spherical cell (15 μm in diameter) in subcortex bearing 3−5 smaller spherical or slightly elongate cells. Fusion cells produced by coalescence of proximal portions of supporting cell and subsidiary cells. Fusion cell reaching 150 μm in extent, cutting off small cells (presumptive connecting filaments) from lobes. Mature cystocarps reaching 6 mm in diameter and 2 mm thick, with one ostiole present from beginning of development. Carposporophyte surrounded by dense pericarp of small-celled filaments, some of these penetrating central sporangial mass separating it into distinct clusters. Carposporangia uninucleate, 20 μm in diameter. Tetrasporangia, 10 μm × 40 μm, borne as apical cells of cortical filaments, scattered over both blade surfaces, divided sequentially to produce 4 spores arranged in zig-zag pattern.

Structura vegetativa *Thallus ex haptero parvo discoideo et lamina sessili vel brevistipitata constans. Lamina vivide rubra, elastica mollisque sed non-lubrica, forma valde variabilis: indivisa lacerata perforata lobata vel divisa in segmenta orbiculata obtusa vel acuta, 1 m in latitudine et 300–1300 μm in crassitudine attingens. Cortex e 4–6 stratis cellularum pigmentosarum sphaericarum vel anticlinaliter elongatarum, circa 4 μm ad superficiem, 10 μm ad partem interiorem composita. Cellulae subcorticales sphaericae vel periclinaliter elongatae, conjunctionibus foveatis secundariis multis interjunctae, exteriores 20 μm, interiores 100 μm. Medulla e grege denso cellularum filiformium, irregulariter elongatarum prolongatarum, et sphaericarum parietibus tenuibus in catenis brevibus composita. Cellulae regulatim stellatae et cellulae contento refractivo nullae. Laminae juvenes (~5 cm) margine angustato et medulla ad marginem e strato singulo filamentorum composita, maturae margine truncato et medulla ad marginem polystromatica.*

Structura reproductiva *Spermatangia, carpogonia, tetrasporangia in thallis separatis formata. Plantae masculae magnitudine plantibus tetrasporangialibus carpogonialibusque similes sed minus robustae et medulla minus densa. Spermatangia in superficie utraque laminae portata, ad soros marginales 1–4 cm latos pallidos limitata. Fasciculi ramorum carpogonialium cellulis subcorticalibus portati, e cellulis 7 constantes: cellulo sustinente ramum carpogonialem 3-cellulatam et 3 cellulas subsidiariis ferente. Omnes cellulae fasciculi primum sphaericae (8–10 μm in diametro), praeter carpogonium a conjunctionibus foveatis in medullam exteriorem extendentes, demum substantia densa refractiva impletae, cellulo sustinente et cellulis subsidiariis in lobos cornibus cervi similes distaliter ramificantibus. Fasciculi cellularum auxiliarium e cellula sphaerica (15 μm in diametro) subcorticali cellulas minores sphaericos vel parum elongatos ferente compositi. Cellulae fusionis per coalescentiam partium proximalium cellulae sustinentis et cellularum subsidiarium factae, 150 μm in longitudine attingentes, cellulas parvas (filamenta conjunctiva) e lobis abscindentes. Cystocarpia matura 6 mm in diametro, 2 mm crassa, ostiolo uno ab initio maturationis praesente. Carposporophyta pericarpio denso e filamentis parvicellulatis composito circumcincta, aliquot ex his filamentis massam centralem sporangialem penetrantibus et in greges distinctos illam separantibus. Carposporangia uninucleata 20 μm in diametro. Tetrasporangia cellulae apicales filamentorum corticalium, 10 μm × 40 μm, supra superficiem utraque laminae dispersa, in sequentia divisa ut sporas 4 in ordine flexuosa ("zigzag") disposita prodeant.*

Holotype UC1935499: Moe 379; 5 April 1974; south side Surge Rocks, Arthur Harbor, Anvers Island, Antarctic Peninsula (64°47.5′ S, 64°05′ W); –22 m; cystocarpic.

Distribution Antarctic Peninsula and South Shetland Islands.

Etymology A reference to the pox-like appearance of the cystocarps.

Habitat Clear, offshore water, usually 20 m or deeper.

Representative specimens Arthur Harbor, Anvers Island, Antarctic Peninsula, –22 m, Moe 379, 5 April 1974, UC1934599; Arthur Harbor, –22 m, Moe 815, 21 November 1974, Moe in UC; Arthur Harbor, –28 m, Moe 891, 27 December 1974, Moe in UC.

Note This species has been incorrectly identified as *Pachymenia orbicularis* in Amsler et al. (2005a,b) and Peters et al. (2005).

Family Phyllophoraceae
***Gymnogongrus antarcticus* Skottsberg 1953**

Syntype localities "Ins. Wiencke, Palmer Archipelago", Antarctic Peninsula.

Representative specimens Arthur Harbor, Anvers Island, Antarctic Peninsula, –10 m, Moe 487, 5 June 1974, UC1571400; Arthur Harbor, –10 m, Moe 933A, 5 January 1975, UC1571404; Vestfold Hills, Davis, East Antarctica, –5 m, Ricker 2717, 12 December 1981, UC1497967.

Note Along the Peninsula, this species can be locally common at depths of 10 m or shallower.

***Gymnogongrus turquetii* ("*turqueti*") Hariot 1907**

Type locality Booth Wandel Island, Antarctic Peninsula.

Taxonomic synonym *Phyllophora appendiculata* Skottsberg in Kylin and Skottsberg 1919; type locality: South Georgia.

Representative specimens Arthur Harbor, Anvers Island, Antarctic Peninsula, intertidal, Moe 781, 13 November 1974, UC1541768; Arthur Harbor, intertidal, Moe, 5 January 1988, UC1541769; South Bay, Livingston Island, South Shetland Islands, Gallardo 5171, 30 December 1990, UC1471751.

Note The basal crust (and possibly the tetrasporophyte) can be extensive. Plants are usually found in the shallow subtidal or in intertidal pools. This name has also apparently been misapplied to specimens of *Hymenocladiopsis* that have simple blades; see comments under *Hymenocladiopsis*.

***Phyllophora ahnfeltioides* Skottsberg in Kylin and Skottsberg 1919**

Type locality South Georgia.

Representative specimens Arthur Harbor, Anvers Island, Antarctic Peninsula, –22 m, Moe 741, 24 October 1974, UC1571399; Arthur Harbor, Anvers Island, Antarctic Peninsula, –6 m, Moe 758, 27 October 1974, UC1571463; Lavebrua Island, Deception Island, South Shetland Islands, –13 m, Moe 969, 1 January 1975, *Synarthrophyton*, *Porphyra plocamiestris* epiphytic, UC1469676.

Note This species is very common in the shallow subtidal of the Antarctic Peninsula; it grows from a fairly extensive pinkish basal crust (possibly also the tetrasporophyte).

Phyllophora antarctica **A. Gepp *et* E.S. Gepp 1905**

Type locality Victoria Land, Antarctica.

Representative specimens Showa Base, East Antarctica, −9 m, Ohno, 19 January 1975, UC1457208; Terra Firma Island, Marguerite Bay, Antarctic Peninsula, −22 m, Moe 1024, 10 February 1975, UC1571421; Cape Evans, McMurdo Sound, Ross Sea, −21 m, K.A. Miller, 12 November 1999, UC1557937.

Note This taxon is found in very deep water along the northern Antarctic Peninsula and in shallower water towards the south. It can be very abundant at sites in the Ross Sea.

Family Gigartinaceae
Sarcothalia papillosa **(Bory) Leister in Hommersand et al. 1993**

Basionym *Iridaea papillosa* Bory 1826.

Type locality Falkland Islands.

Note We have not seen any plants from Antarctica that correspond to this species.

Sarcothalia circumcincta **(J. Agardh) Hommersand in Hommersand et al. 1993**

Basionym *Gigartina circumcincta* J. Agardh 1876.

Type locality Dunedin, New Zealand.

Note We have not examined the specimens on which Antarctic records are based.

Gigartina skottsbergii **Setchell *et* N.L. Gardner 1936**

Type locality Fuegia (Tierra del Fuego).

Representative specimens Arthur Harbor, Anvers Island, Antarctic Peninsula, −20 m, Moe 332, 25 March 1974, UC1557742; Arthur Harbor, −13 m, Moe 728, 24 October 1974, UC1557746; Arthur Harbor, intertidal pool, Moe 824, 16 November 1974, UC1571482.

Note Collections of "*Gigartina skottsbergii*" from Antarctica are morphologically similar to plants from South America and the Falkland Islands but probably represent a separate species. Antarctic plants are separated from South American plants by more than a 2% *rbc*L base pair distance (Hommersand and Fredericq 2003). Furthermore, neither is closely related to the type of *Gigartina* (P. Gabrielson and M. Hommersand, unpublished observations). Along the Antarctic Peninsula, the species can be locally abundant across a wide depth range.

Iridaea cordata **(Turner) Bory 1826**

Basionym *Fucus cordatus* Turner 1809.

Type locality Isla de los Estados, Argentina.

Representative specimens Arthur Harbor, Anvers Island, Antarctic Peninsula, Moe 399, 11 April 1974, UC1557669; Bahia Elefante, King George Island, South Shetland Islands, MER(2-5-94-1-2), 5 February 1994, MER in NC; Cape Evans, McMurdo Sound, Ross Sea, −7 m, K.A. Miller, 10 November 1999, UC1557933.

Note Antarctic plants are morphologically similar to plants from South America and the Falkland Islands but are separated by greater than a 3% *rbc*L base pair distance (Hommersand and Fredericq 2003) and probably belong in a separate species. They are very common in the shallow subtidal of West and East Antarctica.

***Iridaea mawsonii* A.H.S. Lucas 1919**

Type locality Commonwealth Bay (George V Coast).

Note The identity of the Lucas plant is uncertain.

Family Cystocloniaceae

***Acanthococcus antarcticus* J.D. Hooker *et* Harvey 1845**

Syntype localities Cape Horn and the Falkland Islands.

Note All of the specimens we have examined correspond to forms of the plant known as *Cystoclonium obtusangulum* in Antarctica (see below). Carpospores germinating *in situ* and producing spiny outgrowths may be responsible for the misidentification. *Acanthococcus* appears to be absent from Antarctica.

"*Cystoclonium*" *obtusangulum* Kützing 1849

Type locality Cape Horn.

Representative specimens Byers Peninsula, Livingston Island, South Shetland Islands, in drift, Moe 87, January 1974, UC1571703; Arthur Harbor, Anvers Island, Antarctic Peninsula, −25 m, Moe 670, 7 October 1974, UC1557735; English Strait, Greenwich Island, South Shetland Islands, −7 m, Moe 1011, 26 January 1975, UC1469702.

Note Antarctic plants that have been assigned to this species are related to "*Cystoclonium*" *obtusangulum* from South America and the Falkland Islands but are separated by relatively long *rbc*L base pair distances based on unpublished observations by M. Hommersand and S. Fredericq. They do not belong in the Northern Hemisphere genus *Cystoclonium* but constitute a new genus. They can be locally abundant at depths exceeding 20 m, and are easily confused with *Pantoneura plocamioides*. The plant illustrated in plate 23: 2 of Wiencke and Clayton (2002) as *Pantoneura* appears to be this species since the regularity of ultimate branching that is characteristic of *Pantoneura* is not apparent. The name is usually considered to be based on *Gracilaria obtusangula* J.D. Hooker *et* Harvey 1845, but, according to Index Nominum Algarum, the binomial cannot be considered a combination, since Kützing followed Hooker and Harvey (1847) in merging *Gracilaria obtusangula* with *G. nigrescens* (C. Agardh) J.D. Hooker *et* Harvey 1847.

Family Solieriaceae
***Notophycus fimbriatus* R.L. Moe 1986**

Type locality Argentine Islands, Antarctic Peninsula.

Representative specimens Joubin Islands, Anvers Island, Antarctic Peninsula, between −20 m and −24 m, Moe 951, 16 January 1975, UC1557709; Cape Rey, Antarctic Peninsula, −15 m, Moe 1052, 15 February 1975 (Laine leg.), UC1571503; Argentine Islands, Antarctic Peninsula, −22 m, Moe 1070, 19 February 1975, UC1475018.

Note Records from the South Shetland Islands (Wiencke and Clayton 2002) should be critically compared since the specimens differ from the type in form (reduced proliferations of stipe) and habitat (eulittoral pools vs. deep water). This species has been reported recently from the Ross Sea (Page et al. 2002).

Part 2: Biogeographic relationships of the Antarctic red algae

The distribution of the red algae of Antarctica can be attributed partly to vicariance events resulting from plate tectonics and sea floor spreading and partly to long distance dispersal. Changes in the climate of Antarctica over geological time played an important role in its algal composition and distribution, as did the formation, direction, strengthening and decline of both shallow and deep ocean currents. Along with Australia, Antarctica is the major remnant of the once continuous continent of East Gondwana, which faced the Pacific Ocean to the south and east and the Neotethys Ocean to the north. Africa rifted from East Gondwana in the Mesozoic beginning 165–162 Ma and India, Madagascar and the Kerguelen Plateau split off around 132–130 Ma, at about the same time that seafloor spreading rifted the Falkland Plateau from southern Africa and that separation of southern Africa from South America began with the formation of the South Atlantic Ocean (Lawver et al. 1985, McLoughlin 2001). Ghidella et al. (2007) reviewed models of the tectonic framework for this process, one of which can be seen as a PowerPoint movie by Lawver et al. (1999). Whatever connection there may have been between the flora of Antarctica and that of Africa and India was certainly overwhelmed by species introductions starting in the Late Paleocene around 55 Ma at a time when the late Cretaceous circum-equatorial ocean circulation was diverted around South Africa coupled with commencement of the collapse of the Tethyan Seaway (Lee and Lawver 1995, Lawver and Gahagan 2003).

The closest floristic affinity of the red algae of the Antarctic Peninsula (including the South Shetland Islands) is South Georgia. Were the records more complete, this relationship would certainly include the South Orkney Islands and, perhaps, the South Sandwich Islands as well as South Georgia. These island groups sit on microplates that were rifted from the tip of the Antarctic Peninsula beginning at ca. 31 ± 2 Ma and their separation continues to the present day (Figure 4.6; Lawver and Gahagan 1998, 2003; Lawver et al. 1999). Fourteen species were reliably reported from both South Georgia and the Antarctic Peninsula in Wiencke and Clayton (2002). The morphological similarity of these species is not in question, suggesting that the floristic relationship between the two regions is real. The separation of South Georgia from the Antarctic Peninsula was progressive over time. Since we do not have molecular data for any species from South Georgia, it is not possible to know the base pair distances between species for any gene or to estimate the time when species separation took place. We cannot determine the relative contribution of microplate movements compared to that of currents, such as that postulated to have resulted from a passageway formed by the spreading of the Powell Basin at the northeast end of the Antarctic Peninsula (Figure 4.6; Lawver and Gahagan 1998).

A survey of phylogenetic trees and pairwise base distances determined from analyses of *rbc*L sequences for ~60 species of red algae support the opinion that New Zealand, the Antarctic Peninsula, southern South America and the Falkland Islands,

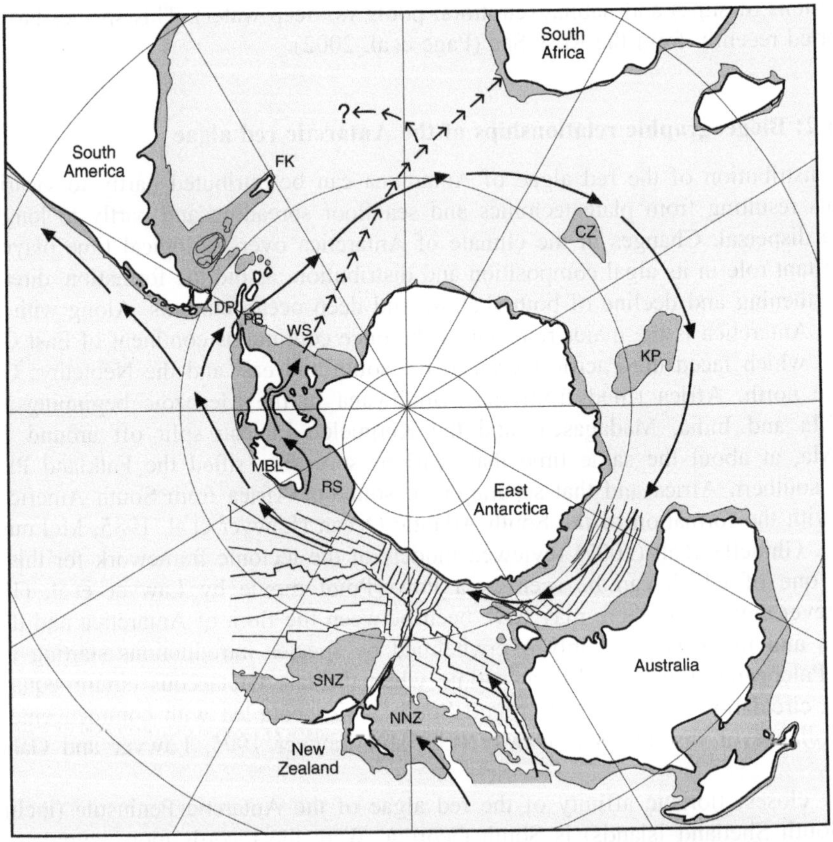

Figure 4.6 Polar oceanographic projection of the Southern Hemisphere at ca. 32 Ma. Exposed land is shown in white, shallow ocean basins in gray, and deep ocean basins in light gray. Selected isochrons that reflect seafloor spreading between Antarctica and Australia or New Zealand were reconstructed from Cande and Stock (2004). Estimated directions of ocean currents at 32 Ma were taken from Lawver and Gahagan (1998, 2003), Exon et al. (2004) and Kennett and Exon (2004).

CZ, Crozet Plateau; DP, Drake Passage; FK, Falkland Islands; KP, Kerguelen Plateau; MBL, Marie Byrd Land; PB, Powell Basin; NNZ, North New Zealand crustal block; RS, Ross Sea; SNZ, South New Zealand crustal block; WS, Weddell Sea. The short arrows indicate the postulated migration of species from West Antarctica to South Africa.

and western South Africa and Namibia form a special biogeographic province containing phylogentically related taxa (M. Hommersand and D.W. Freshwater, unpublished observations). Were more data available, this province would also probably include Tasmania and the islands of the West Wind Drift, as listed in the Introduction. The groups of species involved include: diverse members of the Bangiales (*sensu* Nelson et al. 2006); forms of *Ballia callitricha* and other species of *Ballia* in the Balliales; certain crustose corallines in the Hapalidiaceae; members of the *Delisea/Ptilonia* group in the Bonnemaisoniaceae; certain species of *Plocamium*; the genus *Trematocarpus* in the Sarcodiaceae; *Curdiea* in the Gracilariaceae; *Pachymenia/Aeodes* and an assemblage of related undescribed genera in the Halymeniaceae; species of *Georgiella* and *Falklandiella*, as well as *Dasyptilon* and *Plumariopsis* of the Ceramiaceae; a wide range of genera belonging to the Delesseriaceae including members of the tribes Botryocarpeae, Myriogrammeae, Schizoserideae, some Southern Hemisphere species placed in "*Delesseria*" (=*Paraglossum*) in Apoglosseae, some species of *Phycodrys* in the Phycodryeae and some *Hymenena* in the Cryptopleureae; an assemblage of assorted species in the Kallymeniaceae; many Southern Hemisphere species placed in *Gymnogongrus*, *Ahnfeltiopsis* or *Phyllophora* of the Phyllophoraceae; species of *Sarcothalia* and *Iridaea*, and "*Gigartina*" *skottsbergii* of the Gigartinaceae; and some species placed in *Acanthococcus*, "*Cystoclonium*" or "*Rhodophyllis*" of the Cystocloniaceae.

No genera and species are shared uniquely between the Antarctic Peninsula and southern Africa. The longest base pair distances between sister and nearest-neighbor taxa found in our phylogenetic trees involved tribes and genera distributed between the Antarctic Peninsula and New Zealand. The biogeographic affinity between New Zealand and southern South America and the Falkland Islands is closer, and that between New Zealand and southern Africa may be closer still. The relationships between the red algae of the Antarctic Peninsula and South America is complex, with some groups of species as distant as those distributed between New Zealand and Antarctica, and some much closer, about the same as the most closely related species distributed between New Zealand and South Africa. The interpretation of these complex relationships includes the possibility of separation through plate tectonics, the occurrence of long-distance dispersal at different times during the Cenozoic, and the isolation of Antarctica from the other regions through the formation and strengthening of the Antarctic Circumpolar Current, also known as the West Wind Drift.

The islands of New Zealand were formed from two crustal blocks, both of which split off from Antarctica beginning in the late Cretaceous at a time when that part of Antarctica was close to the South Pole. The southern crustal block rifted from Marie Byrd Land, then a part of West Antarctica but now united with East Antarctica at ca. 84 Ma (Lawver et al. 1987) and the northern block separated from the vicinity of Tasmania beginning at about the same time to produce a wedge-shaped oceanic basin. The two crustal blocks had joined at least by the beginning of the Cenozoic and underwent rearrangements through a complex series of tectonic events to form the present-day North and South Islands of New Zealand. Cande and Stock (2004) published a series of isochrons from 61 Ma to the present that describe this process (see Figure 4.6). Long-range dispersal may have taken place from New Zealand or Tasmania to West Antarctica and South America by the prevailing currents directed from east to west during the first half of the Cenozoic. Our molecular data only support a relationship between New Zealand and Antarctic red algae; however, it is clear that significant information will be added when molecular studies have been carried out on Tasmanian and related southern Australian algae.

Hommersand (1986) proposed that temperate marine algae reached South Africa from New Zealand by long-distance dispersal by means of passageways between the islands of West Antarctica during periods when the region was not fully glaciated and sea levels were high (Figure 4.6). A similar pathway might account for the special relationship between the New Zealand flora and that of southern South America or, more likely, species that reached South America may have come by way of the Antarctic Peninsula. Temperate floras would have only reached South America and South Africa beginning with worldwide cooling of the oceans from the middle (47 Ma) to the end (34 Ma) of the Eocene related to a reduction in pCO_2 concentration (Pearson and Palmer 2000). A shallow (700 m) seaway remained open through West Antarctica from the Ross Sea to the Weddell Sea until at least early (34 Ma) Oligocene (Lawver and Gahagan 2003). The marine transgression across the Tasmanian Land Bridge at 33 Ma that finally separated Australia from East Antarctica with the formation of a deep-water passage by 32 Ma (Exon et al. 2004) led to the extension of the East Antarctic ice sheet to the continental shelf, significantly lowering the water level in the trans-West Antarctic passageways with the formation of Antarctic Bottom Water in the Weddell Sea. The Drake Passage had opened between the southern tip of South America and the northern end of the Antarctic Peninsula by this time; however, deep water flow appears to have been blocked by the South Orkney and South Georgia microcontinents and other continental fragments at the east end of the Scotia Sea until at least 20 Ma (Figure 4.6; Lawver and Gahagan 2003). Antarctic glaciation receded at the end of Oligocene at 25 Ma and the trans-West Antarctic passageways may have been restored and increased with rising sea levels followed by sea level stability during Miocene up to 15 Ma. As Australia moved northward widening its separation from Antarctica, the Australia-New Guinea block collided with Southeast Asia at around 15 Ma substantially blocking the flow of equatorial water from the Pacific Ocean into the Indian Ocean. This event also greatly enhanced the transport and driving force of the Antarctic Circumpolar Current (Lee and Lawver 1995). The East Antarctic ice sheet expanded, again shutting down the trans-West Antarctic passageway with enhancement of Antarctic Bottom Water formation and upwelling in the Weddell Sea. Finally, the closure of the Isthmus of Panama during Pliocene between 4.7 and 3.5 Ma (Coates et al. 1992, Drexler et al. 1998) blocked deep-water circulation between the Atlantic and Pacific Oceans and increased the thermohaline transport of moisture to the polar regions, first leading to warming at the poles followed, ultimately, by the formation of massive polar ice sheets. One consequence of these events was to greatly increase the speed and volume of water circulated by the Antarctic Circumpolar Current to its present levels (Kennett and Exon 2004).

Our analysis of pairwise base distances between species found in New Zealand and those occurring in South Africa (M. Hommersand and D.W. Freshwater, unpublished data) support the opinion held earlier by Hommersand (1986) that a cluster of temperate marine algae was distributed through passageways in West Antarctica from the Pacific side of East Gondwana to the South Atlantic Ocean and from there to South Africa and possibly South America (Figure 4.6). The genera and species clusters distributed between New Zealand, South America and South Africa were essentially the same as those distributed between New Zealand and Antarctica. Only the range of pairwise base distances and the mean base pair distances were less. Uncorrected "p" distances between sister regions calculated using base pair positions 1 and 2 ranged from 0.045 to 0.019 for New Zealand/Antarctic species pairs with a mean distance of 0.025. That between New Zealand and South America or the Falkland Islands ranged from 0.027 to 0.013 with a mean distance of 0.018. That between New Zealand and South Africa ranged from 0.029 to 0.013 with a mean distance of 0.0155. Finally, that between Antarctica and South America ranged from 0.021 to 0.010, with a mean distance of 0.0155. The greater mean base pair distance between species pairs distributed between New Zealand and the Antarctic Peninsula suggest that, either the two regions had been separated for a longer period of time, or that rates of extinction and species evolution were greater in the Antarctic Peninsula following its separation from the New Zealand flora. The latter explanation seems the more likely with the implication that species found in South America passed through the Antarctic Peninsula at some point in time (Figure 4.6).

New Zealand itself and the eastern Tasmanian margin are thought to have been under the influence of a broad warm subtropical gyre from the north that extended to high southern latitudes prior to the opening of the eastern Tasmanian gateway at the Eocene/Oligocene boundary that brought cooler waters from the earliest Antarctic Circumpolar Current. This opinion is based primarily on the composition of the Eocene planktonic and radiolarian assemblages in New Zealand prior to 33 Ma (Kennett and Exon 2004). If New Zealand waters were truly as warm as these authors infer, much of the present Pacific flora of New Zealand would necessarily have been situated further south, possibly on elements of the Campbell Plateau, which maintained its crustal thickness as a unit throughout the break-up of East Gondwana and the assemblage of present-day Zelandia (Grobys 2008).

Hommersand (2007) maintained that the unique elements of the western to southern Australian flora that originated west of the South Tasman Rise did not reach New Zealand as a result of the initial marine transgression across Tasmania. The strong Antarctic Circumpolar Current that developed more recently is largely diverted south of the Campbell Plateau (Lawver and Gahagan 2003) and may never have contributed western and southern Australian species to the New Zealand flora.

The notion that a West Antarctic passageway accounts for seaweed relationships between Pacific Ocean seaweeds in New Zealand and seaweeds in South Africa and possibly South America assumes that surface currents existed in the South Atlantic at various times that would have permitted such distributions to have taken place, either by currents or by rafting. Such pole-to-equator currents were illustrated in models referred to by Hommersand (1986, p. 263, fig. 5) but do not appear in more recent models. This dilemma has not been resolved. The optimum times for such a distribution would have been during windows when passageways through West Antarctica were free of ice, as in the late Eocene or the mid Miocene, or during transition periods of Antarctic Bottom Water formation in the Weddell Sea at the onset of widespread Antarctic glaciation when rafting of algae on or within ice sheets that had separated from the Antarctic continent and drifted northward may have taken place. On the other hand, the sharp cut-off in the minimum distances between base pair separations can only mean that, at some point in time, long-range dispersals among the biogeographic provinces referenced here ceased. This would have taken place when the Antarctic Circumpolar Current became persistent and strong enough to block movement across it, possibly in middle Miocene around 15 Ma when the closure of Australia/New Guinea with Southeast Asia cut off the deep-water equatorial passage between the Pacific and Indian Oceans enhancing the rate of flow of the Antarctic Circumpolar Current.

Even in the absence of molecular data, the morphological evidence establishes that the powerful Antarctic Circumpolar Current distributed South American and Falkland Islands algae eastward and northward to Gough Island (Chamberlain 1965), the Crozet and Kerguelen Islands (Papenfuss 1964) and Macquarie Island (Ricker 1987) reaching, in some instances, as far as the Auckland Islands and Campbell Island south of New Zealand (Papenfuss 1964). Many of the species distributed along the islands of the West Wind Drift are so similar morphologically that their relatedness is not in doubt, even in the absence of molecular data. Indeed, unrelenting population pressure posed by new and persistent introductions coupled with a change in climate has probably replaced or significantly modified any earlier flora that may have existed on these islands. All of the islands north of the Antarctic Polar Front (Moore et al. 1999, Bergstrom and Chown 1999) form a biogeographic province that contains a flora that is distinct from all other floras except that of South America and the Falkland Islands with only a handful of species having reached New Zealand. There may well be remnants of a previous Antarctic flora that may have reached this region during Pleistocene at a time when the Antarctic Polar Front extended further north than at the present time with scouring of the exposed reef by winter ice. In fact, Fraser et al. (2009) have argued that the Southern Bull Kelp, *Durvillaea antarctica* (Chamisso) Hariot, was entirely removed from the subantarctic islands by persistent ice scour during the last glacial maximum only to be replaced through recolonization by a homogeneous population with a single haplotype marker introduced after the ice receeded. A part of this population even reached South Georgia Island.

Pairwise base pair distances between sister and nearest neighbor species calculated from *rbc*L data sets were published for 17 examples by Hommersand and Fredericq (2003). One interesting fact is that the flora of southern Africa and Namibia have been virtually unaffected by an influx of species circulated by the West Wind Drift. This may be attributable to the strength of the barrier imposed by the Benguela upwelling system which causes surface waters to move off shore and be displaced by the deeper waters subducted from the Benguela Current off shore. Storms break up this system and occasionally deposit South American species; however, these seldom become established (Stegenga et al. 1997). The environmental differences between the two regions may simply be too great. Like the Antarctic Circumpolar Current, the Benguela Current and upwelling system is thought to have originated from 12 Ma to the present time and to have strengthened as closure of the Isthmus of Panama was completed. Species endemism is high among the temperate species of South Africa and Namibia, with the greatest morphological similarities and closest base-pair separations being between the South African and New Zealand species, with little or no introductions of species by the West Wind Drift.

By and large, the temperate flora of the Pacific Coast of South America is more closely related to that of New Zealand than is the flora of the Antarctic Peninsula as evidenced by base pair separations. At least half the red algal genera occurring in the

Antarctic Peninsula that exhibit a relationship with New Zealand algae would appear to have reached South America first and only later to have been distributed to Antarctica. The shorter base pair distances between New Zealand and South American red algae may, however, be an artifact. There is a possibility that adaptation to the singularities of the Antarctic climate has caused the *rbc*L gene to evolve more rapidly in Antarctica than in the temperate waters of southern South America. This all remains to be determined.

Some species clearly originated in the north and were distributed southward into Antarctica rather than the reverse. The most striking example is that of *Palmaria decipiens*. An ITS sequence obtained from a collection from the Shetland Islands studied by Lindstrom et al. (1996) came out in the *Palmaria mollis* group and was nearest neighbor to *P. callophylloides* from Alaska. The only reasonable explanation is long-range dispersal, probably by biotic vectors such as birds or whales. *Palmaria decipiens* is treated in the literature as a single species that is widespread in the Southern Hemisphere (Ricker 1987). This may not necessarily be the case. An unpublished sequence of a sample of *Palmaria* from the Falkland Islands (M. Hommersand, personal observations) was entirely unrelated to the plant reported from the Shetland Islands. Many such samples from other localities require reinvestigation.

The observations of Maggs et al. (2007), cited earlier in the species list, suggest that *Rhodochorton purpureum* (Acrochaetiaceae) first evolved in the Northern Hemisphere and was only recently introduced into the Southern Hemisphere, including Antarctica. *Meiodiscus concrescens* (Rhodophysemataceae) exhibits a similar pattern of distribution in the Northern and Southern Hemispheres and would be worth investigating.

An argument might be raised that the affinity between the New Zealand flora and that of South Africa could have come about by distribution through the Western Pacific and Indian Oceans. Indeed this has happened and there are numerous examples that support this pathway. However, most of these species have a more or less continuous distribution in the Indian Ocean and along the coast of eastern and southern Africa. The group of temperate species found along the west coast of South Africa that comprise the special biogeographic assemblage described here appears to be absent in the Indian Ocean.

It might be argued that New Zealand species have reached the west coast of South America by means of the Antarctic Circumpolar and Humboldt Currents. Indeed some may have. A prominent example could be *Durvillaea antarctica*. Those species that might have come across the Pacific Ocean in the direction of the West Wind Drift are all closely related morphologically and, where known, are close to one another genetically as well. This pathway does not account for the distantly related genera and species clusters that are the subject of this discussion.

Conclusion

While the list of species included here is based solely on our personal observations of collections obtained in the Antarctic Peninsula and the Shetland Islands, the identity of the Antarctic red algal flora can be greatly expanded by investigating new collections from other localities within and beyond the Antarctic Polar Front and by detailed morphological studies of each of the groups of red algae. SCUBA collections from each of these localities are needed for such an enterprise. The taxonomic and biogeographical hypotheses presented here, moreover, are all subject to genetic testing using modern molecular methods.

Note

Representative specimens of Antarctic and sub-Antarctic Rhodophyta: The scans of herbarium specimens are available at: http://ucjeps.berkeley.edu/images/antarctic_algae/

References

Adey, W.H., A. Athanasiadis and P.A. Lebednik. 2001. Re-instatement of *Leptophytum* and its type *Leptophytum laeve*: taxonomy and biogeography of the genera *Leptophytum* and *Phymatolithon* (Corallinales, Rhodophyta). *Eur. J. Phycol. 36*: 191–203.

Agardh, J. 1898. *Species genera et ordines algarum*, Vol. 3, Part 3. Gleerup, Lund [VI] + 239.

Alongi, G., M. Cormaci and G. Furnari. 2002. The Corallinaceae (Rhodophyta) from the Ross Sea (Antarctica): a taxonomic revision rejects all records except *Phymatolithon foecundum*. *Phycologia 41*: 140–146.

Amsler, C.D., K. Iken, J.B. McClintock, M.O. Amsler, K.J. Peters, J.M. Hubbard, F.B. Furrow and B.J. Baker. 2005a. Comprehensive evaluation of the palatability and chemical defenses of subtidal macroalgae from the Antarctic Peninsula. *Mar. Ecol. Prog. Ser. 294*: 141–159.

Amsler, C.D., I.N. Okogbue, D.M. Landry, M.O. Amsler, J.B. McClintock and B.J. Baker. 2005b. Potential chemical defenses against diatom fouling in Antarctic macroalgae. *Bot. Mar. 48*: 318–322.

Amsler, C.D., R.J. Rowley, D.R. Laur, L.B. Quetin and R.M. Ross. 1995. Vertical distribution of Antarctic Peninsular macroalgae: cover, biomass, and species composition. *Phycologia 34*: 424–430.

Bergstrom, D.M. and S.L. Chown. 1999. Life at the front: History, ecology and change on southern ocean islands. *Trends Ecol. Evol. 14*: 472–473.

Cande, S.C. and J.M. Stock. 2004. Cenozoic reconstructions of the Australia-New Zealand-South Pacific sector of Antarctica. *In*: (N.F. Exon, J.P. Kennett and M.J. Malone, eds) *The Cenozoic Southern Ocean: tectonics, sedimentation and climate change between Australia and Antarctica. Geophys. Monogr. 151*: 5–17.

Chamberlain, Y.M. 1965. Marine algae of Gough Island. *Bull. Br. Mus. (Nat. Hist.) Bot. 3*: 176–232.

Claydon S.L. and G.W. Saunders, 2010. Recognition of *Rubrointrusa membranacea* gen. et comb. nov., *Rhodonematella subimmersa* gen. et comb. nov. (with a reinterpretation of the life history) and the Meiodiscaceae fam. nov. within the Palmariales (Rhodophyta). *Phycologia 49*: 283–300

Coates, A.G., L.S. Collins, M.-P. Aubry and W.A. Berggren. 1992. Closure of the isthmus of Panama: The near-shore marine record of Costa Rica and western Panama. *Geol. Soc. Amer. Bull. 104*: 814–829.

Cormaci, M., G. Furnari and B. Scammacca. 1992. The benthic algal flora of Terra Nova Bay (Ross Sea, Antarctica). *Bot. Mar. 35*: 541–552.

Drexler, A.W., K.C. Burke, A.D. Cunningham, A.C. Hine, E. Rosencrantz, D.S. Duncan, P. Hallock and E. Robinson. 1998. Caribbean constraints on circulation between Atlantic and Pacific Oceans over the past 40 million years. *In*: (T.J. Crowley and K.C. Burke, eds) *Tectonic boundary conditions for climate reconstructions.* Oxford monographs on geology and geophysics. Oxford University Press, Oxford. pp. 169–191.

Etcheverry, H. 1983. Algas bentónicas de la Antártica Chilena. *Serie Scientifica INACH 30*: 97–124.

Exon, N.F., H. Brinkhuis, C.M. Robert, J.P. Kennett, P.J. Hill and M.K. Macphail. 2004. *In*: (N.F. Exon, J.P. Kennett and M.J. Malone, eds) *The Cenozoic Southern Ocean: tectonics, sedimentation and climate change between Australia and Antarctica. Geophys. Monogr. 151*: 319–344.

Fraser, C.I., Nikula, R., Spencer, H.G. and J.M. Waters. 2009. Kelp genes reveal effects of subantarctic sea ice during the Last Glacial Maximum. *Proc. Nat. Acad. Sci. USA 106*: 3249–3253.

Gain, L. 1912. La flore algologique des regions antarctiques et subantarctiques. *In*: (J. Charcot, ed) *Deuxième expedition Antarctique française (1908–1910). Sciences naturelles.* Documents scientifiques tome 8. Masson et Companie, Paris. pp. 1–218.

Gepp, A. and E. Gepp. 1905. More Antarctic algae. *J. Bot. 43*: 193–196.

Ghidella, M.E., L.A. Lawver and L.M. Gahagan. 2007. Break-up of Gondwana and opening of the South Atlantic: review of existing plate tectonic models. *U.S. Geological Survey and the National Academies: USGS OF-2007-1047. Short Research Paper 055, doi: 10.3133/of2007-1047.srp055.*

Grobys, J.W.G. 2008. Quantitative tectonic reconstructions of Zealandia based on crustal thickness estimates. *Geochem. Geophys. Geosyst.*, an electronic publication. Vol. 9, no. 1. Q01005, Doi:10.1029/2007GC001691.

Gurgel, C.F.D. and S. Fredericq. 2004. Systematics of the Gracilariaceae (Gracilariales, Rhodophyta) a critical assessment based on rbcL sequence analyses. *J. Phycol. 40*: 138–159.

Guiry, M.D. and G.M. Guiry. 2009. *AlgaeBase.* World-wide electronic publication National University of Ireland, Galway. http://www.algaebase.org; searched on 29 April 2009.

Hariot, M. 1907. Algues. *In:* (J. Charcot, ed.) *Expédition antarctique française 1903–1905 commendée par le Dr. Jean Charcot.* Vol. 1(2). Masson et Companie, Paris. pp. 1–9.

Hommersand, M.H. 1986. The biography of the South African marine red algae: a model. *Bot. Mar. 29*: 257–270.

Hommersand, M.H. 2007. Global biogeography and relationships of the Australian marine macroalgae. *In*: (P.M. McCarthy and A.E. Orchard, eds) *Algae of Australia: introduction.* Australian Biological Resources Study and CSIRO Publishing, Melbourne. pp. 511–542.

Hommersand, M.H. and S. Fredericq. 2003. Biogeography of the marine red algae of the South African West Coast: a molecular approach. *In*: (A.R.O. Chapman, R.J. Anderson, V.J. Vreeland and I.R. Davison, eds) *Proceedings of the XVIIth International Seaweed Symposium.* Oxford University Press, Oxford. pp. 325–336.

Hommersand, M.H., D.W. Freshwater, J.M. Lopez-Bautista and S. Fredericq. 2005. Proposal of the Euptiloteae Hommersand et Fredericq, trib. nov. and transfer of some Southern Hemisphere Piloteae to the Callithamnieae (Ceramiaceae, Rhodophyta). *J. Phycol. 42*: 203–225.

Hooker, J.D. 1847. *The botany of the Antarctic voyage of H.M. discovery ships Erebus and Terror, in the years 1839–1843, under the command of Captain Sir James Clark Ross, Vol. I. Flora Antarctica. Pt. II. Botany of Fuegia, the Falklands, Kerguelen's Land, etc. (1845–1847)* Algae. Reeve Brothers, London. pp. 454–502, pls. 165–194.

Kennett, J.P. and N.F. Exon. 2004. Paleooceanograpic evolution of the Tasmanian Seaway and its climatic implications. *In*: (N.F. Exon, J.P. Kennett and M.J. Malone, eds) *The Cenozoic Southern Ocean: tectonics, sedimentation and climate change between Australia and Antarctica. Geophys. Monogr. 151*: 345–367.

Kylin, H. and C. Skottsberg. 1919. Zur Kenntnis der subantarktischen und antarktischen Meeresalgen. II. Rhodophyceen. *In*: (O. Nordenskjöld, ed) *Wissenschaftliche Ergebnisse der schwedischen Südpolarexpedition 1901–1903.* Band IV., Stockholm. 88 pp.

Lawver, L.A. and L.M. Gahagan. 1998. Opening of Drake Passage and its input on Cenozoic ocean circulation. *In*: (T.J. Crowley and K.C. Burke, eds) *Tectonic boundary conditions for climate reconstructions.* Oxford monographs on geology and geophysics. Oxford University Press, Oxford. pp. 212–223.

Lawver, L.A. and L.M. Gahagan. 2003. Evolution of Cenozoic seaways in the circum-Antarctic region. *Palaeogeogr. Paleoclimatol. Paleoecol. 198*: 11–37.

Lawver, L.A., J.G. Sclater and M. Meinke. 1985. Mesozoic and Cenozoic Reconstructions of the South Atlantic. *Tectonophysics 114*: 233–254.

Lawver, L.A., J.Y. Royer, D.T. Sandwell and C.R. Scotese. 1987. Crustal development: Gondwana break-up. *In*: (M.R.A. Thompson, J.A. Cramer and J.W. Thompson, eds) *Geological evolution of Antarctica*. The National Academies Press, Washington, D.C. pp. 533–539.

Lawver, L.A., L.M. Gahagan and W.D. Dalziel. 1999. Antarctica, keystone of Gondwana. http://www.ig.utexas.edu/research/ projects/plates/movies/akog.ppt.

Le Gall, L., J.L. Dalen and G.W. Saunders. 2008. Phylogenetic analyses of the red algal order Rhodymeniales supports recognition of the Hymenocladiaceae fam. nov., Fryeellaceae fam. nov., and *Neogastroclonium* gen. nov. *J. Phycol. 44*: 1556–1571.

Lee, T.-Y. and L.A. Lawver. 1995. Cenozoic plate reconstruction of the Southeast Asia region. *Tectonophysics 251*: 85–138.

Lemoine, M. 1913. Mélobésiées, revision des Mélobésiées antarctiques. *In*: (J. Charcot, ed) *Deuxième expedition Antarctique française (1908–1910). Sciences naturelles.* Documents scientifiques tome 8. Masson et Companie, Paris. pp. 1–64.

Lin S.-M. and W.A. Nelson. 2010. Systematic revision of the genus *Phycodrys* (Delesseriaceae, Rhodophyta) from New Zealand, with the description of *P. novae-zelandiae* sp. nov., *P. franiae* sp. nov. and *P. adamsiae* sp. nov., *Eur. J. Phycol. 45*: 200–214.

Lin, S.-M., S. Fredericq and M. Hommersand. 2001. Systematics of the Delesseriaceae (Ceramiales, Rhodophyta) based on large subunit rDNA and *rbc*L sequences, including the Phycodryoideae, subfam. nov. *J. Phycol. 37*: 881–899.

Lindstrom, S.C., J.L. Olsen and W.T. Stam. 1996. Recent radiation of the Palmariaceae (Rhodophyta). *J. Phycol. 32*: 457–468.

Lucas, A.H.S. 1919. The algae of Commonwealth Bay. *Australasian Antarctic Expedition 1911–14.* Scientific Reports. Series C. Zoology and Botany. VII. W. Applegate Gullick, Sydney. pp. 5–18, 9 pls.

Maggs, C.A., D.P. Fewer and J. Provan. 2007. *The relative roles of dispersal and vicariance in biogeography of marine macroalgae. Symposium 4: Maritime connectivity. Third biennial conference of the International Biogeography Society held in Tenerife, Jan. 9–13, 2007*: 88 (abstract).

McLoughlin, S. 2001. The break-up history of Gondwana and its impact on pre-Cenozoic floristic provincialism. *Aust. J. Bot. 49*: 271–300.

Mendoza, M.L. 1979. *Antarcticophyllum*, nuevo genero para las Corallinaceae. *Boln. Soc. Argent. Bot. 17*: 252–261.

Mendoza, M.L. and J. Cabioch. 1985. Critique et comparaison morphogénétique des genres *Clathromorphum* et *Antarcticophyllum* (Rhodophyta, Corallinaceae). Conséquences biogéographiques et systématiques. *Cah. Biol. mar. 26*: 251–266.

Mendoza, M.L. and J. Cabioch. 1986. Le genre *Hydrolithon* (Rhodophyta, Corallinaceae) sur les côtes subantarctiques et antarctiques d'Argentine et de quelques régions voisines. *Cah. Biol. mar. 27*: 163–191.

Moe, R.L. and T.E. DeLaca. 1976. Occurrence of macroscopic algae along the Antarctic Peninsula. *Antarctic J. 11*: 20–24.

Moe, R.L. and P.C. Silva. 1980. Morphological and taxonomic studies on Antarctic Ceramiaceae (Rhodophyceae). II. *Pterothamnion antarcticum* (Kylin) comb. nov. (*Antithamnion antarcticum* Kylin). *Br. Phycol. J. 15*: 1–17.

Moe, R.L. and P.C. Silva. 1983. Morphological and taxonomic studies on Antarctic Ceramiaceae (Rhodophyta) III. *Georgiella* and *Plumariopsis* (Tribe Ptiloteae). *Br. Phycol. J. 18*: 275–298.

Moore, J.K., Abbott, M.R. and J.G. Richman. 1999. Location and dynamics of the Antarctic Polar Front from satellite sea surface temperature data. *J. Geophys. Res. 104*: 3059–3073.

Nelson, W.A., T.J. Farr and E.S. Broom. 2006. Phylogenetic relationships and generic concepts in the red order Bangiales: challenges ahead. *Phycologia 45*: 249–259.

Page, M.J., N. Alcock, D. Gordon, M. Kelly-Shanks, W. Nelson, K. Neill and J. Watson. 2002. *Preliminary assessment of the biodiversity of benthic macrobiota of the western Ross Sea, Antarctica. Final Research Report for Ministry of Fisheries Research Project ZBD2000/02. Appendix 3: annotated species list of Ross Sea algae collection from the Ross Sea Biodiversity sampling progamme.* NIWA, Wellington. pp. 48 +21 (appendices).

Papenfuss, G.F. 1964. Catalogue and bibliography of Antarctic and Sub-Antarctic benthic marine algae. *In*: (M.O. Lee, ed) *Biology of the Antarctic seas.* Antarctic Research Series Vol. 1. American Geophysical Union, Washington, D.C. pp. 1–76.

Pearson, P.N. and M.R. Palmer. 2000. Atmospheric carbon dioxide concentrations over the past 69 million years. *Nature 406*: 695–699.

Peters, K.J., C.D. Amsler, M.O. Amsler, J.B. McClintock, R.B. Dunbar and B.J. Baker. 2005. A comparative analysis of the nutritional and elemental composition of macroalgae from the western Antarctic Peninsula. *Phycologia 44*: 453–463.

Quartino, M.L., H. Klöser, I.R. Schloss and C. Wiencke. 2001. Biomass and associations of benthic marine macroalgae from the inner Potter Cove (King George Island, Antarctica) related to depth and substrate. *Polar Biol. 24*: 349–355.

Reinsch, P.F. 1890. Zur Meeresalgenflora von Südgeorgien. *In*: (G. von Neumayer, ed.) *Die internationale Polarforschung 1882–1883. Die deutschen Expeditionen und ihre Ergebnisse.* Vol. 2. Beschreibende Naturwissenschaften, Herrmann. Berlin, pp. 366–449, pls. 1–19.

Ricker, R.W. 1987. *Taxonomy and biogeography of Macquarie Island seaweeds.* British Museum (Natural History), London, 344 pp.

Schneider, C.W. and M.J. Wynne. 2007. A synoptic review of the classification of red algal genera a half century after Kylin's "*Die Gattungen der Rhodophyceen*". *Bot. Mar. 50*: 197–249.

Silva, P.C. 1993. Proposal to conserve *Falklandiella* Kylin (Rhodophyceae: Ceramiaceae). *Taxon 42*: 131–132.

Skottsberg, C. 1923. Botanische Ergebnisse der schwedischen Expedition nach Patagonien und dem Feuerlande 1907–1909, IX. Marine Algen, 2. Rhodophyceae. *K. svenska Vetensk.-Akad. Handl. 63*: 1–70.

Skottsberg, C. 1941. Communities of marine algae in subantarctic Antarctic waters. *K. svenska Vetensk.-Akad. Handl. Ser. 3. 19*: 1–92.

Skottsberg, C. 1953. On two collections of Antarctic marine algae. *Arkif för Botanik Ser. 2. 2*: 531–566.

Smith, A.G., A.M. Hurley and J.C. Briden. 1981. *Phanerozoic paleocontinental world maps.* Cambridge University Press, Cambridge. pp. 107.

Stegenga, H., J.J. Bolton and R.J. Anderson. 1997. *Seaweeds of the South African west coast.* Contributions from the Bolus Herbarium Vol. 18. Bolus Herbarium, University of Cape Town, Cape Town. pp. [2] + 1–655.

Taylor, W.R. 1945. Pacific marine algae of the Allan Hancock Expedition to the Galapagos Islands. *Allan Hancock Pacif. Exped. 12*: 1–316, 100 pls.

Thiers, B. 2010. continuously updated. Index herbariorum: a global directory of public herbaria and associated staff. New York Botanical Garden's Virtual Herbarium. http://sweetgum.nybg.org/ih/

Wagner, F.S. 1954. Contributions to the morphology of the Delesseriaceae. *Univ. Calif. Publ. Bot. 27*: [i–ii], 279–345.

Wiencke, C. and M.N. Clayton. 1998. The life history of *Porphyra* from the South Shetland Islands, Antarctica. *Polar Biol. 19*: 257–263.

Wiencke, C. and M.N. Clayton. 2002. Antarctic seaweeds. *In*: (J.W. Wägele, ed) *Synopses of the Antarctic benthos*, A.R.G. Gantner Verlag KG, Ruggell/Lichtenstein. 239 pp.

Woelkerling, W. and D. Lamy. 1998. *Non-geniculate Coralline Red Algae and the Paris Muséum: Systematics and Scientific History.* Publications Scientifiques du Muséum national d'Histoire naturelle/A.D.A.C., Paris. 767 pp.

Wulff, A., K. Iken, M.L. Quartino, A. Al-Handal, C. Wiencke and M.N. Clayton. 2009. Biodiversity, biogeography and zonation of marine benthic micro- and macroalgae in the Arctic and Antarctic. *Bot. Mar. 52*: 491–507.

Wynne, M.J. 1997. Taxonomic and nomenclatural notes on the Delesseriaceae (Rhodophyta) *Contr. Univ. Michigan Herb. 21*: 319–334.

Wynne, M.J. 2007. *Rhodymenia coccocarpa*, a new combination based on *Phyllophora coccocarpa* Mont. *Contr. Univ. Michigan Herb. 25*: 241–247.

Zacher, K., R. Rautenberger, D. Hanelt, A. Wulff and C. Wiencke. 2009. The abiotic environment of polar benthic algae. *Bot. Mar. 52*: 483–490.

Zaneveld, J.S. and R.B. Sanford. 1980. Crustose corallinaceous algae (Rhodophyta) of the New Zealand and United States scientific expedition to the Ross Sea, Balleny Islands and Macquarie Ridge, 1965. *Blumea 26*: 205–231.

Chemical ecology

5 Defenses of polar macroalgae against herbivores and biofoulers

Charles D. Amsler, Katrin Iken, James B. McClintock and Bill J. Baker

Introduction

Macroalgae and other sessile organisms that are unable to avoid predators through behavior or habitat choice must be able either to tolerate damage by predators (Strauss and Agrawal 1999, Stowe et al. 2000) or to resist predation. Although it is possible for some potential prey organisms to be avoided by predators because they are of insufficient nutritional quality (Bullard and Hay 2002), most macroalgae that resist predation do so via physical or chemical defenses (Hanley et al. 2007, Amsler 2008).

Anti-herbivore chemical defenses of macroalgae are relatively well known and play important roles in structuring temperate and tropical marine communities (Paul et al. 2001, Jormalainen and Honkanen 2008, Pereira and da Gama 2008). With the exception of phlorotannins, which can have both primary and secondary metabolic roles (Amsler and Fairhead 2006), most defensive compounds in algae are secondary metabolites without known primary functions (Maschek and Baker 2008). Many such compounds are known from temperate and tropical macroalgae (Harper et al. 2001, Blunt et al. 2006), and chemical defenses are commonly believed to be more widespread and important in macroalgae from the tropics compared to temperate latitudes (Van Alstyne et al. 2001). However, there appears to be a greater diversity of potentially defensive secondary metabolites reported from temperate macroalgae compared to tropical forms because macroalgal species diversity peaks at temperate latitudes (Pereira and da Gama 2008). In contrast, macroalgal species diversity is much lower at polar latitudes (Chapter 3; Wiencke and Clayton 2002, Wulff et al. 2009). Consequently, combined with a lower level of overall research activity in polar communities, we know much less about macroalgal secondary metabolites from polar waters than from lower latitudes (Amsler et al. 2001, 2008, Blunt et al. 2006, Lebar et al. 2007). Nevertheless, recent studies have revealed that chemical defenses are common in Antarctic macroalgae (Wiencke et al. 2007, Amsler et al. 2008, Avila et al. 2008) and are likely to play important roles in Antarctic macroalgal-herbivore relationships, as will be discussed in this chapter.

In addition to defending against herbivores, macroalgae can also defend against biofouling (settlement of propagules from and/or overgrowth by other organisms) via physical or chemical means. In the present context, *biofouling* is equivalent to *epibiosis* as described by Wahl (1989). Continuous sloughing of outer tissue layers can be an effective physical defense by preventing the establishment of epiphytes (Filion-Myklebust and Norton 1981, Moss 1982, Keats et al. 1997, Nylund and Pavia 2005). Macroalgae can chemically defend against biofouling via oxidative burst responses (Weinberger 2007, Potin 2008) or by production of secondary metabolites (Steinberg and de Nys 2002, Lane and Kubanek 2008) that hinder the establishment or growth of epiphytic species. We are aware of no reports on the potential for physical or oxidative burst defenses against biofoulers or pathogens in polar macroalgae. These are important areas for future research. There are many reports of macroalgal secondary metabolites with antifouling bioactivity *in vitro* (Bhadury and Wright 2004). However, using standard crude extract procedures, it is difficult to determine which compounds in what quantities might actually be present at an alga's surface or be released into the surrounding boundary layer. Consequently, it can be very difficult to perform bioassays in an ecologically relevant manner or even to decide which compounds or extract fractions to use, and there have been very few antifouling studies performed with secondary metabolites using ecologically relevant methodology (Steinberg and de Nys 2002, Lane and Kubanek 2008). No such studies have been published to date for polar macroalgae, but initial work has been performed to set the stage for such investigations, as will be discussed below.

Defenses against herbivores

All herbivores are not equal in their modes of feeding, in the resultant impacts of their feeding on macroalgae, or in their responses to algal secondary metabolite defenses. One useful way to differentiate the impacts and responses of marine herbivores is to classify them by size (e.g., Brawley 1992, Hay 1996, Paul et al. 2001) as this often correlates with the magnitude or rate at which they damage their prey. However, exceptions to generalizations by herbivore size certainly occur (Amsler 2001, Paul et al. 2001). Mesoherbivores are commonly defined as being approximately 1 to 25 mm in length (Hay et al. 1987, Brawley 1992), with macroherbivores being larger and microherbivores being smaller than mesoherbivores. Interactions between microherbivores and macroalgae have received very little attention anywhere in the world (Amsler and Fairhead 2006). We are aware of no such studies in polar regions although benthic copepods are known to associate with Antarctic macroalgal detritus (Veit-Köhler 2005) and live macroalgae (C. Aumack, personal communication) as well as with sponges in macroalgal-dominated Antarctic communities (M.O. Amsler, personal communication).

Chemical defenses against macroherbivores

Studies of macroalgal defenses against marine macroherbivores in polar regions have included sea urchins, sea stars, and fish although other large invertebrates are also known to consume macroalgae (e.g., Brand 1980, Iken 1996). In the most extensive such study, Amsler et al. (2005a) reported that thallus fragments of 22 species of subtidal macroalgae from the western Antarctic Peninsula (WAP) were unpalatable to the omnivorous sea star *Odontaster validus* Koehler, while 13 macroalgal species were palatable. Of the species unpalatable as thallus to *O. validus*, chemical defenses explained this unpalatability in 76% of the species used in bioassays offering crude chemical extracts of the algae to *O. validus* in otherwise palatable artificial foods. In parallel studies using the omnivorous fish, *Notothenia coriiceps* Richardson, 24 of 29 macroalgal species assayed were unpalatable as thallus and, of those assayed as crude extracts in artificial foods, chemical defenses explained the unpalatability in 53%

(Amsler et al. 2005a). The large, overstory brown macroalgae that dominate these communities including *Desmarestia anceps* Montagne, *Desmarestia antarctica* Moe *et* Silva, *Desmarestia menziesii* J. Agardh, *Himantothallus grandifolius* (A. *et* E.S. Gepp) Zinova, and *Cystosphaera jacquinotii* (Montagne) Skottsberg, as well as many of the more common red macroalgae, were chemically defended against both the sea stars and fish (Amsler et al. 2005a). From these studies we concluded that chemical defenses are likely to play important roles in mediating macroalgal-macroherbivore interactions in WAP communities.

Some specific secondary metabolites that deter macroherbivore feeding have been detected in WAP macroalgae. *Desmarestia menziesii* produces menzoquinone (Figure 5.1), a hydroquinone that deters feeding in *Odontaster validus* (Ankisetty et al. 2004) but not amphipods (C. Amsler, unpublished data). In addition, the red alga *Delisea pulchra* (Greville) Montagne elaborates six dimeric halogenated furanones (Ankisetty et al. 2004). Although these have not been bioassayed against sympatric Antarctic herbivores, several of these were previously known from Australian *D. pulchra* and are active feeding deterrents against Australian macroherbivores (Wright et al. 2004).

β-Dimethylsulphoniopropionate (DMSP) can function as a defensive compound against sea urchin herbivory in green macroalgae (reviewed by Van Alstyne 2008). Intertidal green macroalgae from the WAP are known to produce DMSP (Karsten et al. 1990, 1992), but to our knowledge, its potential roles in antiherbivore defense against limpets or other intertidal herbivores (Kim 2001) has yet to be investigated.

McMurdo Sound in the Ross Sea is the southernmost geographic limit for benthic algae because it is as far south in the world as open ocean ever occurs. Two species of foliose macroalgae, *Phyllophora antarctica* A. *et* E.S. Gepp and *Iridaea cordata* (Turner) Bory, occur there and can be abundant at some sites (Miller and Pearse 1991). Both species are chemically defended against predation by the abundant, sympatric sea urchin *Sterechinus neumayeri* Meissner (Amsler et al. 1998). However,

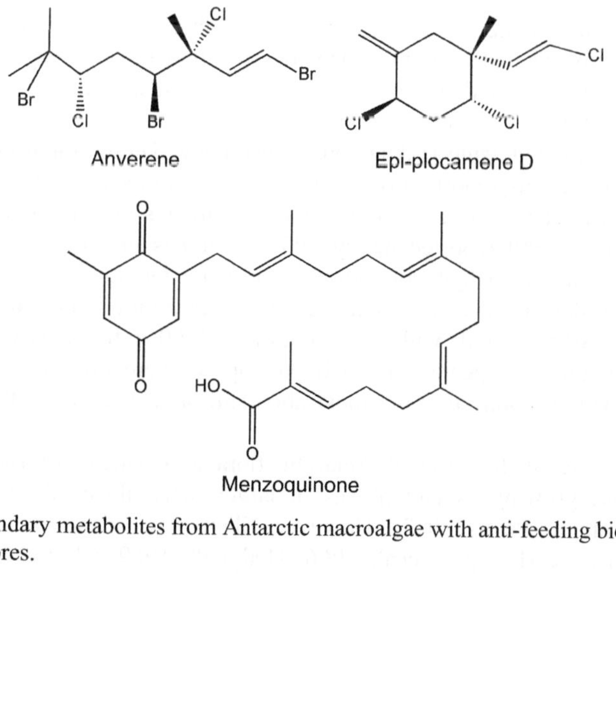

Figure 5.1 Secondary metabolites from Antarctic macroalgae with anti-feeding bioactivity against sympatric herbivores.

S. neumayeri preferentially uses detached individuals of both macroalgal species for protective cover as the algae physically defend them from large sea anemones, which are their primary predators (Amsler et al. 1999). Since most macroalgal biomass in these communities has been detached from the bottom by ice, this also mutualistically benefits the macroalgae because it keeps them in the photic zone where they can continue to photosynthesize (Schwarz et al. 2003) and reproduce (Amsler et al. 1999).

Although macrofauna commonly associate with macroalgae in the Arctic Ocean (Lippert et al. 2001, Wessels et al. 2006), only two of 19 macroalgal-associated animal species from Svalbard were found to directly consume macroalgae (Wessels et al. 2006). One of these was the sea urchin *Strongylocentrotus droebachiensis* (O.F. Müller), which preferred to feed on kelps over smaller, more delicate algal species. The presence of chemical defenses was not directly examined, but of 23 total species of macroalgae used in feeding assays, the authors concluded that only two, *Ptilota gunneri* P.C. Silva, Maggs *et* L.M. Irvine and *Desmarestia viridis* (O.F. Müller) J.V. Lamouroux, were likely to be chemically defended. Molis et al. (2008) demonstrated that the deterrence of *D. viridis* in these communities is likely to be caused by its high acidity, similar to findings of deterrence of *S. droebachiensis* by a temperate species of *Desmarestia* (Pelletreau and Muller-Parker 2002). Molis et al. (2008) further showed that *D. viridis* provides an associational defense for the otherwise relatively palatable kelp *Alaria esculenta* (Linnaeus) Greville.

Chemical defenses against mesoherbivores

Amphipods are very well represented in the Antarctic fauna, particularly in benthic communities (De Broyer and Jažd̶żewski 1993, 1996, De Broyer et al. 1999) and commonly associate with macroalgae in very high densities along the WAP (Richardson 1971, 1977, Huang et al. 2007). For example, *Desmarestia anceps* individuals at Signy Island can support over 10,000 individual amphipods (Richardson 1977). Combining amphipod densities per unit algal biomass (Huang et al. 2007) with raw biomass data previously collected in the same communities (Amsler et al. 1995), Amsler et al. (2008) estimated that amphipod densities per unit sea bottom area range up to greater than 300,000 individuals m^{-2} in monospecific stands of *Desmarestia menziesii,* and up to 32,000 and 26,000 individuals m^{-2} in monospecific stands of *D. anceps* and *Plocamium cartilagineum* (L.) Dixon, respectively. These amphipod densities are very high compared to reports from temperate and tropical waters (Nelson 1980, Wildish 1988, Brawley 1992). Gastropod mesoherbivores are also common on WAP macroalgae. Iken (1999) reported nearly 300 individuals m^{-2} of *Laevilacunaria antarctica* Martens in macroalgal communities on King George Island. Picken (1979) reported a mean density of *L. antarctica* of almost 200 individuals m^{-2} from benthic communities at Signy Island, with a maximum of 2500 individuals m^{-2}. Richardson (1977) reported that 12 species of small gastropods were present on *D. anceps* at Signy Island, with *L. antarctica* the most abundant, and with four other species also common.

A unique feature of the subtidal Antarctic flora is a paucity of filamentous algae other than those growing as endophytes in larger macroalgae, although such endophytes are common (Peters 2003). However, filamentous algae are common in the WAP intertidal zone (Delépine et al. 1966, Hedgpeth 1969, Chung et al. 1994, Kim

2001). Peters (2003) hypothesized that this lack of filamentous algae in the subtidal is because of the very high densities of amphipods and other mesoherbivores found in these communities. He further hypothesized that endophytes are common because they are escaping the mesoherbivores by growing within the larger algae. This hypothesis would predict that filamentous endophytes are relatively palatable compared to macrophytic hosts and that prediction was tested by Amsler et al. (2009). This study compared amphipod feeding rates on nine species of filamentous brown or green endophytes with those on 13 macrophytic species including eight macrophytic species that were commonly observed with endophytes in nature. Feeding rates on the endophytes were two to three orders of magnitude higher than on the host macrophyte species (Amsler et al. 2009). This study also confirmed and extended Peters' (2003) observations that some macrophytic species are usually endophyte-free and that in others, there is an apparent species-specific relationship between hosts and the endophytes that are able to grow to visible densities in them. We hypothesize that this is because of species-specific tolerance of endophytes to secondary metabolite defenses of some of the larger algae but not of others. This hypothesis remains to be tested.

Although filamentous algal species are largely absent in the WAP subtidal, a number of finely branched macroalgal species are common. Aumack et al. (2010) examined the palatability and chemical defenses of five finely-branched macroalgal species to the herbivorous amphipods *Gondogeneia antarctica* (Chevreux) Thurston and *Prostebbingia gracilis* Chevereux. All of the finely-branched algae were unpalatable as fresh thallus to both amphipod species. In all four of the red algal species, this unpalatability could be explained by chemical defenses because organic extracts of the algae in artificial foods were also unpalatable. However, extracts of the single brown alga examined, *Halopteris obovata* (Hooker f. *et* Harvey) Sauvageau, did not deter either amphipod species as was also observed with *G. antarctica* in a previous study (Amsler et al. 2005a).

Huang et al. (2006) examined feeding rates of *Gondogeneia antarctica* and *Prostebbingia gracilis* on eight common species of bladed and branched WAP macroalgae. This study reported that only one of the macroalgal species, *Palmaria decipiens* (Reinsch) Ricker, was consumed at appreciable rates by the two amphipods. However, although subsequent studies by our research group have confirmed the relative palatability of *P. decipiens* to *G. antarctica* (A. Bucolo unpublished, Amsler et al. 2009, Aumack et al. 2010), we have been unable to do so in repeated attempts with *P. gracilis* using similar methodology (A. Bucolo unpublished, Aumack et al. 2010). We do not know the reason for this discrepancy. However, Huang et al. (2006) reported a strong statistical interaction between trial day and feeding rate for *P. gracilis* in addition to high variability in the feeding rate of *P. gracilis* on *P. decipiens*, but not in rates on other macroalgae. It is possible that the high mean feeding rate of *P. gracilis* on *P. decipiens* reported by Huang et al. (2006) was caused by a single unusual trial result. When Huang et al. (2006) presented freeze-dried, ground powders of these eight macroalgal species to the two amphipod species, most of the species that were unpalatable as thallus remained so in the artificial foods, suggesting the presence of chemical defenses. These ground thallus artificial food bioassay results matched well with previous bioassays in which organic extracts of the macroalgae were presented to *G. antarctica* (Amsler et al. 2005a). Amsler et al. (2005a) did not examine the palatability of fresh thallus to amphipods. However, we did perform amphipod feeding bioassays with crude extracts of 27 macroalgal species that were unpalatable as thallus to either the sea star *Odontaster validus* or the fish *Notothenia coriiceps* and observed that 17 (63%) of the species were unpalatable to *G. antarctica*. *Plocamium cartilagineum,* which is unpalatable to *G. antarctica* both as fresh thallus (Huang et al. 2006, Aumack et al. 2010) and crude extract in artificial foods (Amsler et al. 2005a), elaborates the halogenated monoterpenes anverene and epi-plocamene D (Figure 5.1, Ankisetty et al. 2004). Both of these compounds deter feeding by *G. antarctica* but not by sea stars (Ankisetty et al. 2004).

It is of note that *Desmarestia menziesii, D. anceps,* and *Plocamium cartilagineum,* the macroalgal species that Huang et al. (2007) observed with the highest densities of amphipods, were chemically defended and among the most unpalatable species to amphipods (Amsler et al. 2005a, Huang et al. 2006). In addition, *Palmaria decipiens,* which is palatable to the amphipod *Gondogeneia antarctica,* and *Iridaea cordata* (Turner) Bory both supported very low densities of amphipods (Huang et al. 2006, 2007). These observations were made during daylight but may change at night, as has been observed with amphipod densities on other macroalgae (Brawley 1992). *P. decipiens* and *I. cordata* are both preferred food items of the omnivorous fish, *Notothenia coriiceps,* which consumes both macroalgae and amphipods (Iken et al. 1997, 1999). Zamzow et al. (2010) demonstrated that *N. coriiceps* is more likely to capture and consume amphipods associated with *D. menziesii* than with *P. decipiens.* Consequently, this amphipod distribution pattern could well be the result of daylight predator avoidance behavior in the amphipods (cf. Brawley 1992). In this context, it is very possible that amphipod densities on palatable macroalgae like *P. decipiens* and *I. cordata* would be greater at night, as has been observed for amphipod densities on the palatable alga *Mazzaella laminarioides* (Bory) Fredericq (cited as *Iridaea laminarioides*) in Chile (Buschmann 1990).

Iken (1999) examined the feeding ecology and potential for chemical defenses of 12 common macroalgal species from the WAP against the sympatric, herbivorous gastropod *Laevilacunaria antarctica*. *L. antarctica* consumed significant quantities of nine of the 12 macroalgal species but fed at much higher rates on macroalgae with epiphytic diatoms than on those without diatoms. Feeding bioassays with seawater extracts of the unpalatable macroalgae did not reveal the presence of chemical defenses (Iken 1999). However, the experimental design used was exceptionally conservative as only seawater-soluble compounds were included and they were presented in the artificial foods at less than one-twelfth their concentration in the algal thallus.

Wessels et al. (2006) reported that the Arctic amphipod *Gammarellus homari* J.C. Fabricius was one of only two of 19 invertebrates that consumed macroalgae in Svalbard. Unlike the sea urchin *Strongylocentrotus droebachiensis, G. homari* preferred "delicate" red algae over kelps. However, as with *S. droebachiensis,* the only macroalgal species that they concluded might elaborate chemical defenses against the amphipod were *Ptilota gunneri* and *Desmarestia viridis* (Wessels et al. 2006).

The potential roles of phlorotannins in defense against herbivores

Phlorotannins are relatively hydrophilic, polyphenolic metabolites produced by brown algae that have both primary and secondary metabolic roles (Amsler and Fairhead 2006). Although the most commonly studied secondary functions are as defenses against herbivory, phlorotannins also have the potential to serve as defenses against ultraviolet radiation (UVR), pathogens, biofouling, and heavy metal toxicity (Amsler and Fairhead 2006). Some of the potential ecological roles are somewhat controversial, however. It is likely that, in some cases, ecological roles attributed to phlorotannins have in fact been the result of other secondary metabolites, while in other cases it is very likely that phlorotannins are indeed playing important roles in ecological interactions (Amsler and Fairhead 2006). In polar macroalgae, phlorotannins have been studied in the context of defenses against both macroherbivores and mesoherbivores as well as of defenses against UVR. Their potential roles in defense against UVR are treated in a separate chapter (Chapter 13; Karsten et al. 2009).

Antarctic brown algae from the WAP contain phlorotannins at concentrations between 0.5% and 9% of the algal dry weight, which is higher than in most tropical and northern-temperate species, but similar to southern hemisphere species (Iken et al. 2007). In six structurally differentiated brown algae, these phlorotannins were allocated unequally to separate thallus parts, except for *Himantothallus grandifolius*, where thallus allocations were not different (Iken et al. 2007). Conversely, Fairhead et al. (2005a) did not find strong differences in phlorotannin distribution within the two dominant *Desmarestia* species, but patterns in these cases may have been obscured by high variability of phlorotannin concentrations in individuals from different sampling locations and depths. In addition, phlorotannins in different thallus portions were also composed of distinct size classes in different species, but, yet again, in no consistent manner among species (Iken et al. 2007). While holdfasts in all differentiated species were dominated by large phlorotannin size classes (>100 kDa), size class composition of other thallus parts differed strongly among species. If phlorotannins have defensive properties against herbivores, then this differential allocation of phlorotannins to thallus parts may be an adaptive feature to protect more valuable thallus parts from grazing (Chapter 6; Fairhead et al. 2005a, Iken et al. 2007, 2009; also see below tests of chemical defense theories using polar macroalgae section).

Recent investigations of grazer deterrence by phlorotannins from Antarctic brown algae have yielded mixed results (Chapter 6; Iken et al. 2009). Phlorotannin-enriched extracts from five algal species caused significant feeding deterrence against the fish, *Notothenia coriiceps*, but this activity was lost in all but two species (*Desmarestia menziesii* and *Cystosphaera jacquinotii*) when purified phlorotannins of the same species were tested in bioassays with fish. Similarly, out of seven species tested in the same study, only isolated phlorotannins from *D. menziesii* and *Himantothallus grandifolius* deterred feeding in the sea star, *Odontaster validus*. This low proportion of feeding deterrence of phlorotannins was also found for smaller, amphipod mesograzers. Purified *D. anceps* phlorotannins significantly deter feeding by the amphipod *Gondogeneia antarctica* (Chapter 6; Fairhead et al. 2006, Iken et al. 2009). Significant feeding deterrence against *G. antarctica* was also found for *C. jacquinotii* phlorotannins, although phlorotannins of five other brown algal species had no deterrent effects and sometimes even stimulated feeding (Chapter 6; Iken et al. 2009). Such feeding stimulation has been attributed to other, nutritional compounds present in chemical extracts, but may only be a short-term effect seen in bioassays since it does not take into account possible long-term effects of reduced assimilation efficiency in grazers feeding on a high-phlorotannin diet (Fairhead et al. 2006).

Possible roles of nutritional or physical means of resistance to herbivory

Nutritional quality of a food item can interact with other factors to influence palatability. Because herbivores are usually nitrogen-limited (Mattson 1980) the protein or nitrogen content of macroalgae is commonly believed to be the most important nutritional component in determining palatability to herbivores (e.g., Horn and Neighbors 1984, Duffy and Hay 1991, Cruz-Rivera and Hay 2003). Peters et al. (2005) examined multiple nutritional characteristics in 40 WAP macroalgal species and found that, overall, the protein contents were much higher than normally reported from other places in the world. In addition, the elemental C:N ratios were relatively low and nitrogen contents relatively high compared to macroalgae from lower latitudes (Weykam et al. 1996, Peters et al. 2005). This suggests that, in general, and from a purely nutritional standpoint, Antarctic macroalgae should be more desirable to herbivores than temperate or tropical macroalgae. Amsler et al. (2005a) compared data from Peters et al. (2005) to observed differences in palatability to sea stars and fish of WAP macroalgae collected from the same locality. No correlations between protein or any other nutritional component and differences in palatability to sea stars or fish were evident. Huang et al. (2006) performed linear regression analyses comparing amphipod feeding rates on artificial foods made from freeze-dried, ground tissues of eight macroalgae from this same location with protein content and six other nutritional components reported by Peters et al. (2005) and likewise found no significant correlations.

Physical factors, particularly thallus toughness, can also be important parameters in macroalgal resistance to herbivory (e.g., Watson and Norton 1985, Pennings and Paul 1992, Pennings et al. 1996). Amsler et al. (2005a) examined the toughness of 30 species of WAP macroalgae by measuring their resistance to being punctured. Overall, there were no obvious correlations between toughness and palatability to macroherbivores, but toughness may play a role in the unpalatability of some species, particularly to amphipods. For example, the second toughest of the 30 species examined was the red alga *Varimenia macropustulosa* R. Moe [incorrectly identified as *Pachymenia orbicularis* (Zanardini) Setchell *et* Gardner; see chapter 4; Hommersand et al. (2009)]. *V. macropustulosa* was unpalatable as fresh thallus to both sea stars and fish (Amsler et al. 2005a) as well as to amphipods (Amsler et al. 2009), but crude extracts were not deterrent to either sea stars or amphipods (extracts were not bioassayed with fish; Amsler et al. 2005a). Fresh thalli of the first, third, and fourth toughest of these species, the red alga *Gigartina skottsbergii* Setchell *et* Gardner, the brown alga *Ascoseira mirabilis* Skottsberg, and the red alga *Curdiea racovitzae* Hariot, respectively, are not consumed at appreciable rates by the amphipod *Gondogeneia antarctica* (Huang et al. 2006, Amsler et al. 2009). However, Amsler et al. (2005a) reported that crude extracts of all three species are not deterrent to *G. antarctica* [although amphipod feeding results with *G. skottsbergii* extracts were not shown in Amsler et al. (2005a) since it was palatable to both sea stars and fish as fresh thallus]. This suggests that their unpalatability to amphipods may likely be because of their relative toughness. When *G. skottsbergii* thallus was presented in artificial foods freeze-dried and ground to a powder, which should remove most if not all elements contributing to thallus toughness, is was consumed at moderate rates by *G. antarctica* (Huang et al. 2006). On the other hand, the fifth toughest species examined by Amsler et al. (2005a), *Himantothallus grandifolius,* which was not significantly less tough than either *V. macropustulosa, A. mirabilis,* or *C. racovitzae*, is strongly chemically defended against *G. antarctica,* sea stars, and fish (Amsler et al. 2005a). Overall, a majority of the macroalgae that were reported by Amsler et al. (2005a) to be palatable to amphipods as extract, or that were not included in the amphipod bioassays reported there, have subsequently been found to be unpalatable as thallus to the amphipod *Gondogeneia antarctica* (Huang et al. 2006, Amsler et al. 2009, Aumack et al. 2010). Presumably this unpalatability is due to physical toughness.

In the Arctic, Wessels et al. (2006) compared consumption rates on fresh algal thallus from 23 macroalgal species with feeding rates in separate assays with frozen, ground thallus presented in artificial foods. The latter should have reduced or removed thallus physical defenses. From these comparisons, they concluded that physical toughness appeared to play a role in decreasing feeding rates of the amphipod *Gammarellus homari* but not of the sea urchin *Strongylocentrotus droebachiensis* (Wessels et al. 2006).

Tests of chemical defense theories using polar macroalgae

Most theories concerning the evolution or functioning of chemical defenses come from the terrestrial plant literature, but macroalgae can be useful model organisms in testing and extending these theoretical constructs in analogous but, particularly with brown algae, far from homologous systems (Amsler and Fairhead 2006, Pavia and Toth 2008). Tests of three such theories have taken advantage of physical or biotic features of Antarctic benthic macroalgal communities. To date there have been no such published reports from the Arctic.

The resource allocation model (RAM) predicts that on evolutionary time scales, the types of defensive compounds that species produce will be determined in part by their light and nutrient environment (Coley et al. 1985). Species that are primarily carbon-limited (i.e., light-limited) but nitrogen-replete are predicted to preferentially defend themselves with nitrogen-containing metabolites such as alkaloids because these are often bioactive at much lower concentrations than non-nitrogenous defenses such as phenolics or terpenes. So, by investing nitrogen in the secondary metabolites, much less carbon is required for effective defense. Alkaloids and other nitrogen-containing secondary metabolites are common in many groups of marine invertebrates and in marine Cyanobacteria (Baker 1996, Harper et al. 2001, Blunt et al. 2006), but are rare in marine macroalgae. Several authors have suggested that the reason macroalgae make few such compounds is because nitrogen is commonly the major limiting factor for eukaryotic photoautotrophs in marine environments (Hay and Fenical 1988, Hay and Steinberg 1992, Cronin 2001). This idea is very much in agreement with predictions of the RAM. However, one place in the world where that is not true is the coastal waters of Antarctica, where benthic macroalgal habitats typically have high inorganic nitrogen concentrations throughout the year (Gordon et al. 1982, Harrison and Cota 1991, Drew and Hastings 1992). As noted previously, compared to algae from lower latitudes Antarctic macroalgae have relatively high tissue nitrogen and relatively low C:N ratios indicating that they are not growth-limited by nitrogen (Weykam et al. 1996, Dunton 2001, Peters et al. 2005). Paleoceanographic data from the Southern

Ocean indicate that the circulation patterns and other factors responsible for high nitrogen conditions have been present in coastal Antarctica for at least the past 15 million years (Kennett 1977, Kennett and Barker 1990, J. Kennett personal communication), so Antarctic macroalgal species have had ample time to evolve metabolically to the high nitrogen environment. Moreover, although Antarctic macroalgae are usually well adapted to low light conditions (Weykam et al. 1996, Gómez et al. 1997), light is commonly considered the major determinant of overall productivity in Antarctic algae (Holm-Hansen et al. 1977, Heywood and Whitaker 1984, Boyd 2002). Consequently, these communities are an ideal place to test the RAM prediction that selection will have favored the evolution of nitrogen-containing secondary metabolites in nitrogen-replete but carbon-limited species. Amsler et al. (2005a) tested this prediction by targeting isolations of nitrogen-containing metabolites in 23 common WAP macroalgae. Acid extracts were prepared with aqueous-HCl. The extracts were adjusted to pH 8.0 and then repeatedly re-extracted and partitioned with chloroform. The chloroform fraction should have quantitatively contained all alkaloids and similar nitrogen-containing secondary metabolites (Davidson et al. 1994). However, no such compounds were identified in these fractions by thin layer chromatography (Amsler et al. 2005a). In addition, other than nitrogen-containing UVR sunscreen compounds, which may or may not be considered "secondary metabolites" and which are commonly found in macroalgae from Antarctica and elsewhere (Chapter 13; Karsten 2008, Karsten et al. 2009), none of approximately 64 known secondary metabolites from Antarctic macroalgae (Blunt et al. 2006, Lebar et al. 2007) contains nitrogen. Although this does not mean that no nitrogen-containing secondary metabolites will be identified from Antarctic macroalgae in the future, it does indicate that they are not more common than in other, nitrogen-limited regions in contradiction to predictions of the RAM.

The optimal defense theory (ODT) makes a series of predictions, all based around the idea that organisms will allocate defenses in ways that maximize fitness, and all assuming that defenses come with some associated cost (Rhoades 1979). Reproductive structures are of obvious importance with respect to fitness and a number of studies have tested the ODT-based hypothesis that macroalgae devote relatively higher levels of defenses to reproductive tissues relative to somatic tissues (Amsler and Fairhead 2006, Pavia and Toth 2008). However, many macroalgae have a thallus construction in which reproductive and photosynthetic lateral branches are supported by one or a few main axes. Should a main axis break because of damage from herbivores, all the lateral branches supported above the break would be lost. Therefore, one might postulate under the ODT that these main axes should be relatively more heavily defended than any lateral branch. Taylor et al. (2002) and Pavia et al. (2002) tested this hypothesis in two species of temperate brown algae and reported that chemical or physical defenses were indeed stronger in the main axes. In Antarctica, the two brown algal species that dominate in shallower waters along the WAP, *Desmarestia anceps* and *D. menziesii,* are morphologically quite similar, but differ in that *D. anceps* usually has only a single main axis while *D. menziesii* usually has multiple main axes supporting the lateral branches (Wiencke and Clayton 2002). Based on the previous observations of allocation in other brown algae (Pavia et al. 2002, Taylor et al. 2002), Fairhead et al. (2005b) hypothesized that *D. anceps* should similarly devote relatively more resources to defense of its single main axis than of individual lateral branches.

Furthermore, they hypothesized that ODT would predict that there should be relatively less differences in allocation to defense between the multiple main axes of *D. menziesii* and its lateral branches. These hypotheses were supported as *D. anceps* had relatively higher levels of chemical defenses in its main axis than in its lateral branches, but *D. menziesii* did not (Fairhead et al. 2005b). In addition, in both species, the main axes were relatively tougher than the lateral branches. This additional toughness would likely have made the axes relatively less palatable to mesoherbivores than the lateral branches (cf. discussion above) although there are biomechanical reasons why an axis should be tougher than a lateral branch that are completely independent of herbivore defense (Fairhead et al. 2005b).

The induced defense model (IDM; Karban and Meyers 1989, Harvell 1990) is often considered a corollary of the ODT and predicts that production of defenses should be directly correlated with the risk of herbivore attack. If the likelihood of herbivory is high, the model predicts that defenses should be expressed constitutively. However, if herbivory is variable in time or space, then inducing the production of defenses only when herbivores are present may be advantageous. However, for macroalgae this is likely to be true only with mesoherbivores that generally cannot consume an entire alga over a short period as this allows time for an effective defense to be produced (Hay 1996). Fairhead et al. (2006) examined the potential for induced defenses against mesoherbivores in the Antarctic brown alga *Desmarestia menziesii* as well as the potential for induction of defenses against UVR in *D. anceps* and *D. menziesii.* A laboratory mesocosm experiment did not detect increases in the unpalatability of *D. menziesii* relative to ungrazed controls when it was held for 30 days with the amphipods *Gondogeneia antarctica* or *Prostebbingia gracilis*, or with the snail *Laevilacunaria antarctica* (Fairhead et al. 2006). Artificial wounding over the course of the experiment similarly did not induce an increase in unpalatability of *D. menziesii* relative to the ungrazed controls, and no differences in phlorotannin concentrations were detected in any of the treatments (Fairhead et al. 2006). Considering the high densities of mesograzers in communities dominated by *Desmarestia* spp. (cf. Chemical defenses against mesoherbivores section), this constitutive expression of defenses is consistent with IDM predictions. Indeed, the true "experiment" was probably looking for a relaxation of defenses in the ungrazed control as they were released from natural levels of mesoherbivory in the mesocosms.

Fairhead et al. (2006) also manipulated levels of UVR in an *in situ* experiment with *Desmarestia anceps* and *D. menziesii.* As noted in the section titled The potential roles of phlorotannins in defense against herbivores, phlorotannins can serve as ultraviolet screens (Amsler and Fairhead 2006) and this is also the case for polar macroalgae (Chapter 13; Karsten et al. 2009), and so it is possible that recent increases in UVR caused by stratospheric ozone depletion (Frederick et al. 1998) could be inducing higher levels of phlorotannins in Antarctic brown algae. Phlorotannins can serve as feeding deterrents against Antarctic herbivores but this varies on a species-specific basis both with herbivores and macroalgae (Chapter 6; Iken et al. 2009). Consequently, any UVR-induced increases in the phlorotannin contents of either of the dominant, shallow-growing *D. anceps* or *D. menziesii* would have the potential to alter trophic dynamics in these communities. However, no UVR-related changes in palatability or phlorotannin contents were observed in either *Desmarestia* sp. over the 6-week *in situ* experiment (Fairhead et al. 2006).

Defenses against biofoulers

As noted previously, to date there have been no studies of anti-biofouling defenses of polar macroalgae that have presented potential defensive metabolites in a manner that was known to be ecologically relevant. However, there have been efforts to identify Antarctic macroalgae that produce compounds that have the potential to act as antifoulants. Amsler et al. (2005b) screened extracts of two macroalgae from McMurdo Sound and 22 species from the WAP for toxicity to sympatric benthic diatoms. All the macroalgal species had at least one extract type that was bioactive when presented in solution at three-times the tissue-level concentration on a volumetric basis (Amsler et al. 2005b). Both McMurdo Sound species as well as 15 of the WAP species had extracts that were significantly bioactive at the tissue-level concentration or lower, including seven species in which extracts had previously been shown to have no antiherbivore activity (Amsler et al. 2005a,b). Sevak (2010) tested the bioactivity of MPLC-derived fractions of crude extracts from four of these species that had not displayed antiherbivore activity but did have anti-diatom activity. Although a few individual fractions had significant anti-diatom bioactivity at a concentration 10-fold higher than calculated tissue-levels, none had activity at the calculated tissue-level concentration. However, when the fractions were run together in combination, there was strong anti-diatom bioactivity at tissue-level concentrations in one macroalgal species and at 10-times tissue-level in two others (Sevak 2010). This indicates that the bioactivity observed by Amsler et al. (2005b) in the original crude extracts was likely the result of more than one secondary metabolite working additively or synergistically.

Phlorotannins in brown algae can have functions in antimicrobial and antifouling defense (Wilkström and Pavia 2004, Amsler and Fairhead 2006). Phlorotannin extracts from Antarctic brown algae were tested for their anti-diatom and antimicrobial effects (Chapter 6; Iken et al. 2009). Phlorotannins from seven brown algae, including all species that are ecologically dominant along the WAP, caused significant mortality in an Antarctic benthic diatom when bioassayed at natural or higher than natural tissue-level concentrations on a volumetric basis. At 10% of natural concentrations, of the seven species, only phlorotannins from *Desmarestia anceps* and *D. menziesii* failed to significantly reduce diatom survival (Chapter 6; Iken et al. 2009). The ecological significance of this seemingly strong antifouling activity against diatoms remains unclear; however, as the surface concentrations of phlorotannins that would be experienced by diatoms fouling these brown algae in nature are likely to be lower than the lowest whole-tissue level concentrations tested in this study (Chapter 6; Iken et al. 2009). Phlorotannins tested in the same study for antimicrobial activity against sympatric bacteria did not elicit any effects at natural tissue-level concentrations. Only ten-fold natural phlorotannin concentrations of some algal species caused bacterial growth inhibition. Antimicrobial properties were strongly algal- and strain-specific (Chapter 6; Iken et al. 2009). Only phlorotannins from *Desmarestia anceps* and *Phaeurus antarcticus* Skottsberg inhibited growth (at 10-times tissue concentration) in three and four of the four tested bacterial strains, respectively. The ecological significance of this antimicrobial activity at high phlorotannin concentrations is unclear. However, it may be ecologically relevant during wound healing (when phlorotannins can occur in high concentrations) to assist in closing of the wounded tissue (Lüder and Clayton 2004, Amsler and Fairhead 2006), and may thus also decrease the possibility of bacterial infection.

Conclusion

Chemical defenses against herbivory appear to be very important in the Antarctic communities studied to date but not in the only Arctic community examined. The reasons for this potential discrepancy are unknown, but, if the pattern holds when additional Arctic communities are studied, it could be a result of the very different evolutionary history of the Arctic and Antarctic (Smith 1990, Dayton et al. 1994). Clearly, more research into macroalgal-herbivore relationships is needed in the Arctic. Antarctic macroalgae are commonly defended against herbivory by both macroherbivores and mesoherbivores, and almost all the large, ecologically dominant macroalgal species elaborate chemical defenses against herbivory. It appears that chemical defenses play important roles in structuring Antarctic macroalgal communities, and the roles of macroalgal secondary metabolites in the ecology of these systems certainly deserves further attention. Mesoherbivores such as amphipods appear to be very important consumers in benthic Antarctic communities, and in this sense Antarctic macroalgal communities may be more analogous to terrestrial communities dominated by insects than to many other marine communities (cf. Hay et al. 1987). Antarctic macroalgae have proven to be useful models for testing and extending plant defense theories, and studies of their chemical defenses should continue to further our global understanding of the ecology and evolution of algal and plant resistance to herbivores and biofoulers. Antarctic macroalgae elaborate compounds that have the potential to function as defenses against biofouling diatoms and there is evidence in some species that if so, this may require multiple compounds working in combination.

References

Amsler, C.D. 2001. Induced defenses in macroalgae: the herbivore makes a difference. *J. Phycol. 37*: 353–356.

Amsler, C.D. (ed). 2008. *Algal chemical ecology.* Springer-Verlag, Berlin/Heidelberg. pp. xviii +313.

Amsler, C.D. and V.A. Fairhead. 2006. Defensive and sensory chemical ecology of brown algae. *Adv. Bot. Res. 43*: 1–91.

Amsler, C.D., R.J. Rowley, D.R. Laur, L.B. Quetin and R.M. Ross. 1995. Vertical distribution of Antarctic Peninsular macroalgae: cover, biomass, and species composition. *Phycologia 34*: 424–430.

Amsler, C.D., J.B. McClintock and B.J. Baker. 1998. Chemical defense against herbivory in the Antarctic marine macroalgae *Iridaea cordata* and *Phyllophora antarctica* (Rhodophyceae). *J. Phycol. 34*: 53–59.

Amsler, C.D., J.B. McClintock and B.J. Baker. 1999. An antarctic feeding triangle: defensive interactions between macroalgae, sea urchins, and sea anemones. *Mar. Ecol. Prog. Ser. 183*: 105–114.

Amsler, C.D., K.B. Iken, J.B. McClintock and B.J. Baker. 2001. Secondary metabolites from antarctic marine organisms and their ecological implications. *In*: (J.B. McClintock and B.J. Baker, eds), *Marine chemical ecology.* CRC Press, Boca Raton, FL. pp. 267–300.

Amsler, C.D., K. Iken, J.B. McClintock, M.O. Amsler, K.J. Peters, J.M. Hubbard, et al. 2005a. Comprehensive evaluation of the palatability and chemical defenses of subtidal macroalgae from the Antarctic Peninsula. *Mar. Ecol. Prog. Ser. 294*: 141–159.

Amsler, C.D., I.N. Okogbue, D.M. Landry, M.O. Amsler, J.B. McClintock and B.J. Baker. 2005b. Potential chemical defenses against diatom fouling in Antarctic macroalgae. *Bot. Mar. 48*: 318–322.

Amsler, C.D., J.B. McClintock and B.J. Baker. 2008. Macroalgal chemical defenses in polar marine communities. *In*: (C.D. Amsler, ed) *Algal chemical ecology*. Springer-Verlag, Berlin/ Heidelberg. pp. 91–103.

Amsler, C.D., M.O. Amsler, J.B. McClintock and B.J. Baker. 2009. Filamentous algal endophytes in macrophytic Antarctic algae: prevalence in hosts and palatability to mesoherbivores. *Phycologia 48*: 324–334.

Ankisetty, S., S. Nandiraju, H. Win, Y.C. Park, C.D. Amsler, J.B. McClintock, et al. 2004. Chemical investigation of predator-deterred macroalgae from the Antarctic Peninsula. *J. Nat. Prod. 67*: 1295–1302.

Aumack, C.F., C.D. Amsler, J.B. McClintock,and B.J. Baker. 2010. Chemically mediated resistance to mesoherbivory in finely branched macroalgae along the western Antarctic Peninsula. *Eur. J. Phycol. 45:* 19-26.

Avila, C., S. Taboada and L. Núñez-Pons. 2008. Antarctic marine chemical ecology: what is next? *Mar. Ecol. 29*: 1–71.

Baker, B.J. 1996. Carboline and isoquinoline alkaloids from marine organisms. *In*: (S.W. Pelletier, ed) *Alkaloids: chemical and biological perspectives, Volume 10*. Pergamon Press, Oxford. pp. 357–407.

Bhadury, P. and P.C. Wright. 2004. Exploitation of marine algae: biogenic compounds for potential antifouling applications. *Planta 219*: 561–578.

Blunt, J.W., B.R. Copp, M.H.G. Munro, P.T. Northcote and M.R. Prinsep. 2006. Marine natural products. *Nat. Prod. Rep. 23*: 26–78.

Boyd, P.W. 2002. Environmental factors controlling phytoplankton processes in the Southern Ocean. *J. Phycol. 38*: 844–861.

Brand, T.E. 1980. *Trophic interactions and community ecology of the shallow-water marine benthos along the Antarctic Peninsula*. PhD Dissertation, University of California Davis, Davis, CA, USA. pp. 220.

Brawley, S.H. 1992. Mesoherbivores. *In*: (D.M. John, S.J. Hawkins and J.H. Price, eds) *Plant-animal interactions in the marine benthos*. Clarendon Press, Oxford. pp. 235–263.

Bullard, S.B. and M.E. Hay. 2002. Palatability of marine macro-holoplankton: nematocysts, nutritional quality, and chemistry as defenses against consumers. *Limnol. Oceanogr. 47*: 1456–1467.

Buschmann, A.H. 1990. Intertidal macroalgae as refuge and food for amphipoda in central Chile. *Aquat. Bot. 36*: 237–245.

Chung, H., Y.S. Oh, I.K. Lee and D.-Y. Kim. 1994. Macroalgal vegetation of Maxwell Bay in King George Island, Antarctica. *Korean J. Phycol. 9*: 47–58.

Coley, P.D., J.B. Bryant and F.S. Chapin III. 1985. Resource availability and plant antiherbivore defense. *Science 230*: 895–899.

Cronin, G. 2001. Resource allocation in seaweeds and marine invertebrates. *In*: (J.B. McClintock and B.J. Baker, eds) *Marine chemical ecology*. CRC Press, Boca Raton, FL. pp. 325–363.

Cruz-Rivera, E. and M.E. Hay. 2003. Prey nutritional quality interacts with chemical defenses to affect consumer feeding and fitness. *Ecol. Monogr. 73*: 483–506.

Davidson, S., J.B. Hobbs, D.V. Banthorpe and J.B. Harbourne. 1994. *Natural Products, their chemistry and biological significance*. John Wiley and Sons, New York. pp. 455.

Dayton, P.K., B.J. Mordida and F. Bacon. 1994. Polar marine communities. *Am. Zool. 34*: 90–100.

De Broyer, C. and K. Jaždžewski. 1993. Contribution to the marine biodiversity inventory: a checklist of the Amphipoda (Crustacea) of the Southern Ocean. *Doc. Trav. Inst. R. Sci. Nat. Belg. 73*: 1–154.

De Broyer, C. and K. Jaždžewski. 1996. Biodiversity of the Southern Ocean: towards a new synthesis for the Amphipoda (Crustacea). *Bull. Mus. Cw. St. nat. Venova 20*: 547–568.

De Broyer, C., M. Rauschert and Y. Scailteur. 1999. Structural and ecofunctional biodiversity of the benthic amphipod taxocoenoses. (In: *The expedition ANT XV/3 (EASIZ II) of RV Polarstern in 1998*). *Ber. Polarforsch. 301*: 163–174.

Delépine, R., I.M. Lamb and M.H. Zimmerman. 1966. Preliminary report on the marine vegetation of the Antarctic Peninsula. *Proc. Int. Seaweed Symp. 5*: 107–116.

Drew, E.A. and R.M. Hastings. 1992. A year-round ecophysiological study of *Himantothallus grandifolius* (Desmarestiales, Phaeophyta) at Signy Island, Antarctica. *Phycologia 31*: 262–277.

Duffy, J. and M. Hay. 1991. Food and shelter as determinants of food choice by an herbivorous marine amphipod. *Ecology 72*: 1286–1298.

Dunton, K. 2001. $\delta^{15}N$ and $\delta^{13}C$ measurements of Antarctic Peninsula fauna: trophic relationships and assimilation of benthic seaweeds. *Am. Zool. 41*: 99–112.

Fairhead, V.A., C.D. Amsler, J.B. McClintock and B.J. Baker. 2005a. Variation in phlorotannin content within two species of brown macroalgae (*Desmarestia anceps* and *D. menziesii*) from the Western Antarctic Peninsula. *Polar Biol. 28*: 680–686.

Fairhead, V.A., C.D. Amsler, J.B. McClintock and B.J. Baker. 2005b. Within-thallus variation in chemical and physical defenses in two species of ecologically dominant brown macroalgae from the Antarctic Peninsula. *J. Exp. Mar. Biol. Ecol. 233*: 1–12.

Fairhead, V.A., C.D. Amsler, J.B. McClintock and B.J. Baker. 2006. Lack of defense or phlorotannin induction by UV radiation or mesograzers in *Desmarestia anceps* and *D. menziesii* (Phaeophyceae). *J. Phycol. 42*: 1174–1183.

Filion-Myklebust, C. and T.A. Norton. 1981. Epidermis shedding in the brown seaweed *Ascophyllum nodosum* (L.) Lejolis, and its ecological significance. *Mar. Biol. Lett. 2*: 45–51.

Frederick, J.E., Z. Qu and C.R. Booth. 1998. Ultraviolet radiation at sites on the Antarctic coast. *Photochem. Photobiol. 68*: 183–190.

Gómez, I., G. Weykam, H. Klöser and C. Wiencke. 1997. Photosynthetic light requirements, metabolic carbon balance and zonation of sublittoral macroalgae from King George Island (Antarctica). *Mar. Ecol. Prog. Ser. 148*: 281–293.

Gordon, A.L., E.J. Molinelli and T.N. Baker. 1982. *Southern Ocean atlas: thermohaline and chemical distributions and the atlas data set*. Columbia University Press, New York.

Hanley, M.E., B.B. Lamont, M.M. Fairbanks and C.M. Rafferty. 2007. Plant structural traits and their role in anti-herbivore defence. *Perspect. Plant Ecol. 8*: 157–178.

Harper, M.K., T.S. Bubni, B.R. Copp, R.D. James, B.S. Lindsay, A.D. Richardson, et al. 2001. Introduction to the chemical ecology of marine natural products. *In*: (J.B. McClintock and B.J. Baker, eds) *Marine chemical ecology*. CRC Press, Boca Raton, FL. pp. 3–69.

Harrison, W.G. and G.F. Cota. 1991. Primary production in polar waters: relation to nutrient availability. *Polar Res. 10*: 87–104.

Harvell, C.D. 1990. The ecology and evolution of inducible defenses. *Q. Rev. Biol. 65*: 323–340.

Hay, M.E. 1996. Marine chemical ecology: what's known and what's next? *J. Exp. Mar. Biol. Ecol. 200*: 103–134.

Hay, M.E. and W. Fenical. 1988. Marine plant-herbivore interactions: the ecology of chemical defense. *Annu. Rev. Ecol. Syst. 19*: 111–145.

Hay, M.E. and P.D. Steinberg. 1992. The chemical ecology of plant-herbivore interactions in marine versus terrestrial communities. *In*: (G.A. Rosenthal and M.R. Berenbaum, eds) *Herbivores: their interactions with secondary plant metabolites. Volume II. Evolutionary and ecological processes*. Academic Press, New York. pp. 371–413.

Hay, M.E., J.E. Duffy, C.A. Pfister and W. Fenical. 1987. Chemical defense against different marine herbivores: are amphipods insect equivalents? *Ecology 68*: 1567–1580.

Hedgpeth, J.W. 1969. Preliminary observations of life between tidemarks at Palmer Station, 64°45′S. 64°05′W. *Antarctic J. US 4*: 106–107.

Heywood, R.B. and T.M. Whitaker. 1984. The antarctic marine flora. *In*: (R.M. Laws, ed) *Antarctic ecology*. Vol. 2. Academic Press, London. pp. 373–419.

Holm-Hansen, O., S.Z. El-Sayed, G.A. Franceschini and R.L. Cuhel. 1977. Primary production and the factors controlling phytoplankton growth in the Southern Ocean. *In*: (G.A. Llano, ed.) *Adaptations within Antarctic ecosystems*. Smithsonian Institution, Washington, DC. pp. 11–50.

Hommersand, M.H., R.L. Moe, C.D. Amsler and S. Fredericq. 2009. Notes on the systematics of Antarctic and sub-Antarctic Rhodophyta with descriptions of four new genera and five new species. *Bot. Mar. 52*: 509–534.

Horn, M. and M. Neighbors. 1984. Protein and nitrogen assimilation as a factor in predicting the seasonal macroalgal diet of the monkeyface prickleback. *Trans. Am. Fish. Soc. 113*: 388–396.

Huang, Y.M., J.B. McClintock, C.D. Amsler, K.J. Peters and B.J. Baker. 2006. Feeding rates of common Antarctic gammarid amphipods on ecologically important sympatric macroalgae. *J. Exp. Mar. Biol. Ecol. 329*: 55–65.

Huang, Y.M., M.O. Amsler, J.B. McClintock, C.D. Amsler and B.J. Baker. 2007. Patterns of gammarid amphipod abundance and species composition associated with dominant subtidal macroalgae along the western Antarctic Peninsula. *Polar Biol. 30*: 1417–1430.

Iken, K. 1996. Trophic relations between macroalgae and herbivores in Potter Cove (King George Island, Antarctica). *Ber. Polarforsch. 201*: 1–206.

Iken, K. 1999. Feeding ecology of the Antarctic herbivorous gastropod *Laevilacunaria antarctica* Martens. *J. Exp. Mar. Biol. Ecol. 236*: 133–148.

Iken, K., E.R. Barrera-Oro, M.L. Quartino, R.J. Casaux and T. Brey. 1997. Grazing in the Antarctic fish *Notothenia coriiceps*: evidence for selective feeding on macroalgae. *Antarctic Sci. 9*: 386–391.

Iken, K., M. Quartino and C. Wiencke. 1999. Histological identification of macroalgae from stomach contents of the Antarctic fish *Notothenia coriiceps* using semi-thin sections. *Mar. Ecol. 20*: 11–17.

Iken, K., C.D. Amsler, J.M. Hubbard, J.B. McClintock and B.J. Baker. 2007. Allocation patterns of phlorotannins in Antarctic brown algae. *Phycologia 46*: 386–395.

Iken, K., C.D. Amsler, M.O. Amsler, J.B. McClintock and B.J. Baker. 2009. Field studies on deterrent properties of phlorotannins in Antarctic brown algae. *Bot. Mar. 52*: 547–557.

Jormalainen, V. and T. Honkanen. 2008. Macroalgal chemical defenses and their roles in structuring temperate marine communities. *In*: (C.D. Amsler, ed) *Algal chemical ecology*. Springer-Verlag, Berlin/Heidelberg. pp. 57–89.

Karban, R. and J.H. Meyers. 1989. Induced plant responses to herbivory. *Annu. Rev. Ecol. Syst. 20*: 331–348.

Karsten, U. 2008. Defensive strategies of algae and cyanobacteria against solar ultraviolet radiation. *In*: (C.D. Amsler, ed) *Algal chemical ecology*. Springer-Verlag, Berlin/Heidelberg. pp. 273–296.

Karsten, U., C. Wiencke and G.O. Kirst. 1990. The effect of light-intensity and daylength on the β-dimethylsulphoniopropionate (DMSP) content of marine green macroalgae from Antarctica. *Plant Cell Environ. 13*: 989–993.

Karsten, U., G.O. Kirst and C. Wiencke. 1992. Dimethylsulphoniopropionate (DMSP) accumulation in green macroalgae from polar to temperate regions: interactive effects of light versus salinity and light versus temperature. *Polar Biol. 12*: 603–607.

Karsten, U., A. Wulff, M.Y. Roleda, R. Müller, F.S. Steinhoff, J. Fredersdorf and C. Wiencke. 2009. Physiological responses of polar benthic micro- and macroalgae to ultraviolet radiation. *Bot. Mar. 52*: 639–654.

Keats, D.W., M.A. Knight and C.M. Pueschel. 1997. Antifouling effects of epithallial shedding in three crustose coralline algae (Rhodophyta, Corallinales) on a coral reef. *J. Exp. Mar. Biol. Ecol. 213*: 281–293.

Kennett, J. 1977. Cenozoic evolution of antarctic glaciation, the circum-antarctic ocean, and their impact on global paleoceanography. *J. Geophys. Res. 82*: 3843–3860.

Kennett, J. and P. Barker. 1990. Latest Cretaceous to Cenozoic climate and oceanographic developments in the Weddell Sea, Antarctica: an ocean-drilling perspective. *Proc. Ocean Drill. Program Sci. Results 113*: 937–960.

Kim, D. 2001. Seasonality of marine algae and grazers of an Antarctic rocky intertidal, with emphasis on the role of the limpet *Nacella concinna* Strebel (Gastropoda: Patellidae). *Ber. Polarforsch. Meeresforsch. 397*: 1–120.

Lane, A.L. and J. Kubanek. 2008. Secondary metabolite defenses against pathogens and biofoulers. *In*: (C.D. Amsler, ed) *Algal chemical ecology.* Springer-Verlag, Berlin/Heidelberg, pp. 229–243.

Lebar, M.L., J.L. Heimbegner and B.J. Baker. 2007. Cold-water marine natural products. *Nat. Prod. Rep. 24*: 774–797.

Lippert, H., K. Iken, E. Rachor and C. Wiencke. 2001. Macrofauna associated with macroalgae in the Kongsfjord (Spitsbergen). *Polar Biol. 24*: 512–522.

Lüder, U.H. and M.N. Clayton. 2004. Induction of phlorotannins in the brown macroalga *Ecklonia radiata* (Laminariales, Phaeophyta) in response to simulated herbivory – the first microscopic study. *Planta 218*: 928–937.

Maschek, J.A. and B.J. Baker. 2008. The chemistry of algal secondary metabolism. *In*: (C.D. Amsler, ed) *Algal chemical ecology.* Springer-Verlag, Berlin/Heidelberg. pp. 1–24.

Mattson, W.J. 1980. Herbivory in relation to plant nitrogen. *Annu. Rev. Ecol. Syst. 11*: 119–161.

Miller, K.A. and J.S. Pearse. 1991. Ecological studies of seaweeds in McMurdo Sound, Antarctica. *Am. Zool. 31*: 35–48.

Molis, M., H. Wessels, W. Hagen, U. Karsten and C. Wiencke. 2008. Do sulphuric acid and the brown alga *Desmarestia viridis* support community structure in Arctic kelp patches by altering grazing impact, distribution patterns, and behaviour of sea urchins? *Polar Biol. 32*: 71–82.

Moss, B. 1982. The control of epiphytes by *Halidrys siliquosa* (L.) Lyngb. (Phaeophyta, Cystoseiraceae). *Phycologia 21*: 185–191.

Nelson, W.G. 1980. A comparative study of amphipods in seagrasses from Florida to Nova Scotia. *Bull. Mar. Sci. 30*: 80–89.

Nylund, G.M. and H. Pavia. 2005. Chemical versus mechanical inhibition of fouling in the red alga *Dilsea carnosa. Mar. Ecol. Prog. Ser. 299*: 111–121.

Paul, V.J., E. Cruz-Rivera and R.W. Thacker. 2001. Chemical mediation of macroalgal-herbivore interactions: ecological and evolutionary perspectives. *In*: (J.B. McClintock and B.J. Baker, eds), *Marine chemical ecology.* CRC Press, Boca Raton, FL. pp. 227–265.

Pavia, H. and G.B. Toth. 2008. Macroalgal models in testing and extending defense theories. *In*: (C.D. Amsler, ed) *Algal chemical ecology.* Springer-Verlag, Berlin/Heidelberg. pp. 147–172.

Pavia, H., G.B. Toth and P. Åberg. 2002. Optimal defense theory: elasticity analysis as a tool to predict intraplant variation in defenses. *Ecology 83*: 891–897.

Pelletreau, K.N. and G. Muller-Parker. 2002. Sulfuric acid in the phaeophyte alga *Desmarestia munda* deters feeding by the sea urchin *Strongylocentrotus droebachiensis. Mar. Biol. 141*: 1–9.

Pennings, S.C. and V.J. Paul. 1992. Effect of plant toughness, calcification, and chemistry on herbivory by *Dolabella auricularia. Ecology 73*: 1606–1619.

Pennings, S.C., M.P. Puglisi, T.J. Pitlik, A.C. Himaya and V.J. Paul. 1996. Effects of secondary metabolites and $CaCO_3$ on feeding by surgeonfishes and parrotfishes: within-plant comparisons. *Mar. Ecol. Prog. Ser. 134*: 49–58.

Pereira, R.C. and B.A.P. da Gama. 2008. Macroalgal chemical defenses and their roles in structuring tropical marine communities. *In*: (C.D. Amsler, ed) *Algal chemical ecology.* Springer-Verlag, Berlin/Heidelberg. pp. 25–55.

Peters, A.F. 2003. Molecular identification, taxonomy and distribution of brown algal endophytes, with emphasis on species from Antarctica. *Proc. Int. Seaweed Symp. 17*: 293–302.

Peters, K.J., C.D. Amsler, M.O. Amsler, J.B. McClintock, R.B. Dunbar and B.J. Baker. 2005. A comparative analysis of the nutritional and elemental composition of macroalgae from the western Antarctic Peninsula. *Phycologia 44*: 453–463.

Picken, G.B. 1979. Growth, production and biomass of the antarctic gastropod *Laevilacunaria antarctica* Martens 1885. *J. Exp. Mar. Biol. Ecol. 40*: 71–79.

Potin, P. 2008. Oxidative burst and related responses in biotic interactions of algae. *In*: (C.D. Amsler, ed) *Algal chemical ecology*. Springer-Verlag, Berlin/Heidelberg. pp. 245–271.

Rhoades, D. 1979. Evolution of plant chemical defenses against herbivores. *In*: (G.A. Rosenthal and D.H. Janzen, eds) *Herbivores*. Academic Press, New York. pp. 4–54.

Richardson, M.G. 1971. The ecology and physiological aspects of Antarctic weed dwelling amphipods (Preliminary report, II). *British Antarctic Survey Report N9/1971(-72)/H*: 1–16.

Richardson, M.G. 1977. *The ecology including physiological aspects of selected Antarctic marine invertebrates associated with inshore macrophytes*. PhD Dissertation, University of Durham, Durham, UK. pp. 165.

Schwarz, A.M., I. Hawes, N. Andrew, A. Norkko, V. Cummings and S. Thrush. 2003. Macroalgal photosynthesis near the southern global limit for growth; Cape Evans, Ross Sea, Antarctica. *Polar Biol. 26*: 789–799.

Sevak, H.P. 2010. *Potential chemical defenses against diatom fouling in macroalgae from the Antarctic Peninsula: insights from bioassay guided fractionation*. MS Thesis, University of Alabama at Birmingham, Birmingham, AL, USA. pp. 40.

Smith, W.O., Jr (ed). 1990. *Polar oceanography. Part B. Chemistry, biology, and geology*. Academic Press, San Diego, CA. pp. 759.

Steinberg, P.D. and R. de Nys. 2002. Chemical mediation of colonization of seaweed surfaces. *J. Phycol. 38*: 621–629.

Stowe, K.A., R.J. Marquis, C.G. Hochwender and E.L. Simms. 2000. The evolutionary ecology of tolerance to consumer damage. *Annu. Rev. Ecol. Syst. 31*: 565–595.

Strauss, S.Y. and A.A. Agrawal. 1999. The ecology and evolution of plant tolerance to herbivory. *Trends Ecol. Evol. 14*: 179–185.

Taylor, R.B., E. Sotka and M.E. Hay. 2002. Tissue-specific induction of herbivore resistance: seaweed response to amphipod grazing. *Oecologia 132*: 68–76.

Van Alstyne, K.L. 2008. Ecological and physiological roles of dimethylsulfoniopropionate and its products in marine macroalgae. *In*: (C.D. Amsler, ed) *Algal chemical ecology*. Springer-Verlag, Berlin/Heidelberg. pp. 173–194.

Van Alstyne, K.L., M.N. Dethier and D.O. Duggins. 2001. Spatial patterns in macroalgal chemical defenses. *In*: (J.B. Mc-Clintock and B.J. Baker, eds) *Marine chemical ecology*. CRC Press, Boca Raton, FL. pp. 301–324.

Veit-Köhler, G. 2005. Influence of biotic and abiotic sediment factors on abundance and biomass of harpacticoid copepods in a shallow Antarctic bay. *Sci. Mar. 69 (Suppl. 2)*: 135–145.

Wahl, M. 1989. Marine epibiosis. I. Fouling and antifouling, some basic aspects. *Mar. Ecol. Prog. Ser. 58*: 175–189.

Watson, D.C. and T.A. Norton. 1985. The physical characteristics of seaweed thalli as deterrents to littorine grazers. *Bot. Mar. 28*: 383–387.

Weinberger, F. 2007. Pathogen-induced defense and innate immunity in macroalgae. *Biol. Bull. 213*: 290–302.

Wessels, H., W. Hagen, M. Molis, C. Wiencke and U. Karsten. 2006. Intra- and interspecific differences in palatability of Arctic macroalgae from Kongsfjorden (Spitsbergen) for two benthic sympatric invertebrates. *J. Exp. Mar. Biol. Ecol. 329*: 20–33.

Weykam, G., I. Gómez, C. Wiencke, K. Iken and H. Kloser. 1996. Photosynthetic characteristics and C:N ratios of macroalgae from King George Island (Antarctica). *J. Exp. Mar. Biol. Ecol. 204*: 1–22.

Wiencke, C. and M.N. Clayton. 2002. *Antarctic seaweeds*. ARG Gantner Verlag, KG Ruggell. pp. 239.

Wiencke, C., M.N. Clayton, I. Gómez, K. Iken, U.H. Lüder, C.D. Amsler, et al. 2007. Life strategy, ecophysiology and ecology of seaweeds in polar waters. *Rev. Environ. Sci. Biotechnol. 6*: 95–126.

Wildish, D.J. 1988. Ecology and natural history of aquatic Talitroidea. *Can. J. Zool. 66*: 2340–2359.

Wilkström, S.A. and H. Pavia. 2004. Chemical settlement inhibition versus post-settlement mortality as an explanation for differential fouling of two congeneric seaweeds. *Oecologia 138*: 223–230.

Wright, J.T., R. de Nys, A.G.B. Poore and P.D. Steinberg. 2004. Chemical defense in a marine alga: heritability and the potential for selection by herbivores. *Ecology 85*: 2946–2959.

Wulff, A., K. Iken, M. Quartino, A. Al-Handal, C. Wiencke and M.N. Clayton. 2009. Biodiversity, biogeography and zonation of benthic micro- and macroalgae in the Arctic and Antarctic. *Bot. Mar. 52*: 491–507.

Zamzow, J.P., C.D. Amsler, J.B. McClintock, and B.J. Baker. 2010. Habitat choice and predator avoidance by Antarctic amphipods: the roles of algal chemistry and morphology. *Mar. Ecol. Prog. Ser. 400:* 155–163.

6 Field studies on deterrent properties of phlorotannins in Antarctic brown algae

Katrin Iken, Charles D. Amsler,
Margaret O. Amsler, James B. McClintock
and Bill J. Baker

Introduction

Benthic communities of the shallow coastal areas along the western Antarctic Peninsula are characterized by large macroalgal stands dominated in biomass by large brown algae (Chapter 3; Klöser et al. 1994, 1996, Amsler et al. 1995, Quartino et al. 2001, 2005, Wulff et al. 2009). These brown algae contribute substantially to nearshore food webs (Dunton 2001, Corbisier et al. 2004), but seem to do so predominantly through detrital pathways. Antarctic brown algae are directly consumed by only a few organisms (Iken 1996, Amsler et al. 2005a), even though a variety of grazers occur associated with the macroalgal stands (Richardson 1975, 1977, Iken 1996, Iken et al. 1997, 1999, Huang et al. 2006, 2007). Several grazers, such as the fish *Notothenia coriiceps* Richardson, feed directly on macroalgae (Richardson 1975, Moreno and Zamorano 1980, Linkowski et al. 1983, Barrera-Oro and Casaux 1990, Iken et al. 1999) and macroalgae can constitute about 40% of its diet (Iken et al. 1997). Nevertheless, direct grazing is low overall and we have recently demonstrated that Antarctic macroalgae are commonly chemically defended against grazing (Chapter 5; Amsler et al. 2005a, 2008, 2009).

The most prominent secondary compounds in brown algae are phlorotannins, polymers of phloroglucinol units (Ragan and Glombitza 1986, Steinberg 1992, Van Alstyne et al. 2001, Amsler and Fairhead 2006). Their typical molecular size ranges between less than 3 kDa up to more than 100 kDa (Van Altena and Steinberg 1992, Boettcher and Targett 1993). Besides primary metabolic functions as cell-wall strengtheners (Schoenwaelder and Clayton 1999) and in wound healing (Lüder and Clayton 2004), phlorotannins are also assumed to have important secondary functions. These include antimicrobial and antifoulant properties (Chapter 5; Ragan and Glombitza 1986, Wikström and Pavia 2004, Jormalainen et al. 2008, Amsler et al. 2009), protection against UV radiation (Chapter 13; Pavia et al. 1997, Karsten et al. 2009), and, most commonly, resistance against herbivores (Chapter 5; Amsler et al. 2009, Hay and Steinberg 1992, Steinberg 1992, Targett and Arnold 1998, 2001).

The results of past studies on the ecological properties of phlorotannins, particularly with regard to grazer defense, have been mixed and are sometimes controversial (Amsler and Fairhead 2006). These inconsistencies have been attributed to the high variability of phlorotannins on multiple spatial levels, ranging from small-scale differences among thallus parts within an alga (e.g., Tugwell and Branch 1989, Pavia et al. 2003, Fairhead et al. 2005a, Iken et al. 2007) up to global latitudinal and hemispherical patterns (Steinberg 1989, Van Alstyne et al. 2001). In addition, qualitative differences in phlorotannins of different algal species are likely to contribute to variable grazer responses (Ragan and Glombitza 1986, Steinberg 1988, Targett and Arnold 2001). In some cases, grazer deterrence has been attributed to compounds other than phlorotannins in brown algal extracts (e.g., Deal et al. 2003). It is clear that observed variations in grazer defense cannot be consistently explained by phlorotannin concentrations or co-evolutionary patterns with herbivores (Targett et al. 1992, Amsler and Fairhead 2006).

We have recently shown that Antarctic brown algae contain phlorotannins at concentrations between 0.5 to 9% of algal dry weight, which is relatively high compared to other cold-temperate or tropical brown algae (Fairhead et al. 2005a, Iken et al. 2007). Our chemical investigations for other secondary metabolites showed that only very few Antarctic brown algae contain any secondary metabolites other than phlorotannins (Ankisetty et al. 2004, Amsler et al. 2005a, Lebar et al. 2007). These phlorotannins are differentially distributed among different thallus parts in terms of concentration and phlorotannin size classes (Iken et al. 2007). Such unequal allocation of phlorotannins to different thallus parts may be an adaptation for increased protection of valuable thallus parts from grazing if phlorotannins have defensive properties against herbivores (Fairhead et al. 2005b). However, so far, very little is known about any possible ecological functions of Antarctic phlorotannins. A recent study could not confirm a UV-protective function of phlorotannins as experimentally manipulated UV levels did not elicit any changes in phlorotannin levels in the Antarctic *Desmarestia anceps* Montagne and *D. menziesii* J. Agardh (Fairhead et al. 2006). Similarly, increased grazer abundance did not induce increased phlorotannin production in *D. menziesii* (Fairhead et al. 2006). Yet, this does not rule out possible grazer deterrence or other defensive functions of Antarctic phlorotannins as constitutive levels may already be sufficiently high to fulfill such functions. The main goal of the present study was to examine the role of Antarctic brown algal phlorotannins as prospective antifeedants in sympatric omnivorous grazers as well as their ability to inhibit growth in sympatric bacteria and diatoms and thus serve as antifoulants. This is part of a larger aim to better understand the competitive success of brown algae in Antarctic shallow-water ecosystems.

Our approach and the obtained results

Brown algal collections

Brown algae were collected in the periods March–May 2000 and November–December 2001 by SCUBA in the vicinity of Palmer Station (US Antarctic Program) on the southwest coast of Anvers Island (64°46' S, 64°04' W; see Amsler et al. 1995 for map). Algal species included *Desmarestia anceps*, *D. menziesii*, *D. antarctica* Moe *et* Silva, *Himantothallus grandifolius* (A. *et* E. Gepp) Zinova, *Phaeurus antarcticus* Skottsberg (all Desmarestiales), *Cystosphaera jacquinotii* (Montagne) Skottsberg (Fucales), *Ascoseira mirabilis* Skottsberg (Ascoseirales), and *Adenocystis utricularis* (Bory) Skottsberg (Chordariales). Sampling depths in the nearshore areas of islands within a 3.5 km radius of the station ranged from 1 to 37 m, depending on the depth distribution range of each species. The influence of spatial and temporal variability in phlorotannin content (Fairhead et al. 2005a) was reduced by collecting replicate individuals from the same location on the same date.

Phlorotannin extractions

Soluble phlorotannins were extracted as described in detail in Iken et al. (2007), following established, standard procedures for phlorotannin extraction (Ragan and Jensen 1977, Ragan and Glombitza 1986, Van Alstyne 1995, Stern et al. 1996a). Phlorotannin-rich extracts (fish assays in 2000) and isolated phlorotannins (all bioassays in 2001) were used in bioassays. Briefly, for phlorotannin-rich extracts, brown algae were extracted in 80% MeOH and these extracts then partitioned three times with hexane to remove most non-phlorotannin components. Koivikko et al. (2005) recently demonstrated that 70% acetone may be a more powerful solvent to extract phlorotannins, so phlorotannin concentrations established in our study are likely conservative. In 2001, all bioassays were conducted with isolated phlorotannins obtained from species-specific standards that were produced for quantifications using the 2,4-dimethoxybenzaldehyde method (DMBA, Stern et al. 1996a). For this, 80% MeOH algal extracts were purified with toluene and acetone on a microcellulose column. The microcellulose treatment is an important step in phlorotannin isolation because few other typical natural products will absorb to this matrix as phlorotannins do. The only other potentially bioactive compounds that we discovered in our extensive chemical analyses of these Antarctic brown algae, menzoquinone in *Desmarestia menziesii* and cystosphaerol in *Cystosphaera jacquinotii* (Ankisetty et al. 2004, Lebar et al. 2007), are excluded by the microcellulose purification. Hence, the dried acetone fraction, which was used in bioassays in 2001, should have contained purified phlorotannins

Table 6.1 Phlorotannin yields (natural concentrations of isolated phlorotannins) used in feeding, fouling and microbial assays and overview of bioassay results with various grazers and of antimicrobial and antifouling activities.

Algal species	Phlorotannin concentration (isolated; mg ml^{-1})	Fish deterrence (enriched)	Fish deterrence (isolated)	Sea star deterrence	Amphipod deterrence	Antimicrobial activity*	Antifouling activity**
Desmarestia anceps	41.3	+	–	–	+	+	–
Desmarestia menziesii	39.7	+	I	+	–	+	–
Desmarestia antarctica	nd	–	nd	nd	nd	nd	+
Himantothallus grandifolius	47.1	+	–	+	–	–	+
Cystosphaera jacquinotii	49.8	+	+	–	+	+	+
Ascoseira mirabilis	23.1	+	–	–	–	+	+
Adenocystis utricularis	23.5	nd	–	–	–	nd	nd
Phaeurus antarcticus	31.9	nd	–	–	–	+	+

* Activity against at least one of the bacterial strains tested at 10 × phlorotannin concentrations. Results for strain 11.5 are excluded because of weak inhibition. For details see Table 6.2.
** Results for 0.1 × phlorotannin concentrations. nd = not determined.

(Stern et al. 1996a). While analytical limitations in the field prevented us from testing the purity of these extracts, NMR analysis of one of the extracts (*Phaeurus antarcticus*) showed that it contained >85% phlorotannins. Similar purity of phlorotannins was assumed for other extracts.

Phlorotannin concentrations used in bioassays were referenced to natural concentrations determined on a yield per volume algal mass basis (see Iken et al. 2007) (Table 6.1). Phlorotannins were dissolved in a small amount of methanol and dried under reduced pressure in a rotary evaporator onto a feeding stimulant (see Amsler et al. 2005a). Feeding stimulant was a dried powder of the palatable green alga *Cladophora repens* (J. Agardh) Harvey, or the red alga *Palmaria decipiens* (Reinsch) Ricker (the latter for fish assays in 2001 because of insufficient *C. repens*). Feeding stimulant algal powder with (treatment) or without (control) phlorotannin coating was then added to warm 2% alginate and gelatinized with cold 1 M $CaCl_2$ (McClintock and Baker 1997) to form pellets.

Feeding assays

Feeding assays were conducted with the sympatric omnivorous fish *Notothenia coriiceps* (in 2000 and 2001), the omnivorous sea star *Odontaster validus* Koehler, and the primarily herbivorous amphipod *Gondogeneia antarctica* Chevreux (both in 2001). All bioassay organisms were collected using SCUBA from the same general environment as the algae. For details on feeding assay designs, see Amsler et al. (2005a) where we report on the palatability of brown algal thalli and other algal extracts using the same bioassay designs.

For fish assays, 12 *Notothenia coriiceps* individuals were acclimated in separate tanks for several days. For feeding trials, fish were offered either a phlorotannin treatment food pellet or a solvent control food pellet, and, subsequently (for both), another control pellet to check satiation of the fish (i.e., to rule out that treatment pellets were rejected because the fish was not hungry). Typically, food items that are distasteful to the fish are mouthed but then rejected immediately (Amsler et al. 2005a). Rejected food pellets were offered again to the fish to ensure that the same reaction occurred. Only those trials where fish accepted the satiation control food pellets were used in statistical analyses. No fish was used more than once for testing the phlorotannin effects of a particular algal species ($n = 12$ per algal species in both years). Differences between phlorotannin treatment and solvent control pellet consumption were analyzed using Fisher's Exact Test (Sokal and Rohlf 1995). In the *Notothenia coriiceps* assays in 2000 with phlorotannin-rich extracts, all extracts except that of *Desmarestia antarctica* elicited a significant ($p \leq 0.001$) rejection of food pellets compared to control pellets (Figure 6.1A). In contrast, isolated phlorotannins in assays in 2001 did not deter fish feeding significantly, except for *Desmarestia menziesii* and *Cystosphaera jacquinotii* (both $p < 0.01$, Figure 6.1B). In both years, irrespective of feeding stimulant used, fish readily consumed control pellets.

Phlorotannin palatability for the sea star *Odontaster validus* was tested according to an assay design described by McClintock and Baker (1997) and Amsler et al. (2005a). Briefly, sea stars tend to crawl up the sides of the holding tanks and stretch out 1–2 arms on the water surface so that the ambulacral groove becomes exposed. Treatment pellets containing isolated phlorotannins or control pellets were placed one at a time within the ambulacral groove of one arm of a sea star and we observed

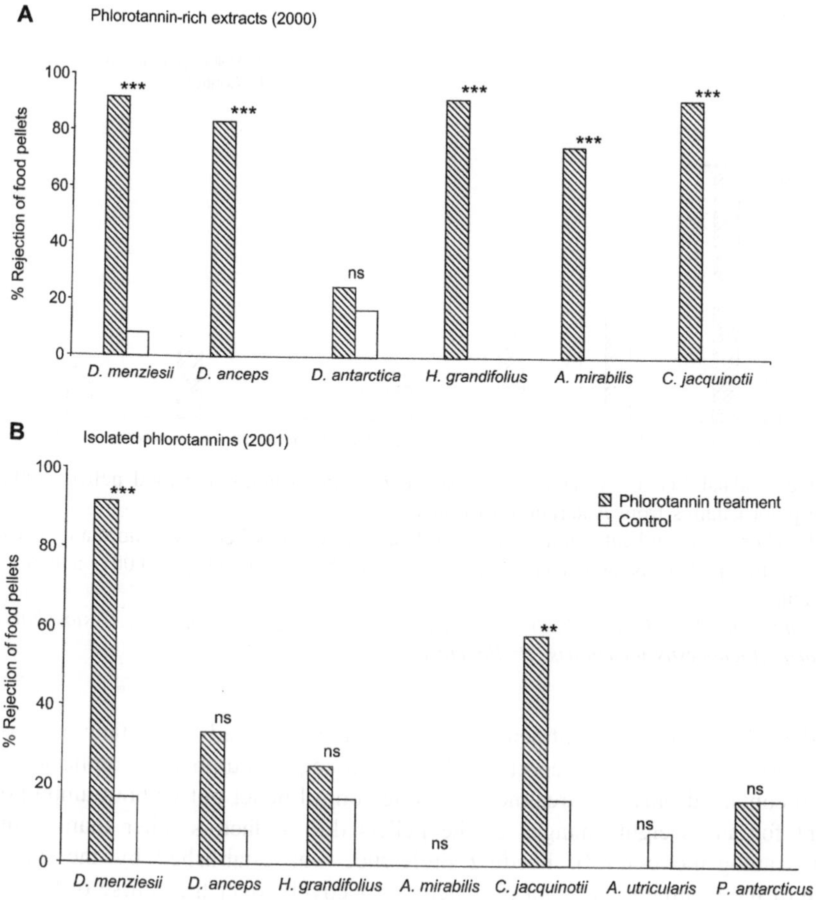

Figure 6.1 Feeding assays with the fish *Notothenia coriiceps* and phlorotannin-rich extracts (A) and isolated phlorotannins (B) of Antarctic brown algae.
Asterisks above bars indicate statistical significance of Fisher's exact test with $^{**}p \leq 0.01$, $^{***}p \leq 0.001$, and ns = not significant.
D.: *Desmarestia, H.*: *Himantothallus, A. mirabilis*: *Ascoseira mirabilis, C.*: *Cystosphaera, A. utricularis*: *Adenocystis utricularis, P.*: *Phaeurus*.

whether the sensory tube feet would pass the offered item towards the mouth (acceptance) or push it off the arm (rejection). Twelve sea stars that subsequently accepted satiation controls were used for phlorotannins and solvent controls from each algal species. Differences between rejection of treatment and solvent control pellets were analyzed using Fisher's exact test (Sokal and Rohlf 1995). The sea star *Odontaster validus* significantly rejected more pellets containing isolated phlorotannins of *Desmarestia menziesii* and *Himantothallus grandifolius* ($p < 0.0001$ and $p = 0.0186$, respectively) than control pellets (Figure 6.2). None of the phlorotannins isolated from the other brown algae had a deterrent effect on the sea stars.

For amphipod assays, 20 *Gondogeneia antarctica* individuals were placed in each of ten 250 ml bottles. Amphipods in each of the ten replicates were simultaneously

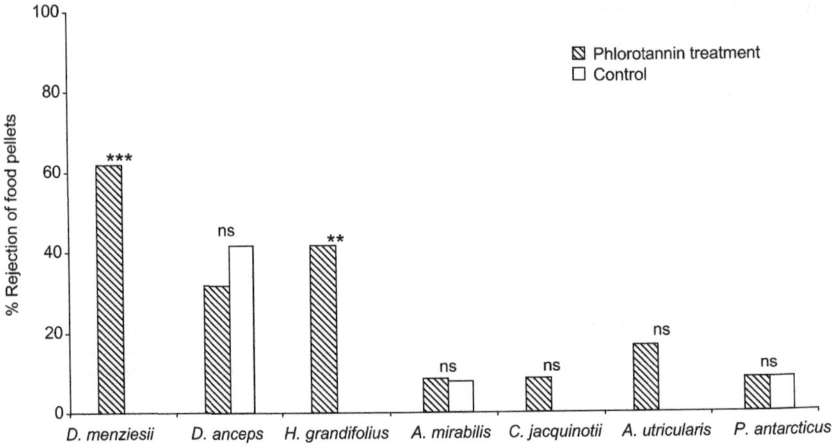

Figure 6.2 Palatability assay with the sea star *Odontaster validus* and food pellets containing isolated phlorotannins from Antarctic brown algae.
Asterisks above bars indicate statistically significant differences between phlorotannin test and solvent control pellets using Fisher's exact test with $^{**}p \leq 0.01$, $^{***}p \leq 0.001$, and ns = not significant.
D.: *Desmarestia, H.*: *Himantothallus, A. mirabilis*: *Ascoseira mirabilis, C.*: *Cystosphaera, A. utricularis*: *Adenocystis utricularis, P.*: *Phaeurus*.

offered a choice of one treatment pellet containing isolated phlorotannins and one solvent control pellet (Amsler et al. 2005a). An equal number of replicate bottles ($n = 10$) contained only one treatment and one control pellet but without amphipods to account for any weight changes of the pellets due to factors other than amphipod feeding (autogenic control). Feeding preference was established by the wet mass change of the treatment and control pellets, respectively, over the course of the experiment. Weight changes of control as well as treatment pellets from bottles with amphipods were adjusted by the percent change in the autogenic control and treatment pellets, respectively. These corrections were treated in a paired fashion, i.e., pellets from one amphipod-containing bottle were adjusted by the autogenic weight changes from one of the bottles without amphipods. Significance of differences between the changes in the ten paired phlorotannin treatments and the ten paired controls was compared using a Wilcoxon signed rank test (SPSS software, SPSS Inc.). Amphipod (*Gondogeneia antarctica*) feeding assays with isolated phlorotannins yielded widely varying results (Figure 6.3). Phlorotannins of *Desmarestia anceps* and *Cystosphaera jacquinotii* significantly deterred amphipod feeding ($p = 0.047$ and $p = 0.005$, respectively). The addition of isolated phlorotannins of *Ascoseira mirabilis*, *Adenocystis utricularis* and *Phaeurus antarcticus* to food pellets, however, elicited significantly higher feeding rates on these treatment than on control pellets ($p = 0.005$ for *A. mirabilis* and *A. utricularis*, and $p = 0.009$ for *P. antarcticus*). Feeding on phlorotannin treatments of the other algal species was not statistically different from that on controls (Figure 6.3).

To test the actual availability of phlorotannins to the bioassay organisms after incorporation into alginate, we re-extracted haphazardly-chosen alginate pellets containing

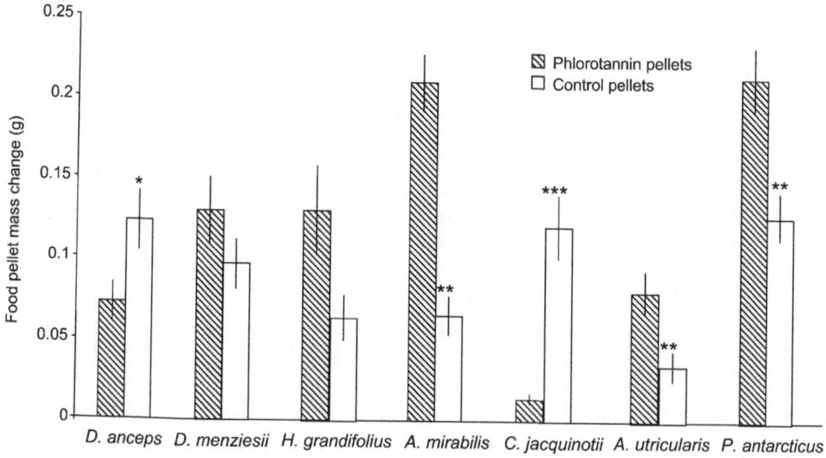

Figure 6.3 Mean (±SD) change in food pellet mass containing either isolated phlorotannins of Antarctic brown algae as treatment or solvent as control after feeding of the amphipod *Gondogeneia antarctica.*
Asterisks above bars indicate statistical difference between treatment and control pellets using Wilcoxon signed rank test with $*p \leq 0.05$, $**p \leq 0.01$, and $***p \leq 0.001$.
D.: *Desmarestia, H.*: *Himantothallus, A. mirabilis*: *Ascoseira mirabilis, C.*: *Cystosphaera, A. utricularis*: *Adenocystis utricularis, P.*: *Phaeurus.*

known amounts of phlorotannins (of *Desmarestia anceps* and *Cystosphaera jacquinotii*) with 80% MeOH. Phlorotannin concentrations were quantified as absorption at 510 nm against known quantities of species-specific standards (DMBA method, Stern et al. 1996a, Iken et al. 2007). Re-extraction of isolated phlorotannins from food pellets indicated an approximately 30% loss of detectable phlorotannins after incorporation into alginate. It is likely that parts of the phlorotannins bound to the alginate matrix or were exuded into the surrounding water and were thus no longer reactive in colorimetric assays.

Antifouling diatom assays

A chain-forming, araphid diatom (McM-C5) of the Class Fragilariophyceae isolated in 1996 in McMurdo Sound, Antarctica was cultured in silicate-enriched Guillard's f/2 medium (Sigma, St. Louis, MO, USA). The diatom anti-fouling bioassay protocol employed in the present study is described in detail in Amsler et al. (2000, 2005b). Isolated phlorotannins were dissolved in f/2 and added to diatom suspensions to yield final concentrations of 5 ×, 1 × and 0.1 × natural phlorotannin concentration. For controls, diatoms were exposed to f/2 medium without phlorotannins. Diatoms exposed to treatments and controls were incubated for three days at −1 °C, after which, aliquots were stained with 50 g ml^{-1} fluoresceine diacetate (10 mg ml^{-1} acetone) and 0.25% Evan's Blue in seawater, and incubated for 30 min in the dark at −1 °C before counting dead and live cells. Living cells stain green with fluoresceine diacetate and are visible with epifluorescence, while dead cells appear blue in brightfield illumination using a compound microscope. A minimum of 100 cells were counted per treatment and control.

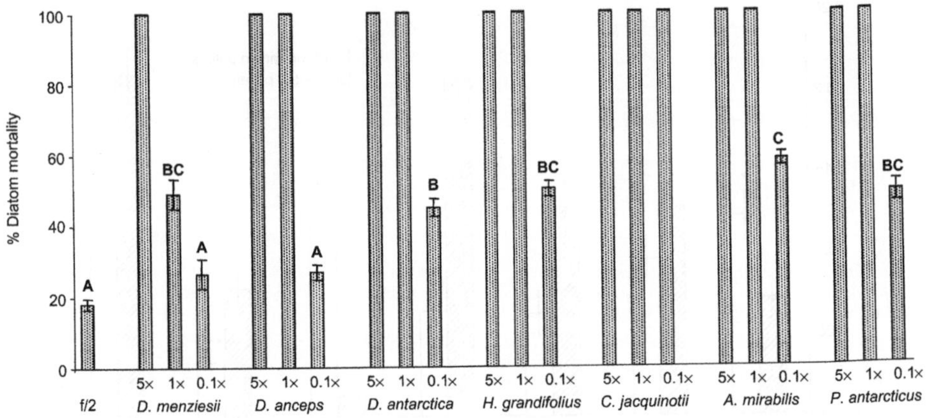

Figure 6.4 Mean (±SE) mortality of diatoms exposed to isolated phlorotannins of Antarctic brown algae at 5 ×, 1 × and 0.1 × natural tissue concentrations, and Guillard's f/2 medium as control.
Letters above bars indicate results of Tukey's HSD post-hoc test with different letters indicating significant differences at $p \leq 0.05$. Replicates without variance (all 5 × concentrations and all 1 × except of *Desmarestia menziesii*) were excluded from analysis (ANOVA, $F_{7,1675} = 35.875$, $p < 0.0001$).
D.: *Desmarestia*, *H.*: *Himantothallus*, *A. mirabilis*: *Ascoseira mirabilis*, *C.*: *Cystosphaera*, *A. utricularis*: *Adenocystis utricularis*, *P.*: *Phaeurus*.

All treatments and the control were tested in triplicate. Results are reported as a mean percent of dead cells of total diatom cell numbers (% diatom mortality); significant differences (= 0.05) between treatment concentrations per algal species and controls were calculated by ANOVA of arcsine-transformed data and Tukey's HSD post-hoc test (SPSS software, SPSS Inc., Chicago, IL, USA). Those treatments where diatom mortality was 100% in all replicates were excluded from statistical analysis because of lack of variance.

Phlorotannins of all algal species caused 100% mortality in diatoms at 5 × natural concentration compared to the control (Figure 6.4). Also, at the 1 × natural concentration, phlorotannins of all species caused 100% mortality, except those of *Desmarestia menziesii*, which still resulted in significant mortality (ca. 50%) over the control (Tukey HSD, $p < 0.001$). At 0.1 × concentration, extracts of several species (*Desmarestia antarctica*, *Himantothallus grandifolius*, *Cystosphaera jacquinotii*, *Ascoseira mirabilis* and *Phaeurus antarcticus*) caused significant mortality compared to controls (Tukey HSD, $p < 0.001$ for all), while others (*Desmarestia menziesii*, *D. anceps*) were not significantly different from the controls at this concentration.

Antimicrobial assays

Antimicrobial activity of phlorotannins was tested against four strains of bacteria previously isolated from McMurdo Sound, Antarctica from seawater and benthic macroinvertebrate surfaces (De Marino et al. 1997). Bacterial strains were grown for 2 days in nutrient broth (DIFCO Marine Broth, Sparks, MD, USA). Of this bacterial suspension,

100 ml were spread evenly onto agar in 100 mm Petri dishes and immediately used in assays. Isolated phlorotannins from the DMBA standards were resuspended in MeOH to a final concentration of 10 × and 1 × natural concentration by volume. Of these, 10 ml were added to each side (20 ml total) of a sterile blank paper disk (6 mm diameter, approximately 20 ml volume, Becton Dickinson, Franklin Lakes, NJ, USA). Control paper disks were laced with 20 ml MeOH. Treatment and control disks were completely dried and then placed on the bacterial spread on the agar. Duplicate paper disks per treatment (algal species and concentration) were used per bacterial strain on separate Petri dishes. Each Petri dish also contained one solvent control paper disk. Bacteria were allowed to grow on plates for 2–3 days. Using the agar diffusion technique (Acar 1980), the diameters of clear zones (halos) around extract and control disks indicating bacterial growth inhibition were measured to the nearest mm.

Table 6.2 Bacterial growth inhibition in agar diffusion assays with Antarctic bacteria. In all cases shown in Table, the concentration was 10 × .

Bacterial strain	Phlorotannin extract	Growth inhibition zone (mm)
11.5	*Desmarestia menziesii*	2 (faint)
	Desmarestia anceps	4 (faint)
	Desmarestia antarctica	–
	Himantothallus grandifolius	2 (faint)
	Ascoseira mirabilis	3 (faint)
	Cystosphaera jacquinotii	4 (faint)
	Phaeurus antarcticus	4 (faint)
13.3	*Desmarestia menziesii*	–
	Desmarestia anceps	1.5
	Desmarestia antarctica	–
	Himantothallus grandifolius	–
	Ascoseira mirabilis	1
	Cystosphaera jacquinotii	–
	Phaeurus antarcticus	3
18.1	*Desmarestia menziesii*	–
	Desmarestia anceps	1
	Desmarestia antarctica	–
	Himantothallus grandifolius	–
	Ascoseira mirabilis	–
	Cystosphaera jacquinotii	1
	Phaeurus antarcticus	1
32.2	*Desmarestia menziesii*	1
	Desmarestia anceps	–
	Desmarestia antarctica	–
	Himantothallus grandifolius	–
	Ascoseira mirabilis	–
	Cystosphaera jacquinotii	–
	Phaeurus antarcticus	2

Inhibition zones are the means of duplicates. None of the 1 × natural concentrations elicited any bacterial growth inhibition and results are not shown in this Table. None of the controls caused any bacterial growth inhibition.

Overall, growth inhibition of Antarctic bacteria by phlorotannin extracts was weak. None of the isolated phlorotannins of any species caused any growth inhibition at natural concentration (1 ×). Phlorotannins of several species had activity against several strains at 10 × natural concentration (Table 6.2). At this higher concentration, phlorotannins of six out of seven algal species caused growth inhibition in bacterial strain 11.5, although zones were very faint, indicating that growth inhibition may not have been complete. Only phlorotannins of *Phaeurus antarcticus* had antibacterial effects against all bacterial strains tested, and *Desmarestia anceps* was active against three out of four strains. *D. antarctica* was the only species whose phlorotannins did not cause any growth inhibition in any of the bacterial strains.

Grazer deterrence of Antarctic brown algal phlorotannins

Feeding deterrence has received the most attention of all possible ecological functions of brown algal phlorotannins (see reviews by Hay and Steinberg 1992, Steinberg 1992, Targett and Arnold 1998, 2001, Amsler and Fairhead 2006). Overall, feeding deterrence of phlorotannins from Antarctic brown algae was specific to individual algal-grazer species combinations. Deterrence patterns in the sympatric fish *Notothenia coriiceps* differed depending on the purification methods used. Phlorotannin-rich extracts of most brown algae (five out of six species tested) elicited a significant rejection response, similar to feeding responses of *N. coriiceps* observed towards general hydrophilic extracts of these algal species (Amsler et al. 2005a). But this effect was only maintained in two algal species (*Desmarestia menziesii* and *Cystosphaera jacquinotii*) when isolated phlorotannins were tested. It is possible that the different feeding stimulant used in the two years when testing the two extract types caused at least some of these differences although this is unlikely because control food pellets were readily consumed in both years. However, we cannot rule out the possibility that the feeding stimulant, when testing isolated phlorotannins (*Palmaria decipiens*), was more attractive to the fish than *Cladophora repens* in the tests of phlorotannin-rich extracts, and hence may have overpowered some of the deterrent effects of the phlorotannins. The loss of activity with higher purification may also support previous suggestions (Deal et al. 2003, Hemmi et al. 2004, Kubanek et al. 2004) that feeding-deterrent effects reported in the literature may not always be caused by phlorotannins but rather by other polar compounds present in phlorotannin-enriched extracts. This could have been the case for some of the phlorotannin-rich extracts although our extensive chemical analyses of Antarctic brown algae yielded only very few secondary metabolites with potential bioactivity (Ankisetty et al. 2004, Amsler et al. 2005a). However, the menzoquinone found in *D. menziesii* and cystosphaerol in *C. jacquinotii* (Ankisetty et al. 2004, Lebar et al. 2007) are removed by the cellulose purification step leading to our isolated phlorotannins. Hence, we are confident that such non-phlorotannin compounds would have been lost with the further purification of the phlorotannin-rich extracts.

Grazer deterrence efficiency of phlorotannins can differ depending on grazer type (Steinberg 1992). Small, less mobile mesograzers that are confined to their algal habitat may be more tolerant of algal chemical defenses than larger, more mobile macrograzers (Hay et al. 1988, 1989). Mesograzers are limited in their range to seek new food sources and also could find protection against predators by residing on a chemically defended host. We showed previously that some Antarctic sedentary mesograzers are not deterred by brown algal homogenates containing phlorotannins (the gastropod *Laevilacunaria antarctica* Martens, Iken 1999), or preferentially reside and feed on Antarctic brown algae (the isopod *Plakarthrium punctatissimum* Pfeffer, Iken 1996), similar to herbivore-algal relationships found for isopods on brown algae in the North Atlantic (Jormalainen et al. 2001). We also found here that isolated phlorotannins of *Desmarestia menziesii* significantly deterred feeding of the fish and sea stars but did not affect the mesograzer amphipods. However, this concept was not supported consistently for phlorotannins of all brown algal species. For example, *Cystosphaera jacquinotii* phlorotannins were significantly distasteful to both the fish and amphipods, but did not elicit a rejection response in sea stars. Similarly, grazing of the same amphipod species (*Gondogeneia antarctica* Chevreux) was deterred by isolated phlorotannins of *Desmarestia anceps* in a previous study (Fairhead et al. 2006). Also overall, the highly mobile macrograzer *Notothenia coriiceps* was similarly little deterred by phlorotannins as were small less mobile mesograzer amphipods, and then only by a few brown algal species. Our results suggest that Antarctic brown algal phlorotannin activities are more species-specific than by categories of meso- or macrograzer.

Another grazer-specific difference that could influence susceptibility to phlorotannins relates to gut structure and the intestinal chemical environment, which may react differently to algal phlorotannins (Horn 1989, Targett and Arnold 1998). For example, an acidic gut environment could maintain the chemical reactivity of phlorotannins in which they can complex proteins and other macromolecules (Tugwell and Branch 1992, Appel 1993, Stern et al. 1996b, Targett and Arnold 2001). The stomach of *Notothenia coriiceps* has a pH of ≤ 3 (Targett and Radtke 1985), which presents a chemical environment in which phlorotannins actively bind proteins such as digestive enzymes. Little is known about the chemical intestinal environment of our invertebrate bioassay species (*Odontaster validus* and *Gondogeneia antarctica*), but data from other echinoderms (pH 6–8, Lawrence 1982) and herbivorous amphipods (pH 5–6.5, Martin 1966) indicate that gut pH may be around neutral, possibly neutralizing the chemical reactivity of phlorotannins. However, differences in gut pH are unlikely to be the sole explanation for the feeding deterrence patterns elicited by phlorotannins of Antarctic brown algae, since fish overall were not more deterred by phlorotannins than the other grazers.

Feeding deterrent patterns can also be related to one or more phlorotannin-specific properties in Antarctic brown algae, including different bulk phlorotannin concentrations in different species (Iken et al. 2007), different molecular structures (Targett and Arnold 2001), or variations in phlorotannin size structure (Boettcher and Targett 1993). Bulk phlorotannin concentrations are relatively high in all investigated Antarctic brown algal species and do not differ much among species (Iken et al. 2007). However, bulk phlorotannin levels did not coincide with overall feeding deterrence patterns, as algal species with higher levels did not always cause stronger feeding deterrence in all grazers. One implication of the generally high phlorotannin levels may be that feeding deterrence against specific grazers would likely be maintained even if phlorotannin concentrations were to fluctuate seasonally or in individual microenvironments (Fairhead et al. 2005a).

Phlorotannins are polymers of phloroglucinol units that can vary species-specifically in structure (Ragan and Glombitza 1986, Targett and Arnold 2001). Phlorotannin molecular structure of Antarctic brown algae is not known but species-specific absorption behavior of isolated standards in spectrophotometric analyses (Iken et al. 2007, DMBA standard curves for phlorotannin quantification) may indicate that phlorotannin structure is different among species. It may be of future interest to analyze phlorotannin structure to evaluate if structure has any influence on grazer deterrence. As polymers, size of the individual molecules increases with age (Ragan and Glombitza 1986, Arnold and Targett 1998) and it has been suggested that larger molecular weight phlorotannins may be ecologically more important than others (Geiselman and McConnell 1981, Boettcher and Targett 1993). We have previously shown that phlorotannin size structure varies by species and thallus parts in many Antarctic brown algae, with most differences between blade and branch portions versus stipes and holdfasts (Iken et al. 2007). Phlorotannins used here in bioassays were isolated from whole plant extractions and thus represent an undetermined mixture of the various size classes. Nevertheless, *Cystosphaera jacquinotii* is dominated in all thallus parts by large molecular size phlorotannins (10–>100 kDa, Iken et al. 2007) and had very strong feeding-deterrent effect against two of three grazers. On the other hand, *Himantothallus grandifolius* also contains mainly large phlorotannin molecules, but had a deterrent effect only against sea stars and neither of the other two grazers. If larger phlorotannin size classes were indeed more active in grazer deterrence, this might explain the allocation of large phlorotannins to attachment structures (holdfasts) in all investigated brown algae (Iken et al. 2007). The holdfast is regarded a specifically important structure for individual survival (Fairhead et al. 2005b).

Deterrence of Antarctic brown algal phlorotannins against biofoulers

Other functions described for brown algal phlorotannins include protection against fouling organisms (Wikström and Pavia 2004) and antibacterial properties (Conover and Sieburth 1964, Ragan and Glombitza 1986, Scalbert 1991). Antifouling and antibacterial properties of brown algal phlorotannins have long been suspected (Conover and Sieburth 1964, Sieburth and Conover 1965), but only few studies have attempted to investigate these properties despite the wealth of studies on brown algal phlorotannins. Fouling organisms can negatively affect growth and reproduction in macroalgae, interfere with nutrient uptake or attract grazers and thereby increase tissue loss of the host (Williams and Seed 1992, de Nys and Steinberg 1999, Littler and Littler 1999). Chemical compounds can reduce attachment of fouling organisms in many seaweeds (Schmitt et al. 1998, Steinberg and de Nys 2002).

Phlorotannin extracts of all species were strongly active against the fouling diatom strain tested in this study. Significant toxicity was detected at as low as 0.1 × natural tissue concentration (ca. 0.6–0.1% DW, Iken et al. 2007) for phlorotannins of all species tested except *Desmarestia menziesii* and *D. anceps*. While these concentration estimates are based on bulk tissue concentrations, fouling diatoms settling on the algal thallus surface would likely experience phlorotannin concentrations resulting from exudation. Few studies have quantified phlorotannin exudation, and rates for uninjured tissue in temperate brown algae range from 5×10^{-4}% DW day^{-1} to 0.01% DW day^{-1} (Ragan and Jensen 1977, Jennings and Steinberg 1994, Koivikko et al. 2005). If exudation rates in Antarctic brown algae are similar to those of temperate species, exudation would still be lower by one to several orders of magnitude than the lowest concentration tested in our assays. While there is a possibility that phlorotannins of some algal species (e.g., *Cystosphaera jacquinotii,* 100% diatom mortality at 0.1 × natural concentration) can cause significant mortality of fouling organisms, surface phlorotannin concentrations will have to be determined to assess whether this could be a major function of phlorotannins in Antarctic macroalgae. It should be noted, however, that settlement of fouling organisms may be deterred at considerably lower phlorotannin concentrations than those causing actual mortality. Some other studies investigating antifouling effects of phlorotannins have shown little effect in the inhibition of multicellular algal and animal foulers (Jennings and Steinberg 1994, Hurd et al. 2000, Honkanen and Jormalainen 2005), although strong negative effects between phlorotannin concentrations and bryozoan epiphytes (Wikström and Pavia 2004) have also been observed, and natural surface concentrations of phlorotannins in *Fucus vesiculosus* L. significantly deterred algal epibionts (Jormalainen et al. 2008).

Conclusion

Overall biological activities of phlorotannins of Antarctic brown algae were species-specific in deterring feeding of a variety of grazers, were moderate in antifouling, and were generally not active as antimicrobial agents. Some of the seemingly conflicting results might be resolved if we knew more about phlorotannin molecular structure in these algae along with interactions of the specific phlorotannin structure and size class with Antarctic grazer digestive systems. For example, phlorotannins may not deter feeding through their immediate noxious taste, but could affect grazers longer-term by binding to digestive enzymes and thus reduce assimilation efficiency in consumers (Hagerman and Butler 1991, Ireland and Horn 1991, Steinberg and Van Altena 1992, Boettcher and Targett 1993). Hence, short-term feeding experiments, as conducted here, may address only part of the ecological significance of phlorotannins. The strong immediate feeding deterrence caused by some but not other phlorotannins, and the inconsistent effects of various algal species' phlorotannins on different grazers suggests that not all phlorotannins are equivalent and that feeding deterrence may at least in part be related to species-specific properties of phlorotannins. Most likely, species-specific characteristics of grazer tolerance and digestion and of phlorotannin quantitative and qualitative properties come together in complex interactions that preclude simple generalizations on the feeding deterrent effects of brown algal phlorotannins in the Antarctic.

References

Acar, J.F. 1980. The disc susceptibility test. *In*: (V. Lorian, ed) *Antibiotics in laboratory medicine.* Williams and Wilkins, Baltimore. pp. 24–54.

Amsler, C.D. and V.A. Fairhead. 2006. Defensive and sensory chemical ecology of brown algae. *Adv. Bot. Res. 43*: 1–91.

Amsler, C.D., R.J. Rowley, D.R. Laur, L.B. Quetin and R.M. Ross. 1995. Vertical distribution of Antarctic peninsular macroalgae: cover, biomass and species composition. *Phycologia 34*: 424–430.

Amsler, C.D., C.B. Moeller, J.B. McClintock, K.B. Iken and B.J. Baker. 2000. Chemical defenses against diatom fouling in Antarctic marine sponges. *Biofouling 16*: 29–45.

Amsler, C.D., K. Iken, J.B. McClintock, M.O. Amsler, K.J. Peters, J.M. Hubbard, F.B. Furrow and B.J. Baker. 2005a. Comprehensive evaluation of the palatability and chemical defenses of subtidal macroalgae from the Antarctic Peninsula. *Mar. Ecol. Prog. Ser. 294*: 141–159.

Amsler, C.D., I.N. Okogbue, D.M. Landry, M.O Amsler, J.B. McClintock and B.J. Baker. 2005b. Potential chemical defense against diatom fouling in Antarctic macroalgae. *Bot. Mar. 48*: 318–322.

Amsler, C.D., J.B. McClintock and B.J. Baker. 2008. Macroalgal chemical defenses in polar marine communities. *In*: (C.D. Amsler, ed) *Algal chemical ecology*. Springer-Verlag, Berlin, Heidelberg. pp. 91–103.

Amsler, C.D., K. Iken, J.B. McClintock and B.J. Baker. 2009. Defenses of polar macroalgae against herbivores and biofoulers. *Bot. Mar. 52:* 535–545.

Ankisetty, S., S. Nandiraju, H. Win, Y.C. Park, C.D. Amsler, J.B. McClintock, J.A. Baker, T.K. Diyabalanage, A. Pasaribu, M.P. Singh, W.M. Maiese, R.D. Walsh, M.J. Zaworotko and B.J. Baker. 2004. Chemical investigation of predator-deterred macroalgae from the Antarctic Peninsula. *J. Nat. Prod. 67*: 1295–1302.

Appel, J. 1993. Phenolics in ecological interactions: the importance of oxidation. *J. Chem. Ecol. 19*: 1521–1552.

Arnold, T.M. and N.M. Targett. 1998. Quantifying in situ rates of phlorotannin synthesis and polymerization in marine brown algae. *J. Chem. Ecol. 24*: 577–595.

Barrera-Oro, E.R. and R.J. Casaux. 1990. Feeding selectivity in *Notothenia neglecta*, Nybelin, from Potter Cove, South Shetland Islands, Antarctica. *Ant. Sci. 2*: 207–213.

Boettcher, A.A. and N.M. Targett. 1993. Role of polyphenolic molecular size in reduction of assimilation efficiency in *Xiphister mucosus*. *Ecology 74*: 891–903.

Conover, J.T. and J.M. Sieburth. 1964. Effect of *Sargassum* distribution on its epibiota and antibacterial activity. *Bot. Mar. 6*: 147–157.

Corbisier T.N., M.A.V. Petti, R.S.P. Skowronski and T.A.S. Brito. 2004. Trophic relationships in the nearshore zone of Martel Inlet (King George Island, Antarctica): $\delta^{13}C$ stable-isotope analysis. *Polar Biol. 27*: 75–82.

Deal, M.S., M.E. Hay, D. Wilson and W. Fenical. 2003. Galactolipids rather than phlorotannins as herbivore deterrents in the brown seaweed *Fucus vesiculosus*. *Oecologia 136*: 107–114.

de Nys, R. and P.D. Steinberg. 1999. Role of secondary metabolites from algae and seagrasses in biofouling control. *In*: (M. Fingerman and R. Nagabhusanam, eds) *Recent advances in marine biotechnology*. Vol. 3. Science Publishers, Enfield. pp. 223–244.

De Marino, S., M. Iorizzi, F. Zollo, L. Minale, C.D. Amsler, B.J. Baker and J.B. McClintock. 1997. Isolation, structure elucidation, and biological activity of the steroid oligoglycosides and polyhydroxysteroids from the Antarctic starfish *Acodontaster conspicuus*. *J. Nat. Prod. 60*: 959–966.

Dunton, K.H. 2001. $\delta^{15}N$ and $\delta^{13}C$ measurements of Antarctic Peninsula fauna: trophic relationships and assimilation of benthic seaweeds. *Am. Zool. 41*: 99–112.

Fairhead, V.A., C.D. Amsler, J.B. McClintock and B.J. Baker. 2005a. Variation in phlorotannin content within two species of brown macroalgae (*Desmarestia anceps* and *D. menziesii*) from the Western Antarctic Peninsula. *Polar Biol. 28*: 680– 686.

Fairhead, V.A., C.D. Amsler, J.B. McClintock and B.J. Baker. 2005b. Within-thallus variation in chemical and physical defenses in two species of ecologically dominant brown macroalgae from the Antarctic Peninsula. *J. Exp. Mar. Biol. Ecol. 233*: 1–12.

Fairhead, V.A., C.D. Amsler, J.B. McClintock and B.J. Baker. 2006. Lack of defense or phlorotannin induction by UV radiation or mesograzers in *Desmarestia anceps* and *D. menziesii* (Phaeophyceae). *J. Phycol. 42*: 1174–1183.

Geiselman, J.A. and O.J. McConnell. 1981. Polyphenols in brown algae *Fucus vesiculosus* and *Ascophyllum nodosum*: chemical defenses against the marine herbivorous snail, *Littorina littorea*. *J. Chem. Ecol. 1*: 1115–1133.

Hagerman, A.E. and L.G. Butler. 1991. Tannins and lignins. *In*: (G.A. Rosenthal and M.R. Berenbaum, eds) *Herbivores: Their interactions with secondary plant metabolites*. Vol. 1. Academic Press, San Diego, CA. pp. 355–388.

Hay, M.E. and P.D. Steinberg. 1992. The chemical ecology of plant-herbivore interactions in marine versus terrestrial communities. *In*: (G.A. Rosenthal and M.R. Berenbaum, eds) *Herbivores: Their interactions with secondary plant metabolites*. Vol. II. Academic Press, New York. pp. 371–413.

Hay, M.E., P.E. Renaud and W. Fenical. 1988. Large mobile versus small sedentary herbivores and their resistance to seaweed chemical defenses. *Oecologia 75*: 246–252.

Hay, M.E., J.R. Pawlik, J.E. Duffy and W. Fenical. 1989. Seaweed-herbivore-predator interactions – host-plant specialization reduced predation on small herbivores. *Oecologia 81*: 418–427.

Hemmi, A., T. Honkanen and V. Jormalainen. 2004. Inducible resistance to herbivory in *Fucus vesiculosus*-duration, spreading and variation with nutrient availability. *Mar. Ecol. Prog. Ser. 273*: 109–120.

Honkanen, T. and V. Jormalainen. 2005. Genotypic variation in tolerance and resistance to fouling in the brown alga *Fucus vesiculosus*. *Oecologia 144*: 196–205.

Horn, M.H. 1989. Biology of herbivorous fishes. *Oceanogr. Mar. Biol. Annu. Rev. 27*: 167–272.

Huang, Y.M., J.B. McClintock, C.D. Amsler, K.J. Peters and B.J. Baker. 2006. Feeding rates of common Antarctic gammarid amphipods on ecologically important sympatric macroalgae. *J. Exp. Mar. Biol. Ecol. 329*: 55–65.

Huang, Y.M., M.O. Amsler, J.B. McClintock, C.D. Amsler and B.J. Baker. 2007. Patterns of gammarid amphipod abundance and species composition associated with dominant subtidal macroalgae along the western Antarctic Peninsula. *Polar Biol. 30*: 1417–1430.

Hurd, C.L., K.M. Durante and P.J. Harrison. 2000. Influence of bryozoan colonization on the physiology of the kelp *Macrocystis integrifolia* (Laminariales, Phaeophyta) from nitrogen-rich and -poor sites in Barkley Sound, British Columbia, Canada. *Phycologia 39*: 435–440.

Iken, K. 1996. *Trophic relations between macroalgae and herbivores in Potter Cove (King George Island, Antarctica)*. PhD Thesis, *Ber. Polarforsch. 201*: 1–206 [in German].

Iken, K. 1999. Feeding ecology of the Antarctic herbivorous gastropod *Laevilacunaria antarctica* Martens. *J. Exp. Mar. Biol. Ecol. 236*: 133–148.

Iken, K., E.R. Barrera-Oro, M.L. Quartino, R.J. Casaux and T. Brey. 1997. Grazing by the Antarctic fish *Notothenia coriiceps*: evidence for selective feeding on macroalgae. *Ant. Sci. 9*: 386–391.

Iken, K., M.L. Quartino and C. Wiencke. 1999. Histological identification of macroalgae from stomach contents of the fish *Notothenia coriiceps* gives new insights in its feeding ecology. *Mar. Ecol. 20*: 11–18.

Iken, K., C.D. Amsler, J.M. Hubbard, J.B. McClintock and B.J. Baker. 2007. Allocation patterns of phlorotannins in Antarctic brown algae. *Phycologia 46*: 386–395.

Ireland, C.D. and M.H. Horn. 1991. Effects of macrophyte secondary chemicals on food choice and digestive efficiency of *Cebidichthys violaceus* (Girard), an herbivorous fish of temperate marine waters. *J. Exp. Mar. Biol. Ecol. 153*: 179–194.

Jennings, J.G. and P.D. Steinberg. 1994. In-situ exudation of phlorotannins by the sublittoral kelp *Ecklonia radiata*. *Mar. Biol. 121*: 349–354.

Jormalainen, V., T. Honkanen and N. Heikkilä. 2001. Feeding preferences and performance of a marine isopod on seaweed hosts: costs of habitat specialization. *Mar. Ecol. Prog. Ser. 220*: 219–230.

Jormalainen, V., S.A. Wikström and T. Honkanen. 2008. Fouling mediates grazing: intertwining of resistances to multiple enemies in the brown alga *Fucus vesiculous*. *Oecologia 155*: 559–569.

Karsten, U., A. Wulff, M.Y. Roleda, R. Müller, F.S. Steinhoff, J. Fredersdorf and C. Wiencke. 2009. Physiological responses of polar benthic algae to ultraviolet radiation. *Bot. Mar. 52*: 639–654.

Klöser, H., G. Mercuri, F. Laturnus, M.L. Quartino and C. Wiencke. 1994. On the competitive balance of macroalgae at Potter Cove (King George Island, South Shetlands). *Polar Biol. 14*: 11–16.

Klöser, H., M.L. Quartino and C. Wiencke. 1996. Distribution of macroalgae and macroalgal communities in gradients of physical conditions in Potter Cove, King George Island, Antarctica. *Hydrobiologia 333*: 1–17.

Koivikko, R., J. Loponen, T. Honkanen and V. Jormalainen. 2005. Contents of soluble, cell-wall-bound and exuded phlorotannins in the brown alga *Fucus vesiculosus*, with implications on their ecological functions. *J. Chem. Ecol. 31*: 195–212.

Kubanek, J., S.E. Lester, W. Fenical and M.E. Hay. 2004. Ambiguous role of phlorotannins as chemical defenses in the brown alga *Fucus vesiculosus*. *Mar. Ecol. Prog. Ser. 277*: 79–93.

Lawrence, J.M. 1982. Digestion. *In*: (J.M. Lawrence and M. Jangoux, eds) *Echinoderm nutrition*. Balkeema, Rotterdam. pp. 39–52.

Lebar, M.L., J.L. Heimbegner and B.J. Baker. 2007. Cold-water marine natural products. *Nat. Prod. Rep. 24*: 774–797.

Linkowski, T.B., P. Presler and C. Zukowski. 1983. Food habits of notothenoid fishes (Nototheniidae) in Admiralty Bay (King George Island, South Shetland Islands). *Pol. Polar Res. 4*: 79–95.

Littler, M.M. and D.S. Littler. 1999. Blade abandonment/proliferation: a novel mechanism for rapid epiphyte control in marine macrophytes. *Ecology 80*: 1736–1746.

Lüder, U.H. and M.N. Clayton. 2004. Induction of phlorotannins in the brown macroalga *Ecklonia radiata* (Laminariales, Phaeophyta) in response to simulated herbivory – the first microscopic study. *Planta 218*: 928–937.

Martin, A.L. 1966. Feeding and digestion in two intertidal gammarids: *Marinogammarus obtusatus* and *M. pirloti*. *J. Zool. 148*: 515–525.

McClintock, J.B. and B.J. Baker. 1997. Palatability and chemical defense in the eggs, embryos and larvae of shallow-water Antarctic marine invertebrates. *Mar. Ecol. Prog. Ser. 154*: 121–131.

Moreno, C.A. and J.M. Zamorano. 1980. Séleccion de los alimentos en *Notothenia coriiceps neglecta* del cinturon de macroalgas de Bahía South Antarctica. *Ser. Cient. INACH 25/26*: 33–43.

Pavia, H., G. Cervin, A. Lindgren and P.Åberg. 1997. Effects of UV-B radiation and simulated herbivory on phlorotannins in the brown alga *Ascophyllum nodosum*. *Mar. Ecol. Prog. Ser. 157*: 139–146.

Pavia, H., G.B. Toth, A. Lindgren and P.Åberg. 2003. Intraspecific variation in the phlorotannin content of the brown alga *Ascophyllum nodosum*. *Phycologia 42*: 378–383.

Quartino, M.L., H. Klöser, I.R. Schloss and C. Wiencke. 2001. Biomass and associations of benthic marine macroalgae from the inner Potter Cove (King George Island, Antarctica) related to depth and substrate. *Polar Biol. 24*: 349–355.

Quartino, M.L., H.E. Zaixso and A.L. Boraso de Zaixso. 2005. Biological and environmental characterization of marine macroalgal assemblages in Potter Cove, South Shetland Islands, Antarctica. *Bot. Mar. 48*: 187–197.

Ragan, M.A. and K.W. Glombitza. 1986. Phlorotannins, brown algal polyphenols. *Prog. Phycol. Res. 4*: 129–241.

Ragan, M.A. and A. Jensen. 1977. Quantitative studies on brown algal phenols. I. Estimation of absolute polyphenol content of *Ascophyllum nodosum* (L.) Le Jol and *Fucus vesiculosus* (L.). *J. Exp. Mar. Biol. Ecol. 30*: 209–221.

Richardson, M.G. 1975. The dietary composition of some Antarctic fish. *Br. Antarct. Surv. Bull. 41/42*: 113–120.

Richardson, M.G. 1977. *The ecology (including physiological aspects) of selected Antarctic marine invertebrates associated with inshore macrophytes*. PhD Thesis, British Antarctic Survey, Cambridge, pp. 296.

Scalbert, A. 1991. Antimicrobial properties of tannins. *Phytochemistry 30*: 3875–3883.

Schmitt, T.M., N. Lindquist and M.E. Hay. 1998. Seaweed secondary metabolites as antifoulants: effects of *Dictyota* spp. diterpenes on survivorship, settlement, and development of marine invertebrate larvae. *Chemoecology 8*: 125–131.

Schoenwaelder, M.E.A. and M.N. Clayton. 1999. The presence of phenolic compounds in isolated cell walls of brown algae. *Phycologia 38*: 161–166.

Sieburth, J.M. and J.T. Conover. 1965. *Sargassum* tannin, an antibiotic which retards fouling. *Nature 208*: 52–53.

Sokal, R.R. and F.J. Rohlf. 1995. *Biometry*. Freeman and Company, New York, pp. 887.

Steinberg, P.D. 1988. Effects of quantitative and qualitative variation in phenolic compounds on feeding in three species of marine invertebrate herbivores. *J. Exp. Mar. Biol. Ecol. 120*: 221–237.

Steinberg, P.D. 1989. Biogeographical variation in brown algal polyphenolics and other secondary metabolites: comparison between temperate Australasia and North America. *Oecologia 78*: 373–382.

Steinberg, P.D. 1992. Geographical variation in the interaction between marine herbivores and brown algal secondary metabolites. *In*: (V.J. Paul, ed) *Ecological roles of marine natural products*. Cornell University Press, Ithaca, London. pp. 51–92.

Steinberg, P.D. and I. Van Altena. 1992. Tolerance of marine invertebrate herbivores to brown algal phlorotannins in temperate Australasia. *Ecol. Monogr. 62*: 189–222.

Steinberg, P.D. and R. de Nys. 2002. Chemical mediation of colonization of seaweed surfaces. *J. Phycol. 38*: 621–629.

Stern, J.L., A.E. Hagerman, P.D. Steinberg, F.C. Winter and J.A. Estes. 1996a. A new assay for quantifying brown algal phlorotannins and comparisons to previous methods. *J. Chem. Ecol. 22*: 1273–1293.

Stern, J.L., A.E. Hagerman, P.D. Steinberg and P.K. Mason. 1996b. Phlorotannin-protein interactions. *J. Chem. Ecol. 22*: 1877–1899.

Targett, N.M. and T.M. Arnold. 1998. Predicting the effects of brown algal phlorotannins on marine herbivores in tropical and temperate oceans. *J. Phycol. 34*: 195–205.

Targett, N.M. and T.M. Arnold. 2001. Effects of secondary metabolites on digestion in marine herbivores. *In*: (J.B. McClintock and B.J. Baker, eds) *Marine chemical ecology*. CRC Press, Boca Raton. pp. 391–411.

Targett, N.M., L.D. Coen, A.A. Boettcher and C.E. Tanner. 1992. Biogeographic comparisons of marine algal phenolics: evidence against a latitudinal trend. *Oecologia 89*: 464–470.

Targett, T.E. and R.L. Radtke. 1985. Growth and feeding ecology studies on coastal Antarctic fishes. *Antarct. J. US. 19*: 147– 148.

Tugwell, S. and G.M. Branch. 1989. Differential polyphenolic distribution among tissues in the kelps *Ecklonia maxima*, *Laminaria pallida* and *Macrocystis angustifolia* in relation to plant-defence theory. *J. Exp. Mar. Biol. Ecol. 129*: 219–230.

Tugwell, S. and G.M. Branch. 1992. Effects of herbivore gut surfactants on kelp polyphenol defenses. *Ecology 73*: 205–215.

Van Alstyne, K.L. 1995. Comparison of three methods for quantifying brown algal polyphenolic compounds. *J. Chem. Ecol. 21*: 45–58.

Van Alstyne, K.L., M.N. Dethier and D.O. Duggins. 2001. Spatial patterns in macroalgal chemical defenses. *In*: (J.B. McClintock and B.J. Baker, eds) *Marine chemical ecology*. CRC Press, Boca Raton. pp. 301–324.

Van Altena, I.A. and P.D. Steinberg. 1992. Are differences in the responses between North American and Australasian marine herbivores to phlorotannins due to differences in phlorotannin structure? *Biochem. Syst. Ecol. 20*: 493–499.

Wikström, S.A. and H. Pavia. 2004. Chemical settlement inhibition versus post-settlement mortality as an explanation for differential fouling of two congeneric seaweeds. *Oecologia 138*: 223–230.

Williams, G.A. and R. Seed. 1992. Interactions between macrofaunal epiphytes and their host algae. *In*: (D.M. John, S.J. Hawkins and J.H. Price, eds) *Plant–animal interactions in the marine benthos*. Systematics Association, Special Vol. 46. Clarendon Press, GB-Oxford. pp. 189–211.

Wulff, A., K. Iken, M.L. Quartino, A. Al-Handal, C. Wiencke and M.N. Clayton. 2009. Biodiversity, biogeography and zonation of marine benthic micro- and macroalgae in the Arctic and Antarctic. *Bot. Mar. 52*: 491–507.

Primary production and ecophysiology

7 Benthic microalgal production in the Arctic: applied methods and status of the current database

Ronnie N. Glud, Jana Woelfel, Ulf Karsten, Michael Kühl and Søren Rysgaard

Introduction

The Arctic oceans, including all marine water bodies at latitudes above the Arctic Circle, cover >20 × 10^6 km^2 and encompass ~25% of the global coastal region (areas with water depths <200 m) (Menard and Smith 1966, Jakobsson et al. 2008). The Arctic coastal regions cover ~5.8–6.1 × 10^6 km^2 with an average water depth of 80 m (Gattuso et al. 2006, Jakobsson et al. 2008) and a significant fraction of this area can be expected to accommodate benthic primary production. Gattuso et al. (2006) estimated that on average, 25% of the Arctic coastal seabed receives >1% of the surface down welling irradiance during the five summer months, with a much larger fraction expected to occasionally experience irradiances of this magnitude.

An increasing number of studies suggest that benthic microalgae contribute significantly to subtidal coastal ecosystem production (Charpy-Roubaud and Sournia 1990, Meyercordt et al. 1999, Nelson et al. 1999, Jahnke et al. 2000, Jahnke 2005, Glud et al. 2008). Furthermore, studies from lower latitudes have documented the importance of benthic microphytes for supporting shallow water food-webs (e.g., Middelburg et al. 2000). In an extensive review of 85 studies, Cahoon (1999) concluded that previous estimates on benthic microalgal productivity had markedly underestimated their importance for marine ecosystem production – especially in oligotrophic systems. Overall, this author provided an annual global estimate of 5 × 10^8 tons C for the neritic microalgal primary production. Cahoon (op. cit.) also showed that, whereas some areas were relatively well studied (i.e., temperate intertidal and eutrophicated systems), others were grossly undersampled, especially the Arctic region for which at that time there were only two studies! The polar coastal areas differ from the better studied temperate regions in being exposed to ice-cover and darkness during extensive periods of the year. Furthermore, they are pristine and experience relatively low temperatures. The seabed is often exposed to ice-mediated erosion and erratic, massive inputs of erosion material during spring (Chapter 2; Rachold et al. 2004, Zacher et al. 2009). Therefore, findings from lower latitude systems cannot be extrapolated *a priori* to polar ecosystems.

Camera-based mapping of the sea bed over larger areas in the Arctic often shows an extensive and dense cover of microalgae that is intensively grazed by macrofauna (Figure 7.1). One striking observation provided by such photo documentation is the pronounced small-scale patchiness, which must be accounted for when evaluating data on benthic productivity. Closer inspection of seabed samples shows that the microalgal cover is often dominated by pennate diatoms and in some cases dinoflagellates or

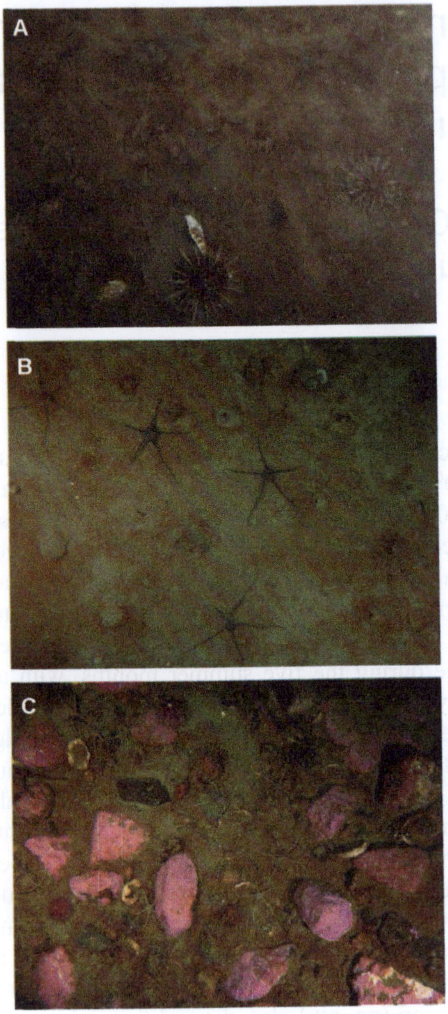

Figure 7.1 *In situ* photographs of benthic microalgal coverage. (A) in NW Greenland (72′22.53″ N, 55′33.37″ W) at 18 m of water depth; (B) in NE Greenland (74′18.59″ N, 20′14.48″ W) in 20 m water depth and (C) in NE Greenland (74′18.59″ N, 20′14.74″ W) in 30 m water depth. In all cases the brown patches indicate the presence of pennate diatoms (the lower panel is dominated by the occurrence of stones covered with coralline red algae). Each image covers an area of 40 × 30 cm.
Photograph made available by P.B. Christensen, M. Sejr, and P. Batty and reproduced by kind permission of the owners.

Cyanobacteria; the first group appears to dominate Arctic microphytobenthos (Vetrov and Romankevich 2004 and references therein). In the following, we have compiled and discussed the available database on benthic microalgal production from the Arctic. In total we have identified 10 peer-reviewed and three unpublished studies of shelf areas off Siberia, Svalbard, Greenland and North America (Figure 7.2). We identified a few more abstracts and book-chapters that assess the benthic productivity

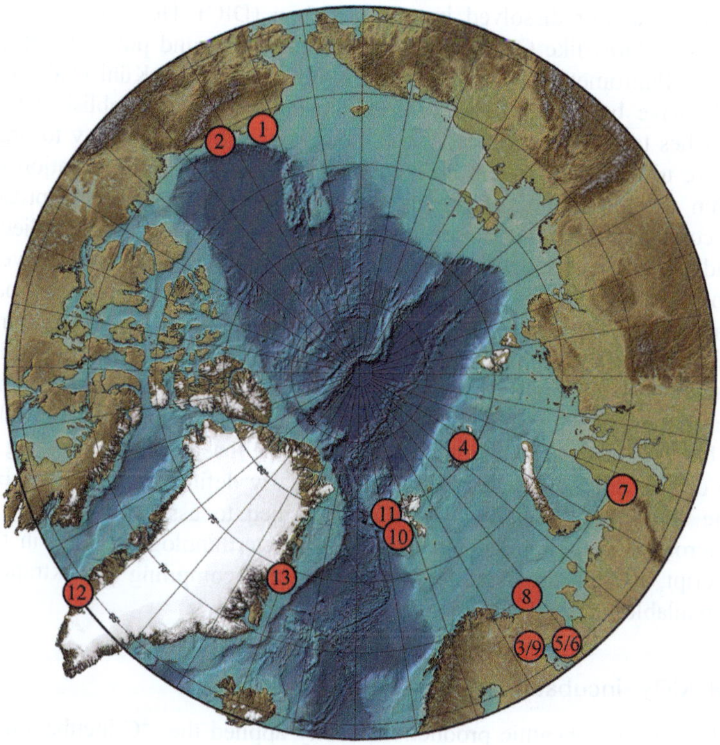

Figure 7.2 Map of the Arctic region indicating positions where benthic microalgal studies have been performed (red dots).
In some instances, the Russian studies did not indicate the exact positions, but the positions are estimated from the context described in the papers. The individual studies are: 1. Matheke and Horner (1974); 2. Horner and Schrader (1982); 3. Kuznetsov (1991); 4. Kuznetsov (2002); 5. Bondarchuk (1974); 6. Bondarchuk 1980; 7. Kuznetsov et al. (1998); 8. Kuznetsov (2005); 9. Kuznetsov and Strogaya (1989); 10. Woelfel et al. (2010) 11. K. Hancke (unpublished data); 12. B. Eyre (unpublished data); 13. Glud et al. (2002).
The Figure is modified from Jakobsson et al. (2008) and published with kind permission from the authors of the quoted article.

in Arctic systems, but in these cases methodology and calculation procedures were often difficult to evaluate. The overall aim of this review is to assess and evaluate the importance of benthic microphytes for primary production in the Arctic coastal region. However, as the current database has been obtained using different measuring approaches that can be confounded by other benthic processes and may express net or gross photosynthesis to differing degrees, it is important to thoroughly evaluate each when compiling the available data.

Applied methods and the current database on Arctic microphytobenthic activity

Most assessments of benthic microalgal productivity are based on core or chamber incubations following the autotrophic incorporation of added $H^{14}CO_3^-$ or the total exchange of O_2 and/or dissolved inorganic carbon (DIC). However, in recent years alternative procedures like O_2-microsensor measurements, and pulse amplitude modulation (PAM) fluorometry (Barranguet and Kromkamp 2000, Kühl et al. 2001, Glud et al. 2002), have been introduced to complement these well-established techniques. The approaches target different steps in the conversion of light energy to organic material and are to a varying extent confounded by the concurrent respiration that takes place within natural benthic communities. The different measuring approaches are, therefore, complementary, but unfortunately most studies have only applied one approach and, as discussed below it is not always straightforward to convert results obtained by one approach to those of another. Furthermore, different approaches define net and gross primary production differently and procedures for extrapolating measurements to *in situ* light levels on daily or seasonal time scales are different. To complicate matters, most published studies do not have sufficient information to recalculate measurements to a standardized procedure. Consequently, evaluating the importance of benthic microalgal photosynthesis on the basis of the existing database requires a certain degree of "creativity" and a clearly defined terminology. Below, we discuss the respective measuring procedures applied to assess the activity of intact benthic microalgal communities and we define the terminology applied in the rest of the manuscript, all with the overall aim of comparing, compiling and extrapolating the presently available Arctic database.

Benthic $H^{14}CO_3^-$ incubation procedures

Early studies on microbenthic production mostly applied the ^{14}C-incubation technique that had been successfully developed to study productivity of planktonic communities (Steeman-Nielsen 1952). In essence, $H^{14}CO_3^-$ is added to benthic chambers placed directly on the seabed or to collected sediment cores exposed to a controlled irradiance in the laboratory (Colijn and de Jonge 1984). The added tracer mixes and equilibrates with the natural DIC-pool of the overlying water and gradually diffuses into the benthic community. In permeable sediments, percolation approaches that facilitate efficient tracer distribution within the photic zone have been adopted (Jönsson 1991). After a given incubation period, the surface sediment is collected and the concentration of radioactively-labeled organic carbon is determined (Matheke and Horner 1974). The technique is very sensitive, but requires the determination of the specific DIC labeling in the microenvironment of the active microalgae. Intact microphytobenthic communities are characterized by extensive temporal and spatial variability and by steep concentration gradients of solutes and light. This makes it very difficult to determine the specific DIC labeling experienced by the active microalgae, which is most certainly not constant during a given incubation (Revsbech et al. 1981, de Beer et al. 1997). There is currently no direct way to determine the specific labeling of inorganic carbon pool at the relevant scale, and estimating an average value of this crucial parameter essentially relies on qualified guessing. Due to this problem, the ^{14}C approach is nowadays rarely used to assess benthic productivity.

In principle, the ^{14}C approach quantifies the gross primary production, but depending on the incubation time, some of the fixed ^{14}C can be respired by the microalgae themselves or by heterotrophic bacteria taking advantage of leaking photosynthetic products and exudates. It is very difficult to assess the extent to which this takes place during a given incubation, but it is well-established that heterotrophic activity in natural benthic communities is markedly stimulated by light due to a rapid turnover of freshly produced organic material (Epping and Jørgensen 1996, Kühl et al. 1996, Fenchel and Glud 2000). Therefore, although dependent on the incubation time, the ^{14}C approach generally underestimates true gross photosynthetic activity. By extending the incubation to 24 h during a natural light/dark cycle it could be argued that the approach expresses the daily net primary production. However, most (and all Arctic) incubations apply an incubation of ~3–5 h and the gross primary production is calculated after subtracting any ^{14}C fixation rate of parallel dark incubations (Matheke and Horner 1974). In the following, we abbreviate gross primary production as determined by this approach as GPPt ("t" for tracer).

While the benthic ^{14}C incubation technique is rarely used anymore, many studies in Arctic waters are based on this approach and represent a significant contribution to the available database. We have identified seven Arctic studies that have applied the ^{14}C approach to *in situ* benthic chambers deployed over a depth range of 3–25 m (Table 7.1). Despite the very limited database and the inherent scatter, maximal rates occurred at very shallow water depths (<6 m), with little activity at the deepest measuring sites (20–25 m), presumably as light availability approached the limit of microalgal photosynthetic activity.

Table 7.1 Benthic microalgal gross primary production in Arctic ecosystems estimated from *in situ* chamber incubations enriched by ^{14}C-labeled DIC (GPPt).

Activity range (mmol C m^{-2} day^{-1})	Mean (mmol C m^{-2} day^{-1})	Chl *a* (mg m^{-2})	Depth (m)	References
37.6–114.0	75.6	~130–380	~5	Matheke and Horner (1974)
11.2–65.6	39.8	~130–380	~5	Matheke and Horner (1974)
0.0–0.2	0.1	1–28	7	Horner and Schrader (1982)
2.0–101.2	32.8	NI	3	Kuznetsov (1991)
0.2–100	30.0	NI	10	Kuznetsov (1991)
0.0–95.2	19.8	NI	15	Kuznetsov (1991)
3.6–19.6	12.0	18–72	20	Kuznetsov (2002)
0.6–4.1	NI	40–940	1–25	Bondarchuk (1974, 1980)
NI	4.6	2–640	5–25	Kuznetsov et al. (1998)

The studies are all seasonal, but the table includes data only from the sea ice-free period.
The Chl *a* data represent measurements for the uppermost centimeter, but in some instances, only mean values were assessable. NI, no information is available.

Total exchange rates of O_2 and DIC

Generally speaking, the most widely applied procedure to assess the benthic production (or degradation) of organic material is the "whole core incubation" approach, wherein chambers or core liners holding an intact sediment core are either incubated in the laboratory under controlled light conditions or *in situ* taking advantage of natural irradiance (Cahoon and Cooke 1992, Jahnke et al. 2000, Glud et al. 2008). The net primary production during illumination can then be inferred from the gradual O_2 accumulation in the overlying water and/or the concurrent decline in the DIC concentration. The approach is relatively simple, and when larger sediment areas are enclosed, the inherent small-scale patchiness in microalgal biomass can, at least to some extent, be accounted for. One concern is, however, that chambers or cores change the local hydrodynamics. Efficient stirring or mixing of the enclosed water volume during incubations is thus essential and has, in various studies, been accommodated by pumps, impellers, discs or stirrers (Pamatmat and Fenton 1968, Cahoon 1988, Huettel and Gust 1992). It is beyond the scope of this review to discuss pros and cons of various chamber and stirring designs or *in situ* vs. laboratory-based measurements (for recent reviews, see Tengberg et al. 2005, Glud 2008), but the effect of any imposed stirring on the measured solute exchange rate must be acknowledged.

In benthic communities with a high O_2 production (or consumption) rate, the thickness of the diffusive boundary layer (DBL) can significantly affect the net exchange of O_2 (Jørgensen and des Marais 1990, Larkum et al. 2003, Glud et al. 2007) and, therefore, affect the derived net photosynthesis. For example, in a benthic microalgal mat, the O_2 exchange increased 5–6-fold as the measured DBL thickness was decreased by a factor of ~4 (Figure 7.3).

Total O_2 or DIC exchange rates measured during illumination quantify the net photosynthesis (NPPc, c for chambers) of the benthic community, including oxygen consumption related to fauna and microbial respiration. The concurrent gross photosynthetic activity can thus be substantially higher. A common procedure for estimating the gross photosynthesis is to add the exchange rates measured in darkness (or in parallel incubations with opaque chambers) to that of the light incubations (e.g., Eyre and Ferguson 2002). This is a pragmatic approach, but the values should be interpreted with caution for at least two reasons: 1) the infauna of shallow water sediments frequently exhibit a daily variation in their activity level and feeding mode that affects the benthic O_2 consumption rate (Rosenberg and Lundberg 2004, Wenzhöfer and Glud 2004), and 2) microbial respiration is typically enhanced in light (Epping and Jørgensen 1996; see also below). Therefore, this approach generally underestimates the true gross photosynthetic activity.

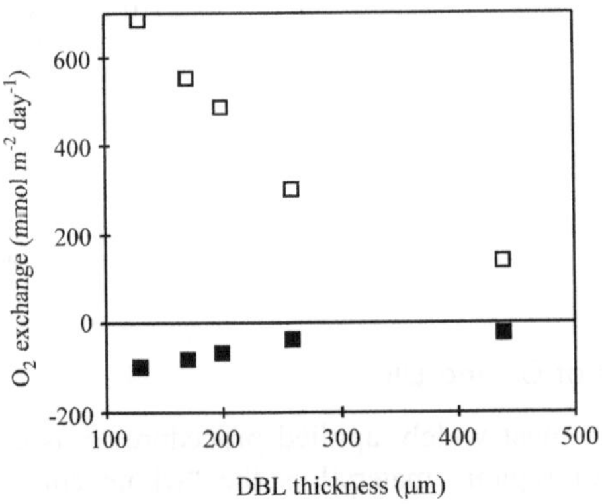

Figure 7.3 The release and uptake of O_2 in a benthic microalgal mat as measured in small benthic chambers (inner diameter 6 cm) kept in light (open symbols) and darkness (closed symbols) with different stirring rates.
Parallel O_2 microelectrode measurements allowed quantification of the actual diffuse boundary layer (DBL) thickness in the periphery of the chambers at the different stirring speeds of a rotating magnet fixed to the chamber wall in the overlying water.

Most studies based on chamber or core incubations infer benthic primary production from change in the O_2 concentration. However, electrons released from the light-induced splitting of water (i.e., the O_2 formation) can be channeled to a range of other processes in cell-housekeeping, rather than CO_2 fixation (Raven and Beardall 1981, Badger et al. 2000, Wagner et al. 2006). Provided any precipitation or dissolution of carbonate can be accounted for, a more direct procedure to quantify primary production could be to measure the daily (24 h) DIC consumption rate. But as O_2 is far easier to measure than DIC, the O_2 exchange measurements remain the preferred procedure and O_2 exchange is subsequently transformed into a CO_2 fixation rate assuming a photosynthetic quotient (PQ). Most measurements of PQ for integrated benthic communities, quantified as the ratio between the exchange of O_2 vs. DIC, range between 0.9 and 1.3, largely depending on the light and nutrient availability (Cammen 1991, Cahoon and Cooke 1992, Longphuirt et al. 2007, Taddei et al. 2008). Generally a value of 1.2 is applied to benthic communities, i.e., 1.0 carbon atom is fixed per 1.2 O_2 molecule produced. Based on 50 incubations of sediment cores collected over a water depth range of 5–30 m and exposed to light levels between 6 and 100 µmol photons m^{-2} s^{-1}, the community PQ of a high Arctic sediment averaged 1.19 ± 0.48 (Glud et al. 2002). Based on these measurements, and in order to be consistent with most studies conducted at lower latitudes, we have chosen to apply a conversion factor of 1.2 to derive the benthic gross primary production (GPPc) from the sum of O_2 exchange measured in darkness and light with benthic incubation chambers [the respiratory quotient (RQ) is thereby assumed to be the reciprocal of PQ].

We have identified two *in situ* studies (Kuznetsov and Strogaya 1989, Kuznetsov 2005) from the Arctic that used chambers for quantifying benthic microphytic production (Table 7.2). The values align with the GPPt data of Table 7.1, but rather than a straightforward depth relationship, maximal phototrophic activity occurred at intermediary water depths.

One published High Arctic study used laboratory core incubations under well-defined irradiance to quantify the productivity of benthic microalgae (Glud et al. 2002). These data are plotted against irradiance along with those from three unpublished studies in Figure 7.4. As expected, the relationship between GPPc and light exposure exhibits significant scatter, mainly reflecting the small- and large-scale variation in microalgal biomass. The cores were sampled at water depths from 2 to 30 m, but

Table 7.2 Benthic microalgal gross primary production in Arctic ecosystems estimated from total O_2 exchange measurements (GPPc) by benthic *in situ* chambers during sea ice-free periods.

Activity range (mmol C m^{-2} day^{-1})	Mean (mmol C m^{-2} day^{-1})	Chl *a* (mg m^{-2})	Depth (m)	References
NI	5.9	30	5	Kuznetsov (2005)
NI	10.1	140	10	Kuznetsov (2005)
NI	16.9	122	15	Kuznetsov (2005)
NI	12.7	126	20	Kuznetsov (2005)
NI	0.8	111	25	Kuznetsov (2005)
8.1–21.6	14.8	42–347	3	Kuznetsov and Strogaya (1989)

The Chl *a* data represent measurements for the uppermost centimeter: in some instances only mean values were available for assessment. NI: no information is available.

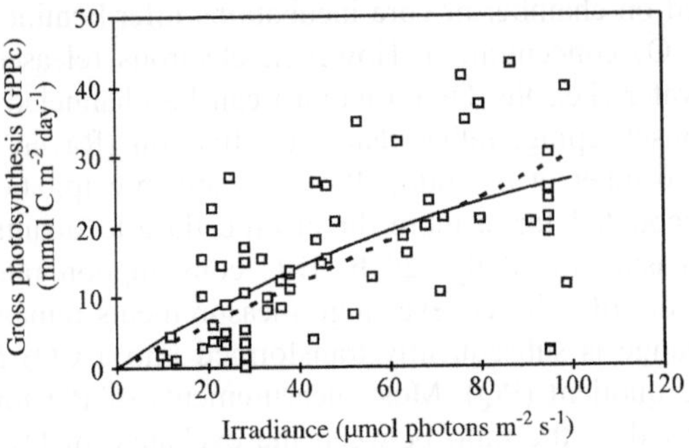

Figure 7.4 Gross primary production calculated from O_2 exchange rates measured in sediment cores exposed to well-defined irradiance levels in the laboratory (GPPc).
The data are extracted from Glud et al. (2002), Woelfel et al. (2010) and two unpublished studies by K. Hancke and B. Eyre, respectively (see also Figure 7.1). The compiled data were fitted by a simple linear equation ($GPP_c = 0.3E$; dotted line), where "E" is the irradiance ($R^2 = 0.29$) and a Platt function [$GPPc = P_m (1\text{-}EXP(\alpha E_z/P_m))$], where "$P_m$" – the photosynthetic capacity – equaled 39 mmol m^{-2} day^{-1}; α – the initial slope of GPPc – equaled 0.45, and "E_z" is the irradiance ($R^2 = 0.34$). Both fits were forced through (0, 0). The index of light adaptation $E_k = P_m/\alpha$ amounted to 86 mmol photons m^{-2} day^{-1}. Two points included in the fit lie outside the boundaries of the axes.

Table 7.3 Benthic microalgal gross primary production in open-water Arctic ecosystems as estimated from total O_2 exchange measurements (GPPc) of laboratory-based core incubations.

Activity range (mmol C m^{-2} day^{-1})	Irradiance (μmol photons m^{-2} s^{-1})	"*In situ*" mean* (mmol C m^{-2} day^{-1})	Chl *a* (mg m^{-2})	Depth (m)	References
36–90	95	112	136	5	Woelfel et al. (2010)
16–46	95	25	30	10	Woelfel et al. (2010)
10–36	95	10	11	15	Woelfel et al. (2010)
10–46	95	6	8	20	Woelfel et al. (2010)
19–26	95	2	2	30	Woelfel et al. (2010)
15–21	95	29	34	5	Woelfel et al. (2010)
19–25	95	16	19	10	Woelfel et al. (2010)
1–3	95	1	1	15	Woelfel et al. (2010)
1–3	95	1	1	20	Woelfel et al. (2010)
0–5	95	<1	<1	30	Woelfel et al. (2010)
18–34	140	36	3	1	K. Hancke (unpublished data)
40–54	200	42	NI	2	B. Eyre (unpublished data)
1–21	21–91	34	75	5	Glud et al. (2002)
1–24	6–69	24	66	10	Glud et al. (2002)
1–43	12–87	12	186	20	Glud et al. (2002)
1–27	11–53	5	92	30	Glud et al. (2002)

The Chl *a* data represent measurements for the uppermost centimeter; in some instances only mean values were available for assessment. NI: no information is available.
* The actual laboratory-based measurements were extrapolated to estimate average *in situ* irradiance at the sampling sites using recorded light levels and light extinction coefficients of the water column during the open-water period.

there is no significant depth trend in the data. This is in accord with the observation that benthic diatoms quickly optimize their photosynthetic apparatus to the available irradiance and that any depth-specific light acclimatization coheres when collected cores are exposed to similar light conditions (Kühl et al. 2001, Glud et al. 2002). Fitting a Platt function to the data (Platt et al. 1980) improved a simple linear fit to the compiled dataset and suggested a photosynthetic capacity (P_{max}) of 39 mmol m^{-2} day^{-1} and an "index of light adaptation" (E_k) of 86 µmol photons m^{-2} s^{-1} (Figure 7.4). The derived Platt equation is by no means universal for Arctic sediments and will, presumably, change as the data base hopefully increases with future studies. It does, however, provide a proxy for extrapolating local production rates to larger regional scales using information on the available irradiance. To extrapolate the studies in Figure 7.4 to *in situ* activity during summer, we assessed the average *in situ* light availability from information on light extinction coefficients of the water column, direct measurements of the down-welling irradiance and day length in the respective studies (Table 7.3).

The recent introduction of the eddy correlation to aquatic biology offers a strong and complementary tool to chamber incubations (Berg et al. 2003, 2007). The technique can be used for *in situ*, non-invasive quantification of the benthic O_2 exchange, integrating the activity of 50–100 m^2 of the sea bed. At present, there are very few studies applying the technique to aquatic sediments (Berg et al. 2003, 2009, Kuwae et al. 2006, Brand et al. 2008) but as yet, no investigations have used eddy correlation to study benthic primary production.

Oxygen microsensor measurements

The prime advantage of using chambers or cores (and eddy correlation) for quantifying benthic exchange rates is that their application is relatively simple and averages out small scale variations. However, such methods represent a "black-box" approach and only provide limited insight into the vertical and horizontal activity distribution or into the microenvironment in which the microbenthic phototrophic activity takes place. Oxygen microelectrode measurements allow a very detailed characterization of the distribution, production and consumption of O_2 at a given point in time (Revsbech et al. 1981, Epping et al. 1999) and multiple measurements thereby complement chamber incubations well. From measured microprofiles, the diffusive export of O_2 from the photic zone can be quantified as the sum of the upward and downward flux, and when converted into carbon equivalents, this represents the net primary production of the community (NPPm) (Figure 7.5). The upward flux is essentially what is quantified in the NPPc measured by chamber or core incubations and generally accounts for 70–90% of the NPPm (Epping and Jørgensen 1996, Kühl et al. 1996, Wenzhöfer et al. 2000, Christensen et al. 2003).

Oxygen microsensors also allow quantification of gross primary production (GPPm) by the light-dark-shift technique (Revsbech and Jørgensen 1983). Basically, this method calculates gross photosynthesis from the rate at which the O_2 concentration at a given point declines immediately after onset of darkness. The approach assumes that the O_2 gradients and respiration remain unaffected during the light/dark shift, and detailed investigations have proven that this is essentially correct, as long as the O_2 decline is determined within the first second of darkness (Glud et al. 1992, Lassen et al. 1998). This approach actually resolves the true gross primary production of the community, provided the applied PQ reflects the relationship between O_2 production and DIC fixation under the given conditions.

Figure 7.5 shows an example of the level of detail that can be obtained by microsensor measurements and degree of complexity in deriving an estimate of benthic productivity from O_2 exchange measurements with benthic chambers or whole cores. The benthic O_2 consumption in darkness amounted to 13.0 mmol m^{-2} day^{-1}, while the O_2 efflux from the photic zone (the upper 1.6 mm) during light equaled 81.7 mmol m^{-2} day^{-1}, of which 74% (60.1 mmol m^{-2} day^{-1}) diffused upwards into the overlying water and the residual 21.6 mmol m^{-2} day^{-1} sustained O_2 consumption in the deeper sediment layers (Figure 7.5). The concurrent depth-integrated gross production of O_2, measured by the light-dark shift technique, was 117.7 mmol m^{-2} day^{-1}, meaning that the respiration within the illuminated photic zone (the upper 1.6 mm) was 36.0 mmol m^{-2} day^{-1} (117.7−81.7). The benthic O_2 consumption during light thus amounted to 57.6 mmol m^{-2} day^{-1} (36.0+21.6), which is >4 times higher than O_2 consumption in darkness. The stimulated activity during light was caused by two factors i) an increased O_2 consumption within the photic zone and ii) a deeper oxic penetration zone within which O_2 was consumed. If we apply the calculation procedure that is generally applied to chamber incubation, the "GPPc" is the sum of the O_2 consumption in darkness and the upward O_2 release in light (converted into C equivalents), which sums to 60.9 mmol C m^{-2} day^{-1} [(13 + 60.1)/1.2]. In reality, the

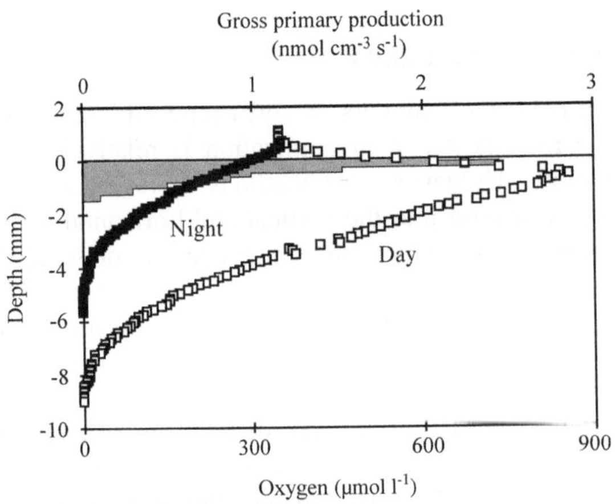

Figure 7.5 Oxygen microprofiles measured in a diatom-dominated community of benthic microalgae.
Black symbols represent measurements in darkness, while open symbols are measurements at a down-welling irradiance of 450 µmol photons m^{-2} s^{-1}. Gray bars represent the gross photosynthesis as measured by the "light-dark-shift" technique. J_{up} represents the upward-directed O_2 flux as calculated from (J_{up} = Do dC/dz), where Do is the molecular diffusion coefficient for O_2 under the experimental conditions, and C is the O_2 concentration at depth Z within the diffusive boundary layer (DBL; the same approach is applied to estimate O_2 consumption from the microprofile measured in darkness). The downward O_2 flux is quantified by the same relation, but this time applying the tortuosity-corrected diffusion coefficient of the sediment matrix. The sum of J_{up} and J_{down} provides the net photosynthetic activity of the photic zone.

NPPm is 68.0 mmol C m^{-2} day^{-1} (81.7/1.2) and the GPPm is 98.0 mmol C m^{-2} day^{-1} (117.7/1.2). This example illustrates how the widely-applied chamber approach underestimates actual benthic productivity and how confusing use of the terms net and gross primary production is in the existing literature. The example also illustrates the detailed insights that can be provided into O_2 turnover at a given spot using microsensor measurement. The approach is, however, very time consuming – a set of microprofiles resolving the O_2 distribution and the GPPm at a given spot for a single level of irradiance typically takes about 40–50 min, and given the natural variability of the seabed it is a non-trivial task to extrapolate the findings from one or a few measurements to a larger area. This remains the prime limitation of the microsensor measuring approach.

To the best of our knowledge, there are only two microsensor studies on benthic microalgal activity in the Arctic (Glud et al. 2002, Hancke and Glud 2004). In the first, sediment cores were recovered from 10 to 30 m water depth and were then incubated under similar light, temperature and salinity conditions in the laboratory for two days. Subsequently, O_2 microprofiles were measured in the diatom cover, and the compiled data exhibited a good PE-relationship, without any light-inhibition at the irradiances applied (Figure 7.6). Benthic microalgal communities rarely exhibit light inhibition because i) a downward migration of motile cells counteracts inhibiting light levels at the surface due to the extremely steep light gradients in such communities (Kühl et al. 1997), and ii) a gradual expansion of the photic zone with increasing irradiance compensates for any potential activity decline at the surface (Epping and Jørgensen 1996, Kühl et al. 1996, Christensen et al. 2003).

The fitted Platt equation of Figure 7.6 shows a photosynthetic capacity (P_{max}) of 67 mmol m^{-2} day^{-1} and an "index of light adaptation" (E_k) of 27 µmol photons m^{-2} s^{-1} for net photosynthesis (NPPc). When comparing with the GPPc-Platt relation (Figure 7.4), it is important to realize that the two relationships are based on different techniques and, as explained above, this causes the otherwise inconsistent observation that P_{max} for gross primary production (GPPc as derived from core incubations) is lower than the P_{max} for the net primary production (NPPm as derived from microprofile measurements). The light compensation point, where the benthic O_2 consumption balances the O_2 production (i.e., the light level at which the net O_2 exchange is zero mmol m^{-2} day^{-1}) amounted to 4.7 µmol photons m^{-2} s^{-1}. This value essentially defines the threshold irradiance required to sustain net primary production of the communities investigated at the given experimental conditions. The data in Figure 7.6 were measured in patches of benthic microphytes and ignored patches of bare sediment. Accounting for *in situ* coverage of benthic microalgae and mean daily light availability at the respective water depths, NPPm for the study was calculated and is presented in Table 7.4.

The only other Arctic microsensor study on benthic microalgae was performed on sediment cores recovered below snow covered sea ice with very low phototrophic biomass (the sediment surface Chl *a* level was 2.7 mg m^{-2}; Hancke and Glud 2004). This study focused on short-term temperature effects on respiration and photosynthesis of benthic communities dominated by diatoms rather than on quantifying benthic productivity. It was shown that both respiration and photosynthesis increased with temperature, but that a stronger Q_{10} response of heterotrophic activity gradually lead to reduced net benthic productivity as the temperature increased (Hancke and Glud 2004). The extent to which this observation is the result of the short-term nature of

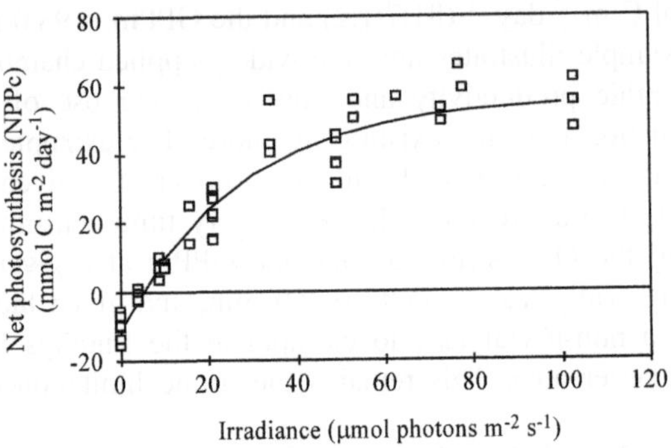

Figure 7.6 Net primary production calculated from O_2 microprofiles measured in sediment cores exposed to well-defined irradiance levels in the laboratory (NPPc).
The data are extracted from Glud et al. 2002 (see also Figure 7.1). The compiled data were fitted by a Platt function including a "respiration" term (R) [$NPPc = P_m (1 - EXP(-\alpha E/P_m))] - R$, where "$P_m$" – the photosynthetic capacity – was 67 mmol m^{-2} day^{-1}; α – the initial slope of GPPc – was 2.5, and R was 10.7 mmol m^{-2} day^{-1}; "E" is the irradiance ($R^2 = 0.70$).

the experiment (days to weeks), and whether seasonal increases in temperature can shift shallow-water Arctic sediments into a more heterotroph-dominated status in late summer remains to be investigated.

Table 7.4 Benthic microalgal net primary production in an open-water Arctic ecosystem as estimated from microsensor measurements (NPPm) in the laboratory and extrapolated to *in situ* activity using information on light availability and benthic microalgal cover (BMC).

Mean (mmol C m^{-2} day^{-1})	Depth (m)	*In situ* BMC (%)
32.0	5	23
31.3	10	26
14.8	20	73
9.3	30	36

Glud et al. (2002).

Relative importance of benthic vs. pelagic microalgal productivity

Growing directly at the sediment surface, benthic microalgae can exploit nutrients released by the underlying biogeochemical mineralization processes and can thus deprive the pelagic community of nutrients. In contrast, pelagic phototrophs can better exploit the down-welling irradiance as compared to communities constrained to a narrow zone on the sediment surface. Thus, nutrient availability often regulates the relative importance of pelagic vs. benthic microalgal productivity. Eutrophic settings favor pelagic productivity, while oligotrophic settings favor benthic productivity (Charpy-Roubaud and Sournia 1990, MacIntyre et al. 1996). Even though rivers can carry nutrient-enriched water and induce plumes of stimulated pelagic production that reduce the benthic light availability locally (Parsons et al. 1988, Springer and McRoy

1993), Arctic coastal waters are pristine with low nutrient levels. Consequently, they can be expected to host a relatively large benthic productivity.

Several of the original benthic studies in Tables 7.1–7.4 performed parallel measurements of the pelagic productivity and have provided estimates on the relative importance of pelagic vs. benthic microalgae (Table 7.5). Except for one study, these estimates show a complete dominance of benthic productivity at shallow depths, which tails off towards deeper waters. But even at the deepest investigated site (30 m), pelagic and benthic microalgal productivity were of similar magnitude, suggesting that benthic microalgal activity dominated in systems shallower than 30 m. Note, however, that Table 7.5 excludes any potential contribution from macrophytes, which may dominate ecosystem production in some Arctic settings (Krause-Jensen et al. 2007).

The relative importance of various photosynthetic communities has also been assessed using diver-operated pulse amplitude modulation (PAM) fluorometry (Schreiber 2005, Diving-PAM, www.walz.com). Measurements in an ice-covered coastal ecosystem off Hokkaido, Japan concluded that benthic microalgae were responsible for 13–66% of the ecosystem production at water depths ranging from 3–9 m (McMinn et al. 2005). However, such indirect approaches involving interpretation of variable chlorophyll fluorescence measurements are based on many assumptions about the photosynthetic apparatus and the coupling between electron transport and carbon fixation of the different organisms – assumptions that are difficult to test or justify with natural samples (Kühl et al. 2001). Therefore, such studies can often at best be regarded as indicative.

The observations in Table 7.5 are in contrast to findings of Kuznetsov (2005), who estimated that the annual pelagic productivity exceeded the benthic microalgal productivity by factors of 1.3–2.8 (average 1.7) when integrated for sediments residing at 0–20 m depth in areas off the Kola Peninsula, Franz Josef Land, eastern Svalbard, West Novaya Zemlya and in the Pechora Sea. These values were, however, presented in abstract format without any explanation on how measurements and extrapolations were performed.

Table 7.5 Relative proportion of benthic microalgal primary production vs. pelagic primary production in Arctic coastal environments.

Benthic contribution (%)	Water depth (m)	Study
76	5	Matheke and Horner (1974)
3	3	Horner and Schrader (1982)
75	3	Kuznetsov (1991)
45	10	Kuznetsov (1991)
40	15	Kuznetsov (1991)
8	0–20	Kuznetsov (2002)
66	0–20	Kuznetsov et al. (1998)
96	5	Glud et al. (2002)
92	10	Glud et al. (2002)
72	20	Glud et al. (2002)
51	30	Glud et al. (2002)

Extrapolating benthic microalgal productivity in time and space

Extrapolations of the limited database on productivity of Arctic benthic microalgae to regional scales requires establishment of relations to simple parameters that are measured more frequently than benthic primary production. The number of studies quantifying benthic Chl *a* concentration is much larger than the number of studies quantifying microalgal activity, and Chl *a* has often been used as an indication of the activity of benthic microalgae (Cahoon 1999, Vetrov and Romankevich 2004). All the original studies of Arctic benthic microalgae (Tables 7.1–7.4) provided estimates on biomass expressed as Chl *a* concentration in the sediment surface layer (mostly 0–1 cm). Minimum or maximum values of Chl *a* and benthic productivity do correlate (Tables 7.1–7.4), but overall there is no clear relationship between these two parameters. A simple linear relationship of the compiled data set expresses an R^2 value of only 0.23, and elimination of some obvious outliers does not markedly improve the relationship (data not shown). However, this may not be too surprising as the photic zone typically extends only a few mm into the sediment, and Chl *a* concentration averaged over the uppermost cm therefore only poorly represents the active phototrophic biomass (Kühl et al. 1997, Kühl 2005). Benthic Chl *a* concentrations may also be confounded by inactive degradation products originating from pelagic microphytes, senescent or saprophytic living microphytes or spores (Sun et al. 1994). Furthermore, it is well-established that one light-adaptive strategy of benthic microalgae is to regulate their cell-specific Chl *a* concentration (Blanchard and Montagna 1992). Clearly extractable Chl *a* concentration is often a poor proxy for extrapolating benthic productivity to wider areas.

Given the fact that benthic microalgae are generally well supplied with nutrients, it seems reasonable to assume that community productivity is light-limited and that light availability could serve as a good proxy for benthic primary production.

The minimum light requirement for benthic microalgae is not well defined, but communities of obligate benthic diatoms have been encountered down to almost 200 m water depth where the maximum light availability was <0.2 µmol photons m^{-2} s^{-1} (McGee et al. 2008). However, a visual coverage of benthic microalgae is rarely observed in water depths >40 m (Cahoon 1999), and the minimal light intensities at which polar and subpolar microphytobenthic activity have been recorded range between 0.5 and 2.5 µmol photons m^{-2} s^{-1} (Palmisano et al. 1985, Grant 1986, Karsten et al. 2006).

Assuming PAR-extinction coefficients of coastal waters ranging between 0.12 and 0.16 m^{-1} for clear and turbid water, respectively (Jerlov 1970), and assuming an average down-welling irradiance during midsummer in the high Arctic of 414 µmol photons m^{-2} s^{-1} (Glud et al. 2002), a diel average of 0.5 µmol photons m^{-2} s^{-1} would reach down to 42 and 52 m in "turbid" and "clear" waters, respectively. The corresponding values for 2.5 µmol photons m^{-2} s^{-1} would be 30 and 46 m. This simple calculation does not account for any changes in spectral composition, but demonstrates that benthic microalgae can be photosynthetically active down to significant water depth in the Arctic region – especially in offshore, clear waters.

Based on remote sensing-derived estimates of light extinction coefficients and PAR distribution in surface waters in the Arctic, Gattuso et al. (2006) proposed that the relative proportion (S) of the coastal seabed receiving light above a given threshold

(E_z mol photons m^{-2} day^{-1}) during the open-water period in summer on average could be estimated as:

$$S = 16.0 - 13.6 \, \text{Log}\,(E_z) + 1.5 \, \text{Log}\,(E_z^2) + 0.7 \, \text{Log}\,(E_z^3).$$

The relationship accounts for the estimated relative distribution of case 1 water (light attenuation due to presence of phytoplankton), which cover ~66% of the investigated area, and case 2 water [light attenuation due to phytoplankton, suspended particles and matter; see Morel and Prieur (1977)]. The relationship predicts that, on average, 35% (~2.1×10^6 km^2) of the coastal Arctic seabed receives a daily average irradiance >0.5 μmol photons m^{-2} s^{-1} during the open-water period. Correspondingly, 25%, 17% and 5% of the coastal Arctic seabed would receive average irradiances above 2.5, 10 and 100 μmol photons m^{-2} s^{-1}, respectively, during the open-water period. Deriving benthic light availability on larger scales from remote sensing is an innovative approach, but it is obviously associated with a number of shortcomings. Measurements can only be performed when cloud and ice cover allow, the spatial and temporal resolution is limited and the distribution of case 1 and case 2 waters can be only crudely assessed at present. Furthermore, the importance of nepheloid layers and high turbidity water, which will both increase light extinction, is poorly defined (Gattuso et al. 2006). Nevertheless such approaches are thus far the best tools available for a first estimation of benthic microalgal productivity over larger scales in the Arctic.

Using the PE relation in Figure 7.4 and the equation for light availability derived by Gattuso et al. (2006), we can extrapolate the estimated gross primary production (GPPc) derived by chamber/core incubations to the entire Arctic region. In essence, the same calculation could be made for the PE relation of NPP_m derived from microsensor measurements. However, as these data only represent conditions within well developed microalgal patches, the small scale variations in biomass have to be accounted for during extrapolating – and, as seen in Figure 7.1, this is a non-trivial task. Such an exercise has, however, been carried out and discussed in a confined area of Young Sound, NE Greenland (Glud et al. 2002).

On average, the daily GPP_c extrapolated to the Arctic coastal region amounts to an average of 1.8×10^5 t C day^{-1} during the open-water period. The average open-water period for the coastal Arctic (water depth 0–50 m) can be estimated from remote sensed sea ice concentrations available at the NSIDC (National Snow and Ice Data Center, USA). The average value amounts to 120 days using a grid size of 12.5 km and a 25% threshold of sea ice concentration (Phil Hwang, unpublished data). However, as grids holding any coastline are masked out by such procedures, areas with land-fast ice are under-represented, and 120 days of open-water period must represent a maximum value. On the basis of selected publications, Cahoon (1999) estimated the average open-water period for the Arctic to be 90 days. Using this value, the annual benthic microalgal primary production amounted to 1.6×10^7 t C $year^{-1}$ in the Arctic. A longer open-water period would increase the estimated primary production proportionally. This estimate does not include potential contributions during sea ice cover, but even though some reports have documented "shade-adapted" benthic photosynthesis below sea ice, reflectance and absorbance in snow-covered sea ice is so high that such contributions must be marginal (Palmisano et al. 1985, Glud et al. 2007b). The integrated benthic primary production may seem marginal as compared to the existing estimates on the pelagic productivity of the Arctic oceans ranging from 21 to

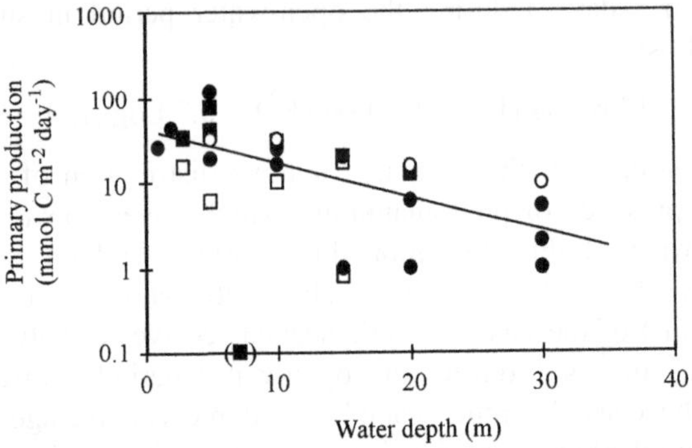

Figure 7.7 Benthic microalgal primary production (PP) derived from Tables 7.1–7.4 plotted against waters depth (z) at the respective study sites (filled squares Table 7.1, open squares Table 7.2, filled circles Table 7.3, open circles Table 7.4). The compiled data (except for point in brackets) were fitted with $PP = 41.5e^{-0.0911z}$ ($R^2 = 0.41$).

42×10^7 t C $year^{-1}$ (Subba-Rao and Platt 1984, Pabi et al. 2008). However, most of the benthic primary production is confined to regions with water depths shallower than 30–40 m, regions that account for only ~10–14% of the Arctic oceans, and in those areas, the relative benthic contribution to the total community primary production is correspondingly higher as also reflected in Table 7.5.

Light availability generally declines exponentially with water depth. Although we used different procedures for quantifying benthic productivity and for extrapolating the data to seasonal time scale, our compiled data (Tables 7.1–7.4) show a quasi exponential decline with increasing water depth (Figure 7.7). The inherent scatter in the data set is partly a result of compiling data obtained with different measuring procedures, but it can also be related to variations in the light extinction coefficients between the different study areas. The natural variability between study areas apparently overrides any systematic bias related to the fact that different measuring procedures have been applied and that the dataset consists of a mixture of NPP and GPP estimates. However, our simple relationship offers another way to extrapolate the limited database to a regional scale by simply multiplying the productivity-depth relation ($42e^{-0.911z}$) to the bathymetry of the coastal Arctic oceans. Such calculation gives an average benthic microalgal primary production of 1.2×10^5 t C day^{-1} for open-water periods, which under the assumption of an average open-water period of 90 days (Cahoon 1999) translates into an annual benthic microalgal primary production in the Arctic coastal ocean of 1.1×10^7 t C $year^{-1}$. This figure is ~30% lower than our estimate of 1.6×10^7 t C $year^{-1}$ using the light relationship above (Figure 7.4), but acknowledging the different procedures and the limited database, the estimates are surprisingly consistent.

Concluding remarks

The limited database (Tables 7.1–7.4) clearly suggests that benthic microalgae contribute significantly to the coastal ecosystem production in Arctic waters. In fact, our compilation indicates that benthic microalgal productivity is of similar magnitude or even exceeds the pelagic productivity in coastal areas with water depths <30 m. Extrapolation from the current data base using empirical relations between light availability, water depth and benthic microalgal activity estimates a contribution of 1.1–1.6×10^7 t C $year^{-1}$ to the coastal Arctic ecosystem.

The entire Arctic region is, however, grossly undersampled and further studies on benthic primary production should be encouraged. As in temperate regions, certain coastal areas are especially underexplored – these include rocky and sandy sediments, and no measurements have been performed in shallow off-shore areas. The introduction of the eddy correlation technique to aquatic biology (Berg et al. 2003) provides an opportunity to improve this situation – especially if linked to benthic observatories – by facilitating *in situ*, non-invasive measurements of large-scale net photosynthesis of benthic algae. Combined with more traditional measuring approaches this could provide an improved insight into benthic microalgal activity and the environmental controls that regulate it. The rapid developments in remote sensing of aquatic light distribution and bathymetric mapping will also facilitate more precise regional extrapolation of specific case studies in the coming years.

The sea ice cover in the Arctic is rapidly declining (Serreze et al. 2007) and given the tight coupling between sea ice cover and marine primary production (Rysgaard et al. 1999), this is expected to increase Arctic productivity. Arrigo et al. (2008) estimated that the pelagic productivity of the Arctic oceans has increased by 5–6% annually in recent years as a consequence of the increased light availability. Increased light availability is expected to further increase the competition for nutrients and we speculate that benthic primary production may consequently be stimulated significantly more than the pelagic production in the low-nutrient Arctic coastal region. On the other hand, a predicted increase in precipitation and permafrost thawing will increase the nutrient-enriched, turbid freshwater run-off and may locally counteract the expected increase in coastal light availability. The net outcome is hard to predict and can only be elucidated by giving priority to long-term observatory-based measurements of coastal primary production, which forms the foundation for the present and the future Arctic coastal food web.

References

Arrigo, K.R., G. van Dijken and S. Pabi. 2008. Impact of a shrinking Arctic ice cover on marine primary production. *Geophys. Res. Lett. 35*: L19603.

Badger, M.R., S. Von Caemmerer, S. Ruuska and H. Nakano. 2000. Electron flow to oxygen in higher plants and algae: rates and control of direct photoreduction (Mehler reaction) and rubisco oxygenase. *Phil. Trans. Roy. Soc. B 355*: 1433–1446.

Barranguet, C. and J. Kromkamp. 2000. Estimating production rates from photosynthetic electron transport in estuarine microphytobenthos. *Mar. Ecol. Prog. Ser. 204*: 39–52.

Berg, P., H. Røy, F. Janssen, V. Meyer, B.B. Jørgensen, M. Huettel and D. de Beer. 2003. Oxygen uptake by aquatic sediments measured with a novel non-invasive eddy correlation technique. *Mar. Ecol. Prog. Ser. 261*: 75–83.

Berg, P., H. Røy and P.L. Wiberg. 2007. Eddy correlation flux measurements – the sediment surface area that contributes to the flux. *Limnol. Oceanogr. 52*: 1672–1684.

Berg, P., R.N. Glud, A. Hume, H. Stahl, K. Oguri, V. Meyer and H. Kitazato. 2009. Eddy correlation measurements of oxygen uptake in deep ocean sediments. *Limnol. Oceanogr. Meth. 7*: 576–584.

Blanchard, G.F. and P.A. Montagna. 1992. Photosynthetic responses of natural assemblages of marine benthic microalgae to short and long-term variations of incident irradiance in Baffin Bay, Texas. *J. Phycol. 28*: 7–14.

Bondarchuk, L.L. 1974. Characteristic features of the benthic diatoms in the near-shore zone of the White Sea (In Russian). *In*: (A.N. Golokov, ed.) *Hydrobiology and biogeography of the shelf of the temperate and cold waters of the ocean*. Nauka: Leningrad. pp. 93–94.

Bondarchuck, L.L. 1980. Diatoms in near-shore sediments of Kandalaksha Bay (White Sea) (In Russian). *In:* (Z.A. Filatova, ed.) *Benthic flora and production of the marginal sea of USSR*. Nauka: Moskow. pp. 63–73.

Brand, A., D.F. McGinnis, B. Wehrli and A. Wüest. 2008. Intermittent oxygen flux from the interior into the bottom boundary of lakes as observed by eddy correlation. *Limnol. Oceanogr. 53*: 1997–2006.

Cahoon, L.B. 1988. Use of a whirling cup rotor to stir benthic chambers. *Hydrobiologia 160*: 193–198.

Cahoon, L.B. 1999. The role of benthic microalgae in neritic ecosystems. *Oceanogr. Mar. Biol. Annu. Rev. 37*: 47–86.

Cahoon, L.B. and J.E. Cooke. 1992. Benthic microalgal production in Onslow Bay, North Carolina, USA. *Mar. Ecol. Prog. Ser. 84*: 185–196.

Cammen, L. 1991. Annual bacterial production in relation to benthic microalgal production and sediment oxygen uptake in an intertidal sandflat and an intertidal mudflat. *Mar. Ecol. Prog. Ser. 71*: 13–25.

Charpy-Roubaud, C. and A. Sournia. 1990. The comparative estimation of phytoplanktonic, microphytobenthic and macrophytobenthic primary production in the oceans. *Mar. Microb. Food Webs 4*: 31–57.

Christensen, P.B., R.N. Glud, T. Dalsgaard and P. Gillespie. 2003. Implications of long line mussel farming on benthos, oxygen and nitrogen dynamics in coastal sediments. *Aquaculture 218*: 567–588.

Colijn, F. and V.N. de Jonge. 1984. Primary production in the Ems – Dollard Estuary. *Mar. Ecol. Prog. Ser. 14*: 185–196.

de Beer, D., A. Glud, E. Epping and M. Kühl. 1997. A fast responding CO_2 microelectrode for profiling in sediments, microbial mats and biofilms. *Limnol. Oceanogr. 42*: 1590–1600.

Epping, E.H.G. and B.B. Jørgensen. 1996. Light enhanced oxygen respiration in benthic phototrophic communities. *Mar. Ecol. Prog. Ser. 139*: 193–203.

Epping, E.H.G., A. Khalili and R. Thar. 1999. Photosynthesis and dynamics of oxygen consumption in a microbial mat as calculated from transient oxygen microprofiles. *Limnol. Oceanogr. 44*: 1936–1948.

Eyre, B.D. and A.J.P. Ferguson. 2002. Comparison of carbon production and decomposition, benthic fluxes and denitrification in seagrass, phytoplankton, benthic microalgae- and macroalgae- dominated warm-temperate Australian lagoons. *Mar. Ecol. Prog. Ser. 229*: 43–59.

Fenchel, T. and R.N. Glud. 2000. Benthic primary production and O_2–CO_2 dynamics in a shallow water sediment: spatial and temporal heterogeneity. *Ophelia 53*: 159–171.

Gattuso, J.-P., B. Gentili, C.M. Duarte, J.A. Kleypas, J.J. Middelburg and D. Antoinne. 2006. Light availability in the coastal ocean: impact on the distribution of benthic photosynthetic organisms and their contribution to primary production. *Biogeosci. 3*: 489–513.

Glud, R.N. 2008. Oxygen dynamics of marine sediments. *Mar. Biol. Res. 4*: 243–289.

Glud, R.N., N.B. Ramsing and N.P. Revsbech. 1992. Photosynthesis and photosynthesis-coupled respiration in natural biofilms quantified with microsensors. *J. Phycol. 28*: 51–60.

Glud, R.N., M. Kühl, F. Wenzhoefer and S. Rysgaard. 2002. Benthic diatoms of a high Arctic fjord (Young Sound, NE Greenland): importance for ecosystem primary production. *Mar. Ecol. Prog. Ser. 238*: 15–29.

Glud, R.N., P. Berg, H. Fossing and B.B. Jørgensen. 2007a. Effect of the diffusive boundary layer (DBL) on the benthic mineralization and O_2 distribution: a theoretical modeling exercise. *Limnol. Oceanogr. 52*: 547–557.

Glud, R.N., S. Rysgaard, M. Kühl and J.W. Hansen. 2007b. The sea-ice in Young Sound: implications for carbon cycling. *In*: (S. Rysgaard and R.N. Glud, eds) *Carbon cycling in Arctic marine ecosystems: case study Young Sound.* Danish Polar Center, Copenhagen. pp. 61–85.

Glud, R.N., B.D. Eyre and N. Patten. 2008. Biogeochemical responses to coral mass spawning at the Great Barrier Reef: effects on respiration and primary production. *Limnol. Oceanogr. 53*: 1014–1024.

Grant, J. 1986. Sensitivity of benthic community respiration and primary production to changes in temperature and light. *Mar. Biol. 90*: 229–306.

Hancke, K. and R.N. Glud. 2004. Temperature effects on respiration and photosynthesis in three diatom dominated benthic communities. *Aquat. Microb. Ecol. 37*: 265–281.

Horner, R. and G.C. Schrader. 1982. Relative contributions of ice algae, phytoplankton and benthic microalgae to primary production in near-shore regions of the Beaufort Sea. *Arctic 35*: 485–503.

Huettel, M. and G. Gust. 1992. Solute release mechanisms from confined cores in stirred benthic chambers and flume flows. *Mar. Ecol. Prog. Ser. 82*: 187–197.

Jahnke, R.A. 2005. Transport processes and organic matter cycling in coastal sediments. *In:* (A.R. Robinson and K.H. Brink, eds) *The global coastal ocean multi scale interdisciplinary processes.* Harvard University Press, Cambridge, MA. pp. 163–191.

Jahnke, R.A., J.R. Nelson, R.L. Marinelli and J.E. Eckman. 2000. Benthic flux of biogenic elements on the southeastern US continental shelf: influence of porewater advective transport and benthic icroalgae. *Cont. Shelf Res. 20*: 109–127.

Jakobsson, M.R., R. Macnab, L. Mayer, R. Andersson, M. Edwards, J. Hatzky, H.W. Schenke and P. Johnson. 2008. An improved bathymetric portrayal of the Arctic Ocean: implications for ocean modeling and geological, geophysical and oceanographic analysis. *Geophys. Res. Lett. 7*: L07602.

Jerlov, N.G. 1970. Light. *In*: (O. Kinne, ed.) *Marine ecology.* Wiley Interscience: London. pp. 95–102.

Jönsson, B. 1991. A C^{14} incubation technique for measuring microphytobenthic primary productivity in intact sediment cores. *Limnol. Oceanogr. 36*: 1485–1492.

Jørgensen, B.B. and D.J. des Marais. 1990. The diffusive boundary layer of sediments: oxygen microgradients over a microbial mat. *Limnol. Oceanogr. 35*: 1343–1355.

Karsten, U., R. Schumann, S. Rothe, I. Jung and L. Medlin. 2006. Temperature and light requirements for growth of two diatom species (Bacillariophyceae) isolated from an Arctic macroalga. *Polar Biol. 29*: 476–486.

Krause-Jensen, D., M. Kühl, P.B. Christensen and J. Borum. 2007. Benthic primary production in Young Sound, Northeast Greenland. *In*: (S. Rysgaard and R.N. Glud, eds) *Carbon cycling in Arctic marine ecosystems: case study Young Sound.* Danish Polar Center, Copenhagen. pp. 159–173.

Kühl, M. 2005. Optical microsensors for analysis of microbial communities. *Meth. Enzymol. 397*: 166–199.

Kühl, M., R.N. Glud, H. Ploug and N.B. Ramsing. 1996. Microenvironmental control of photosynthesis and photosynthesis-coupled respiration in an epilithic cyanobacterial biofilm. *J. Phycol. 32*: 799–812.

Kühl, M., C. Lassen and N.P. Revsbech. 1997. A simple light meter for measurements of PAR (400–700 nm) with fiber-optic microprobes: application for P vs. I measurements in microbenthic communities. *Aquat. Microb. Ecol. 13*: 197–207.

Kühl, M., R.N. Glud, J. Borum, R. Roberts and S. Rysgaard. 2001. Photosynthetic performance of surface-associated algae below sea ice as measured with a pulse-amplitude-modulated (PAM) fluorometer and O_2 microsensors. *Mar. Ecol. Prog. Ser. 223*: 1–14.

Kuwae, T., K. Kamio, T. Inoue, E. Miyoshi and Y. Uchiyama. 2006. Oxygen exchange flux between sediment and water in an intertidal sandflat, measured *in situ* by the eddy-correlation method. *Mar. Ecol. Prog. Ser. 307*: 59–68.

Kuznetsov, L.L. 1991. Microphytobenthos production in the Barents Sea coastal zone. *In:* (G.G. Matishoc, ed.) *Production-destruction processes in the Barents sea coastal zone*. KNC RAN: Apatity. pp. 52–62. (in Russian).

Kuznetsov, L.L. 2002. *Production of phytocoenoses and transformation of nutrients in ecosystem of the Barents Sea* (in Russian) Abstract of thesis of doctor of biology. Shirshov Institute of Oceanology, Moscow.

Kuznetsov, L.L. 2005. Comparative characteristics of phytocoe-noses production in the Barents Sea. *In:* (V. Shibanov, ed.) *Ecosystem dynamics and optimal long-term harvest in the Barents Sea fisheries*. PINRO, Murmansk. pp. 228–232.

Kuznetsov, L.L. and T.L. Strogaya. 1989. Microphytobenthos production indices and oxygen uptake by the community of the Dalnezelenetskaya Bay littoral (Barents Sea) – (in Russian). *Hirobiologicheskii Zhurnal 25*: 104–106.

Kuznetsov, L.L, O.N. Baytaz and P.R. Makarevich. 1998. Structural-functional characteristics of the plankton community in the Baidara bay on materials of expeditions in autmn 1991–1992. *In*: *Biology and oceanography of the Kara and Barents Seas (along the Arctic seaway route)*. KNC RAN. Apatity. pp. 88–95. (in Russian).

Larkum, A.W.D., E.M. Koch and M. Kühl. 2003. Diffusive boundary layers and photosynthesis of the epilithic algal community of coral reefs. *Mar. Biol. 142*: 1073–1082.

Lassen, C., R.N. Glud, N.B. Ramsing, N.P. Revsbech. 1998. A method to improve the spatial resolution of photosynthetic rates obtained by oxygen microsensors. *J. Phycol. 34*: 83–93.

Longphuirt, S.N., J. Clavier, J. Grall, L. Chauvaud, F. le Loc'h, I. le Berre, J. Flye-Sainte-Marie, J. Richard and A. Leynaert. 2007. Primary production and spatial distribution of subtidal microphytobenthos in a temperate coastal system, the Bay of Brest, France. *Estuar. Coast. Shelf Sci. 74*: 367–380.

MacIntyre, H.L., R.J. Geider and D.C. Miller. 1996. Microphytobenthic: the ecological role of the "secret garden" of unvegetated, shallow-water marine habitats. I. Distribution, abundance and primary production. *Estuar. Coasts 19*: 186– 201.

Matheke, G.E.M. and R. Horner. 1974. Primary productivity of the benthic microalgae in the Chukchi Sea near Barrow, Alaska. *J. Fish Res. Board Can. 31*: 1779–1786.

McGee, D., R.A. Laws and L.B. Cahoon. 2008. Live benthic diatoms from the upper continental slope: extending the limits of marine primary production. *Mar. Ecol. Prog. Ser. 356*: 103–112.

McMinn, A., T. Hirawake, T. Hamaoka, H. Hattori and M. Fukuchi. 2005. Contribution of benthic microalgae to ice covered coastal ecosystems in northern Hokkaido, Japan. *J. Mar. Biol. Ass. U.K. 85*: 283–289.

Menarad, H. and S.M. Smith. 1966. Hypsometry of ocean basin provinces. *J. Geophys. Res. 71*: 4305–4325.

Meyercordt, J., S. Gerbersdorf and L.-A. Meyer-Reil. 1999. Significance of pelagic and benthic primary production in two shallow coastal lagoons of different degree of eutrophication in the southern Baltic Sea. *Aquat. Microb. Ecol. 20*: 273–284.

Middelburg, J.J., C. Barranguet, H.T.S. Bushker, P.M.J. Herman and T. Moens. 2000. The fate of intertidal microphytobenthos carbon: an *in situ* 13C labeling study. *Limnol. Oceanogr. 45*: 1224–1234.

Morel, A. and L. Prieur. 1977. Analysis of variations in ocean color. *Limnol. Oceanogr. 22*: 709–722.

Nelson, J.R., J.E. Eckman, C.Y. Robertson, R.L. Marinelli and R.A. Jahnke. 1999. Benthic microalgae biomass and irradiance at the sea floor on the continental shelf of the South Atlantic Bight: spatial and temporal variability and storm effects. *Cont. Shelf Res. 19*: 477–505.

Pabi, S., G.L. van Dijken and K.R. Arrigo. 2008. Primary production in the Arctic ocean 1998–2006. *J. Geophys. Res. 113*: C08005.

Palmisano, A.C., J.B. Soohoo, D.C. White, G.A. Smith, G.A. Stanton and G.R. Burckle. 1985. Shade adapted benthic diatoms beneath Antarctic sea ice. *J. Phycol. 21*: 664–667.

Pamatmat, M.M. and D. Fenton. 1968. An instrument for measuring subtidal benthic metabolism *in situ*. *Limnol. Oceanogr. 13*: 357–540.

Parsons, T.R., D.G. Webb, H. Dovey, R. Haigh, M. Lawrence and G.E. Hopky. 1988. Production studies in the Mackenzie River – Beaufort Sea estuary. *Polar Biol. 8*: 235–239.

Platt, T., C.L. Gallegos and W.G. Harrison. 1980. Photoinhibition of photosynthesis in natural assemblages of marine phytoplankton. *J. Mar. Res. 38*: 687–701.

Rachold, V., H. Eicken, V.V. Gordeev, M.N. Grigoriev, H.-W. Hubberten, A.P. Lisitzin, V.P. Shevchenko, L. Schirrmeister. 2004. Modern terrigenous organic carbon input to the Arctic Ocean *In*: (R. Stein and Macdonald, eds). *The organic carbon cycle in the Arctic Ocean.* Springer-Verlag, Berlin. pp. 33–55.

Raven, J.A. and J. Beardall. 1981. Respiration and photorespiration. *In*: (T. Platt, ed.) Physiological bases and phytoplankton ecology. *Can. Bull. Fish. Aquat. Sci. 210*: 55–82.

Revsbech, N.P. and B.B. Jørgensen. 1983. Photosynthesis of benthic microflora measured with high spatial resolution by the oxygen microprofile method: capabilities and limitations of the method. *Limnol. Oceanogr. 28*: 749–756.

Revsbech, N.P., B.B. Jørgensen and O. Brix. 1981. Primary production of microalgae in sediments measured by oxygen microprofile, $H^{14}CO_3^-$ fixation, and oxygen exchange methods. *Limnol. Oceanogr. 26*: 717–730.

Rosenberg, R. and L. Lundberg. 2004. Photoperiodic activity pattern in the brittle star *Amphiura filiformis*. *Mar. Biol. 145*: 651–656.

Rysgaard, S., T.G. Nielsen and B. Hansen. 1999. Seasonal variation in nutrients, pelagic primary production and grazing in a high-arctic coastal marine ecosystem, Young Sound, NE-Greenland. *Mar. Ecol. Prog. Ser. 179*: 13–25.

Schreiber U. 2005. Pulse-amplitude (PAM) fluorometry and saturation pulse method. *In:* (Papageorgiou G., Govindjee, eds) *Chlorophyll fluorescence: a signature of photosynthesis.* Advances in photosynthesis and respiration series, Vol. 19. Kluwer Academic Publishers, Dordrecht. pp. 279–319.

Serreze, M.C., M.M. Holland and J. Stroeve. 2007. Perspectives on the Arctic's shrinking sea ice cover. *Science 315*: 1533–1536.

Springer, A.M. and C.P. McRoy. 1993. The paradox of pelagic food webs in the northern Bering Sea-III. Patterns of primary production. *Cont. Shelf Res. 5–6*: 575–599.

Steemann-Nielsen, E. 1952. The use of radio-active carbon (^{14}C) for measuring organic production in the sea. *J. du Conseil Expl. M. 18:* 117–140.

Subba-Rao, D.V. and T. Platt. 1984. Primary production of Arctic waters. *Polar Biol. 3*: 191–201.

Sun, M.Y., R.C. Aller and C. Lee. 1994. Spatial and temporal distributions of sedimentary chloropigments as indicators of benthic processes in Long Island Sound. *J. Mar. Res. 52*: 149–176.

Taddei D., P. Cuet, P. Frouin, C. Esbelin and J. Clavier. 2008. Low community photosynthetic quotient in coral reef sediments. Comp. *Rendus Biologies 9*: 668–677.

Tengberg, A., P.O.J. Hall, U. Andersson, B. Lidén, O. Styrenius, G. Boland, F. de Bovee, B. Carlsson, S. Ceradini, A. Devol, G. Duineveld, J.U. Friemann, R.N. Glud, A. Khripounoff, J. Leather, P. Linke, L. Lund-Hansen, G. Rowe, P. Santschi, P. de Wilde and U. Witte. 2005. Intercalibration of benthic flux chambers II. Hydrodynamic characterization and flux comparison of 14 different designs. *Mar. Chem. 94*: 147–173.

Vetrov, A.A. and E.A. Romankevich. 2004. Carbon cycle in the Russians Seas. Springer, Berlin. pp. 335.

Wagner, H., T. Jakob and C. Wilhelm. 2006. Balancing the energy flow from captured light to biomass under fluctuating light conditions. *New Phytol. 169*: 95–108.

Wenzhöfer, F. and R.N. Glud. 2004. Small-scale spatial and temporal variability in benthic O_2 dynamics of coastal sediments: Impact of fauna activity. *Limnol. Oceanogr. 49*: 1471–1481.

Wenzhöfer, F., O. Holby, R.N. Glud, H.K. Nielsen and J.K. Gundersen. 2000. *In situ* microsensor studies of a hydrothermal vent at Milos (Greece). *Mar. Chem. 69*: 43–54.

Woelfel J., R. Schumann, F. Peine, A. Flohr, A. Kruss, J. Tegowski, P. Blondel, C. Wiencke and U. Karsten. 2010. Microphytobenthos of Arctic Kongsfjorden (Svalbard, Norway) – Biomass and potential primary production along the shore line. *Polar Biol 33*: 1239–1253.

Zacher, K., R. Rautenberger, D. Hanelt, A. Wulff and C. Wiencke. 2009. The abiotic environment of polar marine benthic algae. *Bot. Mar. 52*: 483–490.

8 Microphytobenthic biomass along gradients of physical conditions in Arctic Kongsfjorden, Svalbard

Jana Woelfel, Rhena Schumann, Peter Leopold, Christian Wiencke and Ulf Karsten

Introduction

Kongsfjorden, a glacial fjord on the north-western coast of Svalbard (Spitsbergen, Norway), has received considerable research attention in recent years as a high latitude model ecosystem and monitoring site for the impact of climate changes (reviewed in Hop et al. 2002, Wiencke 2004). Although the structure of pelagic and benthic food webs in Kongsfjorden are well described for most higher trophic levels, even basic information of the different phototrophic organisms is still lacking for most groups, particularly for microphytobenthos (Hop et al. 2002).

Microphytobenthic biofilms cover extensive sediment areas worldwide in shallow waters and are responsible for high rates of primary production (Cahoon 1999). Consequently, these communities are a key component in trophic webs and function particularly as a major food source for benthic suspension- or deposit-feeders (MacIntyre et al. 1996, Underwood and Kromkamp 1999). In addition, microphytobenthic organisms stabilise sediment surfaces against erosion by excretion of extracellular polymeric substances (Paterson and Black 1999, De Brouwer et al. 2005 and references therein). Thereby, they also affect exchange of oxygen and nutrients across the sediment-water interface (Sundbäck et al. 1991, Risgaard-Petersen et al. 1994, Bartoli et al. 2003). Although microphytobenthic communities have gained increased research attention, most studies have been conducted on tidal flats or non-tidal shallow-water habitats with water depths <5 m (Wulff et al. 2005). Consequently, much less is known about the vertical distribution of these communities on the shelf (Cahoon 1999).

While the ecology of benthic microalgae is well investigated in cold-temperate to warm-temperate and even tropical waters (Cahoon 1999 and references therein), similar studies for polar regions are rare (Dayton et al. 1986, Gilbert 1991a, b, Glud et al. 2002). This is surprising, since the shelf area of the Arctic is large and accounts for ~22% of the global shelf area at depths of <200 m (Menard and Smith 1966). Although sea ice microalgae have been well examined (Grosselin et al. 1997, Mock and Gradinger 1999), there have been only few microphytobenthic studies. The community composition in sea ice seems to be more similar to benthos than to plankton (Horner et al. 1992, Weissenberger et al. 1992). Benthic diatoms make up 42–88% of Antarctic sea ice assemblages (Kang and Fryxell 1992).

Furthermore, in polar areas, only a minor fraction of gross pelagic primary production reaches the sediment, due to efficient carbon and nutrient recycling in the upper water column (Rysgaard et al. 1999). In a more recent study, Glud et al. (2002) were among the first to document high biomass and high primary production of microphytobenthic assemblages in Young Sound, a high Arctic fjord on the north-eastern coast of Greenland. In their study, sediment-dwelling diatoms accounted for about 40% of total benthic primary production under polar conditions, indicating an ecologically important role as a food source for both benthic and pelagic heterotrophs at these high latitudes. Some earlier investigations in Antarctic and Arctic waters also point to an ecological significance of these microalgae, at least at certain times of the year (Chapter 7; Horner and Schrader 1974, Dayton et al. 1986, Gilbert 1991a, b, Glud et al. 2009). However, except for the study of Glud et al. (2002), more recent studies on microphytobenthic biomass in polar, and especially Arctic waters are lacking.

Kongsfjorden is a glacial fjord influenced by Atlantic and Arctic water masses harbouring a variety of boreal and Arctic organisms. Particularly during summer, influxes from large glaciers and terrestrial melt water create steep sedimentation and salinity gradients along the length of this fjord, leading to a reduced biomass and diversity of the zoobenthic community in the inner reaches (Hop et al. 2002). In contrast, similar information on benthic phototrophic biomass in Kongsfjorden is still lacking (Karsten et al. 2006). In temperate intertidal sediments, nutrients are usually not a limiting factor (Barranguet et al. 1998, Migne et al. 2004) because microphytobenthos can take up nutrients from both the water column and sediment pore water (MacIntyre et al. 1996). However, for Kongsfjorden, this question has not yet been addressed.

Therefore, the aim of the present study was to quantify benthic microalgal communities of Kongsfjorden along the prevailing physical gradients and sediment characteristics. Chlorophyll-based biomass was investigated in sediment samples taken throughout the fjord and from different water depths. In addition, organic content and nutrient concentrations were measured. We formulated two hypotheses: (1) higher microphytobenthic biomass occurs in the inner fjord (near the glaciers) because of finer sediments and enhanced nutrient concentrations, and (2) at shallow water stations (3–5 m water depth), due to better light availability. Another postulate was that benthic diatoms colonise large parts of Kongsfjorden in high abundances and, hence, are very important for trophic relationships in the harsh Arctic environment.

Our approach

The study was conducted during the polar summer from June to August in 2006 in Kongsfjorden (79° N, 12° E) off the west coast of Svalbard (Figure 8.1). Kongsfjorden extends 26 km from northwest to southeast and its width ranges from 3 to 8 km. Maximum water depth is about 400 m. Kongsfjorden is strongly influenced by the presence of four tidewater glaciers: Kronebreen and Kongsvegen at the head as well as Conwaybreen and Blomstrandbreen on its northern coast (Figure 8.1). The tidal range is about 2 m and currents are weak (Svendsen et al. 2002). Sediments (fine mud to sand) are present on level surfaces, often near freshwater outflows. Rocky substrata are most common in areas that are exposed to stronger hydrodynamic activity (Hop et al. 2002). Compared to other locations at this high latitude, air temperature is much higher, with an annual mean ranging from −15°C in winter to about 5 °C in summer. The annual mean water temperature is around 0 °C. The inner part of Kongsfjorden usually becomes ice-covered around December/January, but between 2005/2006 this

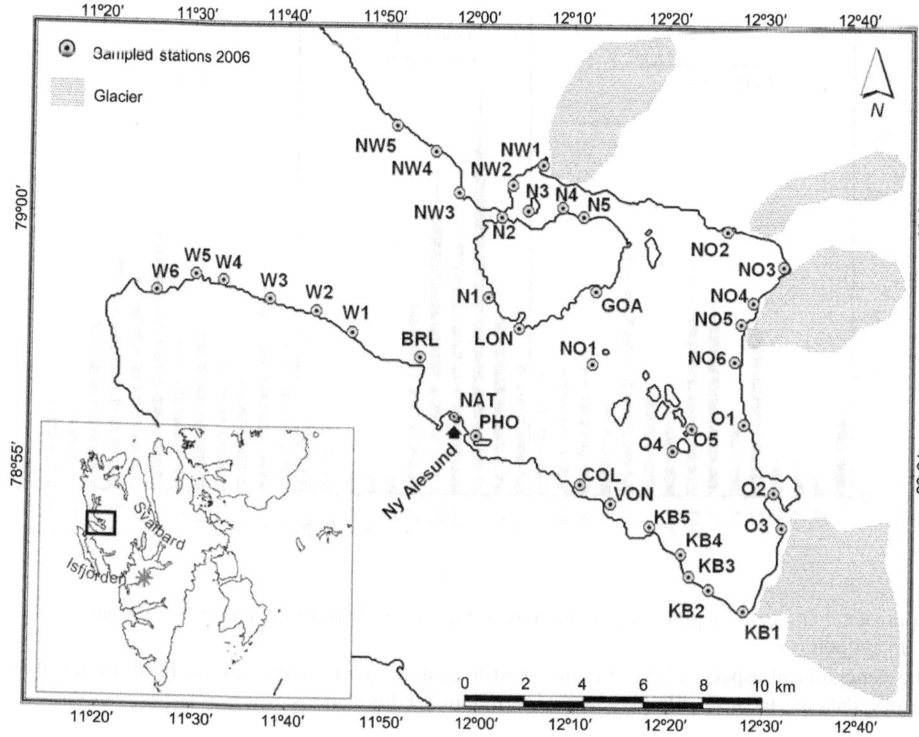

Figure 8.1 Locations of the sampling stations in 2006 in Kongsfjorden, Spitsbergen (Svalbard). The four adjacent glaciers are shown as grey areas.

was not the case. However, the area becomes entirely ice-free during the spring and complete ice cover over the inlet is an exceptional event (Svendsen et al. 2002).

At 41 stations, SCUBA diving was used to collect sediment samples at water depths of 3–5 m in order to monitor the areal distribution of microphytobenthos in the fjord (Figure 8.2, Table 8.1) (average area sampled: 1 m^2). Three replicates were taken with Plexiglas^{TM} sediment tubes (internal diameter 5 cm, 25 cm length) for biomass, cell number, chemical and sedimentary analyses (see below).

More detailed analyses were performed at the three stations "Brandal (BRL)", "Corbel (COL)" and "London (LON)" (Figure 8.1, Table 8.1), because of their good accessibility for diving even under the changing and sometimes difficult meteorological conditions. Two sampling sites, "Brandal" and "Corbel", are located off the northern shore, and one site, "London", is located off the southern shore of Kongsfjorden. At these stations, sediment cores in transect order were taken at water depths of 3–4, 10, 15, 20 and 30 m every month. Due to high patchiness of microphytobenthic biomass, 3–9 cores from each sampling location were used for the determination of biomass (chl*a* and organic matter). Three additional cores were sampled for pore water nutrients and sediment properties. All sediment cores were transported inside their Plexiglas^{TM} tubes to the laboratory under undisturbed, cool and dark conditions. They were kept in a cool room at 2 °C until further processing, which was done as quickly as possible within 24 h after sample transfer from the field to the laboratory.

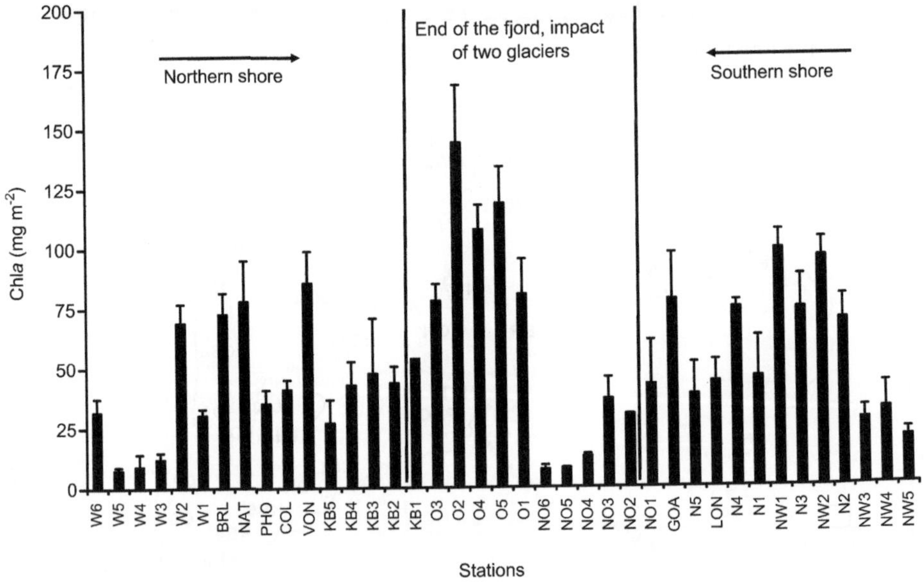

Figure 8.2 Gradient of phototrophic biomass (chl*a* mg m^{-2}) from the coastline of Kongsfjorden in June 2006.
The geographical aspects of the stations, west to east and east to west, in the fjord are given as arrows; data are means ± SD (n = 3 – 9). See Figure 8.1 for stations.

All sediment cores were sectioned horizontally into different layers with razor blades. Each top 5 mm slice of three replicate cores was weighed and homogenised manually before sub-samples were taken and weighed again for photometric pigment and organic content analysis [particulate organic carbon (POC), particulate organic nitrogen (PON)], as well as for cell counting. Three other sediment cores were sliced from the top to 3 cm depth, and used to extract pore water for nutrient analyses and for sediment characterisation (grain size, water content). For analysing vertical distribution patterns of pigments in the sediment (sediment profiles), three replicate cores from station Corbel (taken at water depths of 10 m and 20 m in July 2006) were sliced from the top down to 5.5 cm in steps of 0.5 cm.

Surface water was collected at seven stations (Table 8.2) for nutrient analyses and comparison with the pore water data (Tables 8.4 and 8.5). Dissolved inorganic nitrogen (DIN, as sum of ammonium, nitrate and nitrite concentrations) and soluble reactive phosphorus (SRP) were analysed photometrically (Shimadzu UV 2401, Shimadzu Co., Kyoto, Japan) in Whatman GF/F filtrates (Whatman International Ltd, Maidstone, England) (Rohde and Nehring 1979, Malcolme-Lawes and Wong 1990). The filters with the biomass were used for determination of organic content (PON, POC) in an element analyser "VarioEL III" (Elementar GmbH, Hanau, Germany) (Verardo et al. 1990). The concentration of silicate was measured using the autoanalyser Evolution III (Alliance Instruments GmbH, Salzburg, Austria) according to Hansen and Koroleff (1983). The vertical light field was measured using a Secchi-disc (Hydrobios GmbH, Kiel, Germany) and seston was quantified as a proxy for turbidity. Seston was gravimetrically determined after filtration of 0.5 – 1.5 l of surface water through Whatman

Table 8.1 Locations and water depths of sampling stations for sediments and surface water in Kongsfjorden (June–August 2006)

Sampling stations	Latitude N	Longitude E	Water depths (m) of sediment cores
NAT*	78°55.817′	011°55.236′	5
LON*	78°57.569′	012°2.653′	4, 10, 15, 20, 30
GOA*	78°58.323′	012°11.708′	5
PHO*	78°55.072′	011°59.205′	5
BRL*	78°56.840′	011°51.211′	3, 10, 15, 20, 30
VON*	78°53.829′	012°11.909′	5
KB1	78°52.283′	012°28.004′	5
KB2	78°52.701′	012°24.709′	5
KB3	78°52.957′	012°22.343′	5
KB4	78°53.389′	012°21.426′	5
KB5	78°53.526′	012°19.786′	5
NO1	78°57.119′	012°12.008′	5
NO2	78°57.119′	012°12.008′	5
NO3	78°59.441′	012°23.874′	5
NO4	78°58.972′	012°32.338′	5
NO5	78°58.194′	012°28.955′	5
NO6	78°57.846′	012°27.499′	5
COL*	78°54.307′	012°8.849′	4, 10, 15, 20, 30
O1	78°56.003′	012°26.122′	5
O2	78°54.619′	012°31.144′	5
O3	78°53.924′	012°32.008′	5
O4	78°55.214′	012°19.391′	5
O5	78°55.214′	012°19.391′	5
N1	78°58.238′	011°59.686′	5
N2	78°59.578′	012°2.926′	5
N3	79°0.107′	012°5.232′	5
N4	79°0.172′	012°8.816′	5
N5	78°59.998′	012°11.025′	5
W1	78°57.049′	011°46.795′	5
W2	78°57.584′	011°43.001′	5
W3	78°57.899′	011°38.202′	5
W4	78°58.211′	011°33.263′	5
W5	78°58.368′	011°30.107′	5
W6	78°58.279′	011°26.027′	5
NW1	79°1.275′	012°6.275′	5
NW2	79°0.685′	012°2.039′	5
NW3	79°0.451′	011°57.919′	5
NW4	79°1.258′	011°55,433′	5
NW5	79°1.753′	011°51.395′	5

The sampling stations for surface water are marked *. See Figure 8.1 for locations of sampling stations in the fjord.

GF/F filters and drying the filter for 24 h at 60 °C. Water temperature and salinity were measured with a WTW Salinometer Multiline P4 (WTW GmbH, Weilheim, Germany), the latter as conductivity using the practical salinity scale.

Sediment grain size was determined by wet sieving with defined mesh sizes of 1 mm, 500, 200, 100 and 63 μm, and classified according to Wentworth (1922). Sediment water content was measured as weight loss after drying weighed sediment samples for 24 h at 105 °C.

For concentrations of DIN and SRP in the pore water, the upper 3 cm of sediment cores were used. Pore water was obtained by centrifugation of tubes with embedded sieves (mesh size 1 mm). Afterwards, respective filtrates of pore water passed through Whatman GF/F filters were analysed photometrically (see above).

Pre-weighed subsamples of ca. 2–4 g fresh weight were extracted in 5 ml 90% acetone (v/v) for 24 h in the dark at 4–5 °C with occasional vortex mixing. Afterwards, samples were centrifuged for 5 min at 7200 g at 5 °C, supernatants were collected and absorbance was determined photometrically. Pellets were re-extracted twice in this procedure to improve the extraction efficiency by resuspension in 2.5 ml 90% acetone. All chl*a* values were quantified according to Jeffrey and Humphrey (1975).

POC and PON contents were determined in freeze-dried samples that were homogenised to a fine powder using a mortar and pestle. Samples of 50–100 mg dry weight (DW) were combusted in the element analyser (Verardo et al. 1990) following acidification of the samples in silver cups with 50–100 μl 10% HCl (v/v) to remove inorganic carbonates. The dry and carbonate-free samples were repacked air-tight in tin foil and then measured. However, the "detritus" fraction was not excluded from "photosynthetically active" biomass, thus our C:chl*a* ratios were affected.

Pre-weighed subsamples for cell count (abundance) analyses were suspended in 20 ml 0.2 μm filtered seawater from Kongsfjorden and preserved with 2% glutaraldehyde (v/v) (final concentration). For microscopic analysis, sand particles were removed by sieving through a mesh size of 200 μm. Thus, a fraction of detritus was also removed by this method, affecting our chl*a* cell^{-1} FW ratios. Cell counting chambers with a defined volume of ca. 1 ml were filled with algal samples 20–90 times further diluted. Cells were allowed to sediment for 6–14 h. Counting was performed with an Olympus microscope CH 20 (Olympus Optical Ltd., Tokyo, Japan) equipped

Table 8.2 Water temperature, salinity and Secchi depth in surface samples at different stations (see also Tables 8.1 and 8.4) in Kongsfjorden (May–June 2006).

Station	Date	Water temperature (°C)	Salinity	Secchi depth (m)
PHO	25 May	1.1	34	10
BRL	29 May	1.5	34	9
COL	31 May	3.2	n.d.	n.d.
LON	2 June	3.5	n.d.	6.5
GOA	8 June	3.3	30	n.d.
NAT	12 June	2.5	33	10.5
VON	14 June	2.6	33	4
COL	21 June	4.9	30	2.7
BRL	26 June	5.1	33	7
LON	28 June	6.2	33	4.1

See Figure 8.1 for sampling stations.
n.d., not determined.

with a × 10/0.25 objective. Diatom cells were grouped into empty (dead) and filled (living) valves. The proportion of broken frustules in all samples was always <5%.

Statistical analysis of the data was performed using SPSS software (version 15.0). Three to nine replicates were tested for significant differences using 1-way ANOVA. Homogeneity of variances was tested applying the Levene's test. All data were homogeneous, thus no transformation of data was performed. A critical $p < 0.05$ indicated significant differences. Post-hoc comparisons (Tukey-test) were used to identify pairwise the significantly different averages.

Environmental parameters in surface water and sediment characteristics

Salinity and temperature were relatively stable in the near-shore environment of ≤5 m water depth in Kongsfjorden. With seasonally increasing air temperature, surface water temperature increased from 1 to 6 °C in June, but the salinity remained between 30 and 33 (Table 8.2). Water transparency decreased in early June and was followed by fluctuating values as indicated by changing Secchi depths and PAR data (Tables 8.2 and 8.3). However, small changes in the vertical attenuation coefficients (0.2–0.4; Table 8.3) as well as low seston values (Table 8.4) indicate that irradiance reduction was marginal throughout the study period. Photon fluence rates varied according to weather conditions (May–July 2006) with maximum and minimum values ranging from 56 to 288 µmol m^{-2} s^{-1} at 5 m water depth (R. Müller, unpublished data). Thus, salinity and light availability were rather similar at all sites down to 5 m water depth, except in the near vicinity of glaciers, where more turbid water was observed but not analysed.

In surface water of seven stations, plankton biomass (organic content as well as chl*a* values) and inorganic nutrient values were very low (Table 8.4). However, while SRP values were especially low at 0–0.4 µmol l^{-1}, nutrient concentrations of DIN and silicate were never below 1 µmol l^{-1}. In pore water of the sediments at the same

Table 8.3 Underwater PAR intensities, vertical attenuation coefficients of downward irradiance (Kd) (calculated: 0.5–5 m) and Secchi depths measured close to the Ny-Ålesund harbour in Kongsfjorden (2006) under different weather conditions using a spherical quantum meter (Quantum Li-192 SA. LiCor, Lincoln, USA).

Water depth	PAR (µmol photons m^{-2} s^{-1})			
	23 May 2006	9 June 2006	13 June 2006	21 June 2006
Surface	1200	950	656	650
0.5 m	767 ± 58	500 ± 436	344 ± 4	470 ± 13
1 m	573 ± 110	420 ± 365	276 ± 3	412 ± 14
3 m	400 ± 50	172 ± 149	131 ± 2	258 ± 8
5 m	250 ± 0	85 ± 83	74 ± 3	182 ± 3
Weather	Sunny	Sunny	Overcast	Overcast
Local time	12 : 00	11 : 00	13 : 30	13 : 15
Kd	0.25	0.39	0.34	0.21
Secchi depth (m)	9.5	2.3	3.7	5.5

Data are means ± SD. (R. Müller, unpublished data; Kd values are published in Fricke et al. 2008).

Table 8.4 Abiotic and biotic parameters in surface water at different sampling stations (see also Table 8.1) in Kongsfjorden (June–August 2006).

	Range	Stations
Nutrients		
DIN (μmol l^{-1})	1.0–5.1	7
SRP (μmol l^{-1})	0.0–0.4	7
Silicate (μmol l^{-1})	1.1–3.5	6
Organic content		
Chl*a* (μg l^{-1})	0.3–1.4	6
Seston (mg DW l^{-1})	0.7–7.0	5
POC (mg l^{-1})	0.2–0.5	6
PON (mg l^{-1})	0.0–0.1	6

Data ranges are minimum and maximum values. See Figure 8.1 for sampling stations.
DIN, dissolved inorganic nitrogen; SRP, soluble reactive phosphorus; POC, particulate organic carbon; PON, particulate organic nitrogen.

stations, nutrient concentrations were much higher, with minimum and maximum values of DIN ranging from 1.9 to 36.7 μmol l^{-1} and SRP ranging from 0.9 to 5 μmol l^{-1} (data not shown). At stations Corbel, Brandal and London, DIN:SRP ratios varied from 1 to 30 in relation to transect depth, but no spatial trend was obvious.

At seven stations (all at 5 m depth including the well-investigated BRL, LON, COL stations; see Table 8.1), the predominant sediment type was medium-grained, well sorted, sand with a median grain size of 0.2–0.3 mm (for four stations data not shown). The sediments in deeper waters (30 m) at stations London and Brandal were similar, but with slightly finer sand (modal particle size of 0.16–0.21 mm) was found. Hence, the water content of the sediment varied in relation to grain size between 19 and 36%, but mostly between 21 and 23% of fresh weight (FW).

Comparing biomass variability at different scales

The chl*a* concentrations of the stations in Figure 8.2 followed a geographical trend from west to east off the northern shoreline and from east to west off the southern shoreline, with strong large spatial scale heterogeneity (values ranged from 8 to 145 mg chl*a* m^{-2}) and relatively little variation (range of % SD) within stations. At both western ends of the coast (stations W1–W5 and NW3–NW5), biomass values were lower than in the inner fjord, while concentration increased at stations near the glaciers. Highest biomass values of >100 chl*a* mg m^{-2} were measured at stations O2, O4 and O5 in front of the glacier, respectively.

Data from three intensively investigated stations [two (Brandal and Corbel) located off the northern shore and one (London) off the southern shore] were analysed by 1) sampling month (seasonality), 2) water depth (transects), and 3) sediment depth (profiles). While stations Brandal and Corbel were more exposed to wave action, London was sheltered due to its location in a small inlet.

Monthly differences in biomass values in 10 m water depth are shown in Figure 8.3. A significant continuous seasonal development of microphytobenthos did not occur at this transect depth. Similar high values of 60–130 mg chl*a* m^{-2} were found

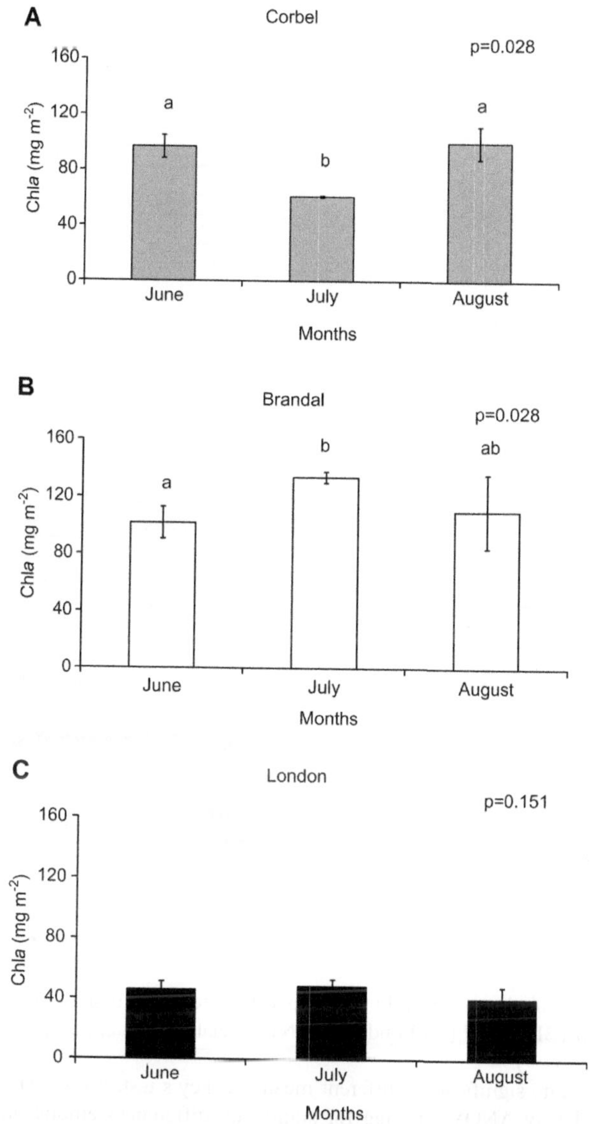

Figure 8.3 Phototrophic biomass in different months of 2006 in Kongsfjorden at stations (A) Corbel (COL), (B) Brandal (BRL) and (C) London (LON) at a water depth of 10 m; data are means ± SD (n = 3–9).
Different letters indicate significantly different means (Tukey's test, $p < 0.05$). p values in Figure indicate results of 1-way ANOVA testing for significant differences among all means (COL: $F_{2,12} = 4.89$; BRL: $F_{2,12} = 5.07$; LON: $F_{2,12} = 2.22$). See Figure 8.1 for stations.

at the exposed stations Brandal and Corbel. At London, values were lower at around 40 mg chl*a* m^{-2}. In contrast to the biomass values at 10 m water depth, microalgal abundance changed seasonally at other transect depths (Figure 8.4). Biomasses at 3–4 m water depth doubled or tripled from June to August at all three stations.

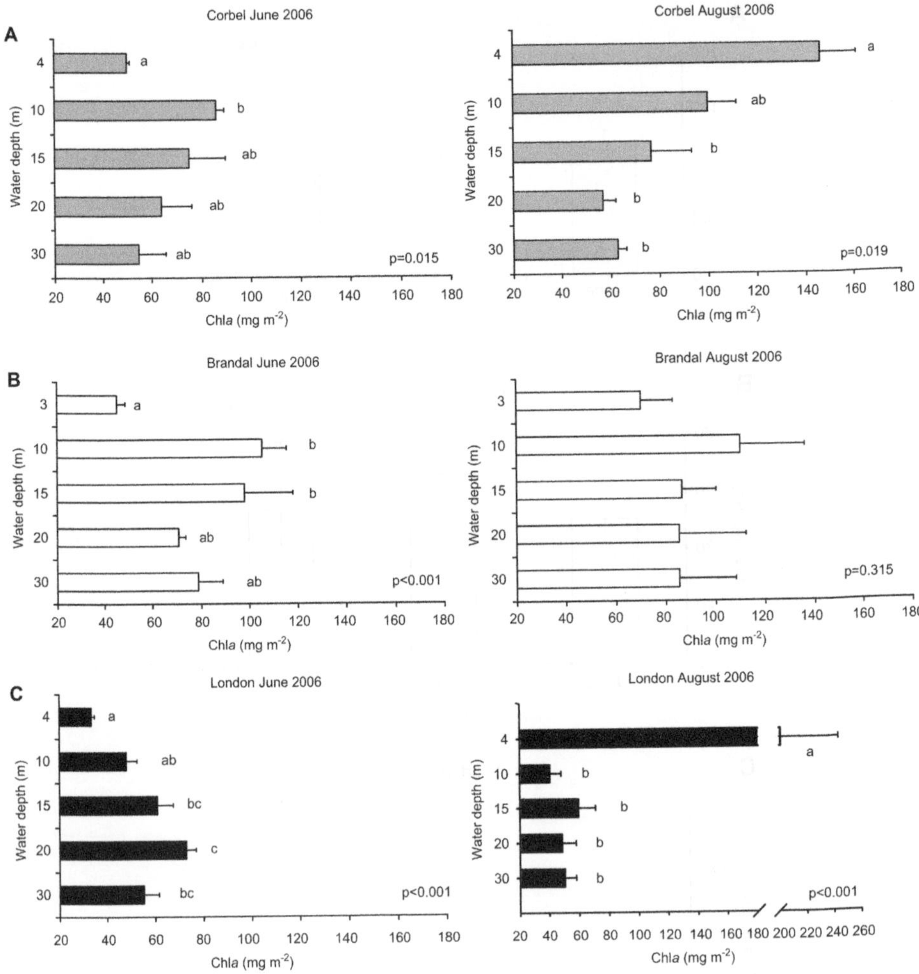

Figure 8.4 Gradient of phototrophic biomass in different water depths at stations (A) Corbel (COL), (B) Brandal (BRL) and (C) London (LON) in relation to sampling months in 2006; data are means ± SD ($n = 3-9$).
Different letters indicate significantly different means (Tukey's test, $p < 0.05$). p values in Figure indicate results of 1-way ANOVA testing for significant differences among all means (COL_{June}: $F_{4,34} = 3.59$; COL_{August}: $F_{4,10} = 1.36$; BRL_{June}: $F_{4,37} = 9.14$; BRL_{August}: $F_{4,10} = 4.87$; LON_{June}: $F_{4,37} = 13.67$; LON_{August}: $F_{4,10} = 31.94$). See Figure 8.1 for stations.

Despite light attenuation with increasing transect depth, an expected parallel decrease in biomass was not evident, and significant differences were absent among different water depths of <10 m at all stations (Figure 8.4). Very high C:N ratios of 28–124 were measured at Brandal and London; ratios at Corbel were moderate (7–8). Furthermore, cell numbers (dominated by diatoms) and the proportion of empty cells varied little by transect depth and geographical location (data not shown). In concordance with these data, we found minor differences in chl*a* values, and no decrease in chl*a*:cell ratios (107–2337 pg chl*a* $cell^{-1}$ FW) in relation to transect depth

Table 8.5 C:N ratio of the organic content, chla:cell ratio (pg cell^{-1} FW) and C:chla ratio [μg C (μg chla)$^{-1}$ DW] of biomass in sediments as well as DIN:SRP ratios (μmol l) (dissolved inorganic carbon:soluble reactive phosphorus) in the pore water determined in different water depths (m) at three sampling stations Corbel (COL), Brandal (BRL) and London (LON) in Kongsfjorden (June–August 2006) (n = 3–4).

Water depth (m)	DIN:SRP ratio of nutrients in the pore water			C:N ratio of organic content			Chl*a* : cell ratio (pg cell^{-1} FW)			C:chl*a* ratio of biomass		
	COL	BRL	LON	COL	BRL	LON	COL	BRL	LON	COL	BRL	LON
3 or 4	2	6	1	7	66	65	107	808	176	75	692	867
10	15	9	7	8	61	124	230	455	564	48	231	1164
15	9	5	9	n.d.	87	112	543	183	542	n.d.	286	1079
20	30	4	9	n.d.	32	52	2337	255	460	n.d.	320	779
30	26	7	28	n.d.	28	35	595	380	1204	n.d.	366	503

See Figure 8.1 for sampling stations.
n.d., not determined.

(Table 8.5). This discontinuity is also evident for C:chl*a* ratios: lowest ratios without a trend were calculated for Corbel at 48–75 μg C (μg chl*a*)$^{-1}$, followed by Brandal at 231–692 μg C (μg chl*a*)$^{-1}$ and lastly London at 503–1164 μg C (μg chl*a*)$^{-1}$.

Pigment concentrations in vertical sediment profiles decreased from 68 mg chl*a* m^{-2} at the top to 12 mg chl*a* m^{-2} at 5.5 cm sediment depth at a transect depth of 10 m, but not as much at a transect depth of 20 m (Figure 8.5). Eighty percent and 63% of biomass in water depths of 10 m and 20 m were concentrated in the upper 1 cm of the sediment, respectively.

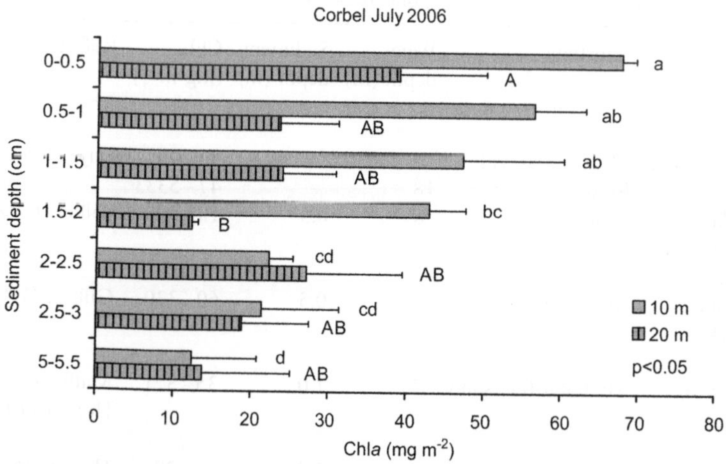

Figure 8.5 Phototrophic biomass in relation to sediment depth in July 2006 in Kongsfjorden at station Corbel (COL) at water depths of 10 m and 20 m; data are means ± SD (n = 3). Different letters indicate significantly different means (Tukey's test, $p < 0.05$). p values in figure indicate results of 1-way ANOVA testing for significant differences among all 3 means (COL_{10m}: $F_{6,12} = 20.07$; COL_{20m}: $F_{6,14} = 2.92$). See Figure 8.1 for location of station.

Distribution of microphytobenthos along physical gradients

The gradients of underwater light climate (transect depth, glacier influence on attenuation) as well as seasonal aspects (daylengths) had a much lower impact on microphytobenthos biomass in the Arctic Kongsfjorden than expected (Figures 8.2–8.4, Table 8.5). There was rather high large-scale patchiness, but only minor changes of biomass with water depth or at stations situated near glaciers (Figures 8.1, 8.2, 8.4). During the Arctic summer 2006, biomass accretion was lower than might have been expected from the observed high population density in June.

Microphytobenthos biomass evaluations in polar regions are still scarce (Table 8.6). Gilbert (1991a, b) emphasised their importance for primary production, with high net productivity rates of 313–701 mg C m^{-2} day^{-1} off Signy Island (South Orkney Islands, Antarctica). Chlorophyll biomass at this shallow water site was very high at 10–500 mg m^{-2}. Other polar locations had similar values (Table 8.6). Microphytobenthic chl*a* concentrations from Kongsfjorden are in good agreement with these data and with those from temperate regions (sandy eulittoral sites), e.g., Baltic Öresund at 1–154 mg m^{-2} (Sundbäck 1983) or German Baltic Sea at 10–142 mg m^{-2} (Woelfel et al. 2007). However, compared to temperate waters, similar biomass values were reached under the much lower Arctic water temperatures, which might indicate 1) a high growth potential under polar conditions and/or 2) a reduced removal of microphytobenthic biomass. Consequently, the important function of microphytobenthic assemblages for the Kongsfjorden benthic ecosystem is substantiated.

No gradient in areal pigment concentration in relation to geographical location was determined (Figure 8.2). Biomass was lower off the western end of the coast (stations W1–W5 and NW–NW5) compared to the inner fjord, most likely caused by higher current rates and coarser sediments. Stations situated near glaciers with lower water

Table 8.6 Microphytobenthic biomass in polar regions.

Site	Substrata	Water depth (m)	Sediment depth (m)	Chl*a* (mg m^{-2})	References
Antarctic					
McMurdo sound	Mud and gravel	6		96–457	Dayton et al. (1986)
	Mud	18		47–5333	
Signy Island, Mooring point, east coats	Sand	8	0.5	10–500	Gilbert (1991a)
Signy Island, Billie rock	Coarse sand	12	0.5	60–330	Gilbert (1991b)
Arctic					
Chukchi Sea,	Mud and fine sand	5	1.0	35–321	Matheke and Horner (1974)
Alaska					
Beaufort Sea	n.d.	7	1.0	1–28	Horner and Schrader (1982)
Kongsfjorden	Sand	4–30	0.5	8–200	This study

n.d., not determined.

transparency had inconsistent chl*a* data; NO had reduced chlorophyll values, while higher concentrations were measured at O2, O4 and O5. Thus, the influence of glaciers on microphytobenthic biomass should be studied in more detail in future.

It is beyond dispute that colonisation and, consequently, structure and functioning of microphytobenthos is highly dependent upon the degree of exposure to water movement, sediment characteristics and environmental stability. Additionally, biomass values are dependent on the "balance" between accumulation (e.g., growth, deposition, migration) and removal (e.g., grazing, resuspension, senescence) processes. Intense movement of sediment grains, especially those that are coarse, inhibits the development of microphytobenthos (particularly epipelic diatoms) (Sundbäck 1983), which probably explains the lowest biomass concentrations at exposed sites such as station Brandal (Figure 8.4B).

In Kongsfjorden, the highest proportion of finer sediments (fine mud to sand) occurs in the deeper water (>50 m), and the dominant bottom types in shallow, better lit regions are rock, boulder substrata and gravel (Hop et al. 2002). Thus, our selection of sampling stations was limited by the lack of sufficient sandy or muddy areas for microphytobenthos, as well as by logistical and meteorological considerations. Hence, more sheltered areas, e.g., the inlet of station London, were chosen for the present study where much organic matter (mostly originating from macroalgal detritus) was deposited. It is most likely that this high amount of detritus distorted our biomass analyses at different scales. Based on the cell counts, extremely high chl*a* concentrations ranging from 107 to 1204 pg chl*a* cell^{-1} FW were calculated, indicating that, in addition to benthic diatoms, there are other chl*a* sources (e.g., macroalgal fragments). Cell-specific chl*a* values of 1.6 and 3.5 pg chl*a* cell^{-1} FW are typical of unialgal diatom cultures of *Nitzschia* cf. *aurariae* Cholnoky and *Cylindrotheca closterium* Reimann *et* Lewin (L. Gustavs, personal communication), which are abundant taxa in sediment samples of Kongsfjorden. Probably, these two cultures do not reflect all cell sizes found in the field, thus higher cell-specific chl*a* values can be expected, but not as high as those we measured. Additionally, high values of C:chl*a* ratios [231–1164 μg C (μg chl*a*)$^{-1}$] at stations Brandal and London confirm this assumption, whereas 80 μg C (μg chl*a*)$^{-1}$ is the average quotient in phytoplankton samples (Garnier and Mourelatos 1991). Without a correction for the detritus contribution, Erga et al. (1989) calculated a quotient of 10–1000 μg C (μg chl*a*)$^{-1}$ for phytoplankton in Boknafjorden (Norway). Furthermore, no spatial trend of C:chl*a* ratios in relation to water depth was determined. However, similar studies that separate microalgae from detritus by sieving or give special emphasis to cell counting on macroalgal remains will be more conclusive.

Low phytoplankton biomass values for seston and POC up to 1 mg DW l^{-1} (Table 8.4) indicate a potential nutrient limitation, which is present when concentrations are below the theoretical half-saturation constant (Ks) for uptake: 2 μM for DIN, 0.2 μM for SRP and 2 μm for DSi (Fisher et al. 1992). But the total range of DIN and silicate over the growing season was high enough to support phytoplankton growth, while low SRP values indicate a phosphorus limitation. However, to estimate a possible nutrient limitation more data on the C:N:P ratio of phytoplankton biomass are needed. In interstitial water, nutrient concentrations of SRP and DIN were never below these threshold values (Table 8.5) making nutrient limitation for benthic microalgae unlikely. Due to naturally higher nutrient concentrations in the sediment, silicate concentrations (measured only in surface water) were most probably sufficient for diatom growth. According to Hillebrand and Kahlert (2002), when *in situ* studies of nutrient enrichment in sediment microalgae result in significant deviations from low N:P quotients <13 and high C:N quotients of >10, there may be nitrogen limitation and, at high N:P quotients >10, potential phosphorus limitation. In pore water at stations Brandal and London, low DIN:SRP values of 1–9 and high C:N values of 32–124 were calculated, indicating nitrogen limitation, whereas biomass at Corbel was not influenced. On the other hand, in the Arctic sediments of Svalbard, nitrogen is mineralised preferentially relative to carbon (Thamdrup and Fleischer 1998). Our very low PON data and at least partly very high C:N ratios up to 1:60 (in maximum 1:125) at Brandal and Corbel (Table 8.5) confirm this study (op. cit.). Probably, low water temperatures prevent rapid degradation of the micro- and macroalgal detritus.

Due to melt water, salinity can decrease from 35 to 23.4, but only at the surface, while in greater depth at around 5 m and deeper, variations are rather small, around 34–35 during summer (Hanelt et al. 2001). Furthermore, water temperature ranged from only 0–5°C. Our measurements confirm these rather small changes in both parameters (Table 8.2), and hence any controlling effect of salinity and/or temperature on microphytobenthos settlement seems to be unlikely.

Water transparency, on the other hand, quickly decreased in early June when ice and snow melt increased (Tables 8.2 and 8.3) and may have reduced the growth of benthic microalgae. Minimal light intensities at which polar and subpolar microphytobenthic activity have been determined range between 0.5 and 2.5 μmol photons m^{-2} s^{-1} (Chapter 6; Glud et al. 2009 and references therein). Although underwater light conditions fluctuated during the investigation period, characterised by changing Secchi depths (Table 8.2) and Kd values (Table 8.3), growth of biomass at greater depths also seemed possible. In addition, benthic microalgae in Arctic waters are well adapted to low light conditions (Karsten et al. 2006). Based on chlorophyll fluorescence measurements of intact communities, these organisms are able to efficiently optimise their photosynthetic apparatus to ambient, often low and fluctuating irradiance changes and can, therefore, inhabit sediments at least down to 30 m water depth (Glud et al. 2002).

However, net accumulation of microphytobenthos was rarely observed and only at 3–4 m water depth from June to August at three stations (Figures 8.3, 8.4). There was no positive biomass accretion at deeper transect sites during the 2006 season. Highest values were measured in upper sediment layers down to 2 cm sediment depth, especially at a transect depth of 10 m (Figure 8.5), and may be explained by better light availability. In conclusion, many more data for shallow water transects and over shorter time intervals from May to September are needed to detect seasonal effects of changing underwater light climate on microphytobenthos performance and productivity.

Sediments of Kongsfjorden are often inhabited by a dense macrofauna including molluscs (e.g., different species of *Yoldiella*), polychaetes and a suite of tanaid crustaceans (Hop et al. 2002). Unfortunately, we do not have sufficient data to estimate the impact of small fauna, but it has been proposed that micro- and meiofauna are able to consume most of microalgae in sediments (Sundbäck et al. 1996). Thus, it seems reasonable to assume that these invertebrates exert high grazing pressure, which could well have been the cause for the low or absent biomass accumulation at most sites and deeper transect depths. Due to this fact, benthic Arctic microalgae could certainly be a driving force in the food web of Kongsfjorden. The following approximate calculation underlies this hypothesis. Assuming a mean chlorophyll concentration of 80 mg chl*a* m^{-2} in sediments down to water depths of 30 m, and a total shore lengths of about 60 km for Kongsfjorden, biomass production would amount to about 143 kg chl*a* (~17,000 kg C) production during summer time, which is considerable. However, more detailed biomass studies on epiphytic and epilithic microalgae must be done in Kongsfjorden as well as similar studies in other Arctic regions to evaluate the ecological significance of microphytobenthic assemblages for polar waters.

References

Barranguet, C., J. Kromkamp and J. Peene. 1998. Factors controlling primary production and photosynthetic characteristics of intertidal microphytobenthos. *Mar. Ecol. Prog. Ser. 173*: 117–123.

Bartoli, M., D. Nizzoli and P. Viaroli. 2003. Microphytobenthos activity and fluxes at the sediment-water interface: interactions and spatial variability. *Aquat. Ecol. 37*: 341–349.

Cahoon, L.B. 1999. The role of benthic microalgae in neritic ecosystems. *Oceanogr. Mar. Biol. Annu. Rev. 37*: 47–86.

Dayton, P.K., D. Watson, A. Palmisano, J.P. Barry, J.S. Oliver and D. Rivera. 1986. Distribution Patterns of Benthic Microalgal Standing Stock at McMurdo Sound, Antarctica. *Polar Biol. 6*: 207–213.

De Brouwer, J.F.C., K. Wolfstein and G.K. Ruddy. 2005. Biogenic stabilization of intertidal sediments: The importance of extracellular polymeric substances produced by benthic diatoms. *Microb. Ecol. 49*: 501–512.

Erga, S.R. 1989. Ecological studies on the phytoplankton of Boknafjorden, Western Norway. 1. The effect of water exchange processes and environmental factors on temporal and vertical variability of biomass. *Sarsia 74*: 161–176.

Fisher, T.R., E.R. Peele, J.W. Ammerman and L.W. Harding. 1992. Nutrient limitation of phytoplankton in Chesapeake Bay. *Mar. Ecol. Prog. Ser. 82*: 51–63.

Fricke, A., M. Molis, C. Wiencke, N. Valdivia and A.S. Chapman. 2008. Natural succession of macroalgal-dominated epibenthic assemblages at different water depths and after transplantation from deep to shallow water on Spitsbergen. *Polar Biol. 31*: 1191–1203.

Garnier, J. and S. Mourelatos. 1991. Contribution of grazing in phytoplankton overall losses in a shallow French lake. *Freshw. Biol. 25*: 515–523.

Gilbert, N.S. 1991a. Primary production by benthic microalgae in nearshore marine sediments of Signy Island, Antarctica. *Polar Biol. 11*: 339–346.

Gilbert, N.S. 1991b. Microphytobenthic seasonality in nearshore marine sediments at Signy Island, South Orkney Islands, Antarctica. *Estuar. Coast. Shelf Sci. 33*: 89–104.

Glud, R.N., M. Kühl, F. Wenzhofer and S. Rysgaard. 2002. Benthic diatoms of a high Arctic fjord (Young Sound, NE Greenland): importance for ecosystem primary production. *Mar. Ecol. Prog. Ser. 238*: 15–29.

Glud, R.N., J. Woelfel, U. Karsten, M. Kühl and S. Rysgaard. 2009. Benthic microalgal production in the Arctic: applied methods and status of the current database. *Bot. Mar. 52*: 559–571.

Grosselin, M., M. Levasseur, P. Wheeler, R. Horner and B. Boot. 1997. New measurements of phytoplankton and ice algae production in the Arctic Ocean. *Deep Sea Res. 44*: 1623–1644.

Hanelt, D., H. Tüg, K. Bischof, C. Groß, H. Lippert, T. Sawall and C. Wiencke. 2001. Light regime in an Arctic fjord: a study related to stratospheric ozone depletion as a basis for determination of UV effects on algal growth. *Mar. Biol. 138*: 649–658.

Hansen, H.P. and F. Koroleff. 1983. Determination of nutrients. In: (K. Grasshoff, M. Ehrhardt and K. Kremling eds.) *Methods of seawater analysis.* Second edition. Verlag Chemie, Weinheim. pp. 159–226.

Hillebrand, H. and M. Kahlert. 2002. Effect of grazing and water column nutrient supply on biomass and nutrient content of sediment microalgae. *Aquat. Bot. 72*: 143–159.

Hop, H., T. Pearson, E.N. Hegseth, K.M. Kovacs, C. Wiencke, S. Kwasniewski, K. Eiane, F. Mehlum, B. Gulliksen, M. Wlodarska-Kowalczuk, C. Lydersen, J.M. Weslawski, S. Cochrane, G.W. Gabrielsen, R.J.G. Leakey, O.J. Lønne, M. Zajaczkowski, S. Falk-Petersen, M. Kendall, S.A. Wängberg, K. Bischof, A.Y. Voronkov, N.A. Kovaltchouk, J. Wiktor, M. Poltermann, G. di Prisco, C. Papucci and S. Gerland. 2002. The marine ecosystem of Kongsfjorden, Svalbard. *Polar Res 21*: 167–208.

Horner, R. and G.C. Schrader. 1982. Relative contributions of ice algae, Phytoplankton, and benthic microalgae to primary production in nearshore regions of the Beaufort Sea. *Arctic 35*: 485–503.

Horner, R., S.F. Ackley, G.S. Dieckmann, B. Gulliksen, T. Hoshiai, L. Legendre, I.A. Melnikov, W.S. Reeburgh, M. Spindler and C.W. Sullivan. 1992. Ecology of sea ice biota.1. Habitat, terminology, and methodology. *Polar Biol. 12*: 417–427.

Jeffrey, S.W. and G.F. Humphrey. 1975. New spectrophotometric equations for determining chlorophylls A, B, C1 and C2 in higher plants, algae and natural phytoplankton. *Biochemie und Physiologie der Pflanzen 167*: 191–194.

Kang, S.H. and G.A. Fryxell. 1992. *Fragilariopsis cylindrus* (Grunow) Krieger: the most abundant diatom in water column assemblages of Antarctic marginal ice-edge zones. *Polar Biol. 12*: 609–627.

Karsten, U., R. Schumann, S. Rothe, I. Jung and L. Medlin. 2006. Temperature and light requirements for growth of two diatom species (Bacillariophyceae) isolated from an Arctic macroalga. *Polar Biol. 29*: 476–486.

MacIntyre, H.L., R.J. Geider and D.C. Miller. 1996. Microphytobenthos: the ecological role of the "secret garden" of unvegetated, shallow-water marine habitats. 1. Distribution, abundance and primary production. *Estuaries 19*: 186–201.

Malcolme-Lawes, D.J. and K.H. Wong. 1990. Determination of orthophosphate in water and soil using a flow analyzer. *Analyst 115*: 65–67.

Matheke, G.E.M. and R. Horner. 1974. Primary productivity of benthic microalgae in Chukchi Sea, near Barrow, Alaska. *J. Fish. Res. Board Can. 31*: 1779–1786.

Menard, H. and S.M. Smith. 1966. Hypsometry of ocean basin provinces. *J. Geophys. Res. 71*: 4305–4325.

Migne, A., N. Spilmont and D. Davoult. 2004. *In situ* measurements of benthic primary production during emersion: seasonal variations and annual production in the Bay of Somme (eastern English Channel, France). *Cont. Shelf Res. 24 (13–14)*: 1437–1449.

Mock, T. and R. Gradinger. 1999. Determination of Arctic ice algae production with a new *in situ* incubation technique. *Mar. Ecol. Prog. Ser. 177*: 15–26.

Paterson, D.M. and K.S. Black. 1999. Water flow, sediment dynamics and benthic biology. *Adv. Ecol. Res. 29*: 155–193.

Risgaard-Petersen, N., S. Rysgaard and L.P. Nielsen. 1994. Diurnal-variation of denitrification and nitrification in sediments colonized by benthic microphytes. *Limnol. Oceanogr. 39*: 573–579.

Rohde, K.H. and D. Nehring (eds.). 1979. *Ausgewählte Methoden zur Bestimmung von Inhaltsstoffen im Meer- und Brackwasser.* Geodätische und Geophysikalische Veröffentlichungen IV H27. Nationalkomitee für Geodäsie und Geophysik bei der Akademie der Wissenschaften der DDR, Berlin. pp. 31–37.

Rysgaard, S., T.G. Nielsen and B.W. Hansen. 1999. Seasonal variation in nutrients, pelagic primary production and grazing in a high-Arctic coastal marine ecosystem, Young Sound, Northeast Greenland. *Mar. Ecol. Prog. Ser. 179*: 13–25.

Sundbäck, K. 1983. *Microphytobenthos on sand in shallow brackish water, Öresund, Sweden.* PhD-thesis, University of Lund. Lund. pp. 209.

Sundbäck, K., V. Enoksson, W. Graneli and K. Pettersson. 1991. Influence of sublittoral microphytobenthos on the oxygen and nutrient flux between sediment and water: a laboratory continuous flow study. *Mar. Ecol. Prog. Ser. 74*: 263–279.

Sundbäck, K., P. Nilsson, C. Nilsson and B. Jönsson. 1996. Balance between autotrophic and heterotrophic components and processes in microbenthic communities of sandy sediments: a field study. *Estuar. Coast. Shelf Sci. 43*: 689–706.

Svendsen, H., A. Beszczynska-Moller, J.O. Hagen, B. Lefauconnier, V. Tverberg, S. Gerland, J.B. Ørbaek, K. Bischof, C. Papucci, M. Zajaczkowski, R. Azzolini, O. Bruland, C. Wiencke, J.G. Winther and W. Dallmann. 2002. The physical environment of Kongsfjorden-Krossfjorden, an Arctic fjord system in Svalbard. *Polar Res. 21*: 133–166.

Thamdrup, B. and S. Fleischer. 1998. Temperature dependence of oxygen respiration, nitrogen mineralization, and nitrification in Arctic sediments. *Aquat. Microb. Ecol. 15*: 191–199.

Underwood, G.J.C. and J. Kromkamp. 1999. Primary production by phytoplankton and microphytobenthos in estuaries. *Adv. Ecol. Res. 29*: 93–153.

Verardo, D.J., P.N. Froelich and A. McIntyre. 1990. Determination of organic carbon and nitrogen in marine sediments using the Carlo Erba NA-1500 Analyser. *Deep Sea Res. 37*: 157–165.

Weissenberger, J., G. Dieckmann and R. Gradinger. 1992. Sea ice – a cast technique to examine and analyze brine pockets and channel structure. *Limnol. Oceanogr. 37*: 179–183.

Wentworth, C.K. 1922. A scale of grade and class terms for clastic sediments. *J. Geol. 30*: 377–392.

Wiencke, C. 2004. The coastal ecosystem of Kongsfjorden, Svalbard. Synopsis of biological research performed at the Koldewey Station in the years 1991–2003. *Ber. Polarforsch. Meeresforsch. 492*: 1–244.

Woelfel, J., R. Schumann, S. Adler, T. Hübener and U. Karsten. 2007. Diatoms inhabiting a wind flat of the Baltic Sea: species diversity and seasonal succession. *Estuar. Coast. Shelf Sci. 75*: 296–307.

Wulff, A., S. Vilbaste and J. Truu. 2005. Depth distribution of photosynthetic pigments and diatoms in the sediments of a microtidal fjord. *Hydrobiologia 534*: 117–130.

9 Phenology and seasonal physiological performance of polar seaweeds

Christian Wiencke, Iván Gómez and Ken Dunton

Introduction

One of the main characteristics of the polar regions that differentiates them from all other marine biogeographic regions is their strong seasonality (Chapter 2; Zacher et al. 2009). Benthic marine algae from the Arctic and Southern Oceans are exposed to long polar nights and sea ice cover in winter that contribute to extended periods of darkness. Following sea ice break-up in spring, the seawater is usually clear and sunlight penetrates deeply into the water column. Later, in summer and autumn, conditions in the sublittoral are again characterised by dim light due to the inflow of turbid meltwater and the development of phytoplankton blooms. Therefore, the periods of favourable light conditions for seaweed growth and development are short. Short days prevail in winter, long days prevail in summer. This seasonal variation of photoperiod is an important factor for triggering seasonal reproduction and growth (Lüning and tom Dieck 1989).

A second important factor for growth is the level of nutrients (Chapman and Craigie 1977). Unlike photoperiod, nutrient availability has a much greater seasonal variation in the Arctic compared to the Southern Ocean (Chapter 2; Zacher et al. 2009). In the Arctic, nutrient levels are usually high only during winter and drop suddenly after sea ice break-up (Chapman and Lindley 1980, Dunton et al. 1982). In a study off Spitsbergen, late winter levels of nitrate varied around 10 μM but decreased down to ≤0.5 μM just before sea ice break-up. Phosphate levels were around 9 μM in late winter and decreased around sea ice break-up down to 0.2 μM (Aguilera et al. 2002). In this respect, the seasonal changes of nutrient concentrations in the Arctic are comparable to those typical of temperate regions (Chapman and Craigie 1977). In contrast, surface nutrient levels off the west Antarctic Peninsula are much higher and persist throughout the year (Ducklow et al. 2007). Levels can range from 30 to 110 μM nitrate and 2 μM phosphate for most of the year, rarely limiting macroalgal growth (Drew and Hastings 1992, Sturz et al. 2003, Ducklow et al. 2007).

The third environmental factor influencing both reproduction and growth is temperature (Lüning 1980a, b). Seawater temperatures in the Antarctic region and in the central Arctic are constantly low throughout the year (Chapter 2; Zacher et al. 2009). The equatorial boundaries of the polar regions are set by the 0 °C winter isotherm. Summer temperatures at this boundary reach nearly 5 °C in the Antarctic and up to 8–10 °C in the Arctic (Chapter 12; Lüning 1990, Müller et al. 2009). In Antarctica, this boundary is located between 48 and 62° S while in the Arctic it is located between 52 and 70° N. Seaweeds from the northern Antarctic Peninsula region are thus exposed to daylengths similar to those in the temperate region of the North Atlantic, a major reason for the similar phenology of species from these regions. In this context it should be noted that the data on *Laminaria solidungula* J. Agardh were obtained at latitudes around 70° N, whereas the phenology of Antarctic species has been studied at daylengths occurring at 60–62° S, which likely explains some of the differences in phenology we describe below.

With the onset of climatic warming, mean monthly sea-water temperatures in the Antarctic Peninsula region rose slowly but steadily by about 1 °C in summer and about 2 °C in winter between 1991 and 2005 (Chapter 2; Schloss et al. 2008, Zacher et al. 2009). However, a considerable rise of the seawater temperatures of up to 4 °C is expected in the Atlantic sector of the Arctic Ocean (Chapter 12; Müller et al. 2009). Such changes will not only change geographic distribution patterns but may also influence seaweed phenology.

Seasonality of reproduction

Typical season responders *sensu* Kain (1989) resemble r-strategists and reproduce in summer when irradiance, temperature and nutrient conditions are most suitable for the species. This concept also applies to species from the polar regions, such as *Adenocystis utricularis* (Bory) Skottsberg, *Geminocarpus geminatus* (Hooker f. *et* Harvey) Skottsberg or *Pylaiella littoralis* (L.) Kjellman (Wiencke and Clayton 2002). In contrast, reproduction in season anticipators *sensu* Kain (1989), whose life strategy is similar to k-strategists, is primarily triggered by daylength. In Antarctic Desmarestiales for example, the formation of gametangia, fertilisation and the development of juvenile sporophytes takes place in winter. In the perennials *Desmarestia anceps* Montagne, *D. menziesii* J. Agardh and *Himantothallus grandifolius* (A. *et* E. Gepp) Zinova the microscopic gametophytes only become fertile under short-day conditions (Wiencke 1990a, Wiencke and Clayton 1990, Wiencke et al. 1995, Wiencke et al. 1996). A night-break regime inhibits induction of fertility in *D. anceps* (Wiencke et al. 1996) and *H. grandifolius* (C. Wiencke, unpublished data). In the annual species *D. antarctica* Moe *et* Silva and *Phaeurus antarcticus* Skottsberg, gametes are formed regardless of daylength. In these species phenology is controlled by sporophyte responses; this life history phase becomes fertile in culture under light:dark cycles between 6:18 h and 8:16 h (Clayton and Wiencke 1990, Wiencke 1990a, Wiencke et al. 1991). The question of whether spore formation in these species is actually under photoperiodic control remains unresolved and requires more detailed studies including night break experiments.

In the endemic Arctic kelp *Laminaria solidungula*, sporangia are present from late autumn to early spring, as shown for populations from Newfoundland and Spitsbergen (Hooper 1984, Wiencke et al. 2004; C. Wiencke, unpublished data). This is similar in most species of *Laminaria sensu lato* (Bartsch et al. 2008) from the temperate regions in the northern hemisphere, which also have their main fruiting period in autumn and winter (Kain 1979). In contrast, there are differences with respect to gametogenesis. Whereas gamete formation is not photoperiodically controlled in temperate species of *Laminaria*, gametogenesis of *L. solidungula* is possible only in short days (tom Dieck 1989).

A further characteristic of Antarctic Desmarestiales is the sensitivity of gametogenesis to photon irradiances $\geq 5\ \mu mol\ m^{-2}\ s^{-1}$ and temperatures $\geq 5\ °C$ (Wiencke 1990a,

Wiencke and Clayton 1990, Wiencke et al. 1991, 1995, 1996). The latter characteristic also has important implications for the geographic distribution of these species (Chapter 12; Müller et al. 2009). A good adaptation to Arctic conditions is found in *Saccharina latissima* (Linnaeus) Lane, Mayes, Druehl *et* Saunders from southwestern Norway compared to *Laminaria digitata* (Hudson) J.V. Lamouroux and *L. hyperborea* (Gunnerus) Foslie from the same locality. Oogenesis and sporophyte formation in *S. latissima* are possible at 0 °C and within a much shorter time period after sporulation compared to either *L. digitata* or *L. hyperborea* (Sjøtun and Schoschina 2002).

The photoperiodic responses produce phenologies of perennial and pseudoperennial seaweeds that are finely tuned to the polar year. Juvenile macrothalli develop predominantly in late winter/spring triggered by the seasonal change of daylength. This strategy of the season anticipators is, however, modified by water depth. For example, sporophytes of *Laminaria solidungula* have sori over a longer time period at greater depth compared to individuals from shallower waters (C. Wiencke, unpublished data). This is probably related to the fact that short days prevail in deep waters for longer durations than in shallow waters (Kain 2006).

Seasonal patterns of growth

As explained above, there are two different strategies by which seaweeds synchronise their development to the seasons. The season anticipators not only reproduce in winter, they also grow under short day or dark conditions in winter-spring, even under sea ice. This was discovered by Chapman and Lindley (1980) in *Laminaria solidungula* from the Canadian High Arctic and later verified by Dunton et al. (1982) and Dunton (1985) for the same species in the Alaskan Beaufort Sea. In this species, new blade formation begins in early winter (mid-November to early December) when the sun remains entirely below the horizon for 24 h day^{-1}. Blade elongation continues through winter with maximal rates in late-winter to early spring under thick ice. Blade expansion is lowest in summer and autumn during the ice-free period while the sun is above the horizon 24 h day^{-1}. In contrast, *Saccharina latissima* completes nearly all of its annual growth during the period when the first light penetrates the water column after sea ice break-up (Dunton 1985). The difference in seasonal growth strategy between the two species is also reflected in peaks of linear growth rates, which range from 1.5 mm day^{-1} for *L. solidungula* (March/April) to 5 mm day^{-1} for *S. latissima* (June/July) in the Beaufort Sea (Dunton 1985). This type of seasonal growth is also typical of temperate populations of the latter species as well as in several North Atlantic species of *Laminaria*, which also exhibit a period of rapid growth in late winter and spring (Kain 1979).

Subsequent laboratory studies have demonstrated typical season anticipator life history strategies in the following mostly endemic Antarctic seaweeds: the brown seaweeds *Himantothallus grandifolius*, *Desmarestia anceps*, *D. antarctica*, *Phaeurus antarcticus* (Wiencke 1990a), *Ascoseira mirabilis* Skottsberg (Gómez et al. 1995a, Gómez et al. 1996), *D. menziesii* (Gómez and Wiencke 1997), and the red seaweeds *Palmaria decipiens* (Reinsch) Ricker (Wiencke 1990b, Weykam et al. 1997), *Delesseria salicifolia* Reinsch, *Gymnogongrus antarcticus* Skottsberg, *G. turquetii* Hariot, *Hymenocladiopsis crustigena* Moe, *Trematocarpus antarcticus* (Hariot) Fredericq *et* R. Moe and *Phyllophora ahnfeltioides* Skottsberg (Dummermuth and Wiencke 2003).

In *H. grandifolius*, this seasonal pattern of growth has also been observed in field experiments (Drew and Hastings 1992).

Representative season responders from the Antarctic are the brown seaweed *Adenocystis utricularis* (Wiencke 1990a), the red seaweeds *Iridaea cordata* (Turner) Bory (Weykam et al. 1997) and *Gigartina skottsbergii* Setchell *et* N.L. Gardner (Wiencke 1990b), and the green seaweeds *Enteromorpha bulbosa* (Suhr) Montagne and *Acrosiphonia arcta* (Dillwyn) J. Agardh (Wiencke 1990b). As most endemic Antarctic and Arctic seaweeds are season anticipators, this life strategy is therefore regarded as a physiological adaptation to the seasonally changing conditions in polar seas. In contrast, season responders mostly have a wider distribution into the adjacent cold-temperate regions. Another difference between both groups is their depth distribution. Whereas polar season anticipators occur in the subtidal only, many polar season responders can grow both in the intertidal and subtidal zones.

As mentioned above, this classification of the Antarctic seaweeds into two life strategy types is based mainly on experimental investigations. How is it possible to study the phenology of seaweeds in the laboratory? Other workers (Chapman and Craigie 1977, 1978, Lüning 1980a, Lüning and tom Dieck 1989) have shown that the seasonal development of seaweeds is generally triggered by light, temperature and/or nutrient conditions. As temperatures and nutrient levels vary only very little over the year in the Southern Ocean (Chapter 2; Zacher et al. 2009), seasonality of Antarctic seaweeds must depend mainly on the light conditions, especially on the seasonal change of daylengths. This conclusion is fully supported by the available fragmentary field observations. In season anticipators from the Antarctic, annual growth and reproduction are most probably based on photoperiodisms and circannual rhythms, triggered or synchronised by daylengths, as demonstrated in several species from other biogeographic regions (Lüning and tom Dieck 1989, tom Dieck 1991, Lüning and Kadel 1993). Thus, assessment of phenology of Antarctic species in long-term culture experiments is possible by simply mimicking the seasonally fluctuating daylengths at the collecting site (Wiencke 1990a,b). This technique allows an easier and better monitoring of seasonal growth, photosynthesis and other physiological processes, including the formation of gametes/spores, than is possible from field observations alone.

Seasonal photosynthesis in Arctic and Antarctic seaweeds

As a consequence of strong seasonality in light availability in polar regions, algal photosynthesis is restricted to the spring-summer period, where both incident irradiance and daylength are optimal. Unlike temperate seaweeds, some of which are able to photosynthesise at considerable rates in the winter season (Gao 1990, Levitt and Bolton 1991), polar algae have to maximise photosynthetic production in a short period of time in order to supply the energy requirements for rapid growth and other metabolic demands. Thus, adaptations of the light harvesting capacities and further carbon assimilation processes are modulated by changes in light availability. For example, polar seaweeds have low light requirements for photosynthesis allowing them to fix carbon and maintain a net assimilatory surplus at photon irradiances as low as 10 μmol m^{-2} s^{-1} (Chapter 10; Gómez et al. 2009), which is a pre-requisite for completion of the life cycle and also allowing algae to occupy locations at great depths.

Major seasonal adjustments of photosynthesis include changes in photosynthetic efficiency (initial slope, α) and particularly the light requirements for saturation (E_k) and compensation (E_c) of photosynthesis (Dunton and Jodwalis 1988, Wiencke et al. 1993, Gómez et al. 1995b). These photosynthetic adaptations resemble those described for light-limited deep seaweeds from subtropical and tropical regions (Drew et al. 1982, Peckol and Ramus 1987, Flores-Moya et al. 1995), but are clearly different from shallow-water seaweeds from temperate regions (Levitt and Bolton 1991, Fairhead and Cheshire 2004). Moreover, seasonal photosynthetic performance is strongly linked to growth patterns and hence the seasonal life strategy of the organisms. In general, these adaptations are species-specific and their expression is a function of local environmental conditions and ecological characteristics of algae. For example, a seasonal comparison of photosynthetic performance in *Laminaria solidungula* and *Saccharina latissima* (Figure 9.1) revealed important differences in their response to light as reflected in P_{max} and I_k (Table 9.1). In contrast to *S. latissima,* photosynthetic efficiency increased and light saturation values decreased in the Arctic endemic *L. solidungula* during the iced-over winter period when *in situ* irradiance values were

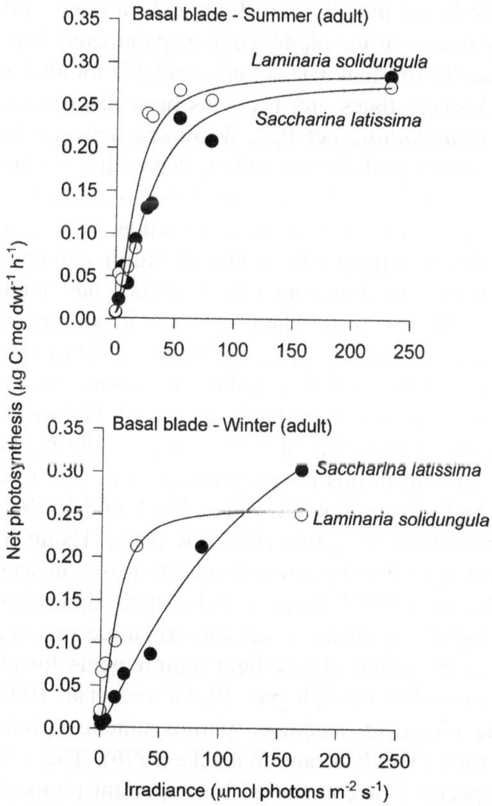

Figure 9.1 Seasonal P vs. I relationships for mature specimens of *Laminaria solidungula* and *Saccharina latissima* from the Beaufort Sea, Alaska.
Entire plants incubated *in situ* during open water (summer, 2–3 °C) and under ice (April, –1.8 °C) conditions. Data summarized from K. Dunton (unpublished data).

Table 9.1 Seasonal photosynthetic characteristics of mature blades of *Laminaria solidungula* and *Saccharina latissima* (see Figure 9.1) measured *in situ* (in the Alaskan Beaufort Sea, 71° N).

Season	*L. solidungula*	*S. latissima*
Summer (open water)		
P_{max} (µg C mg dwt^{-1} h^{-1})	0.3–0.5	0.27
I_k (µmol m^{-2} s^{-1})	30–49	49
Alpha (µg C mg dwt^{-1} h^{-1}/µmol m^{-2} s^{-1})	0.009	0.005
Winter (under ice)		
P_{max} (µg C mg dwt^{-1} h^{-1})	0.25	0.36
I_k (µmol m^{-2} s^{-1})	23	>150
Alpha (µg C mg dwt^{-1} h^{-1}/µmol m^{-2} s^{-1})	0.011	0.003

Data summarised from Dunton and Jodwalis (1988) and K. Dunton (unpublished data).

very low or nearly undetectable (Table 9.1 and Dunton 1990). The lack of a similar response by *S. latissima* reflects its more opportunistic strategy based on a significantly higher P_{max}, enabling the species to achieve maximum production at ice break-up when ambient light levels are initially very high and nutrients still available. Unfortunately, statistical comparisons of the photosynthetic parameters between the two species are not possible because the original data are not available for analysis.

Among Antarctic Desmarestiales, net P_{max} has been determined in the large brown alga *Himantothallus grandifolius*; net P_{max} decreases between March and June (autumn), but exhibits a strong peak in November, decreasing again in December (Drew and Hastings 1992). Increased net P_{max} in November is also observed in *Desmarestia menziesii* (Gómez et al. 1998). The seasonal variation of P_{max} determined in these Antarctic Desmarestiales is comparable to that of North Atlantic *Laminaria digitata*, *L. hyperborea* and *Saccharina latissima* (Drew 1983), but not to that of *Ascoseira mirabilis* (Gómez et al. 1995b). In this latter species, maximum net P_{max} values occur in September and February (no data available between February and September; Gómez et al. 1995b). In contrast, net P_{max} values in *Adenocystis utricularis* gradually decrease between winter and summer (Gutkowski and Maleszewski 1989). High respiration rates in *Ascoseira mirabilis* and *D. menziesii* during late winter-spring are correlated with active growth in this period (Gómez et al. 1995b, Gómez and Wiencke 1997). Interestingly, high respiration rates are also found in *H. grandifolius*, and in some months they exceed net P_{max} (negative P/R ratio). Using models of carbon accretion, Drew and Hastings (1992) predicted more C-losses in winter-early spring than C-gains, but when the annual C balance is calculated, algae show a positive carbon balance. Clearly, during the open-water season, Antarctic macroalgae can maintain a positive carbon balance by virtue of low light requirements for photosynthesis, which set their lower distributional limits (Chapter 10; Gómez et al. 1997, 2009).

For the rhodophyte *Palmaria decipiens*, photosynthetic optima in spring coincide with highest growth rates (Weykam and Wiencke 1996). The seasonal photosynthetic performance in this species is accompanied by important pigment adjustments (Lüder et al. 2001). For example, the initial increase in chl*a* and phycobilins appears to prevent decreases in photosynthesis during the first 2 months of darkness. However, after 6 months exposure to darkness, *P. decipiens* loses its ability to photosynthesise owing to a degradation of the three phycobilins (Weykam et al. 1997, Lüder et al. 2001,

2002). Interestingly, pigments are synthesised rapidly after re-exposure to light and enhance photosynthetic performance to normal values within a week (Lüder et al. 2002). This seasonal pattern of photosynthetic performance contrasts with the pattern found in the season responder *Iridaea cordata*, which maintains a functional photosynthetic apparatus during exposure to darkness (Weykam et al. 1997).

Overall, the seasonal variation in photosynthesis in polar algae is an efficient physiological trait allowing organisms a) to cope with changing light availability and b) to synchronise photosynthesis with biomass formation, which occurs preferentially during conditions of low light in late winter to spring. The well-known shade adaptation of these species is expressed as a strategy based on rapid acclimation to low irradiance conditions or when algae are exposed to excess solar irradiance in summer (Hanelt et al. 2003).

Storage and remobilisation of reserve compounds during the polar year

The capacity of large kelps and kelp-like species for synchronising photosynthesis and growth represents a major adaptation to seasonally changing light conditions that allows algae to optimise available resources, especially nutrients and light. Due to their perennial characteristics, the biomass allocation of these algae (a consequence of the allometric growth and the action of an intercalary meristem) results in different metabolic and physiological characteristics among the various tissue segments. Moreover, ecological constraints (such as light availability, protection against predation, etc.), promote the onset of early and rapid growth in order to reach a critical size or to compensate for tissue losses that have occurred in the previous season.

In Arctic species, synthesis and storage of reserve compounds (e.g., laminaran) occur during spring and summer when light conditions are optimal and nutrients are limiting (Chapman and Craigie 1977). In winter and early spring, when nutrient concentrations increase and photosynthesis ceases due to light limitation, growth of expansion takes place. The high Arctic kelp *Laminaria solidungula* represents an extreme example; the frond of this species elongates appreciably at undetectable levels of irradiance during the polar winter (Dunton and Dayton 1995). About 25% of the original carbon content of the alga is depleted during the dark winter period while it completes nearly 90% of its annual linear growth (Dunton and Schell 1986).

The morpho-functional bases of the storage and further remobilisation of organic compounds of Laminariales lie in the presence of a transport system along the medulla composed of trumpet cells, structures analogous to sieve cells in higher plants (Schmitz 1990). During active growth, sugars and sugar alcohols synthesised and stored in the tissues of the algae are translocated through the trumpet cells as low molecular-weight compounds (e.g., mannitol) to the growth region, as demonstrated in numerous kelps from the temperate region (Kremer 1981a, b). The high growth rates at the meristem during the growth season in winter/spring increase glycolysis and carbon losses due to respiration, and thus carbon from distal regions can supply the increased demands for carbon skeletons and nitrogen, especially in algae exposed to dim light or in darkness (Küppers and Kremer 1978, Johnston and Raven 1986).

One way to replenish carbon intermediates removed from the Krebs cycle is by so-called light-independent carbon fixation (LICF) or β-carboxylation, which is an anaplerotic pathway mediated by the enzymes phosphoenol pyruvate carboxylase (PEPC) or phosphoenol pyruvate carboxykinase (PEP-CK) (Kerby and Evans 1983). Light-independent carbon fixation reactions do not increase the net fixed carbon but are essential for cell metabolism. The pathway provides indispensable C_4 acids that are not synthesised in the Calvin-Benson cycle, but are likely used for formation of new tissue (see below). LICF contributes 13–17% of the total carbon fixed in *Laminaria solidungula* (Dunton and Schell 1986) and *Saccharina latissima* (Küppers and Kremer 1978) and may play an important role in nitrogen metabolism that accounts for the vigorous growth in *L. solidungula* during the dark ice-covered months (Dunton and Schell 1986). Although dark carbon fixation has also been reported in temperate kelps (Kremer 1981a, Cabello-Pasini and Alberte 2001, Gómez et al. 2007), its expression in polar seaweeds is regarded as a key metabolic strategy allowing these species to thrive under harsh environmental conditions that make a temporal decoupling of photosynthesis and biomass formation necessary. In this respect, it may be noted that LICF is not only a characteristic of brown algae, but has been demonstrated to occur also at least in one red alga from the Antarctic, *Palmaria decipiens* (Thomas and Wiencke 1991).

Like species from the Arctic, Antarctic seaweeds also show a marked seasonality in growth and photosynthesis, with most species growing in late winter/spring generally, after sea ice break-up (Wiencke 1990a,b, Weykam and Wiencke 1996, Gómez and Wiencke 1997). However, the relatively high and constant nutrient levels year round (Lipski 1987, Drew and Hastings 1992) suggest that the seasonal growth strategy is primarily triggered by daylength (Gómez et al. 1995a, b, Weykam et al. 1996). Although levels of N in the thallus of some brown algae reflect ambient concentrations, no evidence for storage or remobilisation has been found (Weykam et al. 1997, Gómez and Wiencke 1998). It must be emphasised that growth patterns of large species, such as *Ascoseira mirabilis* and *Himantothallus grandifolius* are complex and resemble those of Arctic/cold-temperate Laminariales, involving re-allocation of biomass along the lamina development (Gómez et al. 1996). In Antarctic species, the mannitol content of the blades shows a clear seasonality. During the period of rapid growth from September to December, mannitol content is low in the different blade regions, but increases significantly towards January and February in the basal and middle blade regions. The low and constant mannitol values measured in the distal region may indicate that this compound is rapidly translocated to basal regions to supply carbon skeletons for biomass formation in the intercalary meristem (Drew and Hastings 1992, Gómez and Wiencke 1998).

Two additional factors support a possible remobilisation of reserve carbohydrates in *Ascoseira mirabilis* and in Antarctic Desmarestiales: a) the existence of "conducting channels" or trumpet hyphae in medullary tissue regions (Moe and Silva 1981, Clayton and Ashburner 1990, Clayton and Wiencke 1990, Wiencke et al. 1991, 1995, Scrosati 1992, Wiencke and Clayton 2002) and b) the patterns of carbon fixation, with higher rates of LICF in the distal thallus regions than in meristematic tissues (Gómez et al. 1995a). Although the morpho-functional mechanisms connecting carbon synthesis and sink regions are still not well understood in Antarctic brown algae, LICF probably plays a role similar to that in *Laminaria* species (Kremer 1981b). Moreover, remobilisation of reserve carbohydrates formed during late summer to power the onset of growth in late winter is not only a characteristic found in large brown algae, but may also appear in some red algae, including *Palmaria decipiens*. In this species, the content of floridean starch decreases gradually in the dark and drops suddenly when new blades are formed in late winter (Weykam and Wiencke 1996, Weykam et al. 1997).

Seaweed phenology in relation to recent climate change

As expected in land plants (Bertin 2008), the phenology and geographic distribution of seaweeds will certainly be influenced by an increase in seawater temperatures that result from global warming (Chapter 12; Müller et al. 2009). Growth of seaweeds is strongly temperature-dependent, as exemplified in numerous species both from the Antarctic (Wiencke and tom Dieck 1989, 1990) and the Arctic (Bischoff and Wiencke 1993). The nearly 30-year continuous record of annual growth measurements in *Laminaria solidungula* in the Beaufort Sea (Dunton et al. 2009) represents an ideal opportunity to utilise the unique phenology of this species to assess the effect of *in situ* temperature changes in relation to polar warming. We hypothesise that if optimum growth temperatures are reached earlier in the year, then development of new thalli will start earlier. Formation of gametes or spores will also respond to temperature, as shown for example in the bipolar species *Desmarestia viridis* (Lüning 1980a). Thus, the timing and formation of juvenile thalli may also be altered.

Such changes in phenology will certainly have implications for various ecological and evolutionary phenomena, including ecosystem productivity, species interactions and community structure (Lotze and Worm 2002). Some of these responses have been observed in polar terrestrial vegetation (summarised by Robinson et al. 2003) but no data on such changes are available for polar seaweeds.

Conclusion

Based on these laboratory and field observations, we conclude that the phenology and seasonal physiological performance of polar seaweeds is finely tuned to the strong seasonal variation in the light conditions in the Arctic and Southern Oceans. Season anticipators, which include most subtidal, perennial and pseudoperennial species, become fertile between autumn and late winter. This strategy is similar to kelps from temperate regions, which mostly release their spores in autumn and winter.

Likewise, biomass formation is mostly confined to the winter half-year. The most extreme example is *Laminaria solidungula*, where blade expansion begins in early winter and continues through the spring, regardless of 24-h periods of continuous darkness. Other season anticipators begin their growth between late winter and late spring, in concert with increasing daylength, as in temperate kelps. Through this strategy, the young photosynthetically active tissue is formed in early spring when the sunlight penetrates deeply into the clear water. For Arctic species, this period of rapid growth corresponds to macroalgal use of the high nutrient levels present only during the darker winter period. We hypothesise that seasonal reproduction and growth are controlled by photoperiodisms and circannual rhythms, as in kelps from the temperate zone. In contrast to most season anticipators, the season responders occur mainly in the eulittoral and upper sublittoral zones and grow from spring to summer, when irradiance and temperature are optimal for the species.

Photosynthetic parameters in polar species are seasonally adjusted to the changing light regime and are in this respect comparable to deep-water kelps from lower latitudes. In Antarctic seaweeds, P_{max} values are highest in spring and lowest in winter, as demonstrated in some temperate kelps. In the red alga *Palmaria decipiens*, the seasonal variation in photosynthetic performance was shown to be based on strong changes in the pigment composition. The photoassimilates formed in the summer-half-year are stored as reserve compounds to support growth in winter and early spring, while the CO_2 released during respiration is partly fixed again through LICF, allowing the rapid synthesis of metabolic intermediates. In Arctic Laminariales, and probably also in Desmarestiales and *Ascoseira mirabilis* from the Antarctic, the reserve carbohydrates are translocated from the distal thallus parts to the meristems in specialised conducting tissues, as has been shown also in kelps from temperate regions. In parallel, respiration rates increase in Antarctic Desmarestiales and *A. mirabilis* to power meristematic activity, as shown for some temperate kelps. These particular morpho-functional processes were first demonstrated in temperate kelps, but are also important components of the life strategies of polar kelps and kelp-like species.

We conclude that phenology and physiology of polar seaweeds have many similarities with seaweeds from the temperate region. We note the existence of important and sometimes critical adaptations to life in an extreme polar environment by some species, while many others exhibit a gradation of responses that often reflects no major differences from species from temperate regions. No specific or unique mechanism occurring only in polar species has been found. Nevertheless, the demonstrated adaptations allow polar species to thrive with great success in polar waters. Whether global climate changes will have an impact on seasonal development and physiology of polar species may be shown in future studies.

References

Aguilera, J., K. Bischof, U. Karsten, D. Hanelt and C. Wiencke. 2002. Seasonal variation in ecological patterns in macroalgae from an Arctic fjord. II. Pigment accumulation and biochemical defence systems against high light stress. *Mar. Biol. 140*: 1087–1095.

Bartsch, I., C. Wiencke, K. Bischof, C.M. Buchholz, B.H. Buck, A. Eggert, P. Feuerpfeil, D. Hanelt, S. Jacobsen, R. Karez, U. Karsten, M. Molis, M.Y. Roleda, H. Schubert, R. Schumann, K. Valentin, F. Weinberger and J. Wiese. 2008. The genus *Laminaria sensu lato*: recent insights and developments. *Eur. J. Phycol. 43*: 1–86.

Bertin, R.I. 2008. Plant phenology and distribution in relation to recent climate change. *J. Torrey Bot. Soc. 135*: 126–146.

Bischoff, B. and C. Wiencke. 1993. Temperature requirements for growth and survival of macroalgae from Disko Island (Greenland). *Helgol. Mar. Res. 47*: 167–191.

Cabello-Pasini, A. and R.S. Alberte. 2001. Expression of carboxylating enzymes in *Laminaria setchelli* (Phaeophyceae). *Phycologia 40*: 351–358.

Chapman, A.R.O. and J.S. Craigie. 1977. Seasonal growth in *Laminaria longicruris*: relation with dissolved inorganic nutrients and internal reserves of nitrogen. *Mar. Biol. 40*: 197–205.

Chapman, A.R.O. and J.S. Craigie. 1978. Seasonal growth in *Laminaria longicruris*: relations with reserve carbohydrate storage and production. *Mar. Biol. 46*: 209–213.

Chapman, A.R.O. and J.E. Lindley. 1980. Seasonal growth of *Laminaria longicruris* in the Canadian High Arctic in relation to irradiance and dissolved nutrient concentration. *Mar. Biol. 57*: 1–5.

Clayton, M.N. and C.M. Ashburner. 1990. The anatomy and ultrastructure of "conducting channels" in *Ascoseira mirabilis* (Ascoseirales, Phaeophyceae). *Bot. Mar. 33*: 63–70.

Clayton, M.N. and C. Wiencke. 1990. The anatomy, life history and development of the Antarctic brown alga *Phaeurus antarcticus* (Desmarestiales, Phaeophyceae). *Phycologia 29*: 303–315.

Drew, E.A. 1983. Physiology of *Laminaria*, II. Seasonal variation of photosynthesis and respiration in *Laminaria digitata* Lamour., *L. hyperborea* (Gunn.) Fosl. and *L. saccharina* (L.) Lamour. and a model for calculation of annual carbon budgets. *P.S.Z.N.I: Mar. Ecol. 4*: 227–250.

Drew, E.A. and R.M. Hastings. 1992. A year-round ecophysiological study of *Himantothallus grandifolius* (Desmarestiales, Phaeophyta) at Signy Island, Antarctica. *Phycologia 31*: 262–277.

Drew, E.A., J.F. Ireland, C. Muir, W.A.A. Robertson and J.D. Robinson. 1982. Photosynthesis, respiration and other factors influencing the growth of *Laminaria ochroleuca* Phyl. below 50 metres in the Strait of Messina. *P.S.Z.N.I. Mar. Ecol. 3*: 335–355.

Ducklow, H.W., K. Baker, D.G. Martinson, L.B. Quetin, R.M. Ross, R.C. Smith, S.E. Stammerjohn, M. Vernet and W. Fraser. 2007. Marine pelagic ecosystems: the west Antarctic Peninsula. *Phil. Trans. R. Soc. B 362*: 67–94.

Dummermuth, A.L. and C. Wiencke. 2003. Experimental investigation of seasonal development in six Antarctic red macroalgae. *Antarct. Sci. 15*: 449–457.

Dunton, K.H. 1985. Growth of dark-exposed *Laminaria saccharina* (L.) Lamour. and *Laminaria solidungula* J. Ag. (Laminariales: Phaeophyta) in the Alaskan Beaufort Sea. *J. Exp. Mar. Biol. Ecol. 94*: 181–189.

Dunton, K.H. 1990. Growth and production in *Laminaria solidungula*: relation to continuous underwater light levels in the Alaskan high Arctic. *Mar. Biol. 106*: 297–304.

Dunton, K.H. and P.K. Dayton. 1995. The biology of high latitude kelps. *In*: (H.R. Skjoldal, C. Hopkins, K.E. Eriktad and H.P. Leinass, eds) *Ecology of fjords and coastal waters*. Elsevier Science B.V., Amsterdam. pp. 499–507.

Dunton, K.H. and C. Jodwalis. 1988. Photosynthetic performance of *Laminaria solidungula* measured *in situ* in the Alaskan High Arctic. *Mar. Biol. 98*: 277–285.

Dunton, K.H. and D.M. Schell. 1986. Seasonal carbon budget and growth of *Laminaria solidungula* in the Alaskan High Arctic. *Mar. Ecol. Prog. Ser. 31*: 57–66.

Dunton, K.H., E. Reimnitz and S.V. Schonberg. 1982. An arctic kelp community in the Alaskan Beaufort Sea. *Arctic 35*: 465–484.

Dunton, K.H., S.V. Schonberg and D.W. Funk. 2009. Interannual and spatial variability in light attenuation: evidence from three decades of growth in the Arctic kelp, *Laminaria solidungula*. *In*: (I. Krupnik, M.A. Lang and S.E. Miller, eds) *Smithsonian at the poles: contributions to international polar year science*. Proceedings of Smithsonian at the Poles Symposium, 3–4 May 2007. Smithsonian Institution, Smithsonian Institution Scholarly Press, Washington, DC. pp. 271–284.

Fairhead, V.A. and A.C. Cheshire. 2004. Rates of primary productivity and growth in *Ecklonia radiata* measured at different depths, over an annual cycle, at West Island, South Australia. *Mar. Biol. 145*: 41–50.

Flores-Moya, A., J.A. Fernández and F.X. Niell. 1995. Seasonal variation of photosynthetic pigments, total C, N, and P content, and photosynthesis in *Phyllariopsis purpurascens* (Phaeophyta) from the Strait of Gibraltar. *J. Phycol. 31*: 867–874.

Gao, K. 1990. Seasonal variation of photosynthetic capacity in *Sargassum horneri*. *Jpn. J. Phycol. (Sôrui). 38*: 25–33.

Gómez, I. and C. Wiencke. 1997. Seasonal growth and photosynthetic performance of the Antarctic macroalga *Desmarestia menziesii* (Phaeophyceae) cultivated under fluctuating Antarctic daylengths. *Bot. Acta 110*: 25–31.

Gómez, I. and C. Wiencke. 1998. Seasonal changes in C, N, and major organic compounds and their significance to morpho-functional processes in the endemic Antarctic brown alga *Ascoseira mirabilis*. *Polar Biol. 19*: 115–124.

Gómez, I., D.N. Thomas and C. Wiencke. 1995a. Longitudinal profiles of growth, photosynthesis and light independent carbon fixation in the Antarctic brown alga *Ascoseira mirabilis*. *Bot. Mar. 38*: 157–164.

Gómez, I., C. Wiencke and G. Weykam. 1995b. Seasonal photosynthetic characteristics of the brown alga *Ascoseira mirabilis* from King George Island (Antarctica). *Mar. Biol. 123*: 167–172.

Gómez, I., C. Wiencke and D.N. Thomas. 1996. Variations in photosynthetic characteristics of the Antarctic marine brown alga *Ascoseira mirabilis* in relation to thallus age and size. *Eur. J. Phycol. 31*: 167–172.

Gómez, I., G. Weykam, H. Klöser and C. Wiencke. 1997. Photosynthetic light requirements, daily carbon balance and zonation of sublittoral macroalgae from King George Island (Antarctica). *Mar. Ecol. Prog. Ser. 148*: 281–293.

Gómez, I., G. Weykam and C. Wiencke. 1998. Seasonal photosynthetic metabolism and major organic compounds in the marine brown alga *Desmarestia menziesii* from King George Island (Antarctica). *Aquat. Bot. 60*: 105–118.

Gómez, I., M. Orostegui and P. Huovinen. 2007. Morpho-functional patterns of photosynthesis in the South Pacific kelp *Lessonia nigrescens*: effects of UV radiation on ^{14}C fixation and primary photochemical reactions. *J. Phycol. 43*: 55–64.

Gómez, I., A. Wulff, M.Y. Roleda, U. Karsten, P. Huovinen, M.L. Quartino, K. Dunton and C. Wiencke. 2009. Light and temperature demands of marine benthic microalgae and seaweeds in polar regions. *Bot. Mar. 52*: 593–608.

Gutkowski, R. and S. Maleszewski. 1989. Seasonal changes of the photosynthetic capacity of the Antarctic macroalga *Adenocystis utricularis* (Bory) Skottsberg. *Polar Biol. 10*: 145–148.

Hanelt, D., C. Wiencke and K. Bischof. 2003. Photosynthesis in marine macroalgae. *In*: (A.W. Larkum, S.E. Douglas and J.A. Raven, eds) *Photosynthesis in algae*. Kluwer Academic Publisher, Dordrecht. pp. 413–435.

Hommersand, M.H., R.L. Moe, C.D. Amsler and S. Fredericq. 2009. Notes on the systematics and biogeographical relationships of Antarctic and sub-Antarctic Rhodophyta with descriptions of four new genera and five new species. *Bot. Mar. 52*: 509–534.

Hooper, R.G. 1984. Functional adaptations to the polar environment by the Arctic kelp, *Laminaria solidungula*. *Br. Phycol. J. 19*: 194.

Johnston, A.M. and J. Raven. 1986. Dark carbon fixation studies on the intertidal macroalga *Ascophyllum nodosum* (Phaeophyta). *J. Phycol. 22*: 483–485.

Kain, J.M. 1979. A view of the genus *Laminaria*. *Annu. Rev. Oceanogr. Mar. Biol. 17*: 101–161.

Kain, J.M. (Jones) 1989. The seasons in the subtidal. *Br. Phycol. J. 24*: 203–215.

Kain, J.M. (Jones). 2006. Photoperiodism in *Delesseria sanguinea* (Ceramiales, Rhodophyta) 2. Daylengths are shorter underwater. *Phycologia 45*: 624–632.

Kerby, W.N. and L.V. Evans. 1983. Phosphoenolpyruvate carboxykinase activity in *Ascophyllum nodosum* (Phaeophyceae). *J. Phycol. 19*: 1–3.

Kremer, B.P. 1981a. Metabolic implications of non-photosynthetic carbon fixation in brown macroalgae. *Phycologia 20*: 242–250.

Kremer, B.P. 1981b. Aspects of carbon metabolism in marine macroalgae. *Oceanogr. Mar. Biol. Annu. Rev. 19*: 41–94.

Küppers, U. and B.P. Kremer. 1978. Longitudinal profiles of CO_2-fixation capacities in marine macroalgae. *Plant Physiol. 62*: 49–54.

Levitt, G.J. and J.J. Bolton. 1991. Seasonal patterns of photosynthesis and physiological parameters and the effects of emersion in littoral seaweeds. *Bot. Mar. 34*: 403–410.

Lipski, M. 1987. Variations of physical conditions, nutrients and chlorophyll *a* contents in Admiralty Bay (King George Island, South Shetland Islands, 1979). *Pol. Polar Res. 8*: 307–332.

Lotze, H.K. and B. Worm. 2002. Complex interactions of climatic and ecological controls on macroalgal recruitment. *Limnol. Oceanogr. 47*: 1734–1741.

Lüder, U.H., J. Knoetzel and C. Wiencke. 2001. Acclimation of photosynthesis and pigments to seasonally changing light conditions in the endemic Antarctic red macroalga *Palmaria decipiens*. *Polar Biol. 24*: 598–603.

Lüder, U.H., J. Knoetzel and C. Wiencke. 2002. Acclimation of photosynthesis and pigments during and after six months of darkness in *Palmaria decipiens* (Rhodophyta) – a study to simulate Antarctic winter sea ice cover. *J. Phycol. 38*: 904– 913.

Lüning, K. 1980a. Control of algal-life-history by daylength and temperature. *In*: (J.H. Price, D. E.G. Irvine and W.F. Farnham, eds) *The shore environment. Vol 2. Ecosystems*. Academic Press, New York. pp. 915–945.

Lüning, K. 1980b. Critical levels of light and temperature regulating the gametogenesis of three *Laminaria* species (Phaeophyceae). *J. Phycol. 16*: 1–15.

Lüning, K. 1990. *Seaweeds. Their environment, biogeography and ecophysiology*. John Wiley and Sons, Inc., New York. pp. 527.

Lüning, K. and P. Kadel. 1993. Daylength range for circannual rhythmicity in *Pterygophora californica* (Alariaceae, Phaeophyta) and synchronisation of seasonal growth by daylength cycles in several other brown algae. *Phycologia 32*: 379–387.

Lüning, K. and I. tom Dieck. 1989. Environmental triggers in algal seasonality. *Bot. Mar. 32*: 389–397.

Moe, R.L. and P.C. Silva. 1981. Morphology and taxonomy of *Himantothallus* (including *Phaeoglossum* and *Phyllogigas*), an Antarctic member of the Desmarestiales (Phaeophyceae). *J. Phycol. 17*: 15–29.

Müller, R., T. Läpple, I. Bartsch and C. Wiencke. 2009. Impact of oceanic warming on the distribution of seaweeds in polar and cold-temperate waters. *Bot. Mar. 52*: 617–638.

Peckol, P. and J. Ramus. 1987. Photosynthetic performance of deep-water macroalgae (Phaeophyta, Dictyotales) off Bermuda. *Hydrobiologia 231*: 93–98.

Robinson, S.A., Wasley, J. and A.K. Tobin. 2003. Living on the edge – plants and global change in continental and maritime Antarctica. *Global Change Biol. 9*: 1681–1717.

Schloss, I., G.A. Ferreyra, O. González, A. Atencio, V.L. Fuentes, G. Tosonotto, G. Mercuri, R. Sahade, M. Tatián and D. Abele. 2008. Long-term hydrographic conditions and climate trends in Potter Cove. *Ber. Polarforsch. Meeresforsch. 571*: 382– 389.

Schmitz, K. 1990. Algae. *In*: (H.-D. Behnke and R.D. Sjolund, eds) *Sieve elements*. Springer Verlag, Berlin. pp. 1–18.

Scrosati, R.A. 1992. Estudio anatómico de *Desmarestia ligulata* de Argentina y *D. menziesii* de Antártida (Phaeophyceae). *Physis A. 113*: 89–98.

Sjøtun, K. and E.V. Schoschina. 2002. Gametophytic development of *Laminaria* spp. (Laminariales, Phaeophyta) at low temperature. *Phycologia 41*: 147–152.

Sturz, A.A., S.C. Gray, K. Dykes, A. King and J. Radtke. 2003. Seasonal changes of dissolved nutrients within and around Port Foster Deception Island, Antarctica. *Deep-Sea Res. Part II 50:* 1685–1705.

Thomas, D.N. and C. Wiencke. 1991. Photosynthesis, dark respiration and light independent carbon fixation of endemic Antarctic macroalgae. *Polar Biol. 11*: 329–337.

tom Dieck, I. 1989. *Vergleichende Untersuchungen zur Ökophysiologie und Kreuzbarkeit innerhalb der digitaten Sektion der Gattung* Laminaria. PhD thesis, University of Hamburg, Hamburg, Germany. 168 pp.

tom Dieck, I. 1991. Circannual growth rhythm and photoperiodic sorus induction in the kelp *Laminaria setchellii* (Phaeophyta). *J. Phycol. 27*: 341–350.

Weykam, G. and C. Wiencke. 1996. Seasonal photosynthetic performance of the endemic Antarctic alga *Palmaria decipiens* (Reinsch) Ricker. *Polar Biol. 16*: 357–361.

Weykam, G., I. Gómez, C. Wiencke, K. Iken and H. Klöser. 1996. Photosynthetic characteristics and C:N ratios of macroalgae from King George Island (Antarctica). *J. Exp. Mar. Biol. Ecol. 204*: 1–22.

Weykam, G., D.N. Thomas and C. Wiencke. 1997. Growth and photosynthesis of the Antarctic red algae *Palmaria decipiens* (Palmariales) and *Iridaea cordata* (Gigartinales) during and following extended periods of darkness. *Phycologia 36*: 395–405.

Wiencke, C. 1990a. Seasonality of brown macroalgae from Antarctica – a long-term culture study under fluctuating Antarctic daylengths. *Polar Biol. 10*: 589–600.

Wiencke, C. 1990b. Seasonality of red and green macroalgae from Antarctica – a long-term culture study under fluctuating Antarctic daylengths. *Polar Biol. 10*: 601–607.

Wiencke, C. and M.N. Clayton. 1990. Sexual reproduction, life history, and early development in culture of the Antarctic brown alga *Himantothallus grandifolius* (Desmarestiales, Phaeophyceae). *Phycologia 29*: 9–18.

Wiencke, C. and M.N. Clayton. 2002. *Antarctic seaweeds*. Synopses of the Antarctic benthos 9. Gantner, Ruggell. 239 pp.

Wiencke, C. and I. tom Dieck. 1989. Temperature requirements for growth and temperature tolerance of macroalgae endemic to the Antarctic region. *Mar. Ecol. Prog. Ser. 54*: 189–197.

Wiencke, C. and I. tom Dieck. 1990. Temperature requirements for growth and survival of macroalgae from Antarctica and Southern Chile. *Mar. Ecol. Prog. Ser. 59*: 157–170.

Wiencke, C., U. Stolpe and H. Lehmann. 1991. Morphogenesis of the brown alga *Desmarestia antarctica* cultivated under seasonally fluctuating Antarctic daylengths. *Ser. Cient. INACH. 41*: 65–78.

Wiencke, C., J. Rahmel, U. Karsten, G. Weykam and G.O. Kirst. 1993. Photosynthesis of marine macroalgae from Antarctica: light and temperature requirements. *Bot. Acta 106*: 78–87.

Wiencke, C., M.N. Clayton and D. Schulz. 1995. Life history, reproductive morphology and development of the Antarctic brown alga *Desmarestia menziesii* J. Agardh. *Bot. Acta 108*: 201–208.

Wiencke, C., M.N. Clayton and C. Langreder. 1996. Life history and seasonal morphogenesis of the endemic Antarctic brown alga *Desmarestia anceps* Montagne. *Bot. Mar. 39*: 435–444.

Wiencke, C., M.N. Clayton and M.E.A. Schoenwaelder. 2004. Sensitivity and acclimation to UV radiation of zoospores from five species of Laminariales from the Arctic. *Mar. Biol. 145*: 31–39.

Zacher, K., R. Rautenberger, D. Hanelt, A. Wulff and C. Wiencke. 2009. The abiotic environment of polar marine benthic algae. *Bot. Mar. 52*: 483–490.

10 Light and temperature demands of marine benthic microalgae and seaweeds in polar regions

Iván Gómez , Angela Wulff, Michael Y. Roleda, Pirjo Huovinen, Ulf Karsten, María Liliana Quartino, Ken Dunton and Christian Wiencke

Introduction

Constant low temperatures and extremes in seasonal underwater light conditions are often associated with the characterization of polar regions as hostile environments for living organisms (Chapter 2; Zacher et al. 2009a). However, marine algae thrive in these habitats, forming rich communities dominated mainly by brown algae (Chapter 3; Svendsen 1959, Dunton et al. 1982, Clayton 1988, Wilce 1994, Wiencke et al. 2007, Wulff et al. 2009). Benthic microalgae are also well represented and may account for around 40–50% of the primary production in polar shallow waters (Glud et al. 2002).

Ambient regimes of temperature, light and nutrients trigger the major biological processes of polar algae and define the scenarios for their highly efficient metabolic adaptations, especially very low light and temperature demands for growth and photosynthesis (Kirst and Wiencke 1995, Wiencke et al. 2007). For example, over a broad range of depths, the patterns of shade adaptation in Antarctic seaweeds are often characterized by a high photosynthetic efficiency and low irradiances for saturation of photosynthesis and growth (Wiencke 1996, Weykam et al. 1996, Gómez et al. 1997). On the other hand, polar algae living in shallow waters are exposed to high solar radiation; they have diverse regulative physiological strategies to avoid damage of sensitive molecules and processes, including photoinhibition of photosynthesis or damage resulting from enhanced UV radiation (Hanelt 1998, Roleda et al. 2006a, b, Bischof et al. 2006, Wulff et al. 2008a, and references therein). The influence of UV radiation on polar algae is summarized in chapter 13 and by Karsten et al. (2009).

In relation to cold adaptation, polar algae have a suite of metabolic strategies including a) the maintenance of fluidity of biological membranes by possession of unsaturated fatty acids that prevent rigidification of membrane lipids; b) molecular adaptations of enzymes in order to maintain sufficient rates of enzyme-catalyzed reactions of key metabolic processes; c) the evolution of cold shock and antifreeze proteins, and d) adaptations of the photosynthetic electron transport chain to function at cold temperatures (reviewed by Morgan-Kiss et al. 2006). These physiological characteristics allow Antarctic and Arctic algae to complete their life cycles *in situ* at temperatures close to 0 °C (Breeman 1988, Bischoff and Wiencke 1993, Wiencke et al. 2007). In general, the cold adaptation of polar algae is a consequence of geological processes that occurred during the last glaciations (Antarctic algae have been exposed to cold conditions for longer than their Arctic counterparts) restricting the distribution of several major taxa, with implications for biogeography in both regions (Chapter 2, 3; van den Hoek and Breeman 1989, Wiencke and tom Dieck 1990, Wiencke et al. 1994, Wulff et al. 2009, Zacher et al. 2009).

The remarkable ability of polar seaweeds and microalgae to tolerate extreme environmental conditions places an emphasis on the threshold for physiological adaptation. Polar benthic algae in particular, in terms of their light and temperature requirements, are excellent models for examining the expression of functional responses that reflect resource utilization or tolerance of environmental stress. Here, we summarize the information on light and temperature demands of polar benthic algae in the context of photosynthetic and growth adaptations. Light limitation is examined as a function of photosynthetic characteristics of seaweeds and benthic microalgae with specific reference to the physiological limits that set the vertical distribution of various species in both Arctic and Antarctic coastal systems. With respect to temperature, we outline the physiological bases underlying the ability of polar algae to grow and photosynthesize at temperatures around 0 °C. We discuss how light and cold adaptations of algae may drive the high benthic primary productivity in high latitude ecosystems. Finally, we examine the consequences of the interaction between light and temperature on algal physiology with an emphasis on the complex interplay between the organism and multiple environmental factors.

Light requirements

Seaweeds

Much of the available information on optimal light conditions for polar seaweeds comes from photosynthesis-light (P-I) curves measured using O_2 evolution, ^{14}C fixation or chlorophyll fluorescence, both from cultured and field material. In general, polar seaweeds have high photosynthetic efficiency and hence low light requirements for saturation of photosynthesis, as determined by E_k values (a parameter defined as the intercept between the initial slope of the curve and the saturated photosynthesis) as reflected in different seaweeds and microalgal assemblages from Antarctic and Arctic locations (Table 10.1). E_k values between 11 and 100 µmol photons m^{-2} s^{-1} are clearly lower than those of species from temperate locations (Kirk 1994). For example, in large Antarctic brown algae such as *Ascoseira mirabilis* or members of Desmarestiales (*Desmarestia anceps*, *D. antarctica* and *Himantothallus grandifolius*), E_k values range between 26 and 105 µmol photons m^{-2} s^{-1}, depending on the depth of collection (e.g., *H. grandifolius* is common at 30–40 m depth) (Gómez et al. 1997; Table 10.1). E_k values of Arctic kelps (e.g., *Laminaria solidungula* and *Saccharina latissima*) are comparable to those of Antarctic Desmarestiales and vary between 13 and 170 µmol photons m^{-2} s^{-1}. In Rhodophyta, light requirements for saturation of photosynthesis are particularly low with summer values ranging between 11 and 49 µmol photons m^{-2} s^{-1} for foliose Antarctic genera such as *Palmaria*, *Gigartina* and *Myriogramme*. E_k values between 11 and 40 µmol photons m^{-2} s^{-1} have been measured for the Arctic species *Dumontia contorta* (Latala 1990). However, some algal assemblages, such as coralline algae from ice covered subtidal sites in the Arctic (e.g.,

Leptophytum foecundum) and Antarctic, have average E_k values as low as 11 or 2 μmol photons m^{-2} s^{-1}, respectively, representing extreme cases of shade adaptation (Kühl et al. 2001, Schwarz et al. 2005). Green algal assemblages comprising genera such as *Acrosiphonia*, *Ulva* and *Urospora* for example, which are common in littoral locations, show in general the highest light requirements for photosynthesis (E_k values between 20 and 65 μmol photons m^{-2} s^{-1}). Based on these data, Antarctic members of Phaeophyta have lower average saturating points than their Arctic counterparts, while in Rhodophyta and Chlorophyta no differences between Arctic and Antarctic species are evident (Table 10.1). However, it must be emphasized that the low number of species studied in both regions, particularly of the two latter groups, makes rigorous comparisons difficult.

Kirst and Wiencke (1995) concluded that one of the factors explaining the ability of polar algae to survive under light-limited conditions is the low rate of respiration. In fact, the irradiance by which photosynthesis compensates respiration in the P–I curve, the so-called compensation point for photosynthesis (E_c), is normally <20 μmol photons m^{-2} s^{-1} in polar seaweeds and, based on their low E_k values, E_c values of benthic microalgae are probably also very low (Table 10.1). In Antarctic brown algae, respiration is generally compensated by photosynthesis at irradiances between 6 and 20 μmol photons m^{-2} s^{-1}, while in Arctic kelps E_c values measured in winter can be as low as 2 μmol photons m^{-2} s^{-1}. Among members of Antarctic Rhodophyta with broad depth distribution (e.g., *Palmaria decipiens* and *Gigartina skottsbergii*), E_c values normally range between 2 and 15 μmol photons m^{-2} s^{-1}: however, in deep assemblages of e.g., *Corallina*, photosynthesis compensates respiration at practically undetectable levels of irradiance (<1 μmol photons m^{-2} s^{-1}; Schwarz et al. 2005). In contrast to E_k, E_c values of littoral green algae are in the range of those found in brown and red algae (4–15 photons m^{-2} s^{-1}). This suggests a direct relation to the minimum light requirements for growth; the maintenance of light requirements for photosynthesis that are always below the ambient irradiance would be an advantage in achieving a positive net C assimilation (high P/R ratios) during periods of maximal growth, which is possible only when respiratory activity is low (see below). In general, E_c values of Arctic and Antarctic seaweeds represent <10% of the average *in situ* irradiance measured in shallow waters (Drew 1977, Dunton and Jodwalis 1988, Gómez et al. 1997, Borum et al. 2002) and are significantly lower than compensating irradiances of e.g., subtidal Mediterranean seaweeds (mean 58 μmol photons m^{-2} s^{-1}; Enríquez et al. 1995).

Light requirements for photosynthesis of seaweeds may be related to depth limits for growth and hence, they can be used to explain zonation patterns. Generally, saturation irradiances for photosynthesis of Antarctic and Arctic seaweeds (Table 10.1) are substantially higher than irradiances required for saturation of growth (normally <15 μmol photons m^{-2} s^{-1}; Dunton and Jodwalis 1988, Wiencke 1990a, b, Wiencke and Fischer 1990). Clearly, saturation of photosynthesis is not necessarily a metabolic pre-requisite for growth as considerable net C assimilation still occurs below the E_k point. Thus, photosynthetic E_k values above the light requirements for growth result in ecological advantages for coping with the strong fluctuations of incident irradiance during the open water period. For example, during late winter-spring, Antarctic seaweeds growing at depths below 20 m can be exposed to irradiances around 80 μmol photons m^{-2} s^{-1} (Klöser et al. 1993, Gómez et al. 1997).

Table 10.1 Light requirements for saturation (E_k) and compensation (E_c) of polar benthic algal photosynthesis from different locations in the Antarctic (Ant) and Arctic (Arc).

Species	Location	Depth (m)	Irradiance* (μmol photons m^{-2} s^{-1})	E_k	E_c	Temperature (°C)	Reference
Brown algae							
Ascoseira mirabilis Skottsberg	King George I. (Ant)	3	700	35	13	−1.8–1.2	Weykam et al. (1996)
Desmarestia anceps Montagne	Signy I. (Ant)	1–10	17–200	–	15	−1.5–0	Drew (1977)
	King George I. (Ant)	10	264	32	15	−1.8–1.2	Weykam et al. (1996)
	King George I. (Ant)	10–30	21–264	31–58	8–27	−1.8–1.2	Gómez et al. (1997)
	King George I. (Ant)	5–15	50–400	73–105	–	2	Rautenberger et al. (unpublished data)
Desmarestia antarctica Moe *et* Silva	Signy I. (Ant)	1–10	17–700	–	20	−1.5–0	Drew (1977)
	King George I. (Ant)	2	780	38	11	−1.8–1.2	Weykam et al. (1996)
Himantothallus grandifolius (A. *et* E.S. Gepp) Zinova	Signy I. (Ant)	5–10	17–75	–	30	−1.5–0.5	Drew (1977)
	King George I. (Ant)	10	264	26	8	−1.8–1.2	Weykam et al. (1996)
	King George I. (Ant)	10–30	21–264	22–29	6–10	−1.8–1.2	Gómez et al. (1997)
Laminaria solidungula J. Agardh	Alaskan High Arctic	6–7	100–300	38	–	−1.8–1.5	Dunton and Jodwalis (1988)
Laminaria digitata (Hudson) Lamouroux	Spitsbergen (Arc)	5	–	20–40	4	3	Latala (1990)
Alaria esculenta (Linnaeus) Greville	Spitsbergen (Arc)	5	–	20–40	5	3	Latala (1990)
Saccharina latissima (Linnaeus) Lane, Mayes, Druehl *et* Saunders	Greenland (Arc)	0.5–15	2–30	13	2	–	Kühl et al. (2001)
	Greenland (Arc)	2.5–15	3–200	15–170	2–22	0–1.5	Borum et al. (2002)
Red algae							
Palmaria decipiens (Reinsch) Ricker	Signy I. (Ant)	1–10	17–200	–	15	−1.5–0.5	Drew (1977)
	King George I. (Ant)	8	346	49	6	−1.5–0.5	Weykam et al. (1996)
	King George I. (Ant)	10–30	21–264	19–42	10–6	−1.8–1.2	Gómez et al. (1997)
Gigartina spp.	Signy I. (Ant)	1–10	17–200	–	15	−1.5–0.5	Drew (1977)
Myriogramme manginii (Gain) Skottsberg	Signy I. (Ant)	0.5–25	0–163	11–18	2–3	−1.8–0.3	Brouwer (1996a)
Gigartina skottsbergii Setchell *et* Gardner	King George I. (Ant)	10–30	21–264	16–31	2–4	−1.8–1.2	Gómez et al. (1997)

Phyllophora antarctica A. Gepp *et* E.S. Gepp	McMurdo Sound (Ant)	10–25	0.5–100	8–10	2–4	1.9	Schwarz et al. (2003)
Coralline assemblage	Ross I. (Ant)	15–20	2	2–6	0.04–0.2	–	Schwarz et al. (2005)
Dumontia contorta (S.G. Gmelin) Ruprecht	Spitsbergen (Arc)	0–1	–	20–40	2	3	Latala (1990)
Leptophytum foecundum (Kjellman) Adey	Greenland (Arc)	15–40	2–30	11	1.6	–0.5	Kühl et al. (2001)
Green algae							
Acrosiphonia spp.	Spitsbergen (Arc)	0–1	–	40	6	3	Latala (1990)
Enteromorpha bulbosa (Suhr) Montagne	King George I. (Ant)	0	–	61	11	–1.8–1.2	Weykam et al. (1996)
Ulvaria obscura (Kützing) Gayral *ex* Bliding	Spitsbergen (Arc)	0–1	–	20–40	4	3	Latala (1990)
Urospora penicilliformis (Roth) Areschoug	King George I. (Ant)	0	–	65	15	–1.8–1.2	Weykam et al. (1996)
Reproductive cells							
Adenocystis utricularis (Bory de Saint-Vincent) Skottsberg, zoospores	King George I. (Ant)	0–1	23	64	–	2	Zacher et al. (2007b)
Ascoseira mirabilis, gametes	King George I. (Ant)	2	22	52	–	2	Roleda et al. (2007)
Saccorhiza dermatodea (Bachelot de la Pylaie) Areschoug, meiospores	Spitsbergen (Arc)	0.5–13	22	18	–	2	Roleda et al. (2006a)
Saccharina latissima, meiospores	Spitsbergen (Arc)	1.5–15	22	13	–	2	Roleda et al. (2006a)
Porphyra endiviifolia (A. Gepp *et* E.S. Gepp) Y.M. Chamberlain, monospores	King George I. (Ant)	0	23	33	–	2	Zacher et al. (2007b)
Gigartina skottsbergii, carpospores	King George I. (Ant)	3	22	54	–	1	Roleda et al. (2008a)
Gigartina skottsbergii, tetraspores	King George I. (Ant)	3	22	44	–	1	Roleda et al. (2008a)
Monostroma hariotii Gain, gametes	King George I. (Ant)	0	23	83	–	2	Zacher et al. (2007b)
Microalgae							
Trachyneis aspera (Ehrenberg) Cleve	McMurdo Sound (Ant)	20	0.6	7–16	–	–2	Palmisano et al. (1985a)
	McMurdo Sound (Ant)	18–20	10–15	6	–	–1	Rivkin and Putt (1987)
Community dominated by *Trachyneis aspera*	McMurdo Sound (Ant)	26	0.2–0.6	1–4	–	–	Robinson et al. (1995)
Amphora sp.	McMurdo Sound (Ant)	18–20	10–15	6	–	–1	Rivkin and Putt (1987)
Diatom community	King George I. (Ant)	5–7	–	66–71	–	–1.8–1.2	Wulff et al. (2008c)
Diatom community	Greenland (Arc)	5–35	2–30	7	–	–0.5	Kühl et al. (2001)

Temperature and *in situ* PAR ranges at the time of the measurements are also indicated. E_k and E_c values are as cited in the literature and are ranges or averages of at least three replicates (standard errors were not indicated for clarity).

* Irradiance values for microstages of seaweeds generally represent conditions during incubation in the laboratory, while *in situ* irradiances reported for studies with benthic microalgae are normally from under ice or under very turbid conditions.

Arc, Arctic; Ant, Antarctic; I., Island.

P-I curves of both Antarctic and Arctic seaweeds are characterized by pronounced initial slopes (the parameter α), which is an indicator of the photosynthetic efficiency at low irradiance (Henley 1993). In these algae, high α values have been measured year-round, which suggests a low potential for acclimation that allows these algae to photosynthesize over a broad depth distribution (Henley and Dunton 1995, Gómez et al. 1995b, Weykam et al. 1996, Gómez 2001). Based on 36 species collected from different depths at King George Island, Weykam et al. (1996) tested the hypothesis that photosynthetic efficiency and light requirements of Antarctic algae vary with depth. Although algae from shallow waters (green and some brown algae) had higher net P_{max} and E_k values than species from deeper habitats, low E_k values were determined regardless of algal division and growing depth. This contrasts with the capacity for depth acclimation of photosynthesis found in temperate algae (Mathieson and Norall 1975, Ramus et al. 1976, Glenn and Doty 1981). A reason explaining this pattern is the extremely transparent water column during the open water season in the Antarctic, which allows average irradiances close to 20 μmol photons m^{-2} s^{-1} at 30 m, with 1% surface irradiance at depths more than 40 m (Klöser et al. 1996, Gómez et al. 1997, Klöser 1998). Although these levels are clearly lower than average midday irradiances (325 μmol photons m^{-2} s^{-1}) at 30 m in some clear temperate and tropical waters (Peckol and Ramus 1988), they exceed reported E_k and E_c values in most of the Antarctic macroalgae studied so far (Table 10.1). Thus, although Antarctic species such as *Desmarestia menziesii*, *D. anceps*, *Palmaria decipiens* and *Gigartina skottsbergii* occur at depths between 5 and 20 m, their lower distribution limit is close to 30 m, coinciding with their compensation irradiances (Klöser et al. 1996, Gómez et al. 1997). Similarly, *Himantothallus grandifolius*, although present at depths close to 10 m, becomes dominant at deeper locations (30 m; Zielinski 1990, Klöser et al. 1996). The Arctic seaweed communities, with the exception of some Ceramiales, are rather characterized by defined belts: at the upper sublittoral zone (<5 m), species such as the brown algae *Fucus distichus*, *Chordaria flagelliformis* and *Saccorhiza dermatodea*, the green algae *Urospora penicilliformis* and *Acrosiphonia arcta*, and the red alga *Devaleraea* are common. Key species in the mid subtidal (3–15 m) are the brown algae *Alaria esculenta*, *Laminaria digitata*, *Saccharina latissima*, *Desmarestia aculeata* and *Sphacelaria plumosa* and the red alga *Callophyllis cristata*. Finally, in the lower sublittoral (30 m) red algae such as *Phycodrys rubens*, *Ptilota gunneri* and *Corallina* assemblages dominate (Chapter 3; Svendsen 1959, Wiencke et al. 2004, Wulff et al. 2009).

The difference in water transparency between Arctic and Antarctic regions can be considerable. The high water transparency of Antarctic coastal waters in late summer, characterized by attenuation coefficients (k_d) ranging from 0.2 to 0.3 m^{-1}, greatly contrasts to that of e.g., nearshore waters in the Arctic Beaufort Sea, which are more estuarine and characterized by muddy bottoms (k_d from 0.5 to 3 m^{-1}) in late summer (Dunton et al. 2009). Seaweeds in Antarctic waters are exposed to nearly 20% of surface irradiance at 8 m depths during spring-summer, which explains both the luxuriant growth of attached species and their penetration to great depths. Overall, the low light conditions in polar regions as a result of extreme seasonality and processes related to coastal water turbidity (Chapter 2; Zacher et al. 2009a) impose severe constraints on seaweed photosynthesis (Chapter 9; Dunton and Dayton 1995, Dunton et al. 2009, Wiencke et al. 2009), which have important consequences for metabolic carbon balance, annual growth and primary productivity (see below).

Microscopic stages of seaweeds

Information on light requirements for photosynthesis and growth of microscopic life history stages (spores, gametes and microthalli) of polar seaweeds is lacking. Basically, reproductive processes and the further viability of morphological components within the life-history are strongly triggered by light (Kain 1964, Lüning 1980a,b, Wiencke 1990a, b). In general, microscopic stages of seaweeds have an ability to grow and photosynthesize under low light, which allows the algae to complete the life cycles in the seasonally changing light environment. Under dim light and short daylength, metabolic performance of spores and gametophytes are favored by diminutive size, higher surface/volume ratio, higher pigment content per weight unit and a more efficient light use compared to adult macrothalli (Wiencke et al. 1995, Gómez and Wiencke 1996).

Propagules released into the water column have a short-lived planktonic phase within the euphotic layer for a period of time when they can be exposed to high solar radiation. Phototaxis can influence the duration of motile spore stages, directing them to areas that support their respective photosynthetic activities (Greer and Amsler 2004). Positive phototaxis may augment propagule dispersal and settlement at higher shoreline levels, while negative phototaxis may direct settlement of algal propagules towards seabed substrata (Greer and Amsler 2004). Active downward swimming and sedimentation of propagules to depths, under algal canopies and boulders, and interstitial spaces between algal fronds and turfs provide low-light microenvironments suitable for germination and growth (Roleda et al. 2006a). Due to their small size and the viscosity of water, they may be able to vertically swim at a speed of 120 $\mu m\ s^{-1}$ covering a distance of 1 m in 2 h (Roleda et al. 2006a). Therefore, zoospores released (normally as a plume of spore cloud) from a 2–4 m large sporophyte should be able to touchdown and settle into a low-light environment under algal canopies within 8 h. Even with extended exposure of Arctic *Saccorhiza dermatodea, Alaria esculenta* and *Laminaria digitata* zoospores to 24 h of polar day ambient solar radiation at different depths in Kongsfjorden, zoospores were to some extent able to resume their physiological functions and germinate under low PAR in the laboratory (Wiencke et al. 2006).

Photosynthesis of freshly released propagules of polar seaweeds is characterized by low E_k values, generally lower than the parental sporophytes in the case of Arctic kelps, but similar to macrothalli of Antarctic species (Table 10.1). The efficient use of low irradiance shows that shade adaptation of propagules is probably related to the chlorophyll antenna size and the number of chloroplasts present in spores and gametes compared to macrothalli (Roleda et al. 2008b). For example, spore release of *Nereocystis luetkeana* (Mertens) Postels *et* Ruprecht at dawn is suggested to be a mechanism to maximize photosynthetic potential of the spores (Amsler and Neushul 1989). Although light harvesting and photosynthetic conversion efficiency parameters vary between propagule cell type, taxonomic group and depth of collection, we suspect that Arctic algal spores have lower E_k values compared to propagules from Antarctica. This is related to propagule size and probably to the high solar radiation during the Antarctic spring (Gómez et al. 1997, Roleda et al. 2008b). E_k values of Arctic kelp zoospores range between 9–18 µmol photons $m^{-2}\ s^{-1}$ (Roleda et al. 2006a, Roleda 2009) compared to the E_k of different Antarctic seaweed propagules ranging between 33–87 µmol photons $m^{-2}\ s^{-1}$ (Roleda et al. 2007, Zacher et al. 2007b, 2009b, Roleda et al. 2008a, b).

Spores of polar algae are extremely photosensitive. Exposure of 1 h to 22 µmol photons m^{-2} s^{-1} of PAR can cause significant reduction in optimum quantum yield of fluorescence (F_v/F_m, Roleda et al. 2006a, b, 2007, 2008a, Roleda 2009, Zacher et al. 2007b, 2009b). This reduction is attributable to the total PAR dose received by the propagules. After exposure to increasing doses of PAR as a function of exposure time, recovery of photoinhibition and germination of spores are observed during post-cultivation under low white light (4–10 µmol photons m^{-2} s^{-1}) in the laboratory. Exposure to blue light (1.8 mol m^{-2}) corresponding to 1% of the spring-time surface irradiance is necessary for gametogenesis of temperate kelp and development of microscopic sporophytes at 45 m depth (Ødegaard et al. 1998). Lüning (1990) estimated that a photon dose of ~70 mol m^{-2} $year^{-1}$ is needed for development of some kelps from Helgoland (North Sea), e.g., *Laminaria hyperborea* (Gunnerus) Foslie. The minimum light requirement for completion of Antarctic Desmarestiales life cycles lies between 31–47 mol photons m^2 $year^{-1}$ (Wiencke 1990a). Although Antarctic brown algae are not phylogenetically closely related to Laminariales, they show remarkable morpho-functional similarities (Chapter 9; Gómez et al. 1995a, Wiencke et al. 2009), which may explain such comparable light requirements. Increasing water turbidity and algal respiration exceeding photosynthesis during winter can, however, decimate the population at deeper locations (Markager and Sand-Jensen 1992, Klöser et al. 1993). On the other hand, kelp gametophytes are able to survive in darkness for 18 months (tom Dieck 1993).

The germination capacity of kelp zoospores exposed to ambient solar radiation in Kongsfjorden (Arctic) is dependent on water transparency and the corresponding PAR dose received at different depths (Wiencke et al. 2006). Viability of spores after exposure to high PAR can vary seasonally and is species-specific (Wiencke et al. 2000, 2006, Roleda et al. 2006b). Thus, the fate of early stages determines the further development of the macroalgal population. Given that development, viability and germination capacity of propagules are strongly regulated by daylength and photon fluence rate (summarized by Lüning 1990), questions related to potential acclimation capacity of spores and gametophytes or the metabolic machinery to withstand darkness or alternatively high solar radiation need to be further addressed.

Benthic diatoms

Pennate diatoms are the dominant benthic microalgal assemblages, and in polar regions they have been shown to very efficiently adjust their photosynthetic activity to current radiation conditions; for Antarctic waters see the studies by Palmisano et al. (1985a), McMinn and Hegseth (2004), Wulff et al. (2008b,c) and for Arctic waters see Kühl et al. (2001) and Glud et al. (2002). Like various seaweeds, many benthic microalgal communities can live up to two months in darkness (Wulff et al. 2008c). Furthermore, the benthic diatom species *Cylindrotheca closterium* (Ehrenberg) Lewin *et* Reimann and *Nitzschia aurariae* Cholnoky from Arctic waters survived in laboratory experiments mimicking polar winter over five months of darkness (U. Karsten, R. Schumann and M. Rickler, unpublished results).

Support for the extreme shade adaptation of benthic diatoms in polar areas comes from studies on the light requirements for photosynthesis. In the Antarctic study by Palmisano et al. (1985a), the benthic diatom *Trachyneis aspera* grew at ambient radiation of <0.6 μmol photons m^{-2} s^{-1} (the limit of detection for the radiometer used), with E_k values between 7 and 16 μmol photons m^{-2} s^{-1} (Table 10.1). Because the diatoms grow between and attached to spicules of a siliceous sponge, the ambient radiation can probably be even lower. Although the diatoms are light-saturated at 11 μmol photons m^{-2} s^{-1}, no photoinhibition has been observed at irradiances as high as 300 μmol photons m^{-2} s^{-1} (Palmisano et al. 1985a). However, this species can also be strongly photoinhibited at 6–10 μmol photons m^{-2} s^{-1} (Rivkin and Putt 1987). In the case of Antarctic freshwater environments (e.g., Lake Hoare), under-ice benthic mats of pennate diatoms have typical E_k values of 2 μmol photons m^{-2} s^{-1} (Hawes and Schwarz 1999). With increasing water depth, the proportion of diatoms to Cyanobacteria increases, indicating that diatoms are indeed strong competitors in cold low-light environments. The ability of polar benthic diatoms to acclimate to extreme low-light conditions, but also to high-light conditions has been shown in a number of studies (Sundbäck et al. 1997, Zacher et al. 2007a, Wulff et al. 2008c). These results indicate that these organisms are very well adapted to fluctuating radiation conditions that result from rapid physical changes (such as ice break up in spring). For example, shallow bottom diatom communities from King George Island can have E_k values between 66 and 77 μmol photons m^{-2} s^{-1} (Wulff et al. 2008; Table 10.1), which are up to 10 times higher than those determined in assemblages adapted to lower irradiances (e.g., Palmisano et al. 1985a).

The low E_k values of benthic diatoms enable them to colonize the deeper ocean. For example, viable Antarctic benthic diatoms (defined as cells with intact chloroplasts) were present at 30 m water depth in Potter Cove, King George Island (Klöser 1998, A. Wulff, personal observation). It must be mentioned that photosynthesis can be well supported at these depths, as during the spring-summer period, maximal irradiances can be close to 50 μmol photons m^{-2} s^{-1} (Gómez et al. 1997). A recent study by McGee et al. (2008) reporting living diatoms at 191 m suggests that photosynthesis of these organisms can (at least temporarily) be supported at PAR intensities as low as 0.1 μmol photons m^{-2} s^{-1} (<0.03% surface irradiance). The pennate diatom *Gyrosigma subsalinum* var. *antarcticum* Frenguelli *et* Orlando typically grows at depths >10 m (Klöser 1998) and was shown to be well adapted to the radiation conditions prevailing in this habitat (Longhi et al. 2003). In general, benthic diatoms grow well below average values of <10 μmol photons m^{-2} s^{-1} (Matheke and Horner 1974, Kühl et al. 2001, Glud et al. 2002) (Table 10.1). Microalgae found in ice-algal communities are a special case; they very often have benthic origins and are transported to sea ice through different bottom-ice interactive processes (e.g., Hawes and Schwarz 1999). These communities, generally dominated by pennate diatoms, are well adapted to low light conditions (Smith et al. 1994, Robinson et al. 1995, McMinn and Hegseth 2004, Lee et al. 2008) and thus, there is no *a priori* reason to believe that diatoms incorporated into ice-algal communities function differently from corresponding species inhabiting soft bottom sediments.

Temperature demands

Seaweeds

Polar seaweeds are well adapted to the low temperatures of their environment. The seaweeds with the lowest determined temperature demands are species endemic to the Antarctic. For example, some endemic Desmarestiales grow up to 5 °C and exhibit upper survival temperatures (USTs) of 11–13 °C. Antarctic cold-temperate species, especially from the intertidal zone, are characterized by higher temperature requirements. They grow up to 10–15 (or 20) °C and have USTs between 13.5 and 19 °C (Wiencke and tom Dieck 1990). The red algae *Georgiella confluens* (Reinsch) Kylin, *Gigartina skottsbergii* Setchell *et* N.L. Gardner and *Plocamium cartilagineum* (Linnaeus) P.S. Dixon grow at 0 °C, but not at 5 °C and have USTs as low as 7–11 °C (Bischoff-Bäsmann and Wiencke 1996).

Seaweeds from the Arctic are comparatively less adapted to low temperatures as shown for example for the kelp *Laminaria solidungula* and the red alga *Devaleraea ramentacea*. Sporophytes of the endemic Arctic *L. solidungula* grow at temperatures up to 15 °C with optimum growth rates between 5 and 10 °C and an UST of 16 °C (tom Dieck 1992). *D. ramentacea* (Linnaeus) Guiry, whose distribution considerably extends into the cold-temperate region, grows at temperatures up to 10 °C with optimum growth rates at 0 °C and an UST of 20 °C (Novaczek et al. 1990, Bischoff and Wiencke 1993). Species with a more prominent distribution in the cold-temperate region are mostly characterized by even higher temperature demands. More data on the temperature demands of endemic Arctic species are urgently needed.

The strong adaptation of Antarctic seaweeds to low temperatures is also reflected in photosynthesis. For example, P_{max} values of a number of cultured endemic Antarctic seaweeds measured at temperatures close to 0 °C (Wiencke et al. 1993) are comparable to P_{max} values reported for temperate seaweeds measured at higher temperatures (0.3–38 µmol O_2 g^{-1} FW^{-1} at 20 °C; e.g., Enríquez et al. 1995). However, optimum temperatures for photosynthesis are clearly above the temperatures of the natural environment. Lowest temperature optima for photosynthesis were recorded in the brown alga *Ascoseira mirabilis* (1–10 °C). The red algae *Trematocarpus antarcticus* (Hariot) Fredericq *et* R. Moe, *Gymnogongrus antarcticus* Skottsberg, *Phyllophora ahnfeltioides* Skottsberg and *Ballia callitricha* (C. Agardh) Kützing as well as the brown alga *Himantothallus grandifolius* are characterized by broad optima between 10 and 20 °C, whereas the red algae *Palmaria decipiens* and *Gigartina skottsbergii* exhibit a strong optimum at 15 °C (Drew 1977, Wiencke et al. 1993, Eggert and Wiencke 2000). The few macroalgal species from the Arctic studied so far have temperature optima around 20 °C (Healey 1972).

The rates of dark respiration in species from the Antarctic increase above the temperature optima for photosynthesis (Drew 1977, Wiencke et al. 1993, Eggert and Wiencke 2000), indicating that the temperature optima of dark respiration are higher than those of photosynthesis, but exact values are not yet available. Photosynthesis/respiration (P:R) ratios are highest at the lowest experimental temperature (0 °C), and decrease strongly with increasing temperatures due to different Q_{10} values for photosynthesis (1.4–3.5) and dark respiration (2.5–5). The high P:R ratios at low temperatures explain the high growth rates and high primary productivity of polar species at low temperatures (see below). Thus, at both the whole organism and physiological levels, there is a clear difference between the temperature demands of seaweeds from the Antarctic and the Arctic.

The different adaptation to low temperatures is an expression of the different cold-water history of these polar regions (Chapter 2; Lüning 1990, Crame 1993, Zachos et al. 2001, Zacher et al. 2009a). In the Southern Ocean, a sharp demarcation of the continent by the Polar Frontal Zone (PFZ) exposed organisms to low temperatures by about 14 Ma, whereas in the Arctic, the cold adaptation of algae occurred in a much shorter time span (about 2 Ma). The temperature requirements for sporophyte growth could explain the high occurrence of endemic species in the Antarctic, e.g., many members of Desmarestiales. Therefore, these species can only occur in areas south of the Antarctic convergence, with winter temperatures below 5 °C allowing sporophyte growth (Wiencke and tom Dieck 1989). In contrast, the distribution of Arctic-North Atlantic species (e.g., the brown algae *Laminaria digitata* and *Halosiphon tomentosus* (Lyngbye) Jaasund is often limited by both the USTs and the upper limit for gametogenesis (ULGLs) (van den Hoek 1982a,b, Breeman 1988, Lüning 1990).

Microscopic stages of seaweeds

As for macrothalli, the optimal temperature for photosynthesis and development of microscopic stages of polar seaweeds is higher than *in situ* temperatures (Kirst and Wiencke 1995). In general, most of the studies on thermal demands of the microscopic life history stages in polar seaweeds have been done on Arctic kelps. Photosynthetic performance of zoospores of species such as *Alaria esculenta, Laminaria digitata* and *Saccharina latissima* are optimum between 7–13 °C, temperatures higher than in the natural environment, while the F_v/F_m measured at 2 °C is compromised but sustained at a minimum (Roleda 2009). Short-term photosynthetic responses (1–4 h) to increasing temperature (13–19 °C) are normally non-lethal. Above the temperature optima, continued cultivation under high temperature has a negative impact on photosynthesis related to the lability of PSII. Species-specific responses to the combined effect of light and temperature stress showed sensitivity in the order *S. latissima* > *L. digitata* > *A. esculenta* (Roleda 2009). In Antarctic species, studies on optimal temperature for photosynthesis are not available, however, comparative measurements carried out at 0 °C, indicate that juvenile sporophytes and gametophytes of e.g., *Desmarestia menziesii* have higher rates of photosynthesis compared to adult sporophytes (Gómez and Wiencke 1996). In the case of spores, a study based on three species (*Adenocystis utricularis, Monostroma hariotii* and *Porphyra endiviifolia*) revealed that photosynthesis can be essential to support repair and photoprotective mechanisms, especially when propagules are exposed to high solar radiation in the water column (Zacher et al. 2007b).

Like photosynthesis, temperature affects a series of reproductive parameters and sporeling development in seaweeds. For example, maturation of gametangia is strongly temperature-dependent. E.g., in *Saccorhiza dermatodea* gametogenesis occurs only at temperatures <14 °C (Henry 1987). In Antarctic seaweeds, reproductive structures have been observed in Austral spring and summer (e.g., Clayton and Wiencke 1986, Wiencke 1988, Roleda et al. 2007, 2008a, Zacher et al. 2007b) when *in situ* water temperature increases to a maximum of 5 °C. Interestingly, microscopic stages of Antarctic seaweeds show a capacity to tolerate temperatures well above those measured *in situ*. For example, microthalli of Antarctic cold-temperate Phaeophyceae have USTs between 21 and 25 °C, and upper limits for growth between 13 and 15 °C (Wiencke and tom Dieck 1990, Peters and Breeman 1993). In the case of gametophytes of endemic Antarctic Desmarestiales, growth occurs at temperatures up to 10 or 15 °C with USTs between 15 and 18 °C (Wiencke and tom Dieck 1989, 1990). The LST of the gametophytes goes down to at least −1.5 °C (Bolton and Lüning 1982, tom Dieck 1993).

Maturation of Arctic brown algal gametophytes and formation of sporophytes is observed at 0 °C (Chapman and Lindley 1980, tom Dieck 1993); however, the upper temperature limit for gametogenesis is yet to be determined. In general, gametophytes of *Laminaria solidungula* have lower and upper survival temperatures of –1.5 and 20 °C, respectively (Bolton and Lüning 1982, tom Dieck 1993). Gametophytes of *Alaria esculenta* and *Agarum clathratum* Dumontier, however, have USTs of 19–21 °C, as low as those of the more Arctic species *L. solidungula* (Wiencke et al. 1994). Zoospore germination in *Alaria esculenta*, *Laminaria digitata* and *Saccharina latissima* has optimal ranges between 2 and 12 °C, but is impaired at 18 °C (Müller et al. 2008).

In Rhodophyta, temperature demands of different stages in the life cycle have been less studied. In the Antarctic, species such as *Gigartina skottsbergii* (Roleda et al. 2008a) and *Iridaea cordata* (Turner) Bory de Saint-Vincent (Zacher et al. 2009b) have maximum survival temperatures of 15–16 °C. In the Arctic, survival, growth and development of different spore ploidies of the red alga *Phycodrys rubens* (Linnaeus) Batters in the Barents Sea may be possible at an upper limit of about 14–15 °C. In this species, asexual reproduction occurs year round, with an end of summer to autumn peak, while sexual reproduction is initiated in spring (Schoschina 1996). Overall, although growth and development of microscopic life history stages are enhanced at temperatures higher than *in situ*, both in Antarctic and Antarctic seaweeds, the initiation of reproductive process in winter may require low temperature or is an endogenous circannual rhythm and adaptive process to ensure that spores and gametes are released in a timely manner and are ready for the start of a new growing season in spring (Molenaar and Breeman 1997). The ability for complex metabolic regulation to optimize photosynthesis over the wide range of temperatures and irradiance levels encountered in nature determines the biogeographic distribution (Roleda 2009).

Benthic diatoms

Polar benthic diatoms preferentially grow in the subtidal at relatively constant low temperatures. For example, microalgae associated with sea ice are able to withstand and even to grow to some extent at extremely low temperatures of –15 °C (Mock and Gradinger 1999). However, some assemblages growing in shallow water or tide pools may be exposed to temperatures higher than in the water column during summer. In Potter Cove (King George Island), temperature in some tide pools may rise from 2 to 14 °C within 8 h (Chapter 2; Klöser et al. 1994, Zacher et al. 2009a).

As far as we know, the temperature demands for growth of Arctic benthic diatoms are somewhat higher than those of endemic Antarctic species. Two *Fragilaria* species from Arctic Kongsfjorden (Spitsbergen) growing as epiphytes on seaweeds with optimum growth rates at 12–14 °C grow well at a reduced rate at 0 °C, but do not survive 20 °C (Karsten et al. 2006), suggesting that Arctic benthic diatoms can be characterized as eurythermal microalgae that are psychrotolerant (organisms tolerant of low growing temperatures). This is in sharp contrast to related Antarctic taxa, which have maximum growth at 0 °C and full inhibition of cell division at 7–9 °C (Longhi et al. 2003). Consequently, Antarctic benthic diatoms can be characterized as polar stenothermal microalgae that are psychrophilic (organisms with a requirement for low growth temperatures). In general, the relatively higher temperature demands for growth in Arctic benthic diatoms compared to Antarctic pendants is indeed documented so far in only very few publications. However, various diatom species from Kongsfjorden (Spitsbergen) show the same high temperature requirements, i.e., optimum growth between 10 and 15 °C (U. Karsten, unpublished data). These differences in the temperature requirements for growth in Arctic and Antarctic benthic diatoms are likely related to the geological cold-water history of both regions, as in the case for seaweeds. While the Antarctic benthic diatoms investigated by Longhi et al. (2003) are indeed characterized as endemic taxa, which well explains their low temperature demands for growth, the respective information on strains from the Arctic is still missing. It is not clear whether an endemic Arctic benthic diatom population actually exists.

To successfully colonize low-temperature environments, polar benthic diatoms have developed mainly physiological adaptive strategies. For example, cold adaptation is required for the activity of one of the key enzymes in photosynthesis and the most abundant enzyme in plants, ribulose-1,5-bisphosphate carboxylase/oxygenase (RUBISCO). This enzyme has poor catalytic efficiency at low temperatures in isolated psychrophilic microalgae. Nevertheless, its activity is maintained, as temperatures drop by an increase in the concentrations of the enzyme (Devos et al. 1998). In contrast to temperature adaptation, acclimation to transient changes in the thermal environment is dependent upon sensor/ signal pathways, and involves, for example, modulations in rates of transcription or translation of enzymes (Morgan-Kiss et al. 2006). Nevertheless, the expression of many genes encoding proteins involved in photosynthesis is up-regulated under decreasing temperatures (Mock and Valentin 2004).

Other adaptive mechanisms to low temperature include the evolution of cold shock and antifreeze proteins, the modulation of the key enzyme kinetics, and the development of more fluid biological membranes through the accumulation of polyunsaturated fatty acid chains (Morgan-Kiss et al. 2006). Many benthic diatoms release so-called extracellular polymeric substances (EPS) that consist of polysaccharides and proteins, which can help to preserve liquid conditions during freezing (Krembs et al. 2002). In addition, extracellular, secreted proteins are described as having an affinity for ice crystals have been described in some sea ice diatoms such as *Navicula glaciei* Van Heurck (Raymond and Fritsen 2001, Janech et al. 2006). Ice-binding proteins do not lower the freezing point, they rather seem to prevent damage to membranes by the inhibition of the recrystallisation of ice and hence may act as effective structural cryoprotectants (Raymond and Knight 2003). Various benthic diatoms from polar waters are known to contain high concentrations of low-molecular weight solutes acting as organic antifreezing substances, e.g., the amino acid proline and the tertiary sulfonium compound dimethylsulfoniumpropionate (DMSP) (Thomas and Dieckmann 2002, Kasamatsu et al. 2004). Although unknown yet for polar microalgae, at least in Antarctic green seaweeds, decreasing temperatures strongly stimulate the biosynthesis and accumulation of DMSP (Karsten et al. 1996).

Benthic primary productivity in polar regions

Seaweeds

Despite the environmental constraints in polar regions, benthic algae in some locations can have high primary productivity similar to or higher than phytoplankton production (Fogg 1977). The contribution of benthic primary productivity has been difficult to quantify, mainly because we cannot provide accurate real estimations of coastal waters exposed to irradiances > E_c (Gattuso et al. 2006). By means of analysis of satellite data (SeaWiFS), these authors (op. cit.) determined that only 28 and 13% of the coastal surface area in the Arctic may potentially support net primary production of benthic microalgae and seaweeds, respectively. In the case of some Antarctic coastal systems of King George Island, seaweeds can cover up to 64% of the total surface of phytobenthos between 10 and 60 m (Zielinski 1990), suggesting that primary production can be well supported in deeper zones.

In general, primary production estimates have been reported mostly for Arctic kelps, while information available for Antarctic seaweeds is based on standing crop data. Arctic kelps such as *Saccharina latissima*, which comprises only 5–10% of the macroalgal biomass, can reach 1–2 g C m^{-2} $year^{-1}$ (Borum et al. 2002), while grand estimates for the High Arctic *Laminaria solidungula* can approach 20 g C m^{-2} $year^{-1}$ (Chapman and Lindley 1981), but generally range 6–10 g C m^{-2} $year^{-1}$ as has been estimated for forests in the Beaufort Sea (Dunton and Schell 1986, Dunton and Dayton 1995; Table 10.2). In the case of Antarctic brown algae, although biomass per area is considerable (Zielinski 1990, Klöser et al. 1994, Brouwer et al. 1995), few estimations of primary production and standing crops are available. Values of primary productivity of *Himantothallus grandifolius* at Signy Island range between 16 and 56 g C m^{-2} $year^{-1}$ (Drew and Hastings 1992), while standing crops can vary between 1 and 5 kg FW m^{-2}. Other algae such as *Desmarestia menziesii* and *D. anceps* can have crops of 0.2 and 1.5 kg DW m^{-2} (Quartino and Boraso de Zaixso 2008; Table 10.2). Production estimates from Rhodophyta (e.g., Arctic coralline assemblages) indicate values close to 2.6 g C m^{-2} $year^{-1}$, and standing crops of some foliose Antarctic species can be close to 0.05 kg DW m^{-2} (Table 10.2). Considering that much of the available information comes from measurements carried out in summer, extrapolations to annual yields or comparisons with other geographic areas or pelagic productivity must be treated with caution. For example, annual values of productivity of polar seaweeds (Table 10.2) are generally lower than estimations for temperate kelps (e.g., *Macrocystis*, *Laminaria* sp.) or Fucales, which can be as high as 1900 g C m^{-2} $year^{-1}$ (summarized by Lüning 1990). Similarly, seaweed production in polar regions is lower than phytoplankton production measured in some productive zones around the Antarctic Peninsula (182 g C m^{-2} $year^{-1}$; Vernet and Smith 2006) or "hotspots" such as the upwelling zones in sub-tropical areas (500–800 g C m^{-2} $year^{-1}$; reviewed by Reynolds 2006).

An essential prerequisite for primary production under low light conditions is a low respiration rate and a positive carbon balance (Markager and Sand-Jensen 1992). With increasing depth, the daily metabolic carbon balance (a parameter relating daily incident radiation, dark respiration and E_k for photosynthesis) decreases due to the fact that the daily period in which algae remain under irradiances above compensation or saturation (H_{comp} and H_{sat}, respectively) becomes shorter. In polar regions, seaweeds exposed to marked seasonality in daylength only exhibit $H_{sat} > 0$ h during the open water season in spring-summer. Annual values for H_{sat} range from 63 to 126 h for *Laminaria solidungula* (Dunton and Jodwalis 1988, Dunton 1990), compared to 43–72 h for *Himantothallus grandifolius* and *Desmarestia antarctica*, recorded only for one month (Gómez et al. 1997). Thus, if the daily carbon balance is negative, then growth of seaweeds can be limited. In the Alaskan Beaufort Sea, *Laminaria solidungula* may only receive 45 mol photons m^{-2} $year^{-1}$, clearly the lowest photon flux of any deep water kelp, which represents a paltry 0.2% of total surface PAR (Dunton 1990). Consequently, its linear growth can sometimes decrease to 5 cm annually (Dunton et al. 2009). In contrast, the clearer waters of the Antarctic Peninsula will allow deep-water species, such as *Himantothallus grandifolius* or several understory red algae, to maintain a positive carbon balance and colonize depths down to at least 40 m (Chapter 3; Gómez et al. 1997, Wulff et al. 2009). However, in other species such as *Desmarestia anceps*, a negative carbon balance coincides with its distribution limit at 30 m. These values estimated using light requirement for photosynthesis are supported by light requirements for growth, which can be <1 μmol photons m^{-2} s^{-1} both in Antarctic (Wiencke and Fischer 1990, Wiencke 1990a) and Arctic species (Chapman and Lindley 1980, Dunton and Schell 1986).

The fate of seaweed primary production regulates much of the energy flux in coastal waters of Arctic and Antarctic regions. In Potter Cove (King George Island), the summer production of the most abundant species (*Desmarestia anceps, D. menziesii* J. Agardh, *Himantothallus grandifolius, Iridaea cordata* and *Gigartina skottsbergii*) is 1401 t, equivalent to 1.76 times the mean dry biomass present in the cove during the whole study. The calculated flux of dry biomass to the ecosystem of Potter Cove in the period December-March is 1370 t (Quartino and Boraso de Zaixso 2008). An important fraction of macroalgal carbon production seems to be used as a food source for diverse zoobenthic Antarctic (Iken et al. 1998, Huang et al. 2006) and High Arctic (Dunton and Schell 1987) communities. In the Antarctic benthos, assimilation numbers at different trophic levels estimated using $\delta^{13}C$ and $\delta^{15}N$ measurements can vary between 30 and 100% (Dunton 2001), which suggests consumption levels comparable to those reported for seaweed-herbivore interactions e.g. from coral reefs (Carpenter 1986). Moreover, after detaching, seaweeds can sink to seabed hollows, be washed ashore or out to sea (Klöser et al. 1994, Thiel and Gutow 2005). Piles of drifted seaweeds are used as resting places by seals and sea elephants (Brouwer 1996b). The mix of seaweeds, coastal sediments and seal faeces improve the development of the microorganisms, which decompose organic matter and increase remineralization (Zielinski 1981). Seaweed decomposition processes (physical and microbial) vary in relation to season and type of thallus (Zielinski 1981, Brouwer 1996b). Polar seaweeds are thus not only important as primary producers that supply significant amounts of fixed carbon to marine ecosystems, but as substrates for organisms that inevitably provide a large fraction of the secondary production to the benthos.

Benthic microalgae

Most studies on microbenthic primary productivity have been conducted in temperate ecosystems (reviewed by Cahoon 1999). Primary productivity of polar benthic microalgae is briefly summarized in order to put these data in a perspective comparable with polar seaweeds (Table 10.2). It must be emphasized that values of microbenthic primary production are difficult to compare due to different methods used and because it is not always apparent from the cited studies whether gross or net production is being reported. Arctic microbenthic primary productivity has been studied by Matheke and Horner (1974) and Horner and Schrader (1982), but has more recently received attention in studies around Greenland (Chapter 7; Glud et al. 2009). For example, in

Table 10.2 Polar benthic primary production and standing crop of selected seaweed and microalgal assemblages.

Species/assemblage	Location	Primary production	Standing crop	Reference
Seaweeds				
Laminaria solidungula	Beaufort Sea (Arc)	6–10 g C m^{-2} $year^{-1}$		Dunton and Schell (1986)
	Canadian Archipelago (Arc)	20 g C m^{-2} $year^{-1}$		Chapman and Lindley (1981)
Saccharina latissima	Young Sound, Greenland (Arc)	0.1–2 g C m^{-2} $year^{-1}$		Borum et al. (2002)
Himantothallus grandifolius	Signy I. (Ant)	16–56 g C m^{-2} $year^{-1}$		Drew and Hastings (1992)
	Signy I. (Ant)		4 kg FW m^{-2}	Heywood and Whitaker (1984)
	Signy I. (Ant)		1–2 kg FW m^{-2}	Richardson (1979)
	King George I. (Ant)		0.4 kg DW m^{-2}	Quartino and Boraso de Zaixso (2008)
	Anvers I. (Ant)		4.5–5 kg FW m^{-2}	Amsler et al. (1995)
Desmarestia spp.	Signy I. (Ant)		0.3–1.5 kg FW m^{-2}	Richardson (1979)
Desmarestia anceps	King George I. (Ant)		0.2 kg DW m^{-2}	Quartino and Boraso de Zaixso (2008)
Desmarestia menziesii	King George I. (Ant)		0.4 kg DW m^{-2}	Quartino and Boraso de Zaixso (2008)
Coralline assemblage (17 m)	Young Sound, Greenland (Arc)	2.6 g C m^{-2} $year^{-1}$		Roberts et al. (2002)
June–September		0.2 g C m^{-2} $year^{-1}$		Roberts et al. (2002)
Myriogramme manginii (5 and 20 m)	Signy I. (Ant)	0.01–2 g C g^{-1} DW $year^{-1}$		Brouwer (1996a)
Rhodophyta community	Signy I. (Ant)		0.1–0.5 kg FW m^{-2}	Richardson (1979)
Iridaea cordata	King George I. (Ant)		0.05 kg DW m^{-2}	Quartino and Boraso de Zaixso (2008)
Gigartina skottsbergii	King George I. (Ant)		0.025 kg DW m^{-2}	Quartino and Boraso de Zaixso (2008)
Seaweed community (average)	King George I. (Ant)		2 kg FW m^{-2}	Zielinski (1990)
Benthic microalgae				
Microphytobenthic community gross production (3–11 m)	Kongsfjorden, Svalbard (Arc)	7–16 mg C m^{-2} h^{-1} 15–35 g C m^{-2} $year^{-1}$	14–63 mg chl*a* m^{-2}	J. Woelfel, unpublished data
Microphytobenthic community (5 m)	Chukchi Sea (Arc)	0.5–57 mg C m^{-2} h^{-1} (~16 g C m^{-2} $year^{-1}$)*	35–321 mg chl*a* m^{-2}	Matheke and Horner (1974)
Microphytobenthic community (7 m)	Beaufort Sea, Alaska (Arc)	0.01 mg C m^{-2} h^{-1} (9 g C m^{-2} $year^{-1}$)*	1–28 mg chl*a* m^{-2}	Horner and Schrader (1982)
Benthic diatoms (5–30 m)	Young Sound, Greenland (Arc)	109–387 mg C m^{-2} day^{-1}		Glud et al. (2002)
Benthic diatoms (30 m)	McMurdo Sound (Ant)	0.01 mg C mg chl*a*$^{-1}$ day^{-1}		Palmisano et al. (1985a)
Microphytobenthic community (ice-free season)	Signy I. (Ant)	13–29 mg C m^{-2} h^{-1} (16 g C m^{-2} $year^{-1}$)*		Gilbert (1991)

In some cases, depth is indicated following description in the original source. Note different units for microbenthic primary production.
* Estimated by Cahoon (1999).
Arc, Arctic; Ant, Antarctic; I., Island.

Young Sound, Greenland the microbenthic community dominated by diatoms accounted for up to 40% of the total benthic primary productivity (Glud et al. 2002). Studies on microalgal primary productivity in Antarctic regions have focused on ice-algae, with only few studies related to benthic microalgae over the years (Palmisano et al. 1985a,b, Dayton et al. 1986, Rivkin and Putt 1987, Gilbert 1991; Table 10.2). Thus, production of polar benthic microalgae is comparable to phytoplankton production estimated for the Central Arctic Ocean (~15 g C m^{-2} $year^{-1}$; Gosselin et al. 1997) or some coastal waters e.g., Young Sound, East Greenland (10 g C m^{-2} $year^{-1}$; Rysgaard et al. 1999), but clearly lower than average values reported for soft bottom communities of microalgae from temperate locations (100–400 g C m^{-2} $year^{-1}$; e.g., McIntyre and Moore 1977, Migné et al. 2004). Taking into account the remarkable ecophysiological adaptation of these organisms (e.g., low light requirements for photosynthesis and an ability to survive extremely low temperatures), as well as their broad distribution in polar benthos, the contribution of polar intertidal and subtidal microphytobenthos to total primary production, is underestimated, and probably considerably (Cahoon 1999, Gattuso et al. 2006).

Concluding remarks and future prospects

The information summarized emphasizes the capacity of polar seaweeds and benthic diatoms a) to endure their harsh environment by virtue of their very low light requirements for photosynthesis and growth, b) to complete their life cycles under constant low temperatures based on effective ecophysiological adaptations, and c) to grow, reproduce and survive at temperatures well above those recorded *in situ*. A major conclusion emerging from this review is that there is a need for examination of the metabolic characteristics of algae in defining the threshold for physiological adaptations (i.e., the expression of functional responses to optimize resource utilization or to withstand environmental stress). For example, it is especially remarkable that many convergent ecophysiological adaptations to low light and temperature exhibited by polar algae occur in taxa that are not closely related. This clearly has implications for phylogenetic issues and challenges the paradigms related to the environmental barriers controlling biogeographic distribution of these organisms.

Much of the available information originates from laboratory experiments using different light and temperature conditions, often these factors being studied separately. There are few studies examining the simultaneous effect of light and temperature on photosynthetic or growth physiology of polar algae. For example, interactive light and temperature requirements for growth have been tested for two epiphytic taxa (*Fragilaria, Synedropsis*) from the Arctic Kongsfjorden (Karsten et al. 2006). No effect on growth was observed at low temperature (0 °C) and PAR (2–20 μmol photons m^{-2} s^{-1}), a small positive effect was seen at 5 °C, but a strong positive relationship was observed at light levels up to 15 μmol photons m^{-2} s^{-1} at temperatures of 10 and 15 °C (Karsten et al. 2006). Meiospores of *Laminaria digitata* from Arctic Spitsbergen showed a reduced germination capacity and photosynthetic efficiency at low temperatures under moderate UVB exposure (Müller et al. 2008, Roleda 2009). Interaction of these two factors, irradiance (especially UV wavelengths) and temperature, as well as their interactions with other environmental factors, such as nutrients, aquatic contaminants etc., can further modify the responses of algae in ways that are not fully understood. For example, is the photoinduced toxicity of certain environmental pollutants by UV radiation posing a threat also to polar algae in the context of climate change scenarios (Petersen et al. 2008, Huovinen et al. 2010)? Thus, it will be increasingly important to take into account the interactions of multiple environmental factors when evaluating the fate of the unique polar ecosystems. Although polar seaweeds are conspicuous organisms and have successfully developed life strategies to colonize and dominate different coastal systems (e.g., fjords, estuaries), our understanding of their physiology and molecular mechanisms of stress tolerance is still limited. Knowledge is needed, for example, to predict the direct effects of present and future threats (global warming, ozone depletion, etc., cf. chapters 12, 13; Müller et al. 2009, Karsten et al. 2009) on polar ecosystems, as algae from polar and cold regions can be assumed to be particularly susceptible to these changes. One of the major problems involved in the evaluation of the biological impact of global climate change on marine organisms is the lack of basic knowledge on the species (Harley et al. 2006 and references therein).

For a long time, sensitive methods to quantify physiological processes or mechanisms that underlie the ecological responses of polar benthic seaweeds to stress under field conditions were generally unavailable. In recent years, modern genomic, proteomic and metabolomic approaches accompanied by high-tech instrumentation have allowed better understanding of the mechanisms by which polar organisms respond to rapid environmental changes. Thus, future research should be focused on defining integrative models of aspects related to the action of multiple factors and how environmental constraints modulate the cold-adaptation. Clearly, this must be addressed at molecular, organism and community scales.

References

Amsler, C.D. and M. Neushul. 1989. Diel periodicity of spore release from kelp *Nereocystis luetkeana* (Mertens) Postels et Ruprecht. *J. Exp. Mar. Biol. Ecol. 134*: 117–127.

Amsler, C.D., R.J. Rowley, D.R. Laur, L.B. Quentin and R.M. Ross. 1995. Vertical distribution of Antarctic peninsular macroalgae: cover, biomass and species composition. *Phycologia 34*: 424–430.

Bischof, K., I. Gómez, M. Molis, D. Hanelt, U. Karsten, U. Lüder, M.Y. Roleda, K. Zacher and C. Wiencke. 2006. Ultraviolet radiation shapes seaweed communities. *Rev. Environ. Sci. Biotech. 5*: 141–166.

Bischoff, B. and C. Wiencke. 1993. Temperature requirements for growth and survival of macroalgae from Disko Island (Greenland). *Helgol. Mar. Res. 47*: 167–191.

Bischoff-Bäsmann, B. and C. Wiencke. 1996. Temperature requirements for growth and survival of Antarctic Rhodophyta. *J. Phycol. 32*: 525–535.

Bolton, J.J. and K. Lüning. 1982. Optimal growth and survival temperatures of Atlantic *Laminaria* species (Phaeophyta) in culture. *Mar. Biol. 66*: 89–94.

Borum, J., M.F. Pedersen, D. Krause-Jensen, P.B. Christensen and K. Nielsen. 2002. Biomass, photosynthesis and growth of *Laminaria saccharina* in a high-Arctic fjord, NE Greenland. *Mar. Biol. 141*: 11–19.

Breeman, A.M. 1988. Relative importance of temperatures and other factors in determining geographic boundaries of seaweeds: experimental and phenological evidence. *Helgol. Meeresunters. 42*: 199–241.

Brouwer, P.E.M. 1996a. *In situ* photosynthesis and estimated annual production of the red macroalga *Myriogramme mangini* in relation to underwater irradiance at Signy Island (Antarctica). *Antarct. Sci. 8*: 245–252.

Brouwer, P.E.M. 1996b. Decomposition *in situ* of the sublittoral Antarctic macroalgae *Desmarestia anceps* Montagne. *Polar Biol. 16*: 129–137.

Brouwer, P.E.M., E.F.M. Geilen, N.J.M. Gremmen and F. Van Lent. 1995. Biomass, cover and zonation pattern of sublittoral macroalgae at Signy Island, South Orkney Islands, Antarctica. *Bot. Mar. 38*: 259–270.

Cahoon, L. 1999. The role of benthic microalgae in neritic ecosystems. *Oceanogr. Mar. Biol. Annu. Rev. 37*: 47–86.

Carpenter, R. 1986. Partitioning herbivory and its effects on coral reef algal communities. *Ecol. Monogr. 56*: 345–364.

Chapman, A.R.O. and J.E. Lindley. 1980. Seasonal growth of *Laminaria solidungula* in the Canadian High Arctic in relation to irradiance and dissolved nutrient concentrations. *Mar. Biol. 57*: 1–5.

Chapman, A.R.O. and J.E. Lindley. 1981. Productivity of *Laminaria solidungula* J. Ag. in the Canadian High Arctic: a year round study. *In:* (T. Levring, ed) *Proceedings of the Xth International Seaweed Symposium*. Walter de Gruyter, Berlin. pp. 247–252.

Clayton, M.N. 1988. Evolution and life histories of brown algae. *Bot. Mar. 31*: 379–387

Clayton, M.N. and C. Wiencke. 1986. Techniques and equipment for culturing Antarctic benthic marine algae, and for preparing specimens for electron microscopy. *Ser. Cient. INACH 34*: 93–97.

Crame, J.A. 1993. Latitudinal range fluctuations in the marine realm through geological time, *Trends Ecol. Evol. 8*: 162–266.

Dayton, P.K., D. Watson, A. Palmisano, J.P. Barry, J.S. Oliver and D. Rivera. 1986. Distribution patterns of benthic microalgal standing stock at McMurdo Sound, Antarctica. *Polar Biol. 6*: 207–213.

Devos, N., M. Ingouff, R. Loppes and R.F. Matagne. 1998. RUBISCO adaptation to low temperatures: a comparative study in psychrophilic and mesophilic unicellular algae. *J. Phycol. 34*: 655–660.

Drew, E.A. 1977. The physiology of photosynthesis and respiration in some Antarctic marine algae. *Br. Antarct. Surv. Bull. 46*: 59–76.

Drew, E.A. and R.M. Hastings. 1992. A year-round ecophysiological study of *Himantothallus grandifolius* (Desmarestiales, Phaeophyta) at Signy Island. Antarctica. *Phycologia 31*: 262–277.

Dunton, K.H. 1990. Growth and production in *Laminaria solidungula*: relation to continuous underwater light levels in the Alaskan High Arctic. *Mar. Biol. 106*: 297–304.

Dunton, K.H. 2001. $\delta\ ^{15}N$ and $\delta\ ^{13}C$ measurements of Antarctic peninsular fauna: trophic relationships and assimilation of benthic seaweeds. *Am. Zool. 41*: 99–112.

Dunton, K.H. and P.K. Dayton. 1995. The biology of high latitude kelp. *In:* (H.R. Skjoldal, ed.) *Ecology of fjords and coastal waters: proceedings of the Mare Nor symposium on the ecology of fjords and coastal waters, Tromsø, Norway, 5–9 December 1994*. Elsevier, Amsterdam. pp. 499–507.

Dunton, K.H. and C.M. Jodwalis. 1988. Photosynthetic performance of *Laminaria solidungula* measured *in situ* in the Alaskan High Arctic. *Mar. Biol. 98*: 277–285.

Dunton, K.H. and D.M. Schell. 1986. Seasonal carbon budget and growth of *Laminaria solidungula* in the Alaskan High Arctic. *Mar. Ecol. Prog. Ser. 31*: 57–66.

Dunton, K.H. and D.M. Schell. 1987. Dependence of consumers on macroalgal (*Laminaria solidungula*) carbon in an Arctic kelp community: $\delta\ ^{13}C$ evidence. *Mar. Biol. 93*: 615–625.

Dunton, K.H., E. Reimnitz and S. Schonberg.1982. An Arctic kelp community in the Alaskan Beaufort Sea. *Arctic 35*: 465–484.

Dunton, K.H., S.V. Schonberg and D.W. Funk. 2009. Interannual and spatial variability in light attenuation: evidence from three decades of growth in the arctic kelp, *Laminaria solidungula*. *Proceedings of Smithsonian at the Poles Symposium*, Smithsonian Institution, Washington, DC, 3–4 May 2007. Smithsonian Institute Scholarly Press, Washington, DC. pp. 271–284.

Eggert, A. and C. Wiencke. 2000. Adaptation and acclimation of growth and photosynthesis of five Antarctic red algae to low temperatures. *Polar Biol. 23*: 609–618.

Enríquez, S., C.M. Duarte and K. Sand-Jensen. 1995. Patterns in the photosynthetic metabolism of Mediterranean macrophytes. *Mar. Ecol. Prog. Ser. 119*: 243–252.

Fogg, G.E. 1977. Aquatic primary production in the Antarctic. *Phil. Trans. R. Soc. Lond. B. 279*: 27–38.

Gattuso, J.-P., B. Gentili, C.M. Duarte, J.A. Kleypas, J.J. Middelburg and D. Antoine. 2006. Light availability in the coastal ocean: impact on the distribution of benthic photosynthetic organisms and contribution to primary production. *Biogeosciences Discuss. 3*: 895–959.

Gilbert, N. 1991. Primary production by benthic microalgae in nearshore marine sediments of Signy Island, Antarctica. *Polar Biol. 11*: 339–346.

Glenn, E.P. and M.S. Doty.1981. Photosynthesis and respiration of the tropical red seaweeds, *Eucheuma striatum* (tamba-lang and elkhorn varieties) and *E. denticulatum*. *Aquat. Biol. 10*: 353–364.

Glud, R.N., M. Kühl, F. Wenzhofer and S. Rysgaard. 2002. Benthic diatoms of a high Arctic fjord (Young Sound, NE Greenland): importance for ecosystem primary production. *Mar. Ecol. Prog. Ser. 238*: 15–29.

Glud, R.N., J. Woelfel, U. Karsten, M. Kühl and S. Rysgaard. 2009. Benthic microalgal production in the Arctic: applied methods and status of the current database. *Bot. Mar. 52*: 559–571.

Gómez, I. 2001. Ecophysiology of Antarctic macroalgae: effects of environmental light conditions on photosynthetic metabolism. *Rev. Chil. Hist. Nat. 74*: 251–271.

Gómez, I. and C. Wiencke. 1996. Photosynthesis, dark respiration and pigment contents of gametophytes and sporophytes of the Antarctic brown alga *Desmarestia menziesii*. *Bot. Mar. 39*: 149–157.

Gómez, I., D.N. Thomas and C. Wiencke. 1995a. Longitudinal profiles of growth, photosynthesis and light independent carbon fixation in the Antarctic brown alga *Ascoseira mirabilis*. *Bot. Mar. 38*: 157–164.

Gómez, I., C. Wiencke and G. Weykam. 1995b. Seasonal photosynthetic characteristics of the brown alga *Ascoseira mirabilis* from King George Island (Antarctica). *Mar. Biol. 123*: 167–172.

Gómez, I., G. Weykam, H. Klöser and C. Wiencke. 1997. Photosynthetic light requirements, daily carbon balance and zonation of sublittoral macroalgae from King George Island (Antarctica). *Mar. Ecol. Prog. Ser. 148*: 281–293.

Gosselin, M., M. Levasseur, P.A. Wheeler, R.A. Horner and B.C. Booth. 1997. New measurements of phytoplankton and ice algal production in the Arctic Ocean. *Deep Sea Res. II. 44*: 1623–1644.

Greer, S.P. and C.D. Amsler. 2004. Clonal variation in phototaxis and settlement behaviors of *Hincksia irregularis* (Phaeophyceae) spores. *J. Phycol. 40*: 44–53.

Hanelt, D. 1998. Capability of dynamic photoinhibition in Arctic macroalgae is related to their depth distribution. *Mar. Biol. 131*: 361–369.

Harley, C.D.G., R.A. Hughes, K.M. Hultgren, B.G. Miner, C.J.B. Sorte, C.S. Thornber, L.F. Rodriguez, L. Tomanek and S.L. Williams. 2006. The impacts of climate change in coastal marine systems. *Ecol. Lett. 9*: 228–241.

Hawes, I. and A.M. Schwarz. 1999. Photosynthesis in an extreme shade environment: benthic microbial mats from Lake Hoare, a permanently ice-covered Antarctic lake. *J. Phycol. 35*: 448–459.

Healey, F.P. 1972. Photosynthesis and respiration of some Arctic seaweeds. *Phycologia 11*: 267–271.

Henley, W.J. 1993. Measurement and interpretation of photosynthetic light-response curves in algae in the context of photoinhibition and diel changes. *J. Phycol. 29*: 729–739.

Henley, W.J. and K.H. Dunton. 1995. A seasonal comparison of carbon, nitrogen, and pigment content in *Laminaria solidungula* and *L. saccharina* (Phaeophyta) in the Alaskan Arctic. *J. Phycol. 31*: 325–331.

Henry, E.C. 1987. Primitive reproductive characters and a photoperiodic response in *Saccorhiza dermatodea* (Laminariales, Phaeophyceae). *Br. Phycol. J. 22*: 23–31.

Heywood, R.B. and T.M. Whitaker. 1984. The Antarctic marine flora. *In*: (R.M. Laws, ed), *Antarctic ecology*. Vol. 2. Academic Press, London. pp. 373–419.

Horner, R. and G.C. Schrader. 1982. Relative contributions of ice algae, phytoplankton, and benthic microalgae to primary production in nearshore regions of the Beaufort Sea. *Arctic 35:* 485–503.

Huang, Y.M., J.B. McClintock, C.D. Amsler, K.J. Peters and B.J. Baker. 2006. Feeding rates of common Antarctic gammarid amphipods on ecologically important sympatric macroalgae. *J. Exp. Mar. Biol. Ecol. 329*: 55–65.

Huovinen, P., P. Leal and I. Gómez. 2010. Interacting effects of copper, nitrogen and ultraviolet radiation on the physiology of three south Pacific kelps. *Mar. Freshw. Res. 61:* 330–341.

Iken, K., M.L. Quartino, E. Barrera-Oro, J. Palermo, C. Wiencke and T. Brey. 1998. Trophic relations between macroalgae and herbivores in Potter Cove (King George Island, Antarctica). *In*: (C. Wiencke, G. Ferreyra, W. Arntz and C. Rinaldi, eds) *The Potter Cove coastal ecosystem, Antarctica. Rep. Polar Res. 299*: 258–262.

Janech, M.G., A. Krell, T. Mock, J.S. Kang and J.A. Raymond. 2006. Ice-binding proteins from sea ice diatoms (Bacillariophyceae). *J. Phycol. 42*: 410–416.

Kain (Jones), J.M. 1964. Aspects of the biology of *Laminaria hyperborea*. III. Survival and growth of gametophytes. *J. Mar. Biol. Assoc. UK 44*: 415–433.

Karsten, U., K. Kück, C. Vogt and G.O. Kirst. 1996. Dimethylsulfoniopropionate production in phototrophic organisms and its physiological function as a cryoprotectant. *In*: (R.P. Kiene, ed.) *Biological and environmental chemistry of DMSP and related sulfonium compounds*. Plenum Press, New York. pp. 143–153.

Karsten, U., R. Schumann, S. Rothe, I. Jung and L. Medlin. 2006. Temperature and light requirements for growth of two diatom species (Bacillariophyceae) isolated from an Arctic macroalga. *Polar Biol. 29*: 476–486.

Karsten, U., A. Wulff, M.Y. Roleda, R. Müller, F.S. Steinhoff, J. Fredersdorf and C. Wiencke. 2009. Physiological responses of polar benthic algae to ultraviolet radiation. *Bot. Mar. 52*: 639–654.

Kasamatsu, N., T. Hirano, S. Kudoh, T. Odate and M. Fukuchi. 2004. Dimethylsulfoniopropionate production by psychrophilic diatom species. *J. Phycol. 40*: 874–878.

Kirk, J.T.O. 1994. *Light and photosynthesis in aquatic ecosystems*. 2nd edition. Cambridge University Press, Cambridge. pp. 509.

Kirst, G.O. and C. Wiencke. 1995. Ecophysiology of polar algae. *J. Phycol. 31*: 181–199.

Klöser, H. 1998. Habitats and distribution patterns of benthic diatoms in Potter Cove (King George Island, Antarctica) and its vicinity. *Ber. Polarforsch. 299*: 95–105.

Klöser, H., G. Ferreyra, I. Schloss, G. Mercuri, F. Laturnus and A. Curtosi. 1993. Seasonal variation of algal growth conditions in sheltered Antarctic bays: the example of Potter Cove (King George Island, South Shetlands). *J. Mar. Syst. 4*: 289–301.

Klöser, H., G. Ferreyra, I. Schloss, G. Mercuri, F. Laturnus and A. Curtosi. 1994. Hydrography of Potter Cove, a small fjord-like inlet on King George Island (South Shetlands). *Estuar. Coast. Shelf Sci. 38*: 523–537.

Klöser, H., M.L. Quartino and C. Wiencke. 1996. Distribution of macroalgae and macroalgal communities in gradients of physical conditions in Potter Cove, King George Island, Antarctica. *Hydrobiologia 333*: 1–17.

Krembs, C., H. Eicken, K. Junge and J.W. Deming. 2002. High concentrations of exopolymeric substances in Arctic winter sea ice: implications for the polar ocean carbon cycle and cryoprotection of diatoms. *Deep-Sea Res. Oceanogr. Res. Papers 49*: 2163–2181.

Kühl, M., R.N. Glud, R. Borum, J. Roberts and S. Rysgaard. 2001. Photosynthetic performance of surface-associated algae below sea ice as measured with a pulse-amplitude-modulated (PAM) fluorometer and O_2 microsensors. *Mar. Ecol. Progr. Ser. 223*: 1–14.

Latala, A. 1990. Photosynthesis and respiration of some marine benthic algae from Spitsbergen. *Polar Res. 8*: 303–307.

Lee, S.H., T.E. Whitledge and S.H. Kang. 2008. Carbon uptake rates of sea ice algae and phytoplankton under different light intensities in a landfast sea ice zone, Barrow, Alaska. *Arctic 61*: 281–291.

Longhi, M.L., I.R. Schloss and C. Wiencke. 2003. Effect of irradiance and temperature on photosynthesis and growth of two Antarctic benthic diatoms, *Gyrosigma subsalinum* and *Odontella litigiosa*. *Bot. Mar. 46*: 276–284.

Lüning, K. 1980a. Control of algal-life-history by daylength and temperature. *In*: (J.H. Price, D.E.G. Irvine and W.F. Farnham, eds) *The shore environment*, Vol 2. Ecosystems. Academic Press, New York. pp. 915–945.

Lüning, K. 1980b. Critical levels of light and temperature regulating the gametogenesis of three *Laminaria* species (Phaeophyceae). *J. Phycol. 16*: 1–15.

Lüning, K. 1990. *Seaweeds. Their environment, biogeography and ecophysiology.* John Wiley and Sons, Inc, New York, Chichester, Brisbane, Toronto, Singapore. pp. 527.

Markager, S. and K. Sand-Jensen. 1992. Light requirements and depth zonation of marine macroalgae. *Mar. Ecol. Prog. Ser. 88*: 83–92.

Matheke, G.E.M. and R. Horner. 1974. Primary productivity of the benthic microalgae in the Chukchi Sea near Barrow, Alaska. *J. Fish. Res. Board Canada 31:* 1779–1786.

Mathieson, A.C. and T.L. Norall. 1975. Physiological studies of subtidal red algae. *J. Exp. Mar. Biol. Ecol. 20*: 237–47.

McGee, D., R.A. Laws and L.B. Cahoon. 2008. Live benthic diatoms from the upper continental slope: extending the limits of marine primary production. *Mar. Ecol. Prog. Ser. 356*: 103–112.

McIntyre, C.D. and W.W. Moore. 1977. Marine littoral diatoms: ecological considerations. *In*: (D. Werner, ed.) *The biology of diatoms*. Blackwell Scientific Publications, Oxford. pp. 333–371.

McMinn, A. and E.N. Hegseth. 2004. Quantum yield and photosynthetic parameters of marine microalgae from the southern Arctic Ocean, Svalbard. *J. Mar. Biol. Ass. UK 84*: 865–871.

Migné, A., N. Spilmont and D. Davoult. 2004. *In situ* measurements of benthic primary production during emersion: seasonal variations and annual production in the Bay of Somme (eastern English Channel, France). *Cont. Shelf Res. 24*: 1437–1449.

Mock, T. and R. Gradinger. 1999. Determination of Arctic ice algae production with a new *in situ* incubation technique. *Mar. Ecol. Prog. Ser. 177*: 15–26.

Mock, T. and K. Valentin. 2004. Photosynthesis and cold acclimation – molecular evidence from a polar diatom. *J. Phycol. 40*: 732–741.

Molenaar, F.J. and A.M. Breeman. 1997. Latitudinal trends in the growth and reproductive seasonality of *Delesseria sanguinea*, *Membranoptera alata*, and *Phycodrys rubens* (Rhodophyta). *J. Phycol. 33*: 330–343.

Morgan-Kiss, R.M., J.C. Priscu, T. Pocock, L. Gudynaite-Savitch and N.P.A. Huner. 2006. Adaptation and acclimation of photosynthetic microorganisms to permanently cold environments. *Microbiol. Mol. Biol. Rev. 70*: 222–252.

Müller, R., C. Wiencke and K. Bischof. 2008. Interactive effects of UV radiation and temperature on microstages of Laminariales (Phaeophyceae) from the Arctic and North Sea. *Clim. Res. 37*: 203–213.

Müller, R., T. Laepple, I. Bartsch and C. Wiencke. 2009. Impact of oceanic warming on the distribution of seaweeds in polar and cold-temperate waters. *Bot. Mar. 52*: 617–638.

Novaczek, I., G.W. Lubbers and A.M. Breeman. 1990. Thermal ecotypes in amphi-Atlantic algae. I. Algae of Arctic to cold-temperate distribution (*Chaetomorpha melagonium*, *Devaleraea ramentacea* and *Phycodrys rubens*). *Helgol. Meeresunters. 44*: 459–474.

Ødegaard, S., K. Sjøtun, T.E. Lein and E. Aas. 1998. Sporophyte formation of *Laminaria hyperborea* (Laminariales, Phaeophyceae) related to photon doses of blue light in the sea. *Sarsia 83*: 301–308.
Palmisano, A.C., J.B. SooHoo, D.C. White, G.A. Smith, G.R. Stanton and L.H. Burckle. 1985a. Shade adapted benthic diatoms beneath Antarctic sea ice. *J. Phycol. 21*: 664–667.
Palmisano, A.C., J.B. SooHoo and C.W. Sullivan. 1985b. Photosynthesis-irradiance relationships in sea-ice microalgae from McMurdo Sound, Antarctica. *J. Phycol. 21*: 341–346.
Peckol, P. and J. Ramus. 1988. Abundances and physiological properties of deep-water seaweeds from the Carolina outer continental shelf. *J. Exp. Mar. Biol. Ecol. 115*: 25–39.
Peters, A.F. and A.M. Breeman. 1993. Temperature tolerance and latitudinal range of brown algae from temperate Pacific South America. *Mar. Biol. 115*: 143–150.
Petersen, D., F. Reichenberg and I. Dahlöff. 2008. Phototoxicity of pyrene affects benthic algae and bacteria from the Arctic. *Environ. Sci. Technol. 42:* 1371–1376.
Quartino, M.L. and A.L. Boraso de Zaixso. 2008. Summer macroalgal biomass in Potter Cove, South Shetland Islands, Antarctica: its production and flux to the ecosystem. *Polar Biol. 31*: 281–294.
Ramus, J., S.I. Beale and D. Mauzerall.1976. Correlation of changes in pigment content with photosynthetic capacity of seaweeds as a function of water depth. *Mar. Biol. 37*: 231–238.
Raymond, J.A. and C.H. Fritsen. 2001. Semipurification and ice recrystallization inhibition activity of ice-active substances associated with Antarctic photosynthetic organisms. *Cryobiology 43*: 63–70.
Raymond, J.A. and C.A. Knight. 2003. Ice binding, recrystallization inhibition and cryoprotective properties of ice-active substances associated with Antarctic sea ice diatoms. *Cryobiology 46*: 174–181.
Reynolds, C.S. 2006. Ecology of phytoplankton. Cambridge University Press, Cambridge. pp. 535.
Richardson, M.G. 1979. The distribution of Antarctic marine macroalgae related to depth and substrate. *Br. Antarct. Surv. Bull. 49*: 1–13.
Rivkin, R.B. and M. Putt. 1987. Photosynthesis and cell division by Antarctic microalgae: comparison of benthic, planktonic and ice algae. *J. Phycol. 23*: 223–229.
Roberts, R.D., M. Kühl, R.N. Glud and S. Rysgaard. 2002. Primary production of crustose coralline red algae in a high Arctic fjord. *J. Phycol. 38*: 273–283.
Robinson, D.H., K.R. Arrigo, R. Iturriaga and C.W. Sullivan. 1995. Microalgal light-harvesting in extreme low-light environments in McMurdo Sound, Antarctica. *J. Phycol. 31*: 508–520.
Roleda, M.Y. 2009. Photosynthetic response of Arctic kelp zoospores exposed to radiation and thermal stress. *Photochem. Photobiol. Sci. 8*, 1302–1312.
Roleda, M.Y., D. Hanelt and C. Wiencke. 2006a. Exposure to ultraviolet radiation delays photosynthetic recovery in Arctic kelp zoospores. *Photosynth. Res. 88*: 311–322.
Roleda, M.Y., C. Wiencke and U.H. Lüder. 2006b. Impact of ultraviolet radiation on cell structure, UV-absorbing compounds, photosynthesis, DNA damage and germination in zoospores of Arctic *Saccorhiza dermatodea*. *J. Exp. Bot. 57*: 3847–3856.
Roleda, M.Y., K. Zacher, A. Wulff, D. Hanelt and C. Wiencke. 2007. Photosynthetic performance, DNA damage and repair in gametes of the endemic Antarctic brown alga *Ascoseira mirabilis* exposed to ultraviolet radiation. *Austral Ecol. 32*: 917–926.
Roleda, M.Y., K. Zacher, A. Wulff, D. Hanelt and C. Wiencke. 2008a. Susceptibility of spores of different ploidy levels from Antarctic *Gigartina skottsbergii* (Gigartinales, Rhodophyta) to ultraviolet radiation. *Phycologia 47*: 361–370.
Roleda, M.Y., K. Zacher, G.L. Campana, A. Wulff, D. Hanelt, M.L. Quartino and C. Wiencke. 2008b. Photosynthetic performance and impact of ultraviolet radiation on the reproductive cells of Antarctic macroalgae. *Ber. Polarforsch. Meeresforsch. 571*: 254–262.
Rysgaard, S., T.G. Nielsen and B.W. Hansen. 1999. Seasonal variation in nutrients, pelagic primary production and grazing in a high-Arctic coastal marine ecosystem, Young Sound, Northeast Greenland. *Mar. Ecol. Prog. Ser. 179*: 13–25.

Schoschina, E.V. 1996. Seasonal and age dynamics of growth and reproduction of *Phycodrys rubens* (Rhodophyta) in the Barents and White Seas. *Aquat. Bot. 55*: 13–30.

Schwarz, A., I. Hawes, N. Andrew, A. Norkko, V. Cummings and S. Thrush. 2003. Macroalgal photosynthesis near the southern global limit for growth, Cape Evans, Ross Sea, Antarctica. *Polar Biol. 26*: 789–799.

Schwarz, A.-M., I. Hawes, N. Andrew, S. Mercer, V. Cummings and S. Trush. 2005. Primary production potential of nongeniculate coralline algae at Cape Evans, Ross Sea, Antarctica. *Mar. Ecol. Prog. Ser. 294*: 131–140.

Smith, R.E.H., L.C. Stapleford and R.S. Ridings. 1994. The acclimated response of growth, photosynthesis, composition, and carbon balance to temperature in the psychrophilic ice diatom *Nitzschia seriata*. *J. Phycol. 30*: 8–16.

Sundbäck, K., S. Odmark, A. Wulff, C. Nilsson and S.A. Wängberg. 1997. Effects of enhanced UVB radiation on a marine benthic diatom mat. *Mar. Biol. 128*: 171–179.

Svendsen, P. 1959. The algal vegetation of Spitsbergen. A survey of the marine algal flora of the outer part of Isfjorden. *Norsk Polarinst Skrifter 116*: 1–49.

Thiel, M. and L. Gutow. 2005. The ecology of rafting in the marine environment. I. The floating substrata. *Oceanogr. Mar. Biol. Annu. Rev. 42*: 181–264.

Thomas, D.N. and G.S. Dieckmann. 2002. Antarctic sea ice – a habitat for extremophiles. *Science 295*: 641–644.

tom Dieck, I. 1992. North Pacific and North Atlantic digitate *Laminaria* species (Phaeophyta): hybridisation experiments and temperature responses. *Phycologia 31*: 147–163.

tom Dieck, I. 1993. Temperature tolerance and survival in darkness of kelp gametophytes (Laminariales, Phaeophyta): ecological and biogeographical implications. *Mar. Ecol. Prog. Ser. 100*: 253–264.

van den Hoek, C. 1982a. Phytogeographic distribution groups of benthic marine algae in the North Atlantic Ocean. A review of experimental evidence from life history studies. *Helgol. Meeresunters. 35*: 153–214.

van den Hoek, C. 1982b. The distribution of benthic marine algae in relation to the temperature regulation of their life histories. *Biol. J. Linn. Soc. 18*: 1–144.

van den Hoek, C. and A.M. Breeman. 1989. Seaweed biogeography in the North Atlantic: where are we now? *In*: (D.J. Garbary and G.R. South, eds) *Evolutionary biogeography of the marine algae of the North Atlantic*. NATO ASI Series, Vol. G 22. Springer-Verlag, Berlin/Heidelberg. pp. 57–86.

Vernet, M. and R.C. Smith. 2006. Measuring and modeling primary production in marine pelagic ecosystems. *In*: (J. Fahey and A. Knapp, eds) *LTER Net primary production methods*. Oxford University Press, Oxford.

Weykam, G., I. Gómez, C. Wiencke, K. Iken and H. Klöser. 1996. Photosynthetic characteristics and C:N ratios of macroalgae from King George Island (Antarctica). *J. Exp. Mar. Biol. Ecol. 204*: 1–22.

Wiencke, C. 1988. Notes on the development of some benthic marine macroalgae of King George Island, Antarctica. *Ser. Cient. INACH 37*: 23–47.

Wiencke, C. 1990a. Seasonality of brown macroalgae from Antarctica – a long-term culture study under fluctuating Antarctic daylengths. *Polar Biol. 10*: 589–600.

Wiencke, C. 1990b. Seasonality of red and green macroalgae from Antarctica – a long-term culture study under fluctuating Antarctic daylengths. *Polar Biol. 10*: 601–607.

Wiencke, C. 1996. Recent advances in the investigation of Antarctic macroalgae. *Polar Biol. 16*: 231–240.

Wiencke, C. and G. Fischer. 1990. Growth and stable carbon isotope composition of cold water macroalgae in relation to light and temperature. *Mar. Ecol. Prog. Ser. 65*: 283–292.

Wiencke, C. and I. tom Dieck. 1989. Temperature requirements for growth and temperature tolerance of macroalgae endemic to the Antarctic region. *Mar. Ecol. Prog. Ser. 54*: 189–197.

Wiencke, C. and I. tom Dieck. 1990. Temperature requirements for growth and survival of macroalgae from Antarctica and Southern Chile. *Mar. Ecol. Prog. Ser. 59*: 157–170.

Wiencke, C., J. Rahmel, U. Karsten, G. Weykam and G.O. Kirst. 1993. Photosynthesis of marine macroalgae from Antarctica: light and temperature requirements. *Bot. Acta 106*: 78–87.

Wiencke, C., I. Bartsch, B. Bischoff, A.F. Peters and A.M. Breemann. 1994. Temperature requirements and biogeography of Antarctic, Arctic and amphiequatorial seaweeds. *Bot. Mar. 37*: 247–259.

Wiencke, C., M.N. Clayton and D. Schulz. 1995. Life history, reproductive morphology and development of the Antarctic brown alga *Desmarestia menziesii* J. Agardh. *Bot. Acta 108*: 201–208.

Wiencke, C., I. Gómez, H. Pakker, A. Flores-Moya, M. Altamirano, D. Hanelt, K. Bischof and F.L. Figueroa. 2000. Impact of UV radiation on viability, photosynthetic characteristics and DNA of brown algal zoospores: implications for depth zonation. *Mar. Ecol. Prog. Ser. 197*: 217–229.

Wiencke, C., B. Vögele, N.A. Kovaltchouk and H. Hop. 2004. Species composition and zonation of marine benthic macroalgae at Hansneset in Kongsfjorden, Svalbard. *Ber. Polarforsch. Meeresforsch. 492*: 55–62.

Wiencke, C., M.Y. Roleda, A. Gruber, M.N. Clayton and K. Bischof. 2006. Susceptibility of zoospores to UV radiation determines upper depth distribution limit of Arctic kelps: evidence through field experiments. *J. Ecol. 94*: 455–463.

Wiencke, C., M.N. Clayton, I. Gómez, K. Iken, U. Lüder, C.D. Amsler, U. Karsten, D., Hanelt, K. Bischof and K. Dunton. 2007. Life strategy, ecophysiology and ecology of seaweeds in polar waters. *Rev. Environ. Sci. Biotech. 6*: 95–126.

Wiencke, C., I. Gómez and K. Dunton. 2009. Phenology and seasonal physiological performance of polar seaweeds. *Bot. Mar. 52*: 585–592.

Wilce, R.T. 1994. The Arctic subtidal as habitat for macrophytes. *In*: (C.S. Lobban, P.J. Harrison, eds) *Seaweed ecology and physiology*. Cambridge University Press, Cambridge. pp. 89–92.

Wulff, A., S. Vilbaste and J. Truu. 2005. Depth distribution of photosynthetic pigments and diatoms in the sediments of a microtidal fjord. *Hydrobiologia 534*: 117–130.

Wulff, A., M.Y. Roleda, K. Zacher and C. Wiencke. 2008a. UV radiation effects on pigments, photosynthetic efficiency and DNA of an Antarctic marine benthic diatom community. *Aquat. Biol. 3*: 167–177.

Wulff, A., K. Zacher, D. Hanelt, A. Al-Handal and C. Wiencke. 2008b. UV radiation – a threat to Antarctic benthic marine diatoms? *Antarct. Sci. 20*: 13–20.

Wulff, A., M.Y. Roleda, K. Zacher and C. Wiencke. 2008c. Exposure to sudden light burst after prolonged darkness – a case study on benthic diatoms in Antarctica. *Diatom Res. 23*: 519–532.

Wulff, A., K. Iken, M.L. Quartino, A. Al-Handal, C. Wiencke and M.N. Clayton. 2009. Biodiversity, biogeography and zonation of marine benthic micro- and macroalgae in the Arctic and Antarctic. *Bot. Mar. 52*: 491–507.

Zacher, K., D. Hanelt, C. Wiencke and A. Wulff. 2007a. Grazing and UV radiation effects on an Antarctic intertidal microalgal assemblage: a long-term field study. *Polar Biol. 30*: 1203–1212.

Zacher, K., M.Y. Roleda, D. Hanelt and C. Wiencke. 2007b. UV effects on photosynthesis and DNA in propagules of three different Antarctic seaweeds (*Adenocystis utricularis, Monostroma hariotii and Porphyra endiviifolium*). *Planta 225*: 1505–1516.

Zacher, K., R. Rautenberger, D. Hanelt, A. Wulff and C. Wiencke. 2009a. The abiotic environment of polar marine benthic algae. *Bot. Mar. 52*: 483–490.

Zacher, K., M.Y. Roleda, A. Wulff, D. Hanelt and C. Wiencke. 2009b. Responses of Antarctic *Iridaea cordata* (Rhodophyta) tetraspores exposed to ultraviolet radiation. *Phycol. Res. 57*: 186–193.

Zachos, J., M. Pagani, L. Sloan, E. Thomas and K. Billups. 2001. Trends, rhythms, and aberrations in global climate 65 Ma to present. *Science 292*: 686–693.
Zielinski, K. 1981. Benthic macroalgae of Admiralty Bay (King George Island, South Shetland Islands, Antarctica) and circulation of algal matter between the water and the shore. *Pol. Polar Res. 2*: 71–94.
Zielinski, K. 1990. Bottom macroalgae of the Admiralty Bay (King George Island, South Shetlands, Antarctica). *Pol. Polar Res. 11*: 95–131.

11 Freezing tolerance and photosynthetic performance of polar seaweeds at low temperatures

Susanne Becker, Bettina Walter and Kai Bischof

Introduction

Species inhabiting the intertidal zone cope with drastic changes in the physical environment during the course of the tides. Moreover, in polar regions, intertidal organisms are frequently exposed to atmospheric temperatures substantially below 0 °C (Chapter 2; Zacher et al. 2009). However, as specimens are exposed to a variety of simultaneously changing abiotic factors, it is essential to understand how the combination of (stress-) factors influences organismic performance (Chapter 10; Gómez et al. 2009). For instance, Antarctic macroalgae are generally considered to be strongly shade-adapted (Wiencke 1990, Lüder et al. 2001) and hence more susceptible to increasing light intensities. However, immediately after sea ice break up, radiation may penetrate deeply into transparent waters (Drew and Hastings 1992, McMinn et al. 2004), and even subtidal specimens will be exposed to elevated light intensities. In addition, photosynthetic activity is particularly sensitive to low temperatures, as enzymatic secondary reactions are strictly temperature-dependent, while primary reactions are not. Thus, exposure to high irradiances in combination with low temperatures may result in a surplus of electrons participating in the Mehler reaction and, consequently, in an increased production of reactive oxygen species (ROS), resulting in photooxidation (Asada and Takahashi 1987, Andersson et al. 1992). Furthermore, low temperatures strongly affect membrane fluidity and protein function, for instance the turnover of the D1 protein, which is the central protein within photosystem II (Aro et al. 1993). Freezing has physiological effects similar to those of dehydration (Pearson and Davison 1994), as the water potential within cells may be considerably reduced.

The Antarctic endemic rhodophyte *Palmaria decipiens* (Reinsch) Ricker is an excellent organism for studying the interactive effects of low temperatures and high light intensities; its temperature and light requirements, as well as its growth activity have been studied previously (Wiencke and tom Dieck 1989, Wiencke 1990, Wiencke et al. 1993, Bischoff-Bäsmann and Wiencke 1996, Lüder et al. 2001).

Along the coasts of the North Atlantic Ocean, the brown algal genus *Fucus* forms dense vegetation fringes within the intertidal zone. *F. distichus* Linnaeus is the species with the northernmost distribution and is, thus, the only member of the genus inhabiting the coasts of the Svalbard archipelago (Lüning 1990). As a general pattern in polar regions, only very few algal species populate the intertidal zone, which is mostly attributed to the influence of ice scouring. *F. distichus* is almost the only intertidal macroalgal species along the Svalbard coastline (Kirst and Wiencke 1995,

Wiencke et al. 2004). In winter, it is exposed to subzero temperatures and, in extreme cases, survives being frozen in ice for several months (Kanwisher 1957). However, earlier studies on *F. distichus* from more southerly locations reported that this species is less tolerant to freezing, desiccation or high light than other species of *Fucus* (Denton and Chapman 1991, Collén and Davison 1999). Thus, the ecological success of *F. distichus* and the absence of other species of *Fucus* in the European high Arctic remains an open field of research.

We present here two separate studies on the effects of low temperature, light intensity and freezing on photosynthetic performance. The first was conducted in Antarctica and focused on the combined effects of temperature and radiation conditions on photosynthesis of the Antarctic endemic *Palmaria decipiens*. The second was conducted at Arctic Kongsfjorden, Spitsbergen, where the effects of freezing on photosynthesis of *F. distichus* were studied in a set of repeated freezing and thawing experiments. These two individual studies focus on the adaptability of the photosynthetic apparatus to cold environments in a more eurythermal (*F. distichus*) and a strictly stenothermal species (*P. decipiens*).

Experimental approaches

Palmaria decipiens Experiments were conducted at the German-Argentinean Dallmann Laboratory, Jubany Base, King George Island, South Shetland Islands, Antarctica (62°14′ S, 58°40′ W) during the Antarctic summer. A detailed overview on the physical and biological environment of the Potter Cove ecosystem is given by Wiencke et al. (2008). Specimens of *Palmaria decipiens* were collected by SCUBA diving between 20 January and 20 February 2007 offshore from the Fourcade glacier. Fifteen to 20 individuals 30–50 cm long were collected at 8–10 m water depth and brought to the laboratory in black boxes to avoid light stress during the transfer. Prior to the experiment, algae were kept under low light conditions (15 µmol photons m^{-2} s^{-1}) in a seawater tank at ambient water temperature (~2 °C). Water was aerated and replaced every second day.

Algal material was cleaned, cut into discs approximately 50 mm in diameter taken from the middle parts of the thalli and randomly distributed among three 1-l plastic beakers (Vitalab GmbH, Darmstadt, Germany) filled with ambient seawater. The beakers serving as three different experimental units were placed in temperature-controlled water tanks providing temperatures of 0 °C and 8 °C. Temperature was maintained with cryostats (1160S, VWR International GmbH, Darmstadt, Germany) and controlled by temperature loggers (Testo 175-T1, Testo AG, Lenzkirch, Germany). Two of these tanks were established for each temperature treatment.

Light treatments were applied by six halogen spot lights (EXN-P, 50W, SP581, Conrad, Hirschau, Germany) mounted on an aluminium frame (500 × 300 mm). Light intensity was adjusted by varying the distance to the beakers or by using black net gauze. Light intensity was set to 200 µmol photons m^{-2} s^{-1} or 400 µmol photons m^{-2} s^{-1} PAR. Light measurements were conducted with a LiCor 1400 Data Logger equipped with a flat-head cosine corrected PAR quantum sensor (LICOR 190 SA, Li-Cor, Lincoln, NE, USA). Samples were exposed to the continuous light treatment for three days, after which algae were kept in dim light (15 µmol photons m^{-2} s^{-1}) for 48 h for recovery measurements.

Fucus distichus The study was conducted in July 2004 in Ny-Ålesund (78°56′ N, 11°56′ E), Arctic Kongsfjorden (Spitsbergen, Norway). A detailed overview on the physical and biological environment of Kongsfjorden is provided by Svendsen et al. (2002) and Hop et al. (2002). During low tide conditions, young thalli from the uppermost part of the *Fucus* fringe were collected in Kolhamna and transferred to the laboratory. There, algal material was kept in running seawater at fjord temperature (~4 °C) and under permanent low light conditions (20 μmol photons m^{-2} s^{-1}) for about two days prior to the experiment. The base of the experimental set-up was a custom-made "freezing table" consisting of a Pelletier element controlling the temperature of a steel surface of approximately 450 × 300 mm. The plate of the table can be cooled down to −20 °C. Algal material was placed in Petri dishes (90 mm in diameter, referred to hereafter as "dish") on the surface of the table. For measurements of photosynthetic activity, the fibre optic of a chlorophyll fluorometer (as described below) was fixed above randomly chosen thalli. In this way, the effective quantum yield of photosynthesis (yield, ΔF/Fm′) was determined during the course of each experiment, and, in addition, optimum quantum yield (Fv/Fm) and photosynthesis vs. irradiance curves (PI-curves) were measured at given times.
Two light fields with identical irradiances were arranged (OSRAM HLX 15 V, 150 W, Osram, Munich, Germany) at the surface of the freezing table. We placed the dish for measuring photosynthetic performance in one of the fields; in the second field, we placed a dish containing algal material exposed for later biochemical analysis. The irradiance in these light fields was set to match the I_k-value of experimental specimens, i.e., 100 μmol photons m^{-2} s^{-1}, as determined by previously recorded PI-curves. Inside the dish, an external temperature sensor connected to the fluorometer provided continuous temperature measurements under water, or between immersed thalli in freezing experiments on non-submerged specimens.

Three experiments with different freezing and thawing conditions were conducted:

1. Rapid freezing: algal material was placed in a dish, submerged in seawater and arranged in the light fields on the surface of the table. After 30 min of acclimation at ambient temperatures (~8 °C), the table was cooled down to the minimum temperature (−20 °C). As soon as the temperature stopped falling, the table was held at this lowest temperature for about 10 min, and then reset to initial ambient temperatures. After ambient temperature was reached, the experiment was further continued for another 30 min. During the process of freezing and thawing, yield values were recorded every 2 min, Fv/Fm and PI-curves were measured after 30 min of preacclimation at the lowest temperature, at the end of the experiment, and after 36 h of recovery. As in the following experiments, Fv/Fm determinations were performed on three different thalli, whereas, due to limitations in time, PI-curves were recorded just once at each sampling time.
2. Dry freezing: the setup was the same as in experiment 1, but samples were frozen in a dish without seawater. Fv/Fm and PI-curves were measured at the same time intervals as in the first experiment.
3. Long-term freezing: the algae were placed in a dish (filled with seawater) on the surface of the table. After a short period of acclimation to ambient conditions, the temperature was set to −1.5 °C, which was maintained for about 7 h. After that, the initial temperature was applied again for another 8 h.

In each experiment, reference samples were arranged in dishes as control treatments, which were maintained at ambient temperature and the same irradiance. At each measuring time, algal fragments from the second dish were collected for later biochemical analysis. Three replicates were used in each experiment.

The photosynthetic parameters Fv/Fm, effective quantum yield and relative ETR_{max} were determined by measuring the variable chlorophyll fluorescence of PS II with a pulse-amplitude modulated fluorometer (PAM 2100, Walz, Effeltrich, Germany) as described in detail by Hanelt et al. (1997) and Bischof et al. (1998). PI-curves were conducted using an internal LED as light source at irradiances of 24–809 μmol photons m^{-2} s^{-1} PAR (*Fucus distichus*) and 30–713 μmol photons m^{-2} s^{-1} PAR (*Palmaria decipiens*), respectively. The product of quantum yield and photon fluence rate (PFR) allows calculation of relative electron transport rates (rel.ETR), as described by Schreiber et al. (1994) (rel.ETR = $\Delta F/F'm \times PFR$); maximum relative electron transport rate (ETR_{max}) was determined by curve fitting. Non-photochemical quenching (NPQ) was calculated according to the formula of Bilger et al. (1995):

$$NPQ = Fm - Fm'/Fm'.$$

As a marker of oxidative stress, the degree of lipid peroxidation was measured in *Fucus* samples by determining the concentration of malondialdehyde (MDA), according to Heath and Packer (1968) and Bischof et al. (2003).

Changes in the concentration of D1 protein of photosystem II were determined by SDS-PAGE and subsequent Western blotting, according to Bischof et al. (2000).

The phycobiliprotein content of *Palmaria* samples was determined after Beer and Eshel (1985). Chlorophyll *a* content was determined after Lüder et al. (2002) using the modified protocol of Inskeep and Bloom (1985).

Experiments conducted on *Palmaria* were set-up as split-plot design with repeated measures and three replications. Mean values and standard deviations were calculated from the replicates and for each treatment. Photosynthetic data (Fv/Fm) were arcsine-transformed and a MANOVA was performed for the factors light and temperature and their interactions. Statistically significant differences were compared with Tukey's HSD (honestly significantly different) post-hoc test with $\alpha < 0.1$ (Sokal and Rohlf 1995). Statistical analysis was performed with JMP software (JMP version 6.1., SAS Inc., USA).

Each experiment carried out on *Fucus* was performed three times. For each replicate, one PI-curve and three Fv/Fm values were recorded and mean values as well as standard deviations were calculated. For experiment 2, statistical analysis on ETR_{max} was precluded due to limitation in sample number. An independent Samöels t-test was performed to test treatment differences. The values were accepted as significantly different at $p < 0.05$. Statistic analysis was performed with SPSS 13.0 (SPSS Inc., Chicago, IL, USA).

Interactive effects of irradiance and temperature on the photosynthetic performance of *Palmaria decipiens*

As an indicator of the physiological state of *Palmaria decipiens*, the optimum quantum yield (Fv/Fm) and PI-curves were recorded. The results indicate that particularly the combination of low water temperatures and high PAR irradiances affects photosynthesis of the alga. In the 400 μmol photons m^{-2} s^{-1} PAR treatment, Fv/Fm values decreased significantly at both temperatures ($p = 0.0064$, see Figure 11.1A, 11.1B). At 0 °C and 400 μmol photons m^{-2} s^{-1} PAR, Fv/Fm values decreased from an initial value of 0.601 – 0.237 after three days exposure. Values for specimens exposed to the same temperature, but at 200 μmol m^{-2} s^{-1} remained at an almost constant level of 0.5 (see Figure 11.1A). ANOVA showed a highly significant difference between light treatments on day three ($p = 0.0012$). The combination of temperature, time and radiation level had a negative influence on Fv/Fm values ($p = 0.0945$). For maximum electron transport rate (ETR-max), there was a significant time/temperature interaction between treatments ($p = 0.0605$), but no effects of the interactions time/light ($p = 0.8222$) or time/temperature/light were found ($p = 0.6574$, see also Figure 11.2). In the 0 °C/high PAR-treatment, ETR_{max} decreased within three days to 69% of the initial value, whereas values at 0 °C, but with lower PAR, decreased only slightly to 92% of the initial value.

Comparing the different temperature treatments at PAR intensities of 200 μmol photons m^{-2} s^{-1}, there was only a slight reduction in Fv/Fm values (initial value at

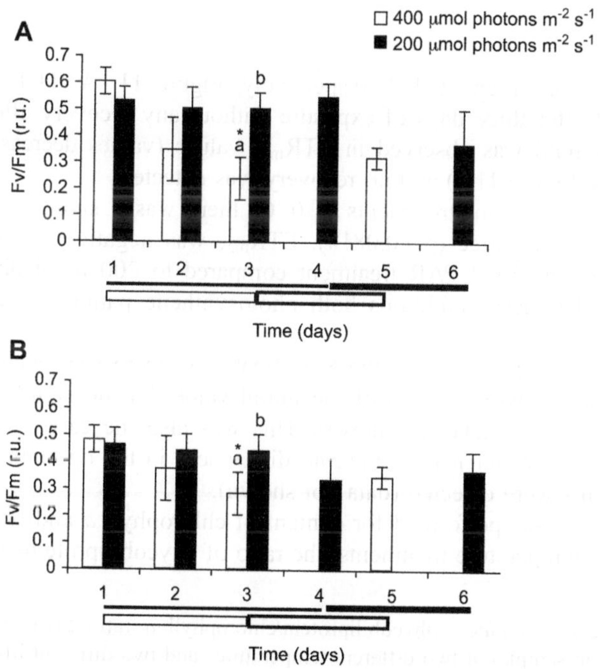

Figure 11.1 *Palmaria decipiens.*
(A) Optimum quantum yield (Fv/Fm) against exposure time (in days) at 0 °C and two light treatments (200 and 400 μmol photons m^{-2} s^{-1}) and (B) at 8 °C and two light treatments as above. Bars beneath the x-axis represent exposure (thin line or thin-framed box) or recovery conditions (bold line or bold framed box) at 400 μmol photons m^{-2} s^{-1} (boxes) or 200 μmol photons m^{-2} s^{-1} (lines) PAR treatments. In both cases, values are means of triplicate measurements, bars show standard deviations. *, significant differences ($p < 0.05$) within a light treatment on day three of exposure. Different letters indicate statistically significant differences between treatments (Tukey's HSD, $p < 0.05$). r.u., relative units.

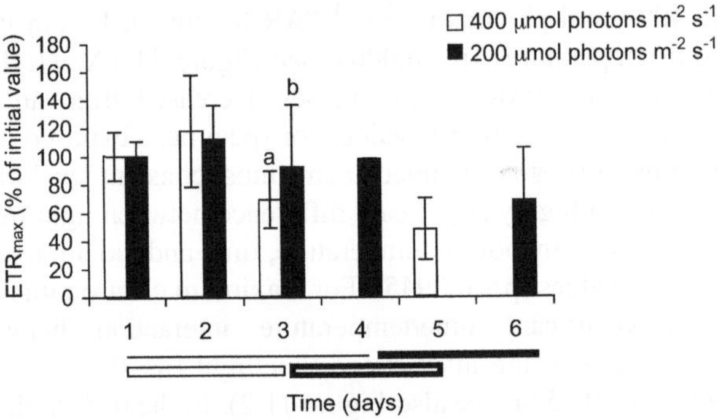

Figure 11.2 *Palmaria decipiens*: maximum electron transport rate (ETR_{max}) as percentage of the initial value against exposure time (in days) at 0 °C.
Bars beneath the x-axis represent exposure (thin line or thin framed box) or recovery conditions (bold line or bold framed box) at 400 µmol photons m^{-2} s^{-1} (boxes) or 200 µmol photons m^{-2} s^{-1} (lines) PAR treatments. Values are means of triplicate measurements, bars show standard deviations. Different letters indicate significant differences (Tukey's HSD) between treatments with $p = 0.0814$.

0 °C and 8 °C of 0.531 and 0.463, respectively, Figure 11.1A, 11.1B) to values of 0.502 and 0.442 after three days of exposure without any recovery after 36 h in dim light. The same trend was observed in ETR_{max} values (values decreasing to 92% of initial values, see Figure 11.2) and no recovery was detected.

Comparing different light treatments at 0 °C, there was a significant difference in ETR_{max} values on day three ($p = 0.0814$). ETR_{max} was negatively influenced by the 400 µmol photons m^{-2} s^{-1} PAR treatment compared to 200 µmol photons m^{-2} s^{-1} PAR, so that under these conditions both photosynthetic parameters decreased (Figure 11.2).

ETR_{max} values for specimens in the 8 °C treatment decreased, even during the recovery period (see above), to 13% of the initial value, but no significant differences were found ($p = 0.6877$, data not shown). This was also the case for the 8 °C treatment with varying light intensities, viz., no differences in the optimum quantum yield between treatments were detected (data not shown).

Pigment analysis was performed for content of chlorophyll *a* and phycobiliproteins. In both light and temperature treatments, the ratio of phycobiliproteins to chlorophyll *a*

Table 11.1 *Palmaria decipiens:* phycobiliprotein/chlorophyll *a* ratio calculated from pigment content measured in samples at two different temperatures and two different irradiances.

Temperature (°C)	Irradiance (200 µmol photons m^{-2} s^{-1})		Irradiance (400 µmol photons m^{-2} s^{-1})	
	Initial	End	Initial	End
0	4.05 ± 1.88	3.07 ± 1.04	3.38 ± 2	4.83 ± 1.74
8	5.35 ± 2.34	4.05 ± 1.36	4.68 ± 1.76	3.13 ± 1.93

Values are means of triplicate measurements ± SD; no statistically significant differences were found (ANOVA $p > 0.1$).

decreased substantially, more so under the lower PAR intensity at both temperatures (see Table 11.1).

The effects of freezing on *Fucus distichus*

In each experiment, there was a pronounced reduction in effective quantum yield with decreasing temperature. A rapid decrease in temperature (experiment 1) from ambient to 0 °C resulted in a marked reduction in yield values (Figure 11.3A). After the water was completely frozen, a further reduction in temperature resulted in continuously decreasing yield values. At −12 °C, quantum yield was reduced to 0.05 units. Subsequently, as soon as temperatures increased again, yield values also increased. The course of NPQ followed a roughly reverse trend, with increasing NPQ at decreasing yield values. Temperature-dependent inhibition of photosynthesis and subsequent recovery were also observed when measuring Fv/Fm and ETR_{max} (Figure 11.3B). Initial Fv/Fm values were 0.68, and at the lowest temperature, Fv/Fm was below 0.2. The recovery of Fv/Fm was not as rapid as the recovery of yield values, and a value of only 0.43 was reached. After a recovery of 36 h in low light and ambient temperature,

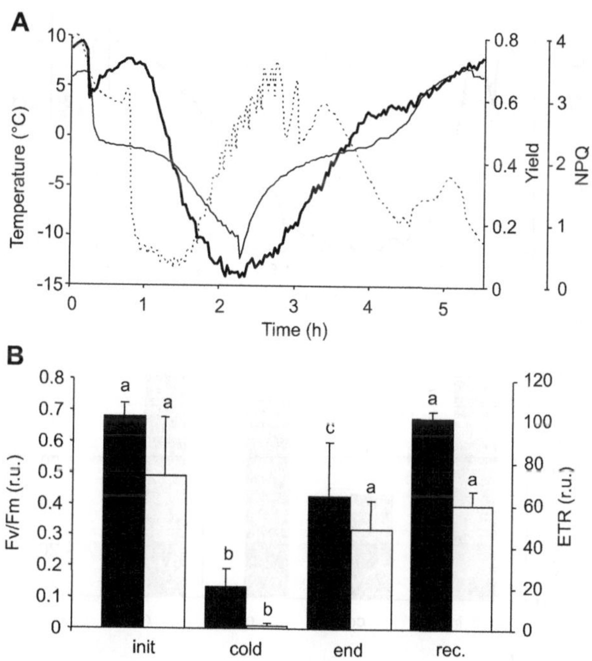

Figure 11.3 *Fucus distichus.*
Experiment 1 (fast and short freezing) with (A) changes in temperature (°C; thin line), effective quantum yield (yield; bold line) and non-photochemical quenching (NPQ; dashed line) in relative units (r.u.) over time; (B) optimum quantum yield (Fv/Fm; black bars) and maximum electron transport rate (ETR_{max}; white bars) measured after a short acclimation (init), at the lowest temperature (cold), at the end of the experiment (end), after a recovery of 36 h in dim white light and at fjord temperature (rec.). Different letters show statistically significant differences (Tukey's HSD $p < 0.05$). Values are means + SD ($n = 3$).

initial values were reached again. Values for ETR_{max} followed an almost identical trend. However, the final value (43 relative units) was lower than the initial value (76 relative units), and had not recovered completely after 36 h, when 52 relative units were reached.

In the second experiment, samples were subjected to freezing without immersion. A rapid and substantial decrease in yield values (Figure 11.4A) was observed as soon as the temperature decreased below 0 °C. When temperature further decreased, the yield dropped to zero. However, yield increased again as soon as the temperature increased above −3 °C, and recovered completely to initial values. Changes in NPQ in the course of the experiment were less pronounced than in the previous experiment; however, the general trend of increasing NPQ with decreasing yield was also observable.

Fv/Fm (Figure 11.4B) decreased almost as strongly at the lowest temperature as in the previous experiment, but recovery proceeded faster than in the submerged specimens. Significant differences were found between the initial, cold and end values. The

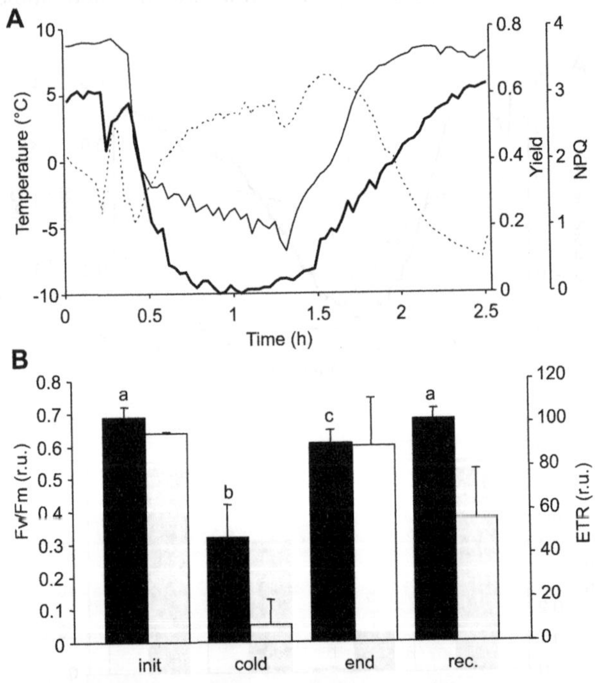

Figure 11.4 *Fucus distichus.*
Experiment 2 (dry freezing) with (A) changes in temperature (°C; thin line), effective quantum yield (yield; bold line) and non-photochemical quenching in relative units (NPQ; dashed line) over time; (B) optimum quantum yield (Fv/Fm; black bars) and maximal electron transport rate (ETR_{max}; white bars) measured after a short acclimation (init), at the lowest temperature (cold), at the end of the experiment (end), after a recovery of 36 h in dim white light and at fjord temperature (rec.). Different letters show statistically significant differences for Fv/Fm values (Tukey's HSD $p < 0.05$). Values are means + SD ($n = 3$). Due to limited sample size, no statistical analysis is available for ETR_{max}. r.u., relative units.

same trend was observed for ETR_{max} (Figure 11.4B), which decreased down to a mean value of <10, and recovery of this parameter was rapid, although the initial value of ETR_{max} was not reached at the end of the experiment.

The results of the third experiment (long-term freezing) are shown in Figure 11.5. In this experiment, the temperature was maintained at −1.5 °C for almost 7 h. Yield values remained nearly constant at 0.6 (Figure 11.5A). After the freezing period, the yield decreased, probably due to an artefact while recording Fv/Fm and a PI-curve before exposure of samples to darkness. Subsequently, with an increase in temperature, the yield also increased. During subsequent exposure for 8 h at ambient temperature (~4 °C), yield values held at around 0.6 over the entire time period. The initial value of Fv/Fm (Figure 11.5B) was 0.7 units and, by the end of the period of freezing, it decreased to 0.25. At the end of the experiment, the value was almost as high as the initial value. After a recovery time of 36 h, the initial value was reached again. ETR_{max} values decreased upon freezing and recovery proceeded more slowly than in experiments 1 and 2 (see Figure 11.5B). However, after a recovery period of 36 h, the values reached the initial level.

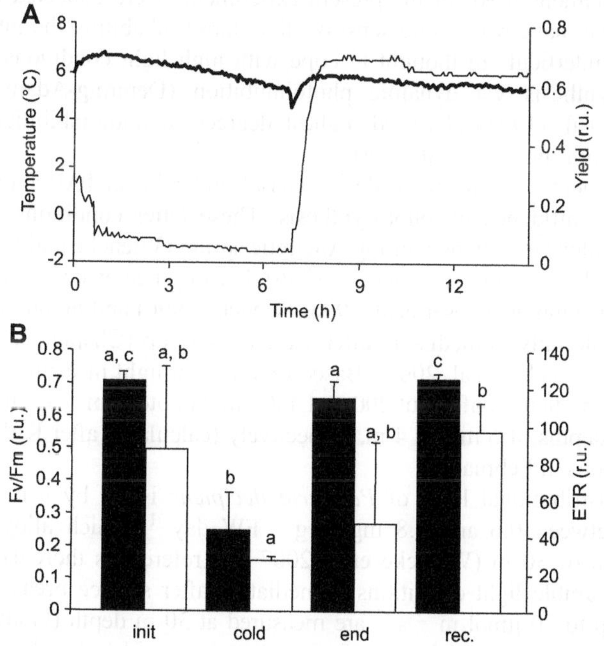

Figure 11.5 *Fucus distichus.*
Experiment 3 (long-term freezing) (A) changes in temperature (°C; thin line) and effective quantum yield (yield; bold line) over time; (B) optimum quantum yield (Fv/Fm) (white bars) and maximal electron transport rate (ETR_{max}) (white bars) measured after a short period of acclimation (init), at the lowest temperature (cold), at the end of the experiment (end), after a recovery of 36 h in dim white light and at fjord temperature (rec.). Different letters show statistically significant differences (Tukey's HSD $p < 0.05$). Values are means + SD ($n = 3$). r.u., relative units.

Significant increases in low temperature-induced oxidative stress as determined by the MDA assay were absent in all experiments (data not shown). Also, temperature-related differences in D1-content were absent in all experiments (data not shown).

Performing photosynthesis at low temperatures

Low water temperatures in combination with high PAR intensities may severely stress algae by inhibiting photosynthetic performance and affecting pigment content. This is thought to be due to the fact that high light and low temperature conditions slow down enzymatic reactions, with a consequently increased generation of reactive oxygen species. Furthermore, *de novo* synthesis of previously degraded D1 protein and its reintegration into the thylakoid membrane may be impaired by low temperatures due to reduced velocity of biosynthetic pathways and membrane fluidity (Aro et al. 1993). As a result, *Palmaria decipiens*' photosynthesis was significantly impaired by high PAR irradiances in combination with low temperatures (see Figures 11.1 and 11.2).

Although *Palmaria decipiens* is thought to cope with high light intensities since it may occur in Antarctic intertidal zones (Wiencke and Clayton 2002), high light intensities under specific experimental conditions were harmful. However, it is important to note that specimens used for the present experiments were collected in the subtidal and therefore may have been more sensitive than those inhabiting the intertidal. Specimens from the intertidal are thought to cope with high light conditions by down-regulating photosynthesis via dynamic photoinhibition (Demmig-Adams and Adams 1992). Hanelt et al. (1994) observed highest degrees of photoinhibition in Antarctic macroalgae floating near the water surface.

In our experiments, treatment of algal material under lower PAR intensities at 0 °C had no negative influence on photosynthesis. These latter conditions are considered similar to those algae experience during Antarctic spring (Wiencke and tom Dieck 1989, Klöser et al. 1993, Weykam and Wiencke 1996). For instance, water temperatures rarely exceed +2 °C in summer (Klöser et al. 1993, Wiencke 1996) and favourable light conditions are available only immediately after sea ice break up (Chapter 2; Wiencke 1996, Lüder et al. 2001, Zacher et al. 2009). Based on our own light measurements conducted in Potter Cove, irradiances of about 200 and 400 µmol photons m^{-2} s^{-1} may be recorded down to water depths of 6 m and 4 m, respectively (calculated after Kirk 1994), during summer conditions in February.

The lower distributional limit of *Palmaria decipiens* is set by a metabolic carbon fixation rate between 0.6 and 0.8 mg C g^{-1} FW day^{-1}, which allows growth to a maximum depth of 30 m (Wiencke et al. 2007 and references therein). Algae benefit from most favourable light conditions immediately after sea ice break up, when light intensities of up to 70 µmol m^{-2} s^{-1} are measured at 30 m depth (Gómez et al. 1997). Like most of the Antarctic endemics, *P. decipiens* is considered to be strongly shade-adapted (Wiencke et al. 1993, Weykam et al. 1996, Weykam and Wiencke 1996, Eggert and Wiencke 2000) and has highest photosynthetic activity during spring (Weykam and Wiencke 1996). However, temperature is thought to be the most important factor controlling the distribution of *P. decipiens* (Wiencke et al. 2007), as growth is suppressed at temperatures >10 °C (Wiencke and tom Dieck 1989), and optimum growth temperatures are between 0 °C and 5 °C (Bischoff-Bäsmann and Wiencke 1996).

This illustrates the importance of our study on interactive effects (compared to the previous single factor studies); even under favourable temperature conditions, high light may easily become a crucial factor. As hypothesised, higher PAR levels generate photoinhibitory conditions at lower temperatures. However, in shallow waters and even in the intertidal, specimens are exposed to higher irradiances. Consequently, there is still a strong need to explore in more detail the underlying acclimation mechanisms under high light and low temperature stress. Our study also supports the concept that higher temperatures, even those close to the tolerance limit, may compensate for photoinhibition. Similar results have been obtained in Antarctic green macroalgae under PAR and UV radiation stress (Rautenberger and Bischof 2006).

Palmaria decipiens is considered to be a season anticipator (Wiencke 1990, Weykam and Wiencke 1996), and thus has a life cycle perfectly adapted to Antarctic conditions. It starts growing under very low light conditions, and daylength seems to be the trigger (Wiencke 1990, 1996, Lüder et al. 2001, Wiencke et al. 2007). Its photosynthetic rhythm coincides with the production of pigments, which increase during spring (Lüder et al. 2001, 2002). Future studies should thus also focus on seasonal variation in species light requirements and temperature tolerance.

In our study on *Fucus distichus*, there were significant reductions in both optimum and effective quantum yield in response to decreasing temperatures. At the lowest temperatures applied, photosynthetic activity was completely halted (Figures 11.3A, 11.4A). However, after release from freezing, quantum yield recovered almost entirely, indicating that no persistent damage to the photosynthetic apparatus was induced. Rather, there may have been a down-regulation of quantum yield in order to protect photosynthesis from increased production of reactive oxygen species (ROS). The ability to recover quickly from photoinhibition is referred to as "dynamic photoinhibition", *sensu* Osmond (1994), or more recently as "photoprotection" (Franklin et al. 2003). This fast response might be part of a physiological protection strategy in a highly variable environment, such as the intertidal zone.

In *Fucus serratus* Linnaeus, Huppertz et al. (1990) demonstrated a decline in Fv/Fm values as well as in oxygen production during low tide accompanied by high light intensities and a fast recovery as soon as thalli were submerged again. Dynamic photoinhibition is usually associated with an increase in NPQ mediated by an increased activity in the xanthophyll cycle (Demmig-Adams and Adams 1992), which is also operative in brown algae (Hanelt et al. 2003). Our results with *F. distichus* also showed an increase in NPQ when low temperatures induced a reduction in quantum yield (Figures 11.3A and 11.4A), which might at least indicate that the above mentioned mechanisms of dynamic photoinhibition may also be deployed upon freezing as well. In accordance with the strict temperature dependence of light-saturated photosynthesis, changes in ETR_{max} are more strongly pronounced than in Fv/Fm, as recovery from cold stress usually took longer.

Apart from the reduced activity of dark reactions, the respective concentration of D1 protein may have affected photosynthetic capacity upon freezing. However, no differences in D1 content were found in our studies. Similarly, only minor changes in ROS-mediated MDA formation were induced by our experimental treatments, which might either indicate the high capacity of *Fucus* to protect itself from radical formation, or that impinging radiation conditions were too low to induce oxidative stress.

In general, frost and drought have similar physiological effects (Pearson and Davison 1994). In our experiments, the combination of drought and frost had no stronger effects on algal physiology than either factor alone. However, the effect of drought was less important at low temperatures because frozen water is not able to evaporate. Overall, the study showed that *Fucus distichus* is able to withstand temperatures down to $-12\,°C$ and also the combination of drought and freezing for about 3 h without long-term damage to photosynthesis. Under field conditions in the intertidal zone of Svalbard, algae are frequently exposed to freezing temperatures during emergence at low tide (Hop et al. 2002, Svendsen et al. 2002).

The ability of *Fucus distichus* to rapidly down-regulate photosynthesis under freezing and desiccating conditions and, subsequently, achieve full photosynthetic performance under submerged conditions is a major component of physiological adaptation to Arctic intertidal conditions. Curiously, in a related study on the effects of freezing, desiccation and high light stress on three different species of *Fucus* (*F. spiralis*, *F. evanescens*, *F. distichus*), *F. distichus* was shown to be the most sensitive species under investigation, responding to the experimental treatments with elevated production of ROS (Collén and Davison 1999). However, in the latter study, an experimental irradiance of about 1600 μmol photons $m^{-2}\,s^{-1}$ was applied, which is above the maximum irradiance encountered at Svalbard during summer (Svendsen et al. 2002). Therefore, it might still be possible that in combination with very high PAR, *F. distichus*, as a species from high northern latitudes, becomes sensitive to drought and freezing.

Summing up, our study indicates that stenothermal *Palmaria decipiens* is well adapted to the current abiotic conditions encountered in Antarctica. Due to the long and stable cold-water history of Antarctica, its endemic species have developed suitable long-term adaptations. However, in combination with the previous findings on the alga's temperature tolerance, we propose that there is a very narrow range for its optimal physiological functioning; this may mean that this species will be sensitive to future environmental change. In contrast, as also indicated by its wide distributional range, *Fucus distichus* is not likely to be impaired by increasing temperatures.

References

Andersson, B., A.H. Salter, I. Virgin and S. Styring. 1992. Photodamage of photosystem II – primary and secondary events. *J. Photochem. Photobiol. B: Biol. 15*: 15–31.

Aro, E.M., I. Virgin and B. Andersson. 1993. Photoinhibition of photosystem II. Inactivation, protein damage and turnover. *Biochim. Biophys. Acta 1143*: 113–134.

Asada, K. and M. Takahashi. 1987. Production and scavenging of active oxygen in photosynthesis. *In*: (D.J. Kyle, C.B. Osmond and C.J. Arntzen, eds) *Photoinhibition*. Elsevier Science, Amsterdam. pp. 89–109.

Beer, S. and A. Eshel. 1985. Determining phycoerythrin and phycocyanin concentrations in aqueous crude extracts of red algae. *Aust. J. Mar. Freshw. Res. 36*: 785–792.

Bilger, W., U. Schreiber and M. Bock. 1995. Determination of the quantum efficiency of photosystem II and of non-photochemical quenching of chlorophyll fluorescence in the field. *Oecologia 102*: 425–432.

Bischof, K., D. Hanelt, H. Tüg, U. Karsten, P.E.M. Brouwer and C. Wiencke. 1998. Acclimation of brown algal photosynthesis to ultraviolet radiation in Arctic coastal waters (Spitsbergen, Norway). *Polar Biol. 20*: 388–395.

Bischof, K., D. Hanelt and C. Wiencke. 2000. Effects of ultraviolet radiation on photosynthesis and related enzyme reactions of marine macroalgae. *Planta 211*: 555–562.

Bischof, K., P.J. Janknegt, A.G.J. Buma, J.W. Rijstenbil, G. Peralta and A.M. Breeman. 2003. Oxidative stress and enzymatic scavenging of superoxide radicals induced by solar UV-B radiation in *Ulva* canopies from southern Spain. *Sci. Mar. 67*: 353–359.

Bischoff-Bäsmann, B. and C. Wiencke. 1996. Temperature requirements for growth and survival of Antarctic Rhodophyta. *J. Phycol. 32*: 525–535.

Collén, J. and I.R. Davison. 1999. Reactive oxygen production and damage in intertidal *Fucus* spp. (Phaeophyceae). *J. Phycol. 35*: 54–61.

Demmig-Adams, B. and W.W. Adams III. 1992. Photoprotection and other responses of plants to high light stress. *Annu. Rev. Plant Physiol. Plant Mol. Biol. 43*: 599–626.

Denton, A.B. and A.R.O. Chapman. 1991. Feeding preferences of gammarid amphipods among four species of *Fucus*. *Mar. Biol. 109*: 503–506.

Drew, E.A. and R.M. Hastings. 1992. A year-round ecophysiological study of *Himantothallus grandifolius* (Desmarestiales, Phaeophyta) at Signy Island, Antarctica. *Phycologia 31*: 262–277.

Eggert, A. and C. Wiencke. 2000. Adaptation and acclimation of growth and photosynthesis of five Antarctic red algae to low temperatures. *Polar Biol. 23*: 609–618.

Franklin, L.A., C.B. Osmond and A.W.D. Larkum. 2003. Photoinhibition, UV-B and algal photosynthesis. *In*: (A.W.D. Larkum, S.E. Douglas and J.A. Raven, eds) *Photosynthesis in algae*. Kluwer Academic Publishers, Dordrecht. pp. 351–384.

Gómez, I., G. Weykam, H. Klöser and C. Wiencke. 1997. Photosynthetic light requirements, metabolic carbon balance and zonation of sublittoral macroalgae from King George Island (Antarctica). *Mar. Ecol. Prog. Ser. 148*: 281–293.

Gómez, I., A. Wulff, M.Y. Roleda, P. Huovinen, U. Karsten, M.L. Quartino, K. Dunton and C. Wiencke. 2009. Light and temperature demands of marine benthic microalgae and seaweeds in polar regions. *Bot. Mar. 52*: 593–608.

Hanelt, D., M.J. Jaramillo, W. Nultsch, S. Senger and R. Westermeier. 1994. Photoinhibition as a regulative mechanism of photosynthesis in marine algae of Antarctica. *Ser. Cient. INACH 44*: 67–77.

Hanelt, D., C. Wiencke and W. Nultsch. 1997. Influence of UV radiation on the photosynthesis of Arctic macroalgae in the field. *J. Photochem. Photobiol. B: Biol. 38*: 40–47.

Hanelt, D., C. Wiencke and K. Bischof. 2003. Photosynthesis in marine macroalgae. *In*: (A.W. Larkum, S.E. Douglas and J.A. Raven, eds) *Photosynthesis in algae*. Kluwer Academic Publishers, Dordrecht. pp. 413–435.

Heath, J.K. and L. Packer. 1968. Photoperoxidation in isolated chloroplasts: Kinetics and stochiometry of fatty acid peroxidation. *Arch. Biochem. Biophys. 125*: 189–198.

Hop, H., T. Pearson, E.N. Hegseth, K.M. Kovacs, C. Wiencke, S. Kwasniewski, K. Eiane, F. Mehlum, B. Gulliksen, M. Wlodarska-Kowalczuk, C. Lydersen, J.M. Weslawski, S. Cochrane, G.W. Gabrielsen, R.J.G. Leakey, O.J. Lønne, M. Zajaczkowski, S. Falk-Petersen, M. Kendall, S.-Å.Wängberg, K. Bischof, A.Y. Voronkov, N.A. Kovaltchouk, J. Wiktor, M. Poltermann, G. di Prisco, C. Papucci and S. Gerland. 2002. The marine ecosystem of Kongsfjorden, Svalbard. *Polar Res. 21*: 167–208.

Huppertz, K., D. Hanelt and W. Nultsch. 1990. Photoinhibition of photosynthesis in the marine brown alga *Fucus serratus* as studied in field experiments. *Mar. Ecol. Prog. Ser. 66*: 175–182.

Inskeep, W.P. and P.R. Bloom. 1985. Extinction coefficients of chlorophyll *a* and *b* in *N,N*-Dimethylformamide and 80% acetone. *Plant Physiol. 77*: 483–485.

Kanwisher, J. 1957. Freezing and drying in intertidal algae. *Biol. Bull. 113*: 275–285.

Kirk, J.O.T. 1994. *Light and photosynthesis in aquatic ecosystems.* Cambridge University Press. pp. 509.

Kirst, G.O. and C. Wiencke. 1995. Ecophysiology of polar algae. *J. Phycol. 31*: 181–199.

Klöser, H., G. Ferreyra, I. Schloss, G. Mercuri, F. Laturnus and A. Curtosi. 1993. Seasonal variation of algal growth conditions in sheltered Antarctic bays: the example of Potter Cove (King George Island, South Shetlands). *J. Mar. Syst. 4*: 289–301.

Lüder, U.H., J. Knoetzel and C. Wiencke. 2001. Acclimation of photosynthesis and pigments to seasonally changing light conditions in the endemic Antarctic red macroalga *Palmaria decipiens*. *Polar Biol. 24*: 598–603.
Lüder, U.H., C. Wiencke and J. Knoetzel. 2002. Acclimation of photosynthesis and pigments during and after six months of darkness in *Palmaria decipiens* (Rhodophyta): a study to simulate Antarctic winter sea ice cover. *J. Phycol. 38*: 904–913.
Lüning, K. 1990. *Seaweeds. Their environment, biogeography, and ecophysiology.* John Wiley and Sons, New York. pp. 527.
McMinn, A., J.W. Runcie and M. Riddle. 2004. Effect of seasonal sea ice breakout on the photosynthesis of benthic diatom mats at Casey, Antarctica. *J. Phycol. 40*: 62–69.
Osmond, C.B. 1994. What is photoinhibition? Insights from comparisons of shade and sun plants. *In:* (N.R. Baker and J.R. Bowyer, eds) *Photoinhibition of photosynthesis from molecular mechanisms to the field*. BIOS Scientific Publishers, Oxford. pp. 1–24.
Pearson, G.A. and I.R. Davison. 1994. Freezing stress and osmotic dehydration in *Fucus distichus* (Phaeophyta): evidence for physiological similarity. *J. Phycol. 30*: 257–267.
Rautenberger, R. and K. Bischof. 2006. Impact of temperature on UV susceptibility of two species of *Ulva* (Chlorophyta) from Antarctic and Subantarctic regions. *Polar Biol. 29*: 988–996.
Schreiber, U., W. Bilger and C. Neubauer. 1994. Chlorophyll fluorescence as a non-intrusive indicator for rapid assessment of in vivo photosynthesis. *In:* (E.D. Schulze and M.M. Caldwell, eds) *Ecophysiology of photosynthesis*. Springer-Verlag, Heidelberg. pp. 49–70.
Sokal, R. and F.J. Rohlf. 1995. *Biometry: the principles and practice of statistics in biological research*. Freeman, New York. pp. 887.
Svendsen, H., A. Beszczynska-Moeller, J.O. Hagen, B. Lefauconnier, V. Tverberg, S. Gerland, J.B. Ørbaek, K. Bischof, C. Papucci, M. Zajaczkowski, R. Attolini, O. Bruland, C. Wiencke, J.G. Winther and W. Dallmann. 2002. The physical environment of Kongsfjorden-Krossfjorden, an Arctic fjord system in Svalbard. *Polar Res. 21*: 133–166.
Weykam, G. and C. Wiencke. 1996. Seasonal photosynthetic performance of the endemic Antarctic red alga *Palmaria decipiens* (Reinsch) Ricker. *Polar Biol. 16*: 357–361.
Weykam, G., I. Gomez, C. Wiencke, K. Iken and H. Klöser. 1996. Photosynthetic characteristics and C:N ratios of macroalgae from King George Island (Antarctica). *J. Exp. Mar. Biol. Ecol. 204*: 1–22.
Wiencke, C. 1990. Seasonality of red and green macroalgae from Antarctica – a long-term culture study under fluctuating Antarctic daylengths. *Polar Biol. 10*: 601–607.
Wiencke, C. 1996. Recent advances in the investigation of Antarctic macroalgae. *Polar Biol. 16*: 231–240.
Wiencke, C. and M.N. Clayton. 2002. *Antarctic seaweeds*. Synopses of the Antarctic benthos 9. Gantner, Ruggell. pp. 239.
Wiencke, C. and I. tom Dieck. 1989. Temperature requirements for growth and temperature tolerance of macroalgae endemic to the Antarctic region. *Mar. Ecol. Prog. Ser. 54*: 189–197.
Wiencke, C., J. Rahmel, U. Karsten, G. Weykam and G.O. Kirst. 1993. Photosynthesis of marine macroalgae from Antarctica: light and temperature requirements. *Bot. Acta 106*: 78–87.
Wiencke, C., M.N. Clayton and M. Schoenwaelder. 2004. Sensitivity and acclimation to UV radiation of zoospores from five species of Laminariales from the Arctic. *Mar. Biol. 145*: 31–39.
Wiencke, C., M.N. Clayton, I. Gómez, K. Iken, U.H. Lüder, C.D. Amsler, U. Karsten, D. Hanelt, K. Bischof and K. Dunton. 2007. Life strategy, ecophysiology and ecology of seaweeds in polar waters. *Rev. Environ. Sci. Biotechnol. 6*: 95–126.
Wiencke, C., G. Ferreyra, D. Abele and S. Marenssi. 2008. The Antarctic ecosystem of Potter Cove, King George Island (Isla 25 de Mayo). Synopsis of research performed 1999–2006 at the Dallmann Laboratory and Jubany Station. *Ber. Polarforsch. Meeresforsch. 571*: 12–47.
Zacher, K., R. Rautenberger, D. Hanelt, A. Wulff and C. Wiencke. 2009. The abiotic environment of polar marine benthic algae. *Bot. Mar. 52*: 483–490.

Polar benthic algae in a changing world

12 Impact of oceanic warming on the distribution of seaweeds in polar and cold-temperate waters

Ruth Müller, Inka Bartsch, Thomas Laepple and Christian Wiencke

Introduction

One major concern in global change scenarios is the actually measured and prospective increase in local and global air and water temperatures and their effects on biological systems. Temperature variation in sea surface waters has been followed since long times and good datasets for marine systems exist from the past decades (Rayner et al. 2005). Increase in mean annual sea surface temperature (SST) has been clearly identified along diverse coastlines in recent years. The global average SST increased by 0.7 °C during the last century (Hulme et al. 2002). Within the last decade, the rate of temperature increase accelerated. One example from the southern hemisphere is King George Island in Antarctica, where mean monthly coastal water temperatures significantly increased by ~1 °C in summer and by ~2 °C in winter between 1991 and 2005 (Schloss et al. 2008). Another example from the northern hemisphere is the island of Helgoland in the North Sea, where mean annual SST has increased by 1.3 °C since 1962; between 1995 and 2005, the increase was 0.5 °C (Wiltshire et al. 2008). The latest report of the Intergovernmental Panel on Climate Change (IPCC 2007a, b) presented long-term projections for climate change into the next century and emphasized that most of the observed warming over the past 50 years is attributable to human activities and that the atmosphere will very likely continue to heat up.

Expansion and retreat of marine species along coastlines during times of temperature change have been documented over long time periods. Many well-preserved fossil records, especially of molluscs, are representative of migration processes over geological time scales (examples given in Lüning 1990). More recently, during the short warm period in the first 50 years of the 20th century in the North Sea/English channel, seaweed migrations along the British and Britanny coastlines were documented (Hiscock et al. 2004, Mieszkowska et al. 2005). Recent publications comparing historical and present datasets partially document distributional changes of seaweed and invertebrate species, indicating the ongoing process and demonstrating a distributional shift of cold- and warm-temperate species to higher latitudes (Mieszkowska et al. 2006, 2007, Lima et al. 2007, Hawkins et al. 2008).

As temperature strongly controls the survivorship, growth, reproduction and recruitment of seaweeds and thereby their distributions (Chapter 10; van den Hoek 1982a, b, Breeman 1988, Lüning 1990, Gómez et al. 2009), the biogeographical distribution patterns of seaweed species are affected by changes in the coastal temperature regime (Breeman 1988, Hiscock et al. 2004). In addition to temperature, photoperiod may trigger crucial steps in the life cycles of seaweeds (Dring 1982). Interaction of both factors narrows seasonal windows for reproduction or growth even more than temperature alone and thereby affects distributional limits as well. Other factors such as ice-scouring or biotic interactions have a modifying influence and will finally determine the local distribution of seaweeds.

Although projections of future distributional shifts of polar and cold-temperate seaweed species have already been calculated on the basis of older sea-surface isotherm and ecophysiological data (Breeman 1988, Hiscock et al. 2004), a large set of coupled climate model simulations prepared for the last IPPC report allows more realistic temperature simulations based on defined greenhouse gas emission scenarios. In the following we will thus predict changes in the latitudinal distribution of ecologically important polar to cold-temperate seaweeds under the anticipated global warming until the end of the 21st century. Comparing modelled SSTs with the temperature requirements of seaweeds, broad biogeographical changes of selected key structural seaweed species in both hemispheres are estimated. Areas of prospective rapid change are identified and the circumstances in the Arctic-Atlantic sector are compared with those in South America and the Antarctic Peninsula. The possible deviation of the prospective broad biogeographical shifts by local abiotic conditions or biotic interactions is discussed in the light of previously published studies.

Expected water temperature changes in polar and cold-temperate regions

An accurate projection of the future climate change in specific regions is possible by using coupled climate model simulations prepared for phase 3 of the Coupled Model Intercomparison Project 3 (CMIP3; http://www-pcmdi.llnl.gov). A multi-model ensemble mean over a large set of models allows a better projection than is possible by focusing on single models (Laepple et al. 2008). To investigate the projected change for the present century, the extraction and analysis of the mean oceanic SST for February and August as representatives of summer and winter extreme temperatures allows an analysis of the differences between different time periods. Müller et al. (2009) investigated the 2080–2099 mean SST using the moderate Special Report on Emission Scenarios – Scenario A1B (SRESA1B) emission scenario and the 1980–1999 mean SST based on historical 20C3M simulations (Figure 12.1; for details of methods see Müller et al. 2009). The SRESA1B emission scenario was based on the assumption that the global human population will reach its peak in the mid 21st century, develop a very rapid economic growth and use balanced fossil-intensive and non-fossil energy sources, which would mean a rapid introduction of new and more efficient technologies. Considering only models in which the ocean model output was available for the historical and the future scenario, 42 simulations from 19 models were used.

According to this projection an overall increase of the North Atlantic and Arctic Ocean February SST is predicted with regional changes ranging from moderate cooling south of Greenland (up to −1.5 °C) to strong warming with maxima in coastal areas of up to 3 °C in northern Norway or the East coast of the United States (Figure 12.1A). The summer pattern (Figure 12.1B) has a similar spatial structure but shows a more pronounced warming up to 4 °C. In contrast to the winter pattern, a strong warming in Baffin and Hudson Bays is predicted in the North Atlantic. Along the Antarctic Peninsula winter temperatures will keep quite stable and summer temperatures will increase by up

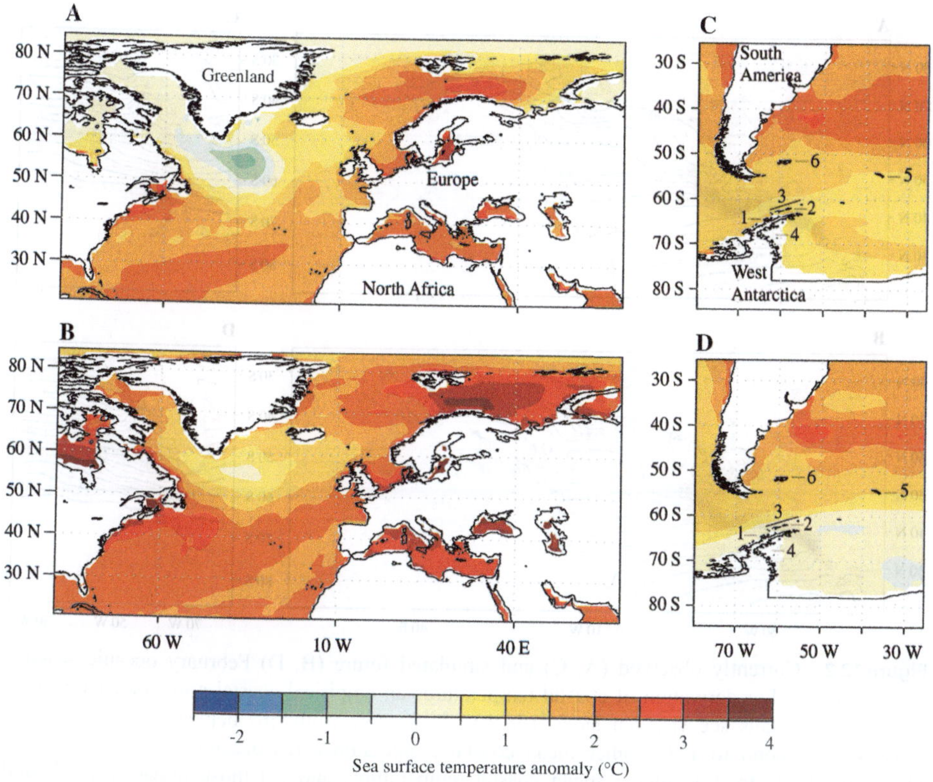

Figure 12.1 Simulated future changes in sea surface temperature. The ensemble mean anomaly for 2080–2099 relative to 1980–1999 is shown for February (A, C) and August (B, D). 1, Anvers Island; 2, King George Island; 3, South Shetland Islands; 4, Larsen shelf; 5, South Georgia; 6, Falklands.

to 1.5 °C. Along the western and eastern coastline of South America the warming trend will be more pronounced with temperature increases of 1.5 to 2.5 °C during austral summer (Figure 12.1C) and austral winter (Figure 12.1D).

Comparing present and future equivalent SST (Figures 12.2, 12.3), the expected summer/winter inequality will increase the amplitude of the seasonal temperature cycle in some sectors while in other sectors the range between winter and summer SST will remain approximately constant.

Looking into this in more detail, in the northern hemisphere, for example, off northwestern Spain, there is a current temperature range between 13 °C in February and 19.5 °C in August, whereas the simulation for 2080–2099 predicts a seasonal range between 14 and 21 °C, i.e., ~7 °C in both cases (Figures 12.2, 12.3). Both, current observations and model predictions show this 7 °C difference between summer and winter temperatures over wide areas along European coastlines. In the western North Atlantic, however, the present seasonal temperature range is much larger (~13–17 °C), and will increase by up to 2 °C in 2080–2099. This increase in the seasonal temperature range will be especially evident in Newfoundland and areas further North such as Baffin Bay or southern and western Greenland.

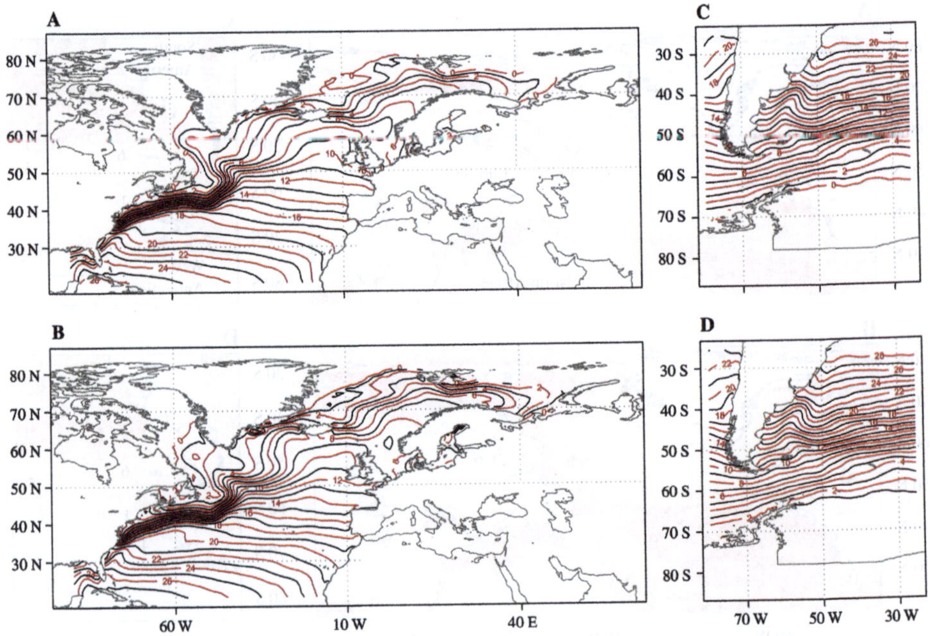

Figure 12.2 Currently observed (A, C) and simulated future (B, D) February oceanic sea-surface isotherms. For derivation of coastal temperatures an empirical coastal correction factor may be used (for details see van den Hoek 1982b): (1) oceanic isotherms between 0 °C and 10 °C actually correspond to 1 °C higher mean coastal temperatures, (2) oceanic isotherms between 11 °C and 20 °C to 2 °C higher mean coastal temperatures and (3) those between 21 °C and 30°C to 3 °C higher mean coastal temperatures.

In Antarctic waters the range between winter and summer SST at specific locations will not change significantly between present and future SST (Figures 12.2, 12.3). For example, off South Georgia, at the northern border of the Antarctic region, the seasonal temperature range between February and August is ~3 °C and is not predicted to change towards the end of the 21st century. In southern South America, however, the range between summer and winter SST will increase slightly. At 50° S, off the west coast of South America, the seasonal difference between winter and summer SST is currently 2.5 °C and will increase to 4.5 °C in future. Off the east coast of the continent at 50° S, the range will also increase by 2 °C, i.e., from 4 °C to 6 °C in future.

The SST changes in the Arctic are related to the patterns of sea ice change, changes in the thermo-haline circulation (THC) and the different land/sea response sensitivities caused by the higher heat capacity of the ocean. As most models predict a reduction in the THC as a response to increases in greenhouse gas concentrations (Schmittner et al. 2005), heat transport to the polar regions will be affected. This will result in a local cooling pattern in the Labrador Sea in winter and a reduced warming in summer. Differences in heat capacity between ocean and land areas will result in a stronger continental warming, which also affects the adjacent water areas by thermal advection (Stouffer and Manabe 1999). As only the water temperature is analyzed, the strong warming caused by positive feedbacks of the sea ice and snow cover known as polar

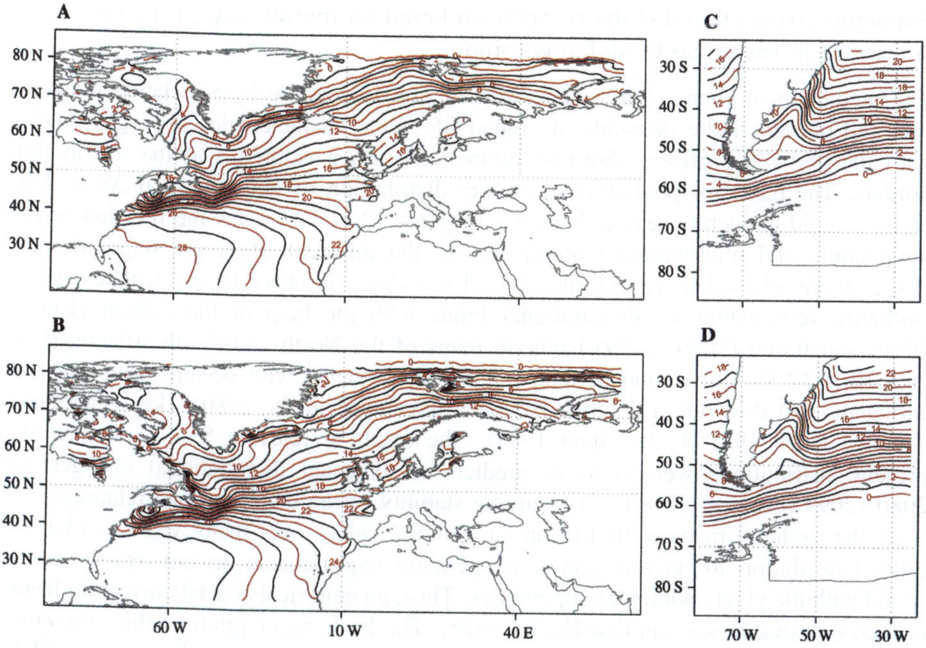

Figure 12.3 Currently observed (A, C) and simulated future (B, D) August oceanic sea-surface isotherms. For further explanation see Figure 12.2.

amplification (e.g., Moritz et al. 2002) is not as apparent as in the air temperature record. Sea ice is thinning, with very rapidly decreasing spatial dissemination due to increasing winter SST in Arctic regions. In regions where the sea ice is projected to persist in future, the SST, which represents the temperature under the sea ice, does not change, even if the air temperature over the ice increases. This is the case for the zonal band north of 80° N as well as in Baffin Bay in February (Figure 12.1A). The overall Arctic warming pattern, including the land/sea contrast, is consistent with changes already observed in the instrumental record (Johannessen et al. 2004), although a direct comparison is difficult due to the natural variability in the record.

For Antarctic areas, similar mechanisms are probably responsible for the projected oceanic warming, although the capacity of the Southern Ocean to absorb large amounts of heat leads to a smaller predicted warming. The region in Antarctica mostly affected by global warming is the Antarctic Peninsula (e.g., Clarke et al. 2007, Barnes and Peck 2008), although the projected SST increase (~1 – 2 °C) is much lower than in Arctic regions (~3 – 4 °C, cf. Figure 12.1A – 12.1D) and the warming is more pronounced in the austral summer. However, the confidence in the future projection around the Antarctic Peninsula is lower than that for the northern hemisphere as the climate models do not reproduce the historical warming observed in this region (King 2003) and the mechanisms for the observed warming are largely unknown (Vaughan et al. 2003). On the other hand, changes in sea ice distribution are more extreme in western Antarctica than in the Arctic. The area covered by sea ice has decreased by ~10% per decade around the Antarctic Peninsula, and the seasonal duration of sea ice formation became shorter during the last century (Clarke et al. 2007).

Expected distributional shifts of polar and cold-temperate key structural seaweeds in response to global warming

The prediction of changes in seaweed distribution is possible by correlating temperature and photoperiodic demands of crucial life cycle characteristics of single species with the prevailing present day and future sea surface isotherms at the distribution borders. The basis for predicting the future distribution of key structural Arctic/Antarctic to cold-temperate seaweed species (Figures 12.4–12.7) was the revision of the temperature and photoperiod requirements of the different life cycle stages or processes of the selected species (Table 12.1). These data were used to re-calculate oceanic isotherms responsible for distributional limits with the help of the current (1980–1999) and future (2080–2099) isotherm maps of the North and South Atlantic (Figures 12.2, 12.3). Correlation of temperature requirements with oceanic isotherms for explanation of distributional limits has been developed in the 1980's (key references: van den Hoek 1982a,b; Breeman 1988). The resulting relevant SST explaining the present distributions were used to predict the expected distributional changes for 2080–2099 (Figures 12.4–12.7) assuming stability of the underlying correlations.

As the isotherm maps exhibit mean temperature values over many years and have a limited resolution for coastal zones, temperature requirements do not exactly correspond with the given isotherm temperatures. Thus, an empirical coastal correction factor is used (for details see van den Hoek 1982b). The basic assumption of this correction factor is that: (1) oceanic isotherms between 0 °C and 10 °C actually correspond to 1 °C higher mean coastal temperatures, (2) oceanic isotherms between 11 °C and 20 °C to 2 °C higher mean coastal temperatures and (3) those between 21 °C and 30 °C to 3 °C higher mean coastal temperatures. This implies, for example, that the distributional limit of a seaweed species with an upper survival temperature of 28 °C is expected to correlate with the 25 °C summer SST. In the following case studies the temperature demands of selected cold-temperate to Arctic/Antarctic key structural seaweed species are described in detail. The reader not interested in these details may directly proceed to the section 'General trends in future seaweed distribution' on page 254.

Case studies of northern-hemispheric seaweed species

***Laminaria solidungula* J. Agardh (Laminariales, Heterokontophyta) (Figure 12.4A, Table 12.1)** Most of the perennial exclusively subtidal kelp populations of *L. solidungula* are truly Arctic, but there are refuge populations in the subarctic, for example within many Newfoundland fjords. Here, the species is either restricted to water masses close to the freezing point that may reach relatively shallow depths (Hooper 1984) or has migrated to greater depths (30 m) where the water is close to 0 °C all year round (A.R.O. Chapman, personal communication, Lüning 1990). Thus, even southern populations are only exposed to small annual temperature ranges. The corresponding 16 °C summer SST does not realistically describe the actual temperature conditions at the southern limit although the sporophytes are able to survive 16 °C (tom Dieck 1992). Truly Arctic populations are exposed to an annual temperature range between 2 and 6 °C (Chapter 2; Zacher et al. 2009; Specific examples: Igoolik, Canadian Arctic, −2–0 °C, Bolton and Lüning 1982; Kongsfjorden, Spitsbergen, up to 5 °C in summer, Svendsen et al. 2002, Hanelt et al. 2004; Disko Island, West

Greenland, up to 6 °C in summer, Jensen and Veland 2006; Stefansson Sound, N-Alaska, −1–4 °C in summer, Dunton 1990). Thus, the 5–6 °C August SST is an appropriate proxy for the southern distribution limit. Hence, a southern lethal or growth boundary for sporophytes or a germination boundary for meiospores can be excluded (cf. Table 12.1). Further data needed to conclusively elucidate the southern limit are lacking. Probably, the southern distribution boundary is determined by the combination of low temperatures and short daylengths needed to induce sporogenesis (I. Bartsch, unpublished data) or a new blade, as in *L. hyperborea* (Lüning 1986).

In future, it is expected that the species will extend its northward distribution to all areas with hard bottom substratum and without a permanent ice cover (northern Greenland, Canadian Arctic Archipelago, north and east of Spitsbergen to Franz Joseph Land, possibly Severnaya Zemlya). In the south, the species will retreat from the southwestern coast of Greenland and from Hudson Bay, southern Baffin Island, Labrador and Newfoundland unless deep cold-water refuges are established. The species will disappear from southern Spitsbergen and most probably also from Novaya Zemlya.

***Saccorhiza dermatodea* (Bachelot de la Pylaie) J.E. Areschoug (Tilopteridales, Heterokontophyta) (Figure 12.4B, Table 12.1)** The annual kelp-like *S. dermatodea* is distributed in Arctic and subarctic regions and extends into cold-temperate areas. Its depth distribution ranges from the lower intertidal/upper subtidal (e.g., at Spitsbergen, Wiencke et al. 2004) to a depth of at least 12 m (Labrador, Hooper and Whittick 1984; Nova Scotia, Novaczek and Mclachlan 1989).

The observed northern distribution limit coincides with the 3–5 °C August SST. This indicates conditions where water temperatures are sufficiently high during summer to enable adequate growth. The annual sporophyte needs to survive summer conditions as reproduction only commences in fall (Keats and South 1985) and growth is faster at 10 °C than at 3 °C (Norton 1977; see also footnote to Table 12.1). In future, once permanent ice cover is considerably reduced in summer, this species will extend northwards at least to 80° N along the Greenland coastline, the complete Spitsbergen coastline and further into the Canadian archipelago.

A composite southern boundary was suggested for *Saccorhiza dermatodea* (Breeman 1988) with lethal limits in northeastern America and a reproduction limit in western Europe without being able to explain the absence of the species in Brittany. In this area, summer temperatures would enable survival and winter temperatures of 10 °C would still permit sexual reproduction. This could be an example of the complex interplay of summer and winter temperatures with the photoperiod. In Brittany, the presumed lethal 17 °C summer SST coincides with the 10 °C winter SST. In February, daylengths increase and range between 9.5 h and 11 h in Brittany. This combination may hinder the completion of the life cycle. Henry (1987) reported predictable gametogenesis only in 8–10 h daylight regimes, while in a 12 h daylight regime, gametogenesis was rare. Thus, there is a mismatch between the photoperiod and the required temperatures in southwestern Europe which would account for its absence. *S. dermatodea* is absent from southwestern Norway where lethal summer temperatures coincide with low winter temperatures and short winter photoperiods but the reasons for its absence are unknown. As the observed southern distribution limit in Europe (northern Norway and Iceland) coincides with a 5–6 °C February SST, future warming may induce the absence of the species from most of northern Norway, but not from Iceland.

***Devaleraea ramentacea* (Linnaeus) Guiry (Palmariales, Rhodophyta) (Figure 12.4C, Table 12.1)** Pseudoperennial *D. ramentacea* is a lower intertidal to mostly subtidal (Spitsbergen, 0.5–7.5 m, Wiencke et al. 2004; Labrador, low intertidal to 10 m, Hooper and Whittick 1984) red algal species and has a circumarctic to cold-temperate distribution. It is able to withstand some ice-scouring or sand coverage as it can vegetatively propagate from the basal parts (Wiencke et al. 2004) and there exist diverse temperature ecotypes (Novaczek et al. 1990, Breeman and Pakker 1994).

The distribution boundaries have been discussed in detail by Bischoff and Wiencke (1993), and Breeman and Pakker (1994). The northern boundary is determined by the ability of the species to withstand ice-scouring and periodic thawing of ice cover during summer. These characteristics translate into a ~1 °C summer isotherm regime (Rayner et al. 2003) and thus, the species is expected to extend its distribution northward, especially along East Greenland coastlines and further north into the Canadian archipelago.

The composite southern distribution boundary has a summer lethal limit on the northeastern American coast coinciding with the 17 °C August SST, and a winter reproduction limit on the European side coinciding with the 6 °C February SST. These delimiting isotherms deviate by 2–3 °C from previously proposed isotherms (20 °C August and 8 °C February SST: Bischoff and Wiencke 1993, Breeman and Pakker 1994). As the current SST database is more reliable, the current assumptions may be treated with more confidence. Reproduction is confined to low temperatures (A. Breeman unpublished data, cited in Bischoff and Wiencke 1993) and to winter (Breeman and Pakker 1994), indicating an additional requirement for short daylengths at the southern distribution limit. As *Devaleraea ramentacea* is subtidal in the Arctic but intertidal to shallow subtidal near the southern distribution range in eastern Canada (A.R.O. Chapman, personal communication) it has to cope with higher and more variable temperatures during summer in this area than in the eastern Atlantic. Future warming will not drastically alter the southern distribution limit along the Canadian coastline as the 17 °C summer SST only shifts northward by about 5° latitude, but the distributional change at European sites will be pronounced. The populations along the whole Norwegian coastline, the Faroes, and southern Greenland will disappear. Although summer temperatures will facilitate survival, winter temperatures will not be low enough to enable reproduction. However, there is the possibility that hidden ecotypes exist as *D. ramentacea* is one of the few seaweed species with pronounced ecotypic variation even between strains at one locality (Novaczek et al. 1990). Unknown ecotypic differentiation may also explain why the isotherms discussed do not correspond to the distribution along the Icelandic coastline.

***Saccharina latissima* (Linnaeus) C.E. Lane, C. Mayes, Druehl *et* G.W. Saunders (Laminariales, Heterokontophyta) (Figure 12.5A, Table 12.1)** In the North Atlantic, the perennial kelp *S. latissima* is distributed from the High Arctic to temperate regions (~40° N). Its depth distribution ranges from shallow subtidal populations (e.g., Helgoland, North Sea, 0–5 m, Pehlke and Bartsch 2008) to depths of 15 to >20 m (e.g., Spitsbergen, 16.5 m, Wiencke et al. 2004; Labrador, 15 m, Hooper and Whittick 1984). Although known temperature optima range between 10 and 15 °C (cf. Table 12.1),

S. latissima extends north to the pack-ice border and hence has truly Arctic populations. The broad distribution pattern coincides with a high degree of polymorphism and plastic physiological responses, suggesting the existence of a large species complex and an opportunistic growth strategy (Bartsch et al. 2008). The existence of temperature ecotypes was proposed by Lüning (1975), Gerard and DuBois (1988) and Müller et al. (2008) although Lüning (1975) and Müller et al. (2008) did not show the persistence of results in the second generation.

As the observed northern distribution boundary extends north of the 1 °C August SST and partially aligns along the pack-ice border (Rayner et al. 2003), the species will become more dominant in these regions in future. As long as substratum is available and light is sufficient to enable growth, it will move northwards with the retreating pack-ice border. In contrast to the Arctic kelp *Laminaria solidungula*, *S. latissima* only commences growth when first light becomes available after the Arctic winter (Dunton 1985).

The southern distribution boundary of *Saccharina latissima* was discussed by van den Hoek (1982a) and Breeman (1988). They suggested that southward expansion is prevented by high lethal summer temperatures, although these permit survival of gametophytes and young sporophytes. Recent data show that in the western Atlantic the distribution limit at the 19–20 °C August SST only corresponds to survival temperatures of the sporophyte (20 to <23 °C; tom Dieck 1993). Additionally, sporophytes become annual at their southern limit (Lee and Brinkhuis 1986), thereby being dependent on regular summer survival of sporophytes for new recruits. In the eastern Atlantic, we presume an additional southern reproduction boundary on the Iberian Peninsula. As sporogenesis only takes place in short daylengths at 15 °C or lower (Lüning 1988) and induction of sori needs at least four weeks (A. Preisler and I. Bartsch, unpublished data), a cool short-day period between autumn to late winter is a pre-requisite for induction of sporangia and, thereby, new recruits. In northern Spain, shortest daylengths of 9 h coincide with a February isotherm of 13–14 °C, while further south, where survival of gametophytes may still be possible, this coincidence disappears. In the Mediterranean, the species is absent due to high lethal summer temperatures of >23 °C. In future, the species will definitely retreat from the Iberian Peninsula and, most probably, from France, The Netherlands, Germany and southern Scandinavia (similar to *Laminaria hyperborea*, Figure 12.4B) moving northwards as determined either by its reproduction or survival boundaries.

***Laminaria hyperborea* (Gunnerus) Foslie (Laminariales, Heterokontophyta) (Figure 12.5B, Table 12.1)** The perennial exclusively subtidal kelp *L. hyperborea* is only present in the temperate eastern Atlantic to subarctic zone. The northern distribution limit of *L. hyperborea* at the 1–2 °C February SST was considered to represent a lethal boundary for gametophytes and young sporophytes by van den Hoek (1982b). Later, it became evident that *L. hyperborea* survives at least two weeks at 2 °C or less (tom Dieck 1992, 1993), but has a reduced sporophyte growth rate of 13% of the maximum rate (tom Dieck 1992) and a retarded fecundity of gametophytes at 0 °C (tom Dieck 1989, Sjøtun and Schoschina 2002). Thus, cold winter conditions determine the limit for sufficient growth and recruitment but are not lethal. This is further substantiated by the observation that recruitment rate decreases significantly with increasing latitude (Rinde and Sjøtun 2005). In future, we expect the species to advance to the west coast of Spitsbergen and perhaps to southern Greenland.

Table 12.1 Temperature demands of the Arctic endemic/cold-temperate species *Laminaria solidungula* (*LamSol*), *Saccorhiza dermatodea* (*SacDer*), *Saccharina latissima* (*SacLat*) and *Devaleraea ramentacea* (*DevRam*), the northern-hemispheric cold-temperate species *Laminaria hyperborea* (*LamHyp*) and *Chondrus crispus* (*ChoCri*), the Antarctic endemic species *Himantothallus grandifolius* (*HimGra*), *Palmaria decipiens* (*PalDec*), *Desmarestia antarctica* (*DesAnt*), *Myriogramme manginii* (*MyrMan*), the Antarctic cold-temperate species *Iridaea cordata* (*IriCor*), *Chordaria linearis* (*ChoLin*), and the southern-hemispheric cold-temperate species *Desmarestia muelleri* (*DesMuel*), and *Lessonia vadosa* (*LesVad*).

Species	Life stage	Process	Temperature (°C) Optimum	Temperature (°C) Range	LTL (°C)	UTL (°C)	PPR/comment	Ecotype	References
Arctic cold-temperate species									
LamSol	Spore	Germination		2–12		<17		ND	1–8
	Gametophytes	Survival			−1.5	19–20			
		Fertility		0–5			SD (1–8 h)		
	Sporophyte	Growth	5–10	0–15	≤0	16			
		Sporogenesis	ND	5	ND	ND	SD		
SacDer	Gametophytes[1]	Survival			< −1.5	21	18 months darkness	ND	1, 7, 9–12
		Fertility	5–10	0–10	<0	<13	SD (<12 h)		
	Sporophyte	Formation	$3–5^2$	<3–10		$<17^2$			
		Survival			−1	$<17^2$			
		Growth	$5–10^2$	<3–10	$<3^2$	$<17^2$			
		Sporogenesis				≥12	SD (<12 h)		
DevRam	Gametophytes/ tetrasporophyte	Survival Canadian isolates	5–10	<−5–22	<−5	$20^{3\ \text{months}}$		Yes	13–15
		Survival European isolates		<−5–20	<−5	$18^{3\ \text{months}}$			
		Survival Greenland isolates			ND	$19^{2\ \text{weeks}}$			
		Growth Canadian isolates	0–5	$-1.5–18^{1\ \text{week}}$	−2	$15\ (18)^{6\ \text{weeks}}$	Promoted in LD	Yes	
		Growth European isolates	5–10	$-1.5–18^{1\ \text{week}}$	−2	$15^{6\ \text{weeks}}$			
		Fertility Canadian isolates	0	0–10		10		Yes	
		Fertility European isolates	5	−2–8					
	Tetrasporophyte	Blade initiation					SD		
Saclat	Spore	Germination Spitsbergen isolate	12	2–12	$<0^{(\text{Norway})}$	<18		Yes	1, 3, 16–22
		Germination Helgoland isolate	2–18	ND		>18			
	Gametophytes	Survival			<−1.5	23–25	18 months darkness		
		Growth Helgoland isolate	10–19	0–21	$<0^{(\text{Norway})}$	>21			
		Fertility Helgoland isolate	10	5–18	$<0^{(\text{Norway})}$	>18			
	Sporophyte	Survival			ND	18–20–<23	SV	Yes	
		Growth diverse isolates^{young}	10–15	0–20	<0	23	SV		
		Growth IOM isolate^{adult}		ND	<5	17	SV		
		Sporogenesis Helgoland isolate	10–15	5–15	2	<20	SD		

Cold-temperate North-Atlantic species									
LamHyp	Spore	Germination	7–12	0–18	0	>18		No	1–4, 7, 16, 18, 19, 21, 23–25
	Gametophytes	Survival			−1.5	21–23			
		Growth	10–15	2–20	<2	21			
		Fertility	7–12	2–18	0	<21 (♀)	BL		
	Sporophyte	Formation	12	2–18	<2	>18			
		Survival			<0	15–20–<23	SV		
		Growth	10–17	0–20	0	>20–<23	SD/stop in LD		
		Sporogenesis					SD probable		
		Blade initiation	ND	−1.5–8	−1.5	<15	SD (≤12 h)		
ChoCri	Gametophytes	Fertility			ND	15–≤20			19, 20, 26–30
	Carpospore	Germination	5–15	5–20	ND	>20	LD (>12 h)		
	Tetrasporophyte	Survival			−5/−20[3]	ND		Yes	
		Growth	10–20	0–25	<0	28–30	SD		
		Sporogenesis	5–15			≥20	Promoted by SD		
Endemic Antarctic species									
HimGra	Gametophyte	Survival			ND	15–16		ND	6, 31–33
		Growth	0–10	0–10					
		Gametogenesis	0	ND			SD (<9)		
	Gametophyte	Survival			ND	15–16			
		Growth	0–10	0–15					
		Gametogenesis	0	ND			SD (<11)		
	Sporophyte	Survival			ND	11–13			
		Growth	0–5	0–5					
PalDec	Gametophyte	Survival			ND	16–17		ND	32
		Growth	0–5	0–10					
DesAnt	Monoecious gametophyte	Growth	0–10	0–10				ND	6, 32, 33
		Fertility	0–5	0–5			No		
	Sporophyte	Survival			ND	13			
		Growth	0–5	0–5					
MyrMan	Tetrasporophyte	Survival			ND	11		ND	34
		Growth	0–5	0–5					

Table continued on next page.

Table 12.1 (Continued)

Species	Life stage	Process	Temperature (°C) Optimum	Temperature (°C) Range	LTL (°C)	UTL (°C)	PPR/comment	Ecotype	References
Antarctic cold-temperate species									
IriCor	Gametophyte	Survival			ND	15–16		ND	35
		Growth	0–5	0–5					
ChoLin	Gametophyte	Survival			ND	19.9–20.5		ND	36, 37
		Fertility			ND	12	SD (≤8)		
Southern-hemispheric cold-temperate species									
DesMuel	Monoecious gametophyte	Survival			−2	22–25.7		No	35, 37–39
		Growth	10–20	0–20					
		Fertility			−2	15			
	Sporophyte	Survival			ND	22–23			
		Growth	10–20	5–20					
LesVad	♂ Gametophyte	Survival			ND	19.9		No	36
		Fertility			−2	12			
	♀ Gametophyte	Survival			ND	21.4			
		Fertility			−2	12			

1 *Saccorhiza dermatodea* has monoecious gametophytes and needs short daylengths for gametogenesis according to Henry (1987).
2 Norton (1977) reported dioecious gametophytes for *Saccorhiza dermatodea* which was disproved by Henry (1987). As adult specimens of *S. dermatodea* are not confusable with other kelp species, the reported dioecious gametophytes, especially the male gametophyte with its high temperature tolerance, probably was a contamination. However, the temperature reaction pattern of the female gametophyte is in the range expected for an Arctic to cold-temperate species. Thus the data of Norton (1977) are used here with some caution. The other argument why Henry (1987) did not believe in Norton's gametophytes was the fact that Norton's gametophytes did not need short daylengths for induction of gametogenesis. This may be explainable by the fact that Norton collected his plants in March when spores probably were already shortday-induced. A similar phenomenon has been observed in *Laminaria solidungula* where cultured gametophytes needed short daylengths for induction of gametogenesis (tom Dieck 1989) but not freshly released meiospores.
3 Survival by vegetative crusts.

References: 1) tom Dieck 1993; 2) tom Dieck 1989; 3) Bolton and Lüning 1982; 4) tom Dieck 1992; 5) I. Bartsch, unpublished data; 6) C. Wiencke, unpublished data; 7) Wiencke et al. 1994; 8) M.Y. Roleda and F.S. Steinhoff, unpublished data; 9) Breeman 1988; 10) Henry 1987; 11) Norton 1977; 12) Keats and South 1985; 13) Bischoff and Wiencke 1993; 14) Breeman and Pakker 1994; 15) Novaczek et al. 1990; 16) Müller et al. 2008; 17) Lüning 1988; 18) Lüning 1980; 19) Lüning 1984; 20) Fortes and Lüning 1980; 21) Sjøtun and Schoschina 2002; 22) Lüning 1975; 23) van den Hoek 1982a; 24) Lüning 1986; 25) Lüning 1982; 26) van den Hoek 1982b; 27) Dudgeon et al. 1989; 28) Dudgeon et al. 1990; 29) Chen and McLachlan 1972; 30) Bird et al. 1979; 31) Wiencke and Clayton 1990; 32) Wiencke and tom Dieck 1989; 33) Wiencke et al. 1991; 34) Bischoff-Bäsmann and Wiencke 1996; 35) Wiencke and tom Dieck 1990; 36) Peters 1992a,b; 37) Peters and Breeman 1993; 38) Ramirez and Peters 1992; 39) Stolpe et al. 1991.

BL, blue light; IOM, Isle of Man; LD, long days; LTL, lower temperature limit; ND, no data; PPR, photoperiodic response; SD, short days; SV, seasonal variation; UTL, upper temperature limit.

The southern distribution boundary of *Laminaria hyperborea* is determined by a winter reproduction (initiation of new blades) and a concurrent summer lethal limit aligning with the 14 °C winter and 20 °C summer SST (Breeman 1988, 1990). In addition, recent evidence suggests that the induction of sporogenesis might be a crucial process limiting distribution. Probably, sporogenesis needs short daylengths, as in other autumn to winter reproducing kelps (Lüning 1988, tom Dieck 1991), and temperatures below 18 °C. Reproduction in kelps is expected to take place several degrees below the upper survival limit of the respective life history phase, as has been shown for *L. digitata* sporogenesis and kelp gametogenesis in general (tom Dieck 1993, tom Dieck and Oliveira 1993, Bartsch et al. 2004). The actual southern boundary is located at the northwestern Iberian Peninsula aligning with the 20 °C August SST. In future, this isotherm will shift further to the north than the 14 °C winter SST. Thus, the distribution will be determined by summer survival and will probably lead to the loss of this habitat-forming key species from coastlines along the whole Iberian Peninsula and most of the French, southern British, Belgian, Dutch, German and Skagerrak coasts, as has already been predicted by Breeman (1990).

***Chondrus crispus* Stackhouse (Gigartinales, Rhodophyta) (Figure 12.5C, Table 12.1)** *C. crispus* is a perennial low intertidal to subtidal cartilagineous red algal species with an amphi-Atlantic cold-temperate to sub-arctic distribution. In the northeastern Atlantic, there is a distributional northern limit at the 3 °C February SST. In the western Atlantic, the northern distribution limit lies north of the 1 °C February SST. As *C. crispus* survives prolonged periods of frost during tidal emersion (but not ice-scouring) and persists through vegetative re-growth from crusts (Dudgeon et al. 1990), the northern boundary of *C. crispus* may be determined by ice-free zones in winter. Summer conditions in the north are characterized by the 9–10 °C August SST, which is neither lethal nor inhibiting for growth or reproduction. Future northward expansion will not be considerable as only the southern tip of Spitsbergen will be colonized, but it is not anticipated that Greenland coastlines will be reached. In the western Atlantic, it will move to zones without ice-scouring in winter. Recent evidence from northern sites (Greenland: P.M. Pedersen, personal communication; Iceland: J. Karlsson, personal communication; Spitsbergen: C. Wiencke, personal communication) suggests no recent change at northern localities.

The southern distribution boundary of *Chondrus crispus* was regarded as a combined reproduction and lethal boundary corresponding to a 17 °C winter and a 24 °C summer SST (van den Hoek 1982a). As occurrence of *C. crispus* off northwestern Africa (Mauritania) is disjunct from southern Portuguese populations and explainable by upwelling (16 °C, van den Hoek 1982a), the southernmost Portuguese localities are more appropriate indicators for a temperature-dependent southern distribution boundary. This assumption is substantiated by the fact that the species becomes rare and plants remain small here (van den Hoek 1982a), an indication for a boundary population. This southernmost site corresponds to a 15 °C winter and a 20–21 °C summer SST. As locations with a 15 °C winter SST (or lower) are also present throughout most of the Mediterranean Sea, where *C. crispus* has never been recorded, the 20–21 °C summer SST seems to act as the effective boundary and also corresponds to the western Atlantic distribution pattern (Lüning 1990). The most probable delimiter is a southern reproduction boundary at the 20–21 °C August SST as fertility of gametophytes is restricted to long daylengths accompanied by temperatures <20 °C, representing pre-summer conditions at the southern limit. In future, the species will probably disappear from the Iberian Peninsula and retreat northwards to Brittany. In the western Atlantic, the species will only change its southern distribution pattern slightly.

Case studies of southern-hemispheric seaweed species

***Himantothallus grandifolius* (A. *et* E.S. Gepp) Zinova (Desmarestiales, Heterokontophyta) (Figure 12.6A, Table 12.1)** *H. grandifolius* is the largest perennial seaweed from the Southern Ocean. It is endemic to East and West Antarctica including South Georgia and grows at depths from about 7 down to 70 m (Wiencke and Clayton 2002).

The southern distribution limit at 72° S in Victoria Land is most certainly determined by the photoperiod, as in the related species *Desmarestia anceps* (Wiencke et al. 1996, C. Wiencke, unpublished data). On the east coast of the Antarctic Peninsula south of 66° S, ice conditions limit the distribution. The southern distribution limit of this species would not be affected by global climate change as temperature is irrelevant in the determination of this boundary but rather it is the photoperiod.

The northern boundary corresponds to the 4 °C winter isotherm as the sporophyte of the species only grows at temperatures ≤5 °C, and as juvenile sporophytes are formed in winter (Wiencke and tom Dieck 1989). In the light of our present knowledge about temperature demands of this species, its northern distribution will not change under the future conditions simulated here.

***Palmaria decipiens* (Reinsch) R.W. Ricker (Palmariales, Rhodophyta) (Figure 12.6A, Table 12.1)** *P. decipiens* is a pseudoperennial alga common throughout Antarctica from the southernmost hard bottom location within the Southern Ocean at Ross Island (77° S) to the Kerguelen Islands, Macquarie Island and the sub-Antarctic islands of New Zealand (Wiencke and Clayton 2002). On the east coast of the Antarctic Peninsula south of 66° S, ice conditions may limit the distribution of the species. In future, we do not expect major shifts of the southern distribution limit of the species with respect to the predicted temperature changes. The northern distribution limit of *P. decipiens* is most probably determined by the 4 °C winter isotherm as growth rates are optimal at temperatures ≤5 °C and approach zero at 10 °C (Wiencke and tom Dieck 1989). Under the simulated moderate warming scenario, the northern distribution limit of *P. decipiens* would not change its position.

***Desmarestia antarctica* R.L. Moe *et* P.C. Silva (Desmarestiales, Heterokontophyta) (Figure 12.6B, Table 12.1)** The annual species *D. antarctica* is endemic to the Antarctic Peninsula region between Anvers Island (65° S) and South Georgia (Wiencke and Clayton 2002). The exact southern boundary is not known due to incomplete surveys. According to the available data, this boundary is probably controlled by neither temperature nor daylength. Off the east coast of the Antarctic Peninsula, ice conditions may limit the distribution south of 66° S. Thus, the southern distribution boundary of this species would only change where ice will retreat by the end of this century.

Desmarestia antarctica cannot occur north of areas with winter temperatures ≥4 °C (like *Himantothallus grandifolius*) as higher temperatures would not allow sufficient growth of the sporophyte during the year (cited as *Desmarestia* sp. in Wiencke and tom Dieck 1989). The boundary is also determined by the temperature requirements for reproduction. Reproduction is only possible at temperatures ≤5 °C (Wiencke et al. 1991). A change of the northern distribution limit is not to be expected under the moderate IPCC global warming scenario.

***Myriogramme manginii* (Gain) Skottsberg (Ceramiales, Rhodophyta) (Figure 12.6B, Table 12.1)** This pseudoperennial species is endemic to the Antarctic Peninsula region between Anvers Island (Chapter 4; 65° S, Amsler et al. 1995, Hommersand et al. 2009) and South Georgia (Wiencke and Clayton 2002). The nature of the southern distribution limit of this species remains unclear. Off the east coast of the Antarctic Peninsula, the ice conditions certainly limit the distribution south of 66° S. The harsh ice-scouring conditions off the east coast of the Antarctic Peninsula might not significantly change until the end of the 21st century, thus a future expansion of *M. manginii* to the south is not very likely. Additionally, no adequate substrata would be available in the southern Weddell Sea.

The northern distribution boundary of the species corresponds to the 4 °C winter isotherm. The species grows between 0 and 5 °C, but not at 10 °C and exhibits an upper temperature limit (UTL) of 12 °C (Bischoff-Bäsmann and Wiencke 1996). Taking these temperature demands into account, the species could at present occur even further north, e.g., on Heard Island or on the Kerguelen Islands, where temperatures vary seasonally between +2.5 and +5.6 °C (20C3M simulation). This indicates incomplete surveys or the presence of reproductive or photoperiodic boundaries. Considering the moderate global warming scenario and the corresponding predicted temperatures, the northern distribution of *Myriogamme manginii* should not change by the end of the present century.

***Iridaea cordata* (Turner) Bory (Gigartinales, Rhodophyta) (Figure 12.6C, Table 12.1)** This annual to biannual species is common throughout Antarctica from 77° S in the Ross Sea to South Georgia, Heard Island, Kerguelen, Macquarie Island, the Falklands, Crozet and southern Tierra del Fuego. It is absent on coasts lacking suitable substrata, e.g., the east coast of the Antarctic Peninsula south of 66° S and the Weddell Sea (Wiencke and Clayton 2002). In future, we expect no major changes in the distribution along the coasts of the Antarctic continent.

Sterile gametophytes raised from tetraspores isolated on King George Island grew only at temperatures ≤5 °C and not at 10 °C (Wiencke and tom Dieck 1990). It is assumed that the Antarctic isolate still grows adequately at 8 °C as the UTL is much higher, i.e., 15–16 °C (Wiencke and tom Dieck 1990). Hence, the northern distribution limit of the isolate tested may imply a growth boundary at the 7 °C summer isotherm. This isotherm includes all populations of the species on South Georgia, and Heard Island, but excludes the population on southern Tierra del Fuego, the Falklands and Crozet (46° S, 51° E). The latter populations of the species may represent ecotypes with a different temperature growth pattern. Under the global warming scenario, the northern distribution limit of the isolate tested may not change, although that prediction is highly uncertain due to the fact that no data on the temperature demands for reproduction are available.

***Chordaria linearis* (Hooker *et* Harvey) Cotton (Chordariales, Heterokontophyta) (Figure 12.6D, Table 12.1)** The annual species *C. linearis* has been recorded in

Antarctica only on Anvers Island (Antarctic Peninsula, 65° S) and on King George Island (South Shetland Islands). However, its main distribution area is in South America along the Pacific coast from 42° S to Cape Horn (56° S) and to northern Patagonia on the Atlantic coast (40°50′ S, Peters 1992b, Peters and Breeman 1993, Wiencke and Clayton 2002). The nature of the southern distribution limit in Antarctica is not clear due to insufficient surveys of this species in Antarctica where it is rare.

On the Pacific coast of South America, the northern boundary is determined by the temperature demands of gametogenesis, as gametes are formed only during short days in winter at temperatures ≤10 °C (Peters 1992a). This should correspond approximately to the 9 °C austral winter isotherm. However, the observed distribution of *Chordaria linearis* in South America correlates with the 10 °C austral winter isotherm. This difference is explained by the fact that the strain used by Peters and Breeman (1993) probably represents a low-temperature ecotype as it came from 54° S, a location very distant from the northern boundary at 42° S. Under the simulated future temperature changes, the northern distribution limit of the species will be shifted from 42° S to 49° S on the Pacific coast of South America. On the Atlantic coast of South America, *C. linearis* populations will disappear between 41° S and 47° S. In addition, the lack of suitable substrata limits the presence of the species along this coastline.

Desmarestia muelleri **Ramirez *et* Peters (Desmarestiales, Heterokontophyta) (Figure 12.7A, Table 12.1)** This annual species occurs off South Georgia, the Falklands, on the Pacific coast of South America up to 33°55′ S, around Tierra del Fuego and on the Atlantic coast of South America up to 47°45′ S. The gametophytes become fertile at −2 °C (Table 12.1; Peters and Breeman 1993) but *D. muelleri* is absent from Antarctica, probably due to the relatively high temperature demands of the sporophyte, which only grows at temperatures between 20 and 5 °C, but not at 0 °C (Wiencke and tom Dieck 1990). Hence, the southern distribution boundary is probably a growth boundary at the 4 °C winter isotherm. Under the future predicted temperature changes, the southern distribution limit of *D. muelleri* will not change.

The northern distribution boundary is determined by the temperature demands for gametogenesis. The gametophytes become fertile only at temperatures ≤15 °C, corresponding to the 13 °C winter isotherm. On the east coast of South America, the discrepancy between observed distribution and the 13 °C winter isotherm might be due to lack of substratum. As a consequence of the global climate change, *Desmarestia muelleri* will most likely disappear along the South American west coast from 33°55′ S to ~40° S by 2089–2099.

Lessonia vadosa **Searles (Laminariales, Heterokontophyta) (Figure 12.7B, Table 12.1)** This perennial species occurs between about 44° S at the Pacific coast around Cape Horn to 47°44′ S on the Atlantic coast and on the Falklands (Ramírez and Santelices 1991, Peters and Breeman 1993). Although the gametophytes become fertile at −2 °C (Peters and Breeman 1993), *L. vadosa* is absent from Antarctica, probably due to higher temperature demands for growth of the sporophyte, as in *Desmarestia muelleri* (see above). Hence, the southern distribution limit is probably determined by the 4 °C August isotherm. In future, the species will not migrate into the Antarctic Peninsula region.

The northern distribution limit of *Lessonia vadosa* is determined by the upper temperature limit for gametogenesis. The gametophytes only become fertile at temperatures ≤12 °C (Peters and Breeman 1993), corresponding to the 10 °C austral winter

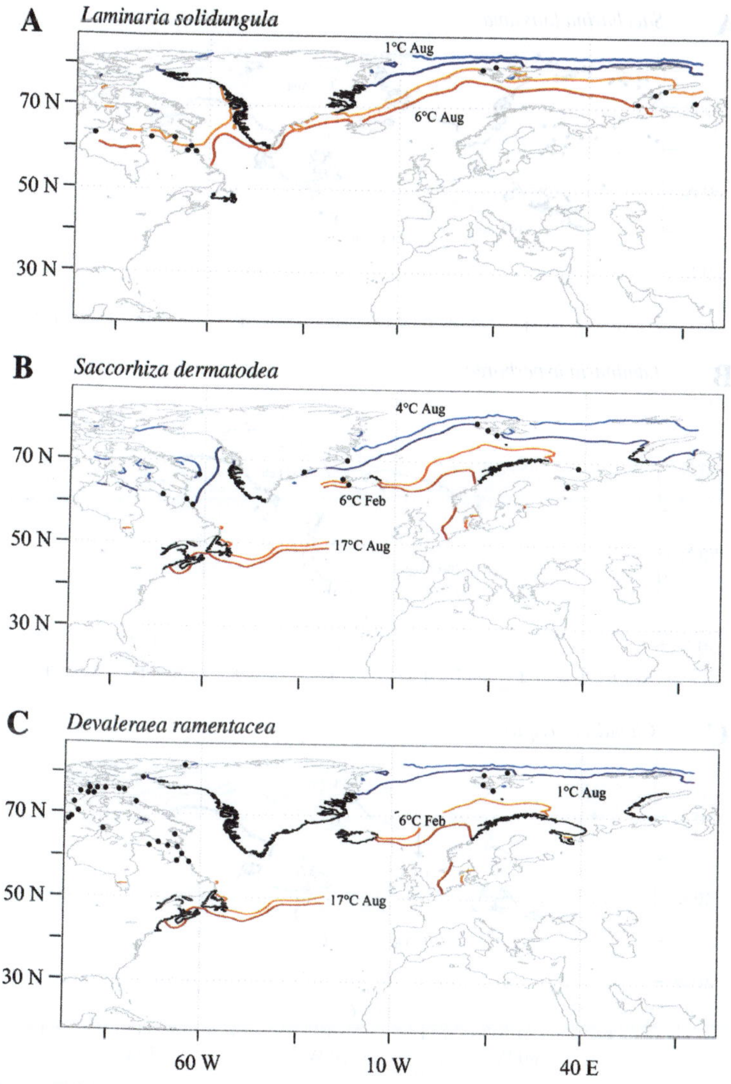

Figure 12.4 Geographic distribution and related sea surface temperature (SST) isotherms for 2080–2099 (summer SST: orange, winter SST: purple) relative to 1980–1999 (summer SST: dark red, winter SST: dark blue) of the endemic Arctic or predominantly Arctic species: (A) *Laminaria solidungula*, (B), *Saccorhiza dermatodea* and (C) *Devaleraea ramentacea*. The present distribution of seaweeds is highlighted by black coastlines or closed circles. The geographic distribution of the species was derived from van den Hoek (1982b), Hooper and Whittick (1984), Bischoff and Wiencke (1993), Lüning (1990). Aug, August; Feb, February.

isotherm. Off the east coast of South America, however, the species is absent from areas north of 48° S due to the absence of adequate substrata. Under the modelled scenario, the northern distribution limit of *L. vadosa* will certainly shift from 44° S to 49° S off the Pacific coast of South America. Off the Atlantic coast of South America, no changes are expected.

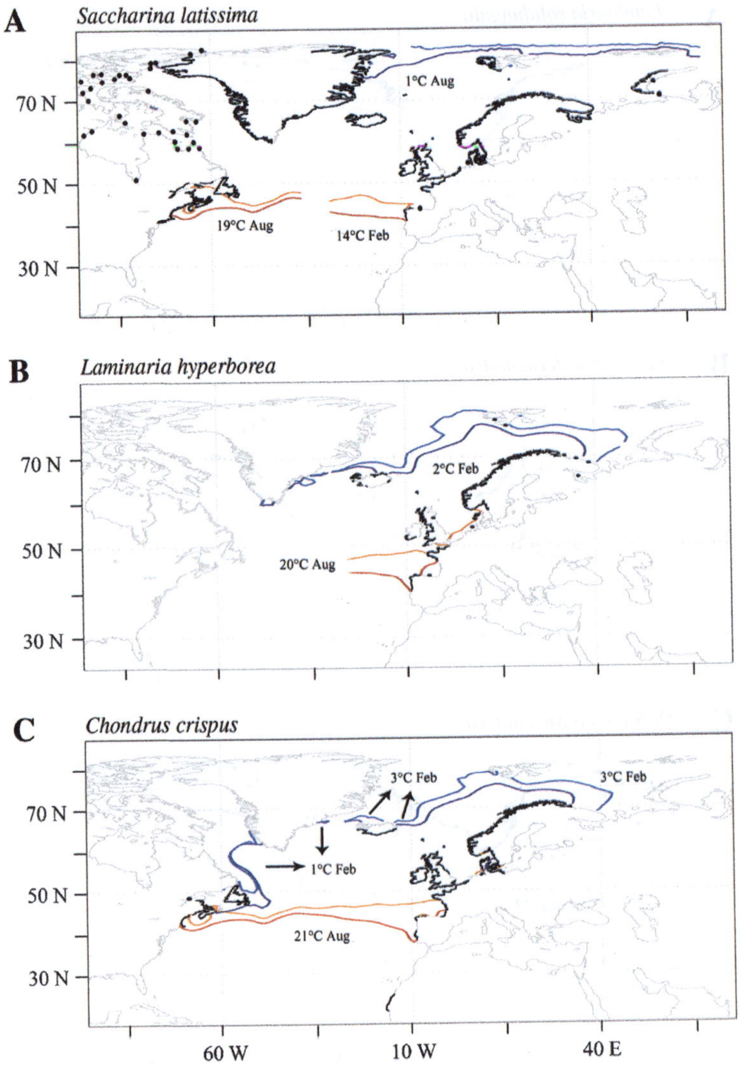

Figure 12.5 Geographic distribution and related sea surface temperature (SST) isotherms for 2080–2099 (summer SST: orange, winter SST: purple) relative to 1980–1999 (summer SST: dark red, winter SST: dark blue) of the Arctic- cold temperate species: (A) *Saccharina latissima*, (B) *Laminaria hyperborea*, and (C) *Chondrus crispus*.
The present distribution of seaweeds is highlighted by black coastlines or closed circles. The geographic distribution of the species was derived from van den Hoek (1982b), Hooper and Whittick (1984), Izquierdo et al. (1993), Lüning (1990), Schoschina (1997), Wiencke and Clayton (2002). Aug, August; Feb, February.

General trends in future seaweed distribution

Polar seaweed distribution is generally restricted to rocky habitats without ice coverage for at least 4–6 weeks during the polar summer (Lüning 1990). Accordingly, the

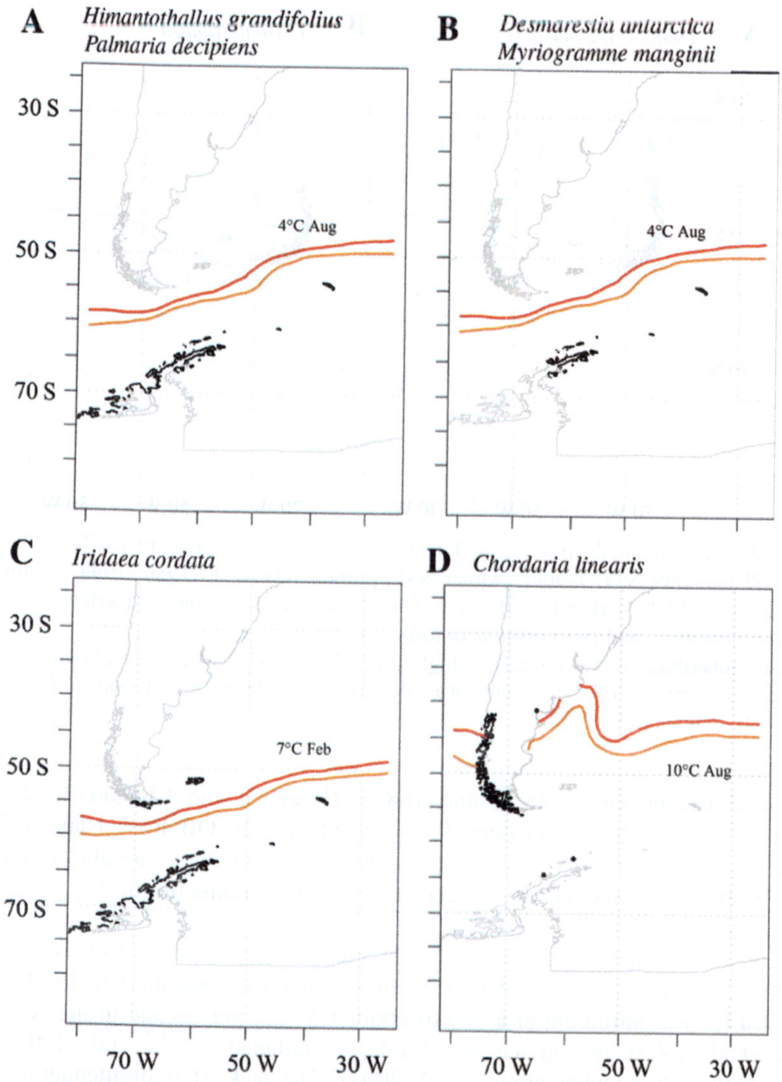

Figure 12.6 Geographic distribution and related sea surface temperature (SST) isotherms for 2080–2099 (summer SST: orange, winter SST: purple) relative to 1980–1999 (summer SST: dark red, winter SST: dark blue) of the endemic Antarctic and Antarctic – cold temperate species: (A) *Himantothallus grandifolius*, *Palmaria decipiens*, (B) *Desmarestia antarctica*, *Myriogramme manginii*, (C) *Iridaea cordata* and (D) *Chordaria linearis*.
The present distribution of seaweeds is highlighted by black coastlines or closed circles. The geographic distribution of the species was derived from Ricker (1987), Wiencke and tom Dieck

actual northern Arctic boundary for seaweeds reaches 80° N but excludes most of the Russian Arctic coastlines, which are mainly composed of soft bottoms and hence do not provide adequate substratum for seaweeds (Widdowson 1971). Seaweeds in Antarctica are restricted to the rocky coastlines of the Antarctic Peninsula and the East

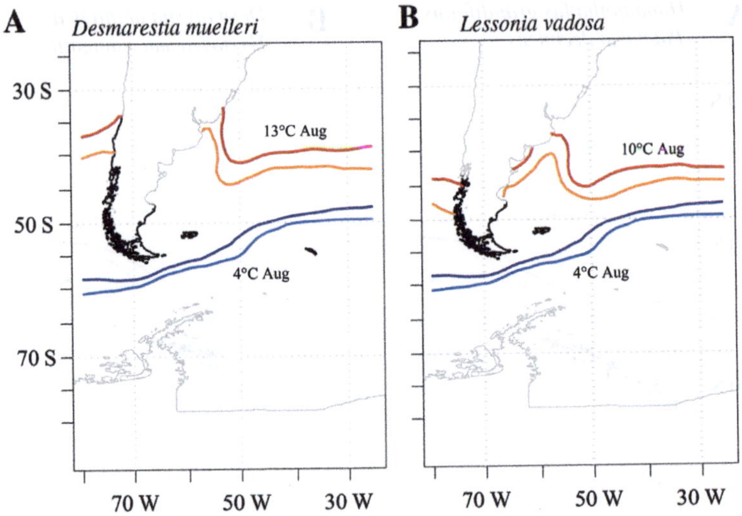

Figure 12.7 Geographic distribution and related sea surface temperature (SST) isotherms for 2080–2099 (summer SST: orange, winter SST: purple) relative to 1980–1999 (summer SST: dark red, winter SST: dark blue) of exclusively cold temperate South American species: (A) *Desmarestia muelleri*, and (B) *Lessonia vadosa*.
The present distribution of seaweeds is highlighted by black coastlines or closed circles. The geographic distribution of the species was derived from Peters and Breeman (1993). Aug, August; Feb, February.

Antarctic coasts between ~45° E and 168° E (Wiencke and Clayton 2002). In the Weddell-Sea, the water is too deep for seaweed growth. Off Ross Island (77°40′ S, 168°00' E), only three red algae and one green alga occur; these are able to cope with about 2 m thick, opaque ice cover occurring for ~10 months during the year (Lüning 1990).

Currently, the sea ice is thinning and its spatial distribution is decreasing very rapidly due to increasing winter SST in Arctic regions. Between the late 1970s and late 1990s, sea ice was shrinking at a rate of about 1.5–2% per decade in the Arctic (e.g., Serreze et al. 2007) and an ice-free Arctic is predicted by the end of the present century at the latest (Johannessen et al. 2004). The lack of light-attenuating sea ice cover will certainly result in new habitats available for seaweed colonization. In particular, the rocky Arctic islands off the Russian mainland, as well as the rocky coastlines of Kola Peninsula, Spitsbergen, Greenland (Baffin Bay and Greenland Sea) and the Canadian Archipelago will experience milder environmental conditions and hence may provide new habitats for retreating polar and progressing temperate seaweeds. Dominant Arctic/cold-temperate species such as *Laminaria solidungula*, *Saccorhiza dermatodea, Devaleraea ramentacea* and *Saccharina latissima* may occupy higher latitudes, in particular along coastlines off Greenland, the Canadian Archipelago, and northern Spitsbergen by 2080–2099 (Figures 12.4 and 12.5A) where rocky substratum becomes available. Arctic species with a southern distribution limit set by winter temperatures between 5 and 7 °C, such as *Saccorhiza dermatodea* and *Devaleraea ramentacea*, will experience considerable northward shifts in the northeastern Atlantic along the Norwegian coastline (Figure 12.4B, 12.4C) as spacing of winter isotherms is especially wide there (Figure 12.2). Some species (e.g. *S. dermatodea*) will additionally colonize the northern part of Novaya Zemlya, and all coasts around Spitsbergen (Figure 12.4B). The northern boundary of cold-temperate species will move northward into the Arctic as higher winter temperatures will enable growth or reproduction during winter. This northward shift will probably not be very pronounced, as exemplified by the cold-temperate species *Laminaria hyperborea* or *Chondrus crispus*.

Similarly, the retreat of the Antarctic ice shelf, especially on the eastern side of the Antarctic Peninsula, may provide new rocky habitats for macroalgae (Clarke et al. 2007). Nevertheless, we do not predict major distributional changes of seaweeds along the coastline of the Antarctic Peninsula until the end of the present century due to anticipated minor increases in SST (Figure 12.1). Shallow marine ecosystems around the Antarctic Peninsula will still be affected by large numbers of persistent icebergs, which have a detrimental effect on the benthic biota. The western Antarctic ice sheet, for instance, had an annual decrease in area of 148 ± 21 km^3 between April 2002 and August 2005 (Velicogna and Wahr 2006). Most prominent was the break-off from the Larsen ice shelves A and B in 1995 and 2002 from the interior ice sheet, which released interior ice into the Antarctic Ocean (Scambos et al. 2004). Moreover, the annual temperature range itself will probably only have a limited effect on seaweed distribution as most brown algae with South American to Antarctic distribution patterns studied so far exhibit northern reproduction boundaries and no lethal boundaries (Peters and Breeman 1993). As shown for *Himantothallus grandifolius*, *Palmaria decipiens*, *Desmarestia antarctica*, and *Myriogramme manginii* expected elevated temperatures under the moderate IPCC scenario are within the temperature range of the key structural endemic Antarctic seaweeds with respect to their growth and survival requirements (Table 12.1, Figures 12.2–12.3). The cold-temperate species *Iridaea cordata*, whose distribution extends from the Ross Sea in East Antarctica to Tierra del Fuego and the Falklands in the cold temperate region, behaves similarly (Figure 12.6C).

The anticipated drastic increase in coastal seawater temperatures involves a high risk of species extinctions in Arctic and both cold-temperate regions. The A1B simulations for 2080–2099 (Figure 12.1) have shown that the phytobenthos off Novaya Zemlya Island, southern to middle Norway, northern Spain and Portugal in the eastern North Atlantic, Connecticut, New Brunswick, Nova Scotia, and Newfoundland in the western North Atlantic, and off Patagonia in the South Atlantic will be most affected by future SST increase. In particular the coastlines along the Iberian Peninsula (Portugal and northern Spain) will be very much affected by species regressions, while Brittany (France) may become a transition region for cold- and warm-temperate floristic elements. According to our model data, cold-temperate seaweed populations may disappear at their southern geographical boundaries in Spain and Portugal and move northwards up to Brittany as the examples of *Saccharina latissima*, *Laminaria hyperborea* and *Chondrus crispus* show (Figure 12.5, Table 12.1). In the northwestern Atlantic, isotherms are generally close together, what will also be the case in future (Figures 12.2–12.3). Hence, northward distributional shifts of seaweeds with summer lethal limits will not be very pronounced. This is exemplified by the southern distribution boundaries of *S. latissima* (Figure 12.5A) and *C. crispus* (Figure 12.5C) which are characterized by relatively high summer SST between 19 and 21 °C (Table 12.1).

These species will only move northward slightly from Cape Cod to approximately Nova Scotia and/or Newfoundland.

A geographical shift of marine species in response to temperature fluctuations has been documented in the past for plankton (e.g., Beaugrand and Reid 2003, Whitehouse et al. 2008), fish (e.g., Sabatés et al. 2006, Rijnsdorp et al. 2009), and invertebrates (e.g., Kröncke et al. 2001), as well as for diverse seaweed species (Lüning 1990, Hiscock et al. 2004, Mieszkowska et al. 2006, overview: Hawkins et al. 2008). Lima et al. (2007) reported a northward distributional shift of seven cold-temperate species between 1950 and 2006 along the Portuguese coastline in conjunction with a SST increase of 0.74 °C during this time. Amongst these species was *Chondrus crispus*, which retreated ~180 km north. However, on a regional scale, the cold-temperate species showed no particular migration trend, in contrast to all migrating warm-temperate species, which extended their ranges northwards (Lima et al. 2007). Similarly, our model data for the cold-temperate species *C. crispus* do not predict a considerable northward extension under the moderate IPCC scenario (Figure 12.5C). Unpublished data from northern sites confirm that no recent invasion of northern regions has occurred yet (Greenland: P.M. Pedersen, personal communication; Iceland: J. Karlsson, personal communication; Spitsbergen: C. Wiencke, personal communication). However, the predicted poleward shift of some cold-temperate species, such as the kelp *Laminaria hyperborea* (Figure 12.5B), seems to be in progress right now as this species has already been recorded in secluded fjords in southern Spitsbergen (Peltikhina 2002, Olsen et al. 2004). Furthermore, there are indications that the predicted retreat from the Iberian Peninsula due to high summer temperatures (Figure 12.5B; Breeman 1990) is in progress as well. Dramatic changes have taken place off the northern Iberian coastline (Asturias region), where the kelp forests disappeared in 2008 over a 150 km stretch of coastline. In two areas (Puerto de Vialvélez and Concha de Artedo) where *L. hyperborea* disappeared together with other abundant kelps, such as *Laminaria ochroleuca* Bachelot de la Pylaie and *Saccharina latissima* in 2008, *Cystoseira* species (mainly *C. baccata* (S.G. Gmelin) P.C. Silva) became dominant. There are indications that the summer upwelling system providing cool temperatures below 20 °C was intermittently disturbed (J. Rico, personal communication). However, the southernmost distribution of *L. hyperborea* in Portugal north of Lisbon (Baleal) has not changed since the 1960s (Lima et al. 2007; R. Raujo, personal communication).

During recent years, the cold-temperate to Arctic kelp *Saccharina latissima* has also exhibited drastic changes in abundance along the European coastline (Moy et al. 2003, Moy and Christie 2007, Pehlke and Bartsch 2008). Similarly, changing abundance patterns have been reported for other kelp species such as *Laminaria digitata* (Hudson) J.V. Lamouroux in Europe (Sivertsen 1997, Cosson 1999, Morizur 2001), but clear evidence for a correlation with changing temperatures is missing in both cases. For all other species discussed above, no information about ongoing distributional shifts is available. In the southern hemisphere, reports about the occurrence of cold-temperate species in Antarctic waters such as *Blidingia minima* (Nägeli *ex* Kützing) Kylin, *Ulva intestinalis* Linnaeus, *Petalonia fascia* (O.F. Müller) Kuntze, *Chordaria linearis*, and *Rhodymenia coccocarpa* (Montagne) M.J. Wynne (Clayton et al. 1997, Wiencke and Clayton 2002, Zacher et al. 2007, Amsler et al. 2009) could be the first sign of migrational shifts. With respect to migration into Antarctic waters, the Scotia Arc with South Georgia as a prominent stepping stone is probably an important migration route. This phenomenon has been observed for animals (Knox and Lowry 1978) and presumably *C. linearis* reached the Antarctic Peninsula via this route.

Deviation of seaweed distribution from isotherms

The existence of ecotypes, which have been described in diverse seaweed species (e.g., Breeman and Pakker 1994, Orfanidis and Breeman 1999), may interfere with the simple projection of future geographical shifts as presented above. As an example, the polar red alga *Devaleraea ramentacea* is one of the few seaweed species with pronounced ecotypic variation even between strains of one locality (Novaczek et al. 1990). Unknown ecotypic differentiation may explain why the distribution of *D. ramentacea* along the Icelandic coastline is not in accordance with the isotherms (Figure 12.4C).

Another example is the kelp *Saccharina latissima*, which extends north to the pack-ice border and thereby has truly Arctic populations, but the known temperature optima between 10 and 15 °C have been derived from cold-temperate populations (cf. Table 12.1). The broad distribution pattern coincides with a high degree of polymorphism and plastic physiological responses suggesting the existence of a large species complex and an opportunistic and fugitive growth strategy (Bartsch et al. 2008). The existence of temperature ecotypes was proposed by Lüning (1975), Gerard and DuBois (1988) and Müller et al. (2008). However, the hybridization experiments of Lüning (1975) and Müller et al. (2008) did not show the persistence of results in the second generation. It is very likely that hidden ecotypes exist in other seaweed species (e.g., *Palmaria decipiens*, *Iridaea cordata*) as well, which would modify present biogeographical distribution. However, a future rapid development of new ecotypes can be excluded, since this process probably takes thousands, if not even millions of years (Breeman and Pakker 1994, Clayton et al. 1997).

The interplay of photoperiodic responses and temperature demands may additionally complicate the forecast of distributional change, as exemplified by *Saccharina latissima* and *Saccorhiza dermatodea* (Figures 12.4B and 12.5A). If critical daylengths needed to initiate the next obligatory life history step do not coincide with the critical temperature demands, the sea surface isotherm alone is not able to explain distributional limits. In *S. latissima*, sporogenesis only takes place in short daylengths at 15 °C or lower (Lüning 1988), and induction of sori needs at least four weeks (A. Preisler and I. Bartsch, unpublished data). Thus, a cool short-day period between autumn to late winter is a prerequisite for induction of sporangia and, thereby, new recruits. At the currently observed southern distribution boundary in northern Spain the shortest daylengths of 9 h coincide with a February isotherm of 13–14 °C, while further south, where survival of sporophytes might still be possible, this coincidence disappears. Thus, the interplay of daylengths and temperature delimit the occurrence of this kelp and not critical temperatures alone.

Breeman (1988) suggested a composite southern boundary for *Saccorhiza dermatodea* with lethal limits in northeastern America, and a reproduction limit in western Europe, without being able to explain the absence of this species in Brittany. This could be another example of the complex interplay of temperatures with the photoperiod. In Brittany, the presumed lethal 17 °C August SST coincides with the 10 °C February SST (cf. Figure 12.2, 12.3, 12.4B). While February temperatures would still permit reproduction (Table 12.1), daylengths range between 9.5 h and 11 h in Brittany.

Henry (1987) reported predictable gametogenesis only in 8–10 h daylight regimes, while in a 12 h daylight regime gametogenesis was rare. This indicates a possible mismatch between photoperiod and required temperatures in southwestern Europe, which would account for the absence of this species. *S. dermatodea* is absent in southwestern Norway, where lethal summer temperatures coincide with low winter temperatures and short winter photoperiods, but the reasons for its absence are unknown.

Reproductive boundaries of seaweeds may be more variable than deduced from unifactorial experiments, which formed the basis of data used for our correlations (Table 12.1). The reproductive period and frequency of individual algal species are highly influenced by the combination of temperature, nutrient status, depth distribution and interspecific relationships, as has been shown for a *Macrocystis* forest (Dayton et al. 1999). Under nutrient-rich, cold-water conditions in La Niña years, *Macrocystis pyrifera* (Linnaeus) C. Agardh forms a dense surface canopy. Thereby, *M. pyrifera* strongly controlled growth and reproduction of the understory kelps *Pterygophora californica* Ruprecht and *Laminaria farlowii* Setchell via light limitation (Dayton et al. 1999). Under nutrient-stressed, warm-water conditions in El Niño years, however, *M. pyrifera* produced soral tissues significantly later, leading to a massive decline in the species' abundance. Due to consequently better light conditions, reproduction (sori development, soral cover) of the two understory algae *P. californica* and *L. farlowii* accelerated, and their abundance significantly increased in El Niño years (Dayton et al. 1999). Such complex effects have the potential to modify boundaries of seaweeds and make it difficult to exactly predict their distributional changes.

Besides the autecological constraints discussed above, local temperature conditions influence species distribution and thereby alter the projections of predicted future geographical distribution. Local deviations from mean annual SSTs may be most marked in enclosed bays and fjords or in shallow waters (Hiscock et al. 2004) or at sites with strong thermal stratification. The influence of local temperatures on distribution has been observed in the cold-temperate/Arctic kelp *Alaria esculenta* (Linnaeus) Greville near Plymouth (SW England). As the species can survive summer temperatures of 17 °C (Sundene 1962, Fredersdorf et al. 2009), it still grew at the offshore Eddystone Reef characterized by temperatures <17 °C during the intermediate warming period between 1950 and 1960, but disappeared from the nearby, but more protected and hence probably warmer coastline (Hiscock et al. 2004).

In areas with strong temperature stratification, subduction of species into colder waters complicates the correlation of thermal demands and isotherms. An example is the occurrence of stenothermic Arctic *Laminaria solidungula* in southeastern Canada. Although most of the perennial kelp populations of *L. solidungula* are truly Arctic, there are refuge populations in the Subarctic, for example within many Newfoundland fjords. Here, the species is either restricted to water masses close to freezing point that may reach relatively shallow depths (Hooper 1984) or has migrated into greater depths (30 m) where the water is close to 0 °C year round (A.R.O. Chapman, personal communication, Hooper and Whittick 1984, Lüning 1990). Thus, the 16 °C summer isotherm describing the southern distribution boundary (cf. Figure 12.1A and Figure 12.2A) does not indicate the southern distribution limit. Comparably, deep-water relict populations of temperate brown kelps have recently been detected in tropical waters around the Galapagos Archipelago (Graham et al. 2007) and are known from other subtropical deep waters and the Mediterranean Sea (overview in Bartsch et al. 2008).

Subduction below the thermocline at distribution boundaries has been observed in several other Arctic seaweeds like the branched coralline alga *Lithothamnion tophiforme* (Esper) Unger, the crustose coralline alga *Leptophytum laeve* (Foslie) Adey and its adelphoparasite *Kvaleya epilaeve* Adey *et* Sperapani, the brown kelp *Agarum clathratum* Dumortier as well as the red algae *Ptilota serrata* Kützing and *Turnerella pennyi* (Harvey) F. Schmitz (W. Adey, personal communication, Adey and Sperapani 1971, Adey and Steneck 2001). All these species not only occurred in greater depths at lower latitudes, but also reduced their abundance or thallus size towards their southern distribution limit (W. Adey, personal communication). *L. tophiforme*, for example, reaches an abundance of 29–40% cover estimated over the total coralline cover on the collected rock and shell samples (Adey and Steneck 2001) in 10–20 m depths in northern Labrador. Further south, in northeastern Newfoundland, the species only covers 6.7% of total coralline cover above the thermocline, but about 22% below the thermocline in depths >40 m (W. Adey, personal communication, Adey and Steneck 2001). Similar phenomena are also expected in Antarctica but have not been documented yet. Empirical assessments of abundance vs. thermal stress across species ranges have been rarely undertaken (Mieszkowska et al. 2005, Sagarin and Somero 2006), but will be a valuable tool to further assess temperature-related distributional changes (e.g., Adey and Hayek 2005) in future.

Finally, it should be noted, that adequate areas for subduction of seaweed species must provide both an acceptable temperature regime and sufficient light conditions. The predicted sea level rise (IPCC 2007a) will reduce light conditions in present-day deep water habitats and thereby affect lower algal depths limits. Nevertheless, on the whole, it is expected that sea level rise will not considerably increase or diminish the size of seaweed habitats/subduction zones unless rocky coastlines adjoin soft sediments.

Effects of sea surface temperature increase on phytobenthic communities

The prediction of seaweed distributional shifts by correlating temperature demands and oceanic isotherms provides a rough indication of future change and the general pattern of extinctions and invasions along specific coastlines. However, our analysis is not able to provide a reliable picture of the changing community structure under a global warming scenario as there are many interacting factors that are not yet understood or predictable (e.g., Beaugrand 2009). Long-term studies investigating the ecological consequences of increasing and long-lasting oceanic warming on phytobenthos are unfortunately scarce. Although there are many studies investigating the impact of the El Niño southern oscillation (ENSO) on marine benthic communities, they may not be useful for long-term projections. ENSO events seem to have a minor impact on the long-term community structure (Sagarin et al. 1999, Halpern and Cottenie 2007), and the future development of the ENSO phenomenon is unclear (Cane 2005).

A valuable case study which may help to generalize ecological consequences on phytobenthos under the impact of rapid and long-lasting SST increase, is an 18-year monitoring program carried out by Schiel et al. (2004). These authors investigated the intertidal and subtidal species composition and abundance at the temperate rocky Diablo Potter Cove, California, which was affected by the thermal outfall of a power-generating station over 10 years. As consequence of the rapid and long-lasting SST increase to ultimately 3.5 °C, remarkable changes occurred. In total, the algal richness was reduced by 40%. Additionally, the abundance of 54% of algae and 27% of the invertebrates fell by at least 50%. The initial dense cover of foliose algae and scattered grazing invertebrates in the intertidal was replaced by bare rocks or transient turf and slender algal crusts (Schiel et al. 2004). Thus, a complete change in the community structure was observed. Although the rate of temperature increase might be of importance for acclimatization or extinction of seaweeds, similar changes were observed after the breakdown of key structural brown seaweeds in the Californian intertidal following a 0.79–1.26 °C mean annual SST increase within 60 years (Sagarin et al. 1999). Once established, turf-forming, filamentous or other ephemeral algae may inhibit the reestablishment of canopy-forming macroalgae (Airoldi 1998). Turf communities and climax communities have been considered as alternate states under different stress regimes (Worm et al. 1999, Airoldi 2003, Connell 2005).

Therefore, under a moderate global warming scenario, the replacement of foliose or leathery perennial algae by turf species and algal crusts is likely to occur in intertidal zones of cold-temperate and polar transition areas. Hawkins et al. (2008) assume that intertidal rocky shores will be characterized by a reduced abundance of key structural fucoids under elevated temperatures in southwestern Europe (Britain and Brittany). As consequence they expect a drastic decline of biodiversity and of organic material transport into coastal systems.

Similarly, in the subtidal of the Californian Diablo Potter Cove, the dominant canopy-forming kelps were almost completely absent after 10 years of exposure to a SST elevation of 3.5 °C, which led to loss of species diversity (Schiel et al. 2004). Only the warm-temperate giant kelp *Macrocystis pyrifera*, which formerly had been rare, and an understory red alga profited extensively from the newly available substratum resources. However, these developments were accompanied by a delayed increase of sea urchin density (Schiel et al. 2004). Increasing sea urchin populations may result in a loss of the giant kelp in future as warmer SST enhances gonad production and larval growth of sea urchin species (Hart and Scheibling 1988, Ling et al. 2008).

Dense sea urchin populations typically destroy canopy-forming subtidal algae and transform macroalgal beds into barren grounds dominated by coralline crusts as has been observed worldwide (e.g., review: Steneck et al. 2002, Ling 2008, Ling and Johnson 2009). The conversion into barren grounds is directly associated with a local loss of biodiversity (Ling 2008). Once the transition to barren landscapes is established, the recolonisation of canopy-forming seaweeds is often inhibited for a long time period (Airoldi 1998, Steneck et al. 2002). The transition from barren grounds into macroalgal beds is usually triggered by the collapse of sea urchin populations due to the outbreak of diseases, at least in cold-temperate areas (summarized in Scheibling and Hennigar 1997).

Pathogens of sea urchins are not yet established in polar areas due to cold SST (Scheibling and Hennigar 1997). Sea-urchin pathogens are regularly transported to the North Atlantic west coast by warm-water influx, due to regional-scale oceanic events such as wind-driven coastal upwelling and cross-shelf advection of water masses (Scheibling and Hennigar 1997). As extreme weather events are forecasted to increase under greenhouse-forcing conditions (IPCC 2007a), a more frequent wind-driven transportation of sea urchin pathogens to higher latitudes and infestation of sea urchins in these regions may be assumed. The re-establishment of canopy-forming macroalgae at barren grounds would thus be enhanced in the respective polar regions.

Alternatively, re-colonisation of barren landscapes by native seaweeds could become hindered by highly competitive range-extending or invasive species, which may take over barren grounds as newly available habitats. For example, the invasive green alga *Codium fragile* subsp. *fragile* (Suringar) Hariot has recently prevented recolonisation of barren grounds by native kelp species off Nova Scotia by forming dense meadows (Scheibling and Anthony 2001, Scheibling and Gagnon 2006). This process was supported by the invasive encrusting bryozoan *Membranipora membranacea* Linnaeus, which regularly causes kelp defoliation and hence reduces competitive strength of kelp species (Schmidt and Scheibling 2007). Establishment of the invader *C. fragile* was enhanced by sea urchin food preference for kelps under increased water temperature and wave action, leading to increased herbivore pressure on local kelp stands (Lyons and Scheibling 2008). In the future, the introduction of invasive species in polar regions by shipping traffic may play an increasing role (Carlton and Hodder 1995, Clayton et al. 1997). Thus, native seaweeds have not only to withstand future temperature changes, but also increasing competition pressure by invading seaweeds, increasing risk of infection by algal diseases, and increasing herbivore pressure.

Summarizing, one may expect major ecological consequences of global warming on local phytobenthic communities in cold-temperate and polar environments. However, the multiple, complex and often unknown biotic relationships of cold-temperate and polar key species under elevated temperatures remain a challenging topic that should receive more attention in future.

Concluding remarks

There is a high uncertainty in forecasting future trends of seaweed distribution, since the current rate of anthropogenically induced global warming far exceeds the rate of climate warming during the last 10,000 years (IPCC 2007b). Moreover, only tentative estimations about the actual changes at the ecosystem level can be made as interactions of abiotic and biotic factors under rapidly increasing and long-lasting oceanic warming conditions are little understood. In addition, temperature demands of early life history stages of key structural Antarctic seaweeds as well as their ecological/ecophysiological responses to multifactorial stress are mostly unknown, which may lead to incorrect conclusions for their future distribution. Lastly, the mismatch of formerly synchronized reproductive events and the impact of altered food quality for herbivores under climate warming are not fully understood. Indisputably, the expected retreat and progression of macroalgal key species as so called "ecological engineers" will provoke substantial and cascading effects in cold-temperate and polar transition areas with substantial consequences for biodiversity and whole ecosystem functioning.

References

Adey, W.H. and L.-A.C. Hayek. 2005. The biogeographic structure of the western North Atlantic rocky intertidal. *Cryptogam. Algol. 26*: 35–66.

Adey, W.H. and C.P. Sperapani. 1971. The biology of *Kvaleya epilaeve*, a new parasitic genus and species of Corallinaceae. *Phycologia 10*: 29–42.

Adey, W.H. and R.S. Steneck. 2001. Thermogeography over time creates biogeographic regions: a temperature/space/time-integrated model and an abundance weighted test for benthic marine algae. *J. Phycol. 37*: 677–698.

Airoldi, L. 1998. Roles of disturbance, sediment stress, and substratum retention on spatial dominance in algal turf. *Ecology 79*: 2759–2770.

Airoldi, L. 2003. Effects of patch shape in intertidal algal mosaics: roles of area, perimeter and distance from edge. *Mar. Biol. 143*: 639–650.

Amsler, C.D., R.J. Rowley, D.R. Laur, L.B. Quetin and R.M. Ross. 1995. Vertical distribution of Antarctic Peninsular macroalgae: cover, biomass and species composition. *Phycologia 34*: 424–430.

Amsler, C.D., M.O. Amsler, J.B. McClintock and B.J. Baker. 2009. Filamentous algal endophytes in macrophytic Antarctic algae: prevalence in hosts and palatability to mesoherbivores. *Phycologia 48*: 324–334.

Barnes, D. and L. Peck. 2008. Vulnerability of Antarctic shelf biodiversity to predicted regional warming. *Clim. Res. 37*: 149–163.

Bartsch, I., K. Lüning and S. Pang. 2004. Kelp reproduction: environmental control of sporogenesis. *Abstracts 18th International Seaweed Symposium*, Bergen, Norway, 20–25 June: 43.

Bartsch, I., C. Wiencke, K. Bischof, C.M. Buchholz, B.H. Buck, A. Eggert, P. Feuerpfeil, D. Hanelt, S. Jacobsen, R. Karez, U. Karsten, M. Molis, M.Y. Roleda, H. Schubert, R. Schumann, K. Valentin, F. Weinberger and J. Wiese. 2008. The genus *Laminaria sensu lato*: recent insights and developments. *Eur. J. Phycol. 43*: 1–86.

Beaugrand, G. 2009. Decadal changes in climate and ecosystems in the North Atlantic Ocean and adjacent seas. *Deep Sea Res. 56*: 656–673.

Beaugrand, G. and P.C. Reid. 2003. Long-term changes in phytoplankton, zooplankton and salmon related to climate. *Glob. Change Biol. 9*: 801–817.

Bird, N.L., L.C.-M. Chen and J. McLachlan. 1979. Effects of temperature, light and salinity on growth in culture of *Chondrus crispus, Furcellaria lumbricalis, Gracilaria tikvahiae* (Gigartinales, Rhodophyta), and *Fucus serratus* (Fucales, Phaeophyta). *Bot. Mar. 22*: 521–527.

Bischoff, B. and C. Wiencke. 1993. Temperature requirements for growth and survival of macroalgae from Disko Island (Greenland). *Helgol. Meeresunters. 47*: 167–191.

Bischoff-Bäsmann, B. and C. Wiencke. 1996. Temperature requirements for growth and survival of Antarctic Rhodophyta. *J. Phycol. 32*: 525–535.

Bolton, J.J. and K. Lüning. 1982. Optimal growth and maximal survival temperatures of Atlantic *Laminaria* species (Phaeophyta) in culture. *Mar. Biol. 66*: 89–94.

Breeman, A.M. 1988. Relative importance of temperature and other factors in determining geographic boundaries of seaweeds: experimental and phenological evidence. *Helgol. Meeresunters. 42*: 199–241.

Breeman, A.M. 1990. Expected effects of changing seawater temperatures on the geographic distribution of seaweed species. *In*: (J.J. Beukema et al., eds). *Expected effects of climatic change on marine coastal ecosystems*. Kluwer Academic, Netherlands. pp. 69–76.

Breeman, A.M. and H. Pakker. 1994. Temperature ecotypes in seaweeds: adaptive significance and biogeographic implications. *Bot. Mar. 37*: 171–180.

Cane, M.A. 2005. The evolution of El Niño, past and future. *Earth. Planet. Sci. Lett. 164*: 1–10.

Carlton, J.T. and J. Hodder. 1995. Biogeography and dispersal of coastal marine organisms: experimental studies on a replica of a 16th century sailing vessel. *Mar. Biol. 121*: 721–730.

Chen, L.C.-M. and J. McLachlan. 1972. The life history of *Chondrus* in culture. *Can. J. Bot. 50*: 1055–1060.

Clarke, A., J. Murphy, M.P. Meredith, J.C. King, L.S. Peck, D.K.A. Barnes and R.C. Smith. 2007. Climate change and the marine ecosystem of the western Antarctic Peninsula. *Phil. Trans. R. Soc. B. 362*: 149–166.

Clayton, M.N., C. Wiencke and H. Klöser. 1997. New records of temperate and sub-Antarctic marine benthic macroalgae from Antarctica. *Polar Biol. 17*: 141–149.

Connell, S.D. 2005. Assembly and maintenance of subtidal habitat heterogeneity: synergistic effects of light penetration and sedimentation. *Mar. Ecol. Prog. Ser. 289*: 53–61.
Cosson, J. 1999. Sur la disparition progressive de *Laminaria digitata* sur les côtes du Calvados (France). *Cryptogam. Algol. 20*: 35–42.
Dayton, P.K., M.J. Tegner, P.B. Edwards and K.L. Riser. 1999. Temporal and spatial scales of kelp demography: the role of oceanographic climate. *Ecol. Monogr. 69*: 219–250.
Dring, M.J. 1982. *The biology of marine plants*. Arnold, London. pp. 199.
Dudgeon, S.R., I.R. Davison and R.L. Vadas. 1989. Effect of freezing on photosynthesis in intertidal macroalgae: relative tolerances of *Chondrus crispus* and *Mastocarpus stellatus* (Rhodophyta). *Mar. Biol. 101*: 107–114.
Dudgeon, S.R., I.R. Davison and R.L. Vadas. 1990. Freezing tolerance in the intertidal red algae *Chondrus crispus* and *Mastocarpus stellatus*: relative importance of acclimation and adaptation. *Mar. Biol. 106*: 427–436.
Dunton, K.H. 1985. Growth of dark-exposed *Laminaria saccharina* (L.) Lamour. and *Laminaria solidungula* J. Ag. (Laminariales, Phaeophyta) in the Alaskan Beaufort Sea. *J. Exp. Mar. Biol. Ecol. 94*: 181–189.
Dunton, K.H. 1990. Growth and production in *Laminaria solidungula*: relation to continuous underwater light levels in the Alaskan High Arctic. *Mar. Biol. 106*: 297–304.
Fortes, M.D. and K. Lüning. 1980. Growth rates of North Sea macroalgae in relation to temperature, irradiance and photoperiod. *Helgol. Meeresunters. 34*: 15–29.
Fredersdorf J, Müller R, Becker S, Wiencke C and K. Bischof. 2009. Interactive effects of radiation, temperature and salinity on different life history stages of Arctic kelp *Alaria esculenta* (brown algae). *Oecologia 160*: 483–492.
Gerard, V.A. and K.R. DuBois. 1988. Temperature ecotypes near the southern edge boundary of the kelp *Laminaria saccharina*. *Mar. Biol. 97*: 575–580.
Gómez, I., M.Y. Roleda, K. Dunton, A. Wulff, U. Karsten and C. Wiencke. 2009. Light and temperature demands of marine benthic microalgae and seaweeds in polar regions. *Bot. Mar. 52*: 593–608.
Graham, M.H., B.P. Kinlan, L.D. Druehl, L.E. Garske and S. Banks. 2007. Deep-water kelp refugia as potential hotspots of tropical marine diversity and productivity. *Proc. Natl. Acad. Sci. USA 104*: 16576–16580.
Halpern, B.S. and K. Cottenie. 2007. Little evidence for climate effects on local-scale structure and dynamics of California kelp forest communities. *Glob. Change Biol. 13*: 236–251.
Hanelt, D., K. Bischof and C. Wiencke. 2004. The radiation, temperature and salinity regime in Kongsfjorden. *Ber. Polarforsch. Meeresforsch. 492*: 14–25.
Hart, M.W. and R.E. Scheibling. 1988. Heat waves, baby booms, and the destruction of kelp beds by sea urchins. *Mar. Biol. 99*: 167–176.
Hawkins, S.J., P.J. Moore, M.T. Burrows, E. Poloczanska, N. Mieszkowska, R.J.H. Herbert, S.R. Jenkins, R.C. Thompson, M.J. Genner and A.J. Southward. 2008. Complex interactions in a rapidly changing world: responses of rocky shore communities to recent climate change. *Clim. Res. 37*: 123–133.
Henry, E.C. 1987. Primitive reproductive characters and a photoperiodic response in *Saccorhiza dermatodea* (Laminariales, Phaeophyceae). *Br. Phycol. J. 22*: 23–31.
Hiscock, K., A. Southward, I. Tittley and S. Hawkins. 2004. Effects of changing temperature on benthic marine life in Britain and Ireland. *Aquat. Conserv. Mar. Freshw. Ecosyst. 14*: 333–362.
Hommersand, M.H., R.L. Moe, C.D. Amsler and S. Fredericq. 2009. Notes on the systematics and biogeographical relationships of Antarctic and sub-Antarctic Rhodophyta with descriptions of four new genera and five new species. *Bot. Mar. 52*: 509–534.
Hooper, R.G. 1984. Functional adaptations to the polar environment by the Arctic kelp, *Laminaria solidungula*. *Br. Phycol. J. 2*: 194.
Hooper, R.G. and A. Whittick. 1984. The benthic marine algae of the Kaipokok Bay, Makkovik Bay and Big River Bay region of the central Labrador coast. *Naturalist can. (Rev. Ecol. Syst.) 111*: 131–138.

Hulme, M., G.J. Jenkins, X. Lu, J.R. Turnpenny, T.D. Mitchell, R.G. Jones, J. Lowe, J.M. Murphy, D. Hassell, P. Boorman, R. McDonald and S. Hill. 2002. *Climate change scenarios for the United Kingdom: the UKCIP02 scientific report*. Tyndall Centre for Climate Change Research, School of Environmental Sciences, University of East Anglia, Norwich, UK. pp. 120.

IPCC [Intergovernmental Panel on Climate Change; core writing team: R.K. Pachauri and A. Reisinger, eds] 2007a. *Climate change 2007: synthesis report. Contribution of Working Groups I, II and III to the 4th assessment report of the Intergovernmental Panel on Climate Change*. IPCC, Geneva. pp. 104.

IPCC. 2007b. *Climate change 2007: the physical science basis. Contribution of Working Group I to the 4th Assessment Report of the Inter-governmental Panel on Climate Change* (S. Solomon, D. Qin, M. Manning, Z. Chen, M. Marquis, K.B. Averyt, M. Tignor and H.L. Miller, eds). Cambridge University Press, Cambridge, UK. pp. 996.

Izquierdo, J.L., M.J. Navarro and T. Gallardo. 1993. Mapas de distibución de algas marinas de la Península Ibérica. IV. *Laminaria ochroleuca* Pylaie, *L. hyperborea* (Gunner.) Foslie y *L. saccharina* (L.) Lamour. (Laminariales, Fucophyceae). *Botánica Complutensis 18*: 291–304.

Jensen, M.H. and I.R. Veland. 2006. A survey of marine dinoflagellates in the waters surrounding Disko Island, Western Greenland. *In:* (K.A. Halberg, ed.) *Arctic biology field course, Qeqertasuaq, 2006*. Arctic Station, University of Copenhagen. pp. 5–85.

Johannessen, O., L. Bengtsson, M. Miles, S. Kuzmina, V. Semenov, G. Alekseev, A. Nagurnyi, V. Zakharov, L. Bobylev and L. Pettersson. 2004. Arctic climate change: observed and modelled temperature and sea ice variability. *Tellus 56(A)*: 328–341.

Keats, D.W. and G.R. South. 1985. Aspects of the reproductive phenology of *Saccorhiza dermatodea* (Phaeophyta, Laminariales) in Newfoundland. *Br. Phycol. J. 20*: 117–122.

King, J., J. Turner, G. Marshall, W. Connolley and T. Lachlan-Cope. 2003. Antarctic Peninsula climate variability and its causes as revealed by analysis of instrumental records. *Antarctic Res. Ser. 79*: 17–30.

Knox, G.A. and J.K. Lowry. 1978. A comparison between the benthos of the Southern Ocean and the North Polar Ocean with special reference to the Amphipoda and the Polychaeta. *In*: (J.M. Dunbar, ed.) *Polar oceans*. Proceedings of the SCOR/SCAR Polar Ocean Conference, Montreal 1974. pp. 423–462.

Kröncke, I., B. Zeiss and C. Rensing. 2001. Long-term variability in macrofauna species composition off the island of Norderney (East Frisia, Germany) in relation to changes in climatic and environmental conditions. *Senckenb. Marit. 31*: 65–82.

Laepple, T., S. Jewson and K. Coughlin. 2008. Interannual temperature predictions using the CMIP3 multi-model ensemble mean. *Geophys. Res. Lett. 35 (L10701)*: doi:10.1029/2008GL033576.

Lee, J.A. and B.H. Brinkhuis. 1986. Reproductive phenology of *Laminaria saccharina* (L.) Lamour. (Phaeophyta) at the southern limit of its distribution in the northwestern Atlantic Ocean. *J. Phycol. 22*: 276–285.

Lima, F.P., P.A. Ribeiro, N. Queiroz, J.H. Hawkins and A.M. Santos. 2007. Do distributional shifts of northern and southern species of algae match the warming pattern? *Glob. Change Biol. 13*: 2592–2604.

Ling, S.D. 2008. Range expansion of a habitat-modifying species leads to loss of taxonomic diversity: a new and impoverished reef state. *Oecologia 156*: 883–894.

Ling, S.D. and C.R. Johnson. 2009. Population dynamics of an ecologically important range-extender: kelp beds versus sea urchin barrens. *Mar. Ecol. Prog. Ser. 374*: 113–125.

Ling, S.D., C.R. Johnson, S. Frusher and C.K. King. 2008. Reproductive potential of a marine ecosystem engineer at the edge of a newly expanded range. *Glob. Change Biol. 14*: 907–915.

Lüning, K. 1975. Kreuzungsexperimente an *Laminaria saccharina* von Helgoland und von der Isle of Man. *Helgol. wiss. Meeresunters. 27*: 108–114.

Lüning, K. 1980. Critical levels of light and temperature regulating the gametogenesis of three *Laminaria* species (Phaeophyceae). *J. Phycol. 16*: 1–15.

Lüning, K. 1982. Seasonality of larger brown algae and its possible regulation by the environment. *In*: (L.M. Srivastava, ed.) *Synthetic and degradative processess in marine macrophytes*. Walter de Gruyter, Berlin. pp. 47–67.

Lüning, K. 1984. Temperature tolerance and biogeography of seaweeds: the marine algal flora of Helgoland (North Sea) as an example. *Helgol. Meeresunters. 38*: 305–317.

Lüning, K. 1986. New frond formation in *Laminaria hyperborea* (Phaeophyta): a photoperiodic response. *Br. Phycol. J. 3*: 269–273.

Lüning, K. 1988. Photoperiodic control of sorus formation in the brown alga *Laminaria saccharina*. *Mar. Ecol. Prog. Ser. 45*: 137–144.

Lüning, K. 1990. *Seaweeds: their environment, biogeography and ecophysiology*. Wiley & Sons, New York. pp. 527.

Lyons, D.A. and R.E. Scheibling. 2008. Context-dependant survival of the invasive seaweed *Codium fragile* ssp. *tomentosoides* in kelp bed and urchin barren habitats off Nova Scotia. *Aquat. Biol. 2*: 17–27.

Mieszkowska, N., R. Leaper, P. Moore, M.A. Kendall, M.T. Burrows, D. Lear, E. Poloczanska, K. Hiscock, P.S. Moschella, R.C. Thompson, R.J. Herbert, D. Laffoley, J. Baxter, A.J. Southward and S.J. Hawkins. 2005. Marine biodiversity and climate change assessing and predicting the influence of climatic change using intertidal rocky shore biota. Final report for United Kingdom funders. *Occas. Publ. Mar. Biol. Assoc. UK. 20*: 1–53.

Mieszkowska, N., M.A. Kendall, S.J. Hawkins, R. Leaper, P. Williamson, N.J. Hardman-Mountford and A.J. Southward. 2006. Changes in the range of some common rocky shore species in Britain – a response to climate change? *Hydrobiologia 555*: 241–251.

Mieszkowska, N., S.J. Hawkins, M.T. Burrows and M.A. Kendall. 2007. Long-term changes in the geographic distribution and population structures of *Osilinus lineatus* (Gastropoda: Trochidae) in Britain and Ireland. *J. Mar. Biol. Ass. UK 87*: 537–545.

Moritz, R., C. Bitz and E. Steig. 2002. Dynamics of recent climate change in the Arctic. *Science 297*: 1497–1502.

Morizur, Y. 2001. Changements climatiques ou surexploitation? Gros temps sur les algues brunes. *Les nouvelles de l'Ifremer 25*: 1.

Moy, F.E. and H. Christie. 2007. Sugar kelp status in Skagerrak. *Abstracts of the 42nd European Marine Biology Symposium*, Kiel, 27–31 August 2007: 295.

Moy, F., J. Aure, E. Dahl, N. Green, T.M. Johnsen, E.R. Lømsland, J. Magnusson, L. Omli, F. Olsgaard, E. Oug, A. Pedersen, B. Rygg and M. Walday. 2003. Landtidsovervå king av miljøkvaliteten i kystområ dene av Norge. *Årsrapport for 2002*: 1–69.

Müller, R., C. Wiencke and K. Bischof. 2008. Interactive effects of UV radiation and temperature on microstages of Laminariales (Phaeophyceae) from the Arctic and North Sea. *Clim. Res. 37*: 203–213.

Müller, R., T. Laepple, I. Bartsch and C. Wiencke. 2009. Impact of oceanic warming on the distribution of seaweeds in polar and cold-temperate waters. *Bot. Mar. 52*: 617–638.

Norton, T.A. 1977. Experiments on the factors influencing the geographical distribution of *Saccorhiza polyschides* and *Saccorhiza dermatodea*. *New Phytol. 78*: 625–635.

Novaczek, I. and J. Mclachlan. 1989. Investigations of the marine algae of Nova Scotia XVII: vertical and geographic distribution of marine algae on rocky shores of the maritime provinces. *Proc. N.S. Inst. Sci. 38*: 91–143.

Novaczek, I., G.W. Lubbers and A. Breeman. 1990. Thermal ecotypes of amphi-Atlantic algae. I. Algae of cold-temperate distribution (*Chaetomorpha melagonium*, *Devaleraea ramentacea* and *Phycodrys rubens*). *Helgol. Meeresunters. 44*: 459–474.

Olsen, B.R., K. Sjøtun, T. Høsæter and O.J. Lønne. 2004. Population structure of *Laminaria digitata* (Hudson) J.V. Lamouroux from three different areas in the Northeast Atlantic ocean. *Abstracts of the 18th International Seaweed Symposium*, Bergen, Norway, 20–25 June 2004: 75.

Orfanidis, S. and A. Breeman. 1999. Geographic variation in thermal traits in *Digenea simplex* and *Champia parvula* (Rhodophyta) in relation to present and glacial temperature regimes. *J. Phycol. 35*: 919–930.

Pehlke, C. and I. Bartsch. 2008. Changes in depth distribution and biomass of sublittoral seaweeds at Helgoland (North Sea) between 1970 and 2005. *Clim. Res. 37*: 135–147.

Peltikhina, T.S. 2002. Distribution peculiarities and stocks of *Laminaria* algae in the area of Isfjord of the Spitsbergen Archipelago. *In*: (G.G. Matishov and G.A. Tarasov, eds) *The complex investigations of the Spitsbergen nature*. Murmansk Biological Institute. pp. 168–174 (in Russian with English abstract).

Peters, A. 1992a. Culture studies on the life history of *Chordaria linearis* (Phaeophyceae) from Tierra del Fuego, South America. *J. Phycol. 28*: 678–683.

Peters, A. 1992b. Distinction of South American *Haplogloia andersonii* and *Chordaria linearis* (Phaeophyceae) based on differences in sexual reproduction, morphology, ecophysiology and distribution. *J. Phycol. 28*: 684–693.

Peters, A.F. and A.M. Breeman. 1993. Temperature tolerance and latitudinal range of brown algae from temperate Pacific South America. *Mar. Biol. 115*: 143–150.

Ramirez, M.E. and A.F. Peters. 1992. The South American species of *Desmarestia* (Phaeophyceae). *Can. J. Bot. 70*: 2430–2445.

Ramírez, M.E. and B. Santelices. 1991. *Catálogo de las algas marinas bentónicas de la costa temperada del Pacífico de Sudamérica*. Monografías Biológicas 5. Facultad de Ciencias Biologicas, Pontificia Universidad Católica de Chile. pp. 385.

Rayner, N.A., D.E. Parker, E.B. Horton, C.K. Folland, L.V. Alexander, D.P. Rowell, E.C. Kent and A. Kaplan. 2003. Global analyses of sea surface temperature, sea ice, and night marine air temperature since the late nineteenth century. *J. Geophys. Res. 108(D14)*: 4407.

Rayner, N.A., P. Brohan, D.E. Parker, C.K. Folland, J.J. Kennedy, M. Vanicek, T.J. Ansell, and S.F.B. Tett. 2005. Improved analysis of changes and uncertainties in sea surface temperature measured *in situ* since the mid-nineteenth century: The HadSST2 dataset. *J. Climate 19*: 446–469.

Ricker, R.W. 1987. *Taxonomy and biogeography of Macquarie Island seaweeds*. British Museum (Natural History), London. pp. 344.

Rijnsdorp, A.D., M.A. Peck, G.H. Engelhard, C. Mollmann and J.K. Pinnegar. 2009. Resolving the effect of climate change on fish populations. *ICES. J. Mar. Sci.* : DOI 10.1093/icesjms/fsp056.

Rinde, E. and K. Sjøtun. 2005. Demographic variation in the kelp *Laminaria hyperborea* along a latitudinal gradient. *Mar. Biol. 146*: 1051–1062.

Sabatés, A., P. Martín, J. Lloret and V. Raya. 2006. Sea warming and fish distribution: the case of the small pelagic fish, *Sardinella aurita*, in the western Mediterranean. *Glob. Change Biol. 12*: 2209–2219.

Sagarin, R.D. and G.N. Somero. 2006. Complex patterns of expression of heat-shock protein 70 across the southern biogeographical ranges of the intertidal mussel *Mytilus californianus* and snail *Nucella ostrina*. *J. Biogeogr. 33*: 622–630.

Sagarin, R.D., J.B. Barry, S.E. Gilman and C.H. Baxter. 1999. Climate-related community over short and long time scales. *Ecol. Monogr. 69*: 465–490.

Scambos, T.A., J.A. Bohlander, C.A. Shuman and P. Skvarca. 2004. Glacier acceleration and thinning after ice shelf collapse in the Larsen B embayment, Antarctica. *Geophys. Res. Lett. 31 (L18402)*: DOI 10.1029/2004GL020670.

Scheibling, R.E. and S.X. Anthony. 2001. Feeding, growth and reproduction of sea urchins (*Strongylocentrotus droebachiensis*) on single and mixed diets of kelp (*Laminaria* spp.) and the invasive alga *Codium fragile* ssp. *tomentosoides*. *Mar. Biol. 139*: 139–146.

Scheibling, R.E. and P. Gagnon. 2006. Competitive interactions between the invasive green alga *Codium fragile* ssp. *tomentosoides* and native canopy-forming seaweeds in Nova Scotia (Canada). *Mar. Ecol. Prog. Ser. 325*: 1–14.

Scheibling, R.E. and A.W. Hennigar. 1997. Recurrent outbreaks of disease in sea urchins *Strongylocentrotus droebachiensis* in Nova Scotia: evidence for a link with large-scale meteorologic and oceanographic events. *Mar. Ecol. Prog. Ser. 152*: 155–165.
Schiel, D.R., J.R. Steinbeck and M.S. Foster. 2004. Ten years of induced ocean warming causes comprehensive changes in marine benthic communities. *Ecology 85*: 1833–1839.
Schloss, I., G.A. Ferreyra, O. González, A. Atencio, V. Fuentes, G. Tosonotto, G. Mercuri, R. Sahade, M. Tatián and D. Abele. 2008. Long-term hydrographic conditions and climate trends in Potter Cove. *Ber. Polarforsch. Meeresforsch. 571*: 382–389.
Schmidt, A.L. and R.E. Scheibling. 2007. Effects of native and invasive macroalgal canopies on composition and abundance of mobile benthic macrofauna and turf-forming algae. *J. Exp. Mar. Biol. Ecol. 341*: 110–130.
Schmittner, A., M. Latif and B. Schneider. 2005. Model projections of the North Atlantic thermohaline circulation for the 21st century assessed by observations. *Geophys. Res. Lett. 32*: DOI 10.1029/2005GL024368.
Schoschina, E.V. 1997. On *Laminaria hyperborea* (Laminariales, Phaeophyceae) on the Murman coast of the Barents Sea. *Sarsia 82*: 371–373.
Serreze, M.C., M.M. Holland and J. Stroeve. 2007. Perspectives on the Arctic's shrinking sea ice cover. *Science 315*: 1533–1536.
Sivertsen, K. 1997. Geographic and environmental factors affecting the distribution of kelp beds and barren grounds and changes in biota associated with kelp reduction at sites along the Norwegian coast. *Can. J. Fish. Aquat. Sci. 54*: 2872–2887.
Sjøtun, K. and E.V. Schoschina. 2002. Gametophytic development of *Laminaria* spp. (Laminariales, Phaeophyta) at low temperature. *Phycologia 41*: 147–152.
Steneck, R.S., M.H. Graham, B.J. Bourque, D. Corbett, J.M. Erlandson, J.A. Estes and M.J. Tegner. 2002. Kelp forest ecosystems: biodiversity, stability, resilience and future. *Environ. Conserv. 29*: 436–459.
Stolpe, U., C. Wiencke, M.N. Clayton and H. Lehmann. 1991. Life history, morphology and development of a ligulate *Desmarestia* species from southernmost Chile. *Br. Phycol. J. 26*: 235–245.
Stouffer, R. and S. Manabe. 1999. Response of a coupled ocean-atmosphere model to increasing atmospheric carbon dioxide: sensitivity to the rate of increase. *J. Clim. 12*: 2224–2237.
Sundene, O. 1962. The implications of transplant and culture experiments on the growth and distribution of *Alaria esculenta*. *Nytt Magasin for Botanikk 9*: 155–174.
Svendsen, H., A. Beszczynska-Møller, J.O. Hagen, B. Lefauconnier, V. Tverberg, S. Gerland, J.B. Ørbæk, K. Bischof, C. Papucci, M. Zajaczkowski, R. Azzolini, O. Bruland, C. Wiencke, J.-G. Winther and W. Dallmann. 2002. The physical environment of Kongsfjorden-Krossfjorden, an Arctic fjord system in Svalbard. *Polar. Res. 21*: 133–166.
tom Dieck, I. 1989. *Vergleichende Untersuchungen zur Ökophysiologie und Kreuzbarkeit innerhalb der digitaten Sektion der Gattung* Laminaria *(Phaeophyceae)*. PhD thesis. University of Hamburg, Hamburg. pp. 168.
tom Dieck, I. 1991. Circannual growth rhythm and photoperiodic sorus induction in the kelp *Laminaria setchellii* (Phaeophyta). *J. Phycol. 27*: 341–350.
tom Dieck, I. 1992. North Pacific and North Atlantic digitate *Laminaria* species (Phaeophyta): hybridization experiments and temperature responses. *Phycologia 31*: 147–163.
tom Dieck, I. 1993. Temperature tolerance and survival in darkness of kelp gametophytes (Laminariales, Phaeophyta): ecological and biogeographical implications. *Mar. Ecol. Prog. Ser. 100*: 253–264.
tom Dieck, I. and E.C. de Oliveira. 1993. The section Digitatae of the genus *Laminaria* (Phaeophyta) in the northern and southern Atlantic: crossing experiments and temperature responses. *Mar. Biol. 115*: 151–160.
van den Hoek, C. 1982a. The distribution of benthic marine algae in relation to the temperature regulation of their life histories. *Biol. J. Linn. Soc. 18*: 81–144.

van den Hoek, C. 1982b. Phytogeographic distribution groups of benthic marine algae in the North Atlantic Ocean. A review of experimental evidence from life history studies. *Helgol. Meeresunters. 35*: 153–214.

Vaughan, D., G. Marshall, W. Connolley, C. Parkinson, R. Mulvaney, D. Hodgson, J. King, C. Pudsey and J. Turner. 2003. Recent rapid regional climate warming on the Antarctic Peninsula. *Clim. Change 60*: 243–274.

Velicogna, I. and J. Wahr. 2006. Measurements of time-variable gravity show mass loss in Antarctica. *Science 311*: 1754–1756.

Whitehouse, M.J., M.P. Meredith, P. Rothery, A. Atkinson, P. Ward and R.E. Korb. 2008. Rapid warming of the ocean around South Georgia, Southern Ocean, during the 20th century: forcings, characteristics and implications for lower trophic levels. *Deep Sea Res. Part I: Oceanogr. Res. Pap. 55*: 1218–1228.

Widdowson, T.B. 1971. A taxonomic revision of the genus *Alaria* Greville. *Syesis 4*: 11–49.

Wiencke, C. and M.N. Clayton. 1990. Sexual reproduction, life history, and early development in culture of the Antarctic brown alga *Himantothallus grandifolius* (Desmarestiales, Phaeophyceae). *Phycologia 29*: 9–18.

Wiencke, C. and M.N. Clayton. 2002. *Antarctic seaweeds. Synopses of the Antarctic benthos 9*. ARG Gantner Verlag KG, Ruggell, Lichtenstein. pp. 239.

Wiencke, C. and I. tom Dieck. 1989. Temperature requirements for growth and temperature tolerance of macroalgae endemic to the Antarctic region. *Mar. Ecol. Prog. Ser. 54*: 189–197.

Wiencke, C. and I. tom Dieck. 1990. Temperature requirements for growth and survival of macroalgae from Antarctica and southern Chile. *Mar. Ecol. Prog. Ser. 59*: 157–170.

Wiencke, C., U. Stolpe and H. Lehmann. 1991. Morphogenesis of the brown alga *Desmarestia antarctica* cultivated under seasonally fluctuating Antarctic daylengths. *Ser. Cientifica INACH 41*: 65–78.

Wiencke, C., I. Bartsch, B. Bischoff, A.F. Peters and A.M. Breeman. 1994. Temperature requirements and biogeography of Antarctic, Arctic and amphiequatorial seaweeds. *Bot. Mar. 37*: 247–259.

Wiencke, C., M.N. Clayton and C. Langreder. 1996. Life history and seasonal morphogenesis of the endemic Antarctic brown alga *Desmarestia anceps* Montagne. *Bot. Mar. 39*: 435–444.

Wiencke, C., B. Vögele, N.A. Kovaltchouk and H. Hop. 2004. Species composition and zonation of marine benthic macroalgae at Hansneset in Kongsfjorden, Svalbard. *Ber. Polarforsch. Meeresforsch. 492*: 55–62.

Wiencke, C., M.N. Clayton, I. Gómez, K. Iken, U.H. Lüder, C.D. Amsler, U. Karsten, D. Hanelt, K. Bischof and K. Dunton. 2008. Life strategy, ecophysiology and ecology of seaweeds in polar waters. *Rev. Environ. Sci. Biotechnol. 6*: 95–126.

Wiltshire, K.H., M. Malzahn, K. Wirtz, W. Greve, S. Janisch, P. Mangelsdorf, D.F.J. Manly and M. Boersma. 2008. Resilience of North Sea phytoplankton spring bloom dynamics: an analysis of long-term data at Helgoland Roads. *Limnol. Oceanogr. 53*: 1294–1302.

Worm, B., H.K. Lotze, C. Boström, R. Engkvist, V. Labanauskas and U. Sommer. 1999. Marine diversity shift linked to interactions among grazers, nutrients and propagule banks. *Mar. Ecol. Prog. Ser. 185*: 309–314.

Zacher, K., A. Wulff, M. Molis, D. Hanelt and C. Wiencke. 2007. Ultraviolet radiation and consumer effects on a field-grown intertidal macroalgal assemblage in Antarctica. *Glob. Change Biol. 13*: 1201–1215.

Zacher, K., R. Rautenberger, D. Hanelt, A. Wulff and C. Wiencke. 2009. The abiotic environment of polar benthic algae. *Bot. Mar. 52*: 483–490.

13 Physiological responses of polar benthic algae to ultraviolet radiation

Ulf Karsten, Angela Wulff, Michael Y. Roleda, Ruth Müller, Franciska S. Steinhoff, Jana Fredersdorf and Christian Wiencke

Introduction

Stratospheric accumulation of ozone is primarily responsible for absorbing parts of the solar ultraviolet radiation (UVR) before it can reach the marine biosphere. Emission of anthropogenic halogenated volatile substances in the past century have – apart from natural sources of these compounds (Laturnus 2001, Laturnus et al. 2002, Gribble 2003) – resulted in a stratospheric enrichment of these compounds, which could persist for many decades. Because of the high chemical reactivity of halogens, they efficiently destroy ozone in the protective layer. This is particularly well reflected in the strong ozone decline over Antarctica each spring, which can amount to more than 75% depletion, a phenomenon known to the public as the 'ozone hole' (Wessel et al. 1998, Whitehead et al. 2000). The Arctic is also currently affected by stratospheric ozone depletion and consequently increasing UVR (Chapter 2; McKenzie et al. 2003, Zacher et al. 2009b).

UVR is differentiated according to the CIE definition (Commission Internationale de l'Eclairage 1935; http://www.cie.co.at) into three wavebands: UVC: 190–280 nm, UVB: 280–315 nm and UVA: 315–400 nm. UVC is strongly mutagenic and lethal to most organisms; however, owing to its complete absorption by the atmospheric ozone layer it does not reach the biosphere. In contrast, UVA is not attenuated by ozone, and hence its fluence will be unaffected by any ozone layer reduction reaching polar organisms. It is the UVB range that increases as a consequence of stratospheric ozone destruction. Although this waveband represents less than 1% of the total solar flux reaching the earth's surface, it is biologically extremely harmful (Franklin and Forster 1997). Calculations indicate that a 10% decline in column ozone would result in an approximate 5% increase in surface irradiance at 320 nm and a 100% increase at 300 nm (Frederick et al. 1989).

Physical and chemical environment in benthic habitats of polar regions

Sea ice is probably a very important factor affecting polar benthic algae, not only with regard to ice scour (Teixido et al. 2007) but also because benthic algae are stressed during ice break-up in early summer when they become suddenly exposed to very high and stressful photosynthetically active radiation (PAR, 400–700 nm) and UVR (Chapter 2; Aguilera et al. 2002a, Bischof et al. 2002a, Zacher et al. 2009b). Although irradiance decreases logarithmically with increasing water depth, the 1% penetration depth of UVB was recorded down to 19 m in Potter Cove, King George Island (Richter et al. 2008). For benthic microalgae, the sediment has been proposed as a refuge to escape harmful radiation by migration (Wulff et al. 2008a), but UVR has been shown to penetrate approximately 0.6 mm (UVB) and 1 mm (UVA) into a sandy sediment (Wulff et al. 1999). In the undergrowth of seaweed forests, UVR levels are lower, but the canopy algae are exposed to full solar radiation (Hanelt 2004).

Structure and function of benthic micro- and macroalgal assemblages

The shallow water zone of polar regions is dominated by two types of algal assemblages. Whereas macroalgal species of the Chlorophyta, Rhodophyta and Phaeophyceae primarily settle on hard bottom substrata, such as rocky shorelines, the microscopic forms preferentially cover extensive sediment areas. At their growth sites, seaweeds often form submerged, complexly structured underwater forests characterised by high primary productivity, thereby providing food, shelter and habitat to many marine invertebrates and fish (Bischof et al. 2006). Macroalgal assemblages typically show distinct vertical zonation patterns, exhibiting a characteristic sequence of species with increasing depth, which is related to the vertically changing environmental parameters (Bischof et al. 2006). Being at the basis of marine food webs, macroalgae are directly consumed by a diversified suite of micro- and macrograzers (Duffy and Hay 2000), whereas algal exudates might fuel the microbial loop if they are used by free-living and alga-associated bacteria. In addition, seaweeds serve many species as habitat, to which sessile forms attach directly, and can host motile animals by provision of shelter from predators.

Microphytobenthic biofilms in shallow waters are typically dominated by Bacillariophyceae. This phototrophic assemblage is best known from temperate marine regions as highly productive, providing a major food source for benthic suspension- or deposit-feeders (Cahoon 1999), a control barrier for oxygen fluxes at the sediment/water interface (Risgaard-Petersen et al. 1994), and a stabiliser of sediment surfaces against erosion by the excretion of extracellular polymeric substances (De Brouwer et al. 2005). Consequently, the microphytobenthos represents a key component in the functioning of trophic webs in coastal zones that are characterised by sediments (Cahoon 1999). However, in contrast to temperate and tropical regions, structure and function of microphytobenthic assemblages are poorly studied in polar waters (Chapter 7, 8; Glud et al. 2009, Woelfel et al. 2009, 2010).

UVR effects on benthic micro- and macroalgae

Solar radiation is essential for life in polar regions. In the process of photosynthesis, the micro- and macrophytobenthic assemblages absorb visible light energy with their respective photosynthetic antennae to chemically fix inorganic carbon into energy-rich organic compounds. However, an increase in UVR can inhibit many biological processes. The major cellular targets of UVB are various biomolecules, which directly absorb this radiation, or which are indirectly affected by various UV-induced photochemical reactions. The biological and, ultimately, ecological consequences are numerous.

Cell biological and ultrastructural changes

The cell biological and ultrastructural changes due to UVR exposure depend on the general physiological constitution, the life history and developmental stage of the organism under study. The most susceptible stages in the life histories of macroalgae are their microscopic stages, their spores, gametes and zygotes. This has been demonstrated in species of the genus *Fucus*, and in various Laminariales and Gigartinales from Spitsbergen, King George Island (Antarctica) and the North Sea (Chapter 2; Schoenwaelder et al. 2003, Roleda et al. 2007a, 2007b, 2008a, 2010b, Müller et al. 2008, Zacher et al. 2009a). As discussed below, multiple cellular processes are negatively affected by UVR, in particular photosynthesis (Roleda et al. 2006d), nuclear division (Huovinen et al. 2000) and motility (Makarov and Voskoboinikov 2001). By contrast, there are repair mechanisms operating which mitigate damage. This is shown by the effective recovery of photosynthesis and the repair of DNA damage (Roleda et al. 2004, 2006a, 2008a, Zacher et al. 2008). Damage can also be prevented by UV-absorbing phlorotannins in brown algae (Swanson and Druehl 2002, Roleda et al. 2005, 2006c) and mycosporine-like amino acids in red algae (Roleda et al. 2008a, Zacher et al. 2008). The balance between the damaging effects of UVR and the repair and protective mechanisms can be measured by the integrative parameter 'germination'. If spores or zygotes germinate after UVR exposure, the repair and protective mechanisms are strong enough to cope with the damaging effects of UVR.

Thus, the effects on the ultrastructure of an algal cell can be quite diverse, ranging from none to very strong damage, the latter usually indicating cell death. But one should always keep in mind that an electron micrograph shows a snapshot of only the moment when the organism was fixed, and hence provides no information about physiological processes, such as UVR acclimation. Ongoing repair or degradation processes, for instance, can be easily overlooked or misinterpreted. Therefore, when studying UVR-induced ultrastructural effects, additional physiological information should always be taken into account.

UVR affects all cellular components, especially the chloroplasts, the mitochondria, the nucleus and also the cytoplasm. Table 13.1 summarises present knowledge of UVR effects on the fine structure of marine algal cells of different systematic positions from polar to cold temperate regions. When comparing the results, it must be noted that the data were obtained from different life history and developmental stages and that different exposure times and radiation conditions were used in the various studies.

Numerous investigations have been made within the Rhodophyta. Poppe et al. (2002, 2003) investigated laboratory-grown *Palmaria decipiens* (Reinsch) R.W. Ricker, *Palmaria palmata* (Linnaeus) Kuntze, *Phycodrys austrogeorgica* Skottsberg and *Bangia atropurpurea* (Roth) C. Agardh, whereas Holzinger et al. (2004) studied field-grown *Odonthalia dentata* (Linnaeus) Lyngbye and *P. palmata*. The most striking effect of UVR in the species studied was the disturbance of chloroplast fine structure. The intrathylakoidal space was generally enlarged and the thylakoid membranes became wrinkled. In UVR-sensitive species or in less sensitive species after strong UVR exposure, the thylakoids became tubular, e.g., in *B. atropurpurea* (Poppe et al. 2003) or disintegrated into "inverted" vesicles, obviously a general phenomenon in red algal chloroplasts. In these vesicles, the phycobilisomes, which are normally attached to the exterior of the thylakoids, were exposed to the inside. In such cases, the photosynthetic apparatus was strongly damaged and photosynthesis was impaired if not fully inhibited. In some cases, these fine-structural changes were, however, partly reversible, reflecting acclimation of photosynthesis to UVR, as shown in *P. decipiens* (Poppe

Table 13.1 Summary of present knowledge of UVR effects on the ultrastructure of algal cells of different systematic position from polar to cold-temperate regions.

Phylum/ Class	Species	Developmental stage	UV exposure	Chloroplast	Nucleus	Mitochondria	Others	Reference
Rhodo-phyta	*Palmaria decipiens*	Male gametophytes	2–23 h 0.37 W m^{-2} UVB 6.84 W m^{-2} UVA 3.35 W m^{-2} PAR	After 2 h: chloroplast thylakoids dilated, vesicle-like formations at the chloroplasts margins; after 6–8 h: thylakoids disrupted and connected to vesicles; after 12–23 h: changes disappeared, thylakoids irregular in shape	–	After 4 h: tubuli swollen, sacculus appearance; after 23 h: changes disappeared	–	Poppe et al. 2002, 2003
	Palmaria palmata	Vegetative thalli	6/24 h 0.20 W m^{-2} UVB 4.7 W m^{-2} UVA 25–30 μmol photons m^{-2} s^{-1}	After both exposure times: thylakoid lumen enlarged, thylakoid membranes wrinkled, outer chloroplast envelope with protrusions into the cytoplasm	–	Cristae visible, swollen	–	Holzinger et al. 2004
	Palmaria palmata	Vegetative thalli	2–48 h 0.6–0.8 W m^{-2} UVB 6.24–8.41 W m^{-2} UVA 3.12–4.66 W m^{-2} PAR	Vesiculation of thylakoids after 2 h, after 6 h increased vesiculation, after 24 h growth of vesicles and expansion of intrathylakoidal space	–	After 2 h: tubuli swollen, inner mitochondrial membrane sacculus type	–	Poppe et al. 2003
	Phycodrys austrogeorgica	Gametophytes	2–12 h 0.6–0.8 W m^{-2} UVB 6.24–8.41 W m^{-2} UVA 3.12–4.66 W m^{-2} PAR	After 2–7 h: dilated thylakoids, formation of inverted thylakoid vesicles; after 12 h: detached phycobilisomes, thylakoid and chloroplast envelope disintegrated	–	–	Crystal-line inclusions in cyto-plasm	Poppe et al. 2003
	Bangia atropurpurea	Gametophytes	2–96 h 0.6–0.8 W m^{-2} UVB 6.24–8.41 W m^{-2} UVA 3.12–4.66 W m^{-2} PAR	Thylakoids change from parallel/central organisation to vesicle-like appearance adjacent to the pyrenoid, tubularly arranged thylakoids	–	–	–	Poppe et al. 2003
	Odonthalia dentata	Vegetative thalli	6/24 h 0.20 W m^{-2} UVB 4.7 W m^{-2} UVA 25–30 μmol photons m^{-2} s^{-1}	After 6 h: irregularities in thylakoid membranes, granular structures and interrupted appearance of thylakoid membranes; after 24 h:	–	–	–	Holzinger et al. 2004

Phaeo-phyceae	*Laminaria hyperborea*	Zoospores	8 h 0.31 ± 0.05 W m^{-2} UVB 5.7 ± 0.7 W m^{-2} UVA 18.9 ± 1.2 µmol photons m^{-2} s^{-1}	Formation of plastoglobuli, disrupted and fragmented thylakoids at 17 °C, vesicle-like structures	Mottled nucleo-plasm, nucleopores varied in size	Tubulus/sacculus type, swollen cristae	–	Steinhoff et al. 2008
	Saccharina-latissima	Gametophytes	8 h 0.35 W m^{-2} UVB 5.5 W m^{-2} UVA 20 µmol photons m^{-2} s^{-1}	Formation of plastoglobuli	–	–	–	R. Müller, unpub-lished results
	Laminaria digitata	Gametophytes	8 h 0.35 W m^{-2} UVB 5.5 W m^{-2} UVA 20 µmol photons m^{-2} s^{-1}	Formation of plastoglobuli	–	–	–	R. Müller, unpub-lished results
Chloro-phyta	*Prasiola crispa*	Vegetative thalli	6/24 h 0.20 W m^{-2} UVB 4.7 W m^{-2} UVA 25–30 µmol photons m^{-2} s^{-1}	No effects after 6 h; after 24 h: slight alterations, reduced number of plastoglobuli, dilatations of thylakoids	–	After 24 h: slight damage		Holzinger et al. 2006
	Urospora penicilliformis	Vegetative fertile gametophytes	16 h 0.40 W m^{-2} UVB 4.34 W m^{-2} UVA 22 µmol photons m^{-2} s^{-1}	No significant effects in chloroplasts	–	–		Roleda et al. 2010a
	Zygnema sp.	Vegetative filaments	24 h 0.42 W m^{-2} UVB 4.00 W m^{-2} UVA 20 µmol photons m^{-2} s^{-1}	No effects	No effects	No effects	–	Holzinger et al. 2009

et al. 2002). Another effect of UVR was formation of protrusions of the chloroplast envelope, as shown in *P. palmata* and *O. dentata* (Holzinger et al. 2004), or disintegration of both envelope membranes, as shown in *P. austrogeorgica* (Poppe et al. 2003). In the latter species, the protein crystals occurring generally in the cytoplasm became corroded after UVR treatment, indicating either damage or remobilisation of the stored protein for repair processes (Poppe et al. 2003). Within the mitochondria, the cristae appeared swollen under UVR and were transformed into sacculi (Poppe et al. 2002, Holzinger et al. 2004).

In contrast, foliose thalli of the green algae *Prasiola crispa* (Lightfoot) Kützing (Holzinger et al. 2006) and *Urospora penicilliformis* (Roth) J.E. Areschoug (Roleda et al. 2010a) showed no effect after mild UVR exposure. After stronger exposure, only slight alterations appeared within the chloroplasts, such as dilatations of thylakoids and a reduced number of plastoglobuli (Holzinger et al. 2006). Under these conditions, photosynthetic efficiency decreased. The mitochondria showed slight damages and cytoplasmic globules increased in size and became more abundant. Thus, in contrast to most of the red algae discussed above, *P. crispa* and *U. penicilliformis* can cope with UV stress relatively well. The same is true for the freshwater green alga *Zygnema* sp., which was collected from a mat on the soil close to a freshwater runlet in Spitsbergen where it grows under ambient polar day solar radiation (Holzinger et al. 2009). These authors showed that natural as well as artificial PAR and UVR conditions did not affect photosynthetic performance and ultrastructure. However, substantial amounts of lipid bodies were observed in all samples. Such lipid accumulations are typically observed when algae are subjected to unfavorable environmental conditions including photo-oxidative stress (Holzinger et al. 2009). *Zygnema* species are capable of forming "akinetes", persisting vegetative stages that are able to sustain stressful abiotic parameters (McLean and Pessoney, 1971).

In brown algae, two different life history stages have been examined by electron microscopy, namely, zoospores and filamentous gametophytes. After UVB exposure, the zoospores of *Laminaria hyperborea* (Gunnerus) Foslie showed an enhanced formation of plastoglobuli in the chloroplast, the nucleoplasma became mottled and the structure of the mitochondria changed from the tubulus type to the sacculus type (Steinhoff et al. 2008). Phlorotannin-containing physodes were present but did not show any change after UVR exposure, and thus their contribution to UV protection is doubtful. The fine structure of gametophytes of *Saccharina latissima* (Linnaeus) C.E. Lane, C. Mayes, Druehl *et* G.W. Saunders and *Laminaria digitata* (Hudson) J.V. Lamouroux changed little under UVR. The only change was increase in size of plastoglobuli in the chloroplast (R. Müller, unpublished results), generally regarded as an indication for the upregulation of plastid lipid metabolism in response to oxidative stress.

Because no data on these UVR effects on benthic diatoms are available, here we summarise results obtained on the three pelagic diatom species *Cyclotella* sp., *Nitzschia closterium* W. Smith (*Cylindrotheca closterium*) and *Thalassiosira nordenskioeldii* Cleve (Buma et al. 1996). Exposure of these microalgae to UVB resulted in increased cell size. Moreover, although chloroplast number increased, this organelle became disoriented and smaller. The nucleus appeared enlarged and the contrast between nucleolus, chromatin and nucleoplasm faded with increasing UVB exposure. These ultrastructural changes mirror the fact that cell division was inhibited, whereas growth still continued as the general cell metabolism was not or only little affected by UVR. Similar results were obtained with the coccolithophore *Emiliana huxleyi* (Lohm.) Hay *et* Mohler (Buma et al. 2000). On the basis of the observed phenomenon, there probably is a disruption of processes in microtubule dynamics. Similarly, in *Macrocystis pyrifera* (Linnaeus) C. Agardh gametophytes, nuclear division and translocation were inhibited after UVR exposure (Huovinen et al. 2000). An often observed response to UVR, which presumably is also related to microtubule function, is the inactivation and/or loss of flagella as shown, e.g., in *Chlamydomonas reinhardtii* P.A. Dangeard (Hessen et al. 1995) and in sperm of the Japanese scallop (Li et al. 2000). Cell motility is affected, as has also been demonstrated in zoospores of *Saccharina latissima* [as *Laminaria saccharina* (L.) J.V. Lamour.; Makarov and Voskovoinikov 2001].

Changes in chloroplast ultrastructure after UV stress also seem to be a general phenomenon in freshwater algae, e.g., *Micrasterias denticulata* Brébisson *ex* Ralfs (Lütz et al. 1997) and in endosymbiotic dinoflagellates (Hannack et al. 1998). Other features include the formation of large endoplasmic reticulum cisternae in *M. denticulata* (Meindl and Lütz 1996) and of lipid globules containing secondary carotenoids in snow algae, e.g., *Chlamydomonas nivalis* (F.A. Bauer) Wille (Remias et al. 2005). For further information on these algal groups, the reader is directed to the review by Holzinger and Lütz (2006).

Molecular targets and repair mechanisms

DNA is one of the most UV-sensitive biomolecules, and UV-induced damage occurs directly through absorption of UVB quanta by aromatic residues. Absorbed energy can be dissipated by various mechanisms involving single bases (e.g., single-strand breaks) or interactions between adjacent bases (e.g., dimerisation) and between non-adjacent bases (i.e., inter- or intrastrand crosslinks) (Karentz et al. 1991). The structural consequences are conformational alterations, such as the frequently observed formation of cyclobutane dimers and pyrimidine (6-4)-pyrimidone (6-4)-photoproducts (CPDs) (Lois and Buchanan 1994). Such UV-induced DNA damage can significantly affect transcription, causing erroneous replication and promoting mutations, which finally increase mortality of algal cells.

Life stage-dependent susceptibility to UVB-induced DNA damage has been observed in several polar macroalgae. Zoospores of the sublittoral kelps from Spitsbergen [*Alaria esculenta* (Linnaeus) Greville, Wiencke et al. 2007; *Laminaria digitata*, Roleda et al. 2010b; *Saccorhiza dermatodea* Bachelot de la Pylaie, Roleda et al. 2006a] are more susceptible to UVB-induced DNA damage than their respective young sporophytes (Roleda et al. 2005, 2006b). Recently, filamentous *Urospora penicilliformis* gametophytes from Antarctica were also reported to sustain significantly less DNA damage compared with propagules exposed to the same UVB dose (Roleda et al. 2009b). In multicellular life stages, the relatively thick cell wall might be able to selectively filter short UV-wavelengths from reaching the UV-sensitive targets (i.e., nucleus) compared with the thin-walled or "naked" propagules. By contrast, an inverse relationship between thallus thickness and remaining DNA damage also occurs in young sporophytes of different kelp species (Roleda et al. 2005, 2006b, 2007a). Aside from UV screening by cell walls, intracellular mechanisms, such as synthesis of

UV-absorbing compounds (see below), are also important for UV protection against DNA damage. Moreover, release of a cloud of spores (e.g., kelp zoospores with phlorotannin-containing physodes) can self-shade, thereby forming a type of "biofilter" against harmful UVR (Roleda et al. 2006c).

Sensitivity of propagules to UVR-induced DNA damage is related to size of the propagules, ploidy levels and depth distribution of the adult plants (Roleda et al. 2007a, 2008a, Zacher et al. 2007a). Propagules of eulittoral *Urospora penicilliformis* (Roleda et al. 2009b) exposed to a comparable UVB dose showed significantly lower DNA damage than gametes of the sublittoral *Ascoseira mirabilis* Skottsberg (Roleda et al. 2007b). Among kelps, the reported prevalence of larger, more UV-tolerant meiospores originating from species or populations from sites exposed to high UV radiation suggest that kelp meiospores are pre-adapted to the UV conditions of the parent plant (Swanson and Druehl 2000, Wiencke et al. 2006). Comparison between spores of different ploidy levels in Antarctic *Gigartina skottsbergii* Setchell *et* N.L. Gardner showed less DNA damage in diploid carpospores compared with haploid tetraspores (Roleda et al. 2008a). Haploid cells are reported to be efficient replicators, whereas diploid cells are resistant to damage (Long and Michod 1995). The higher tolerance of carpospores to UVR confirms the genetic buffering hypothesis suggested for vascular plants which proposes that diploids benefit from better cellular regulation and are, therefore, more vigorous and tolerant to radiation stress (Raper and Flexer 1970). During the diploid state, DNA damage can be repaired, because there are two copies of this biomolecule in each cell and one copy can be presumed to be undamaged.

Repair of UV-induced DNA photoproducts can take place by various mechanisms that include photoreactivation (light-mediated repair) or nucleotide and base excision repair (Britt 1996). Light-induced DNA repair is based on the activity of the key enzyme photolyase, which binds to complementary DNA strands and efficiently breaks pyrimidine dimers. However, this enzyme only functions as a DNA repair mechanism when blue light is available for activation. Significant removal of CPDs, indicating repair of DNA damage, was observed in Arctic seaweeds after recovery in low white light. DNA damage repair could be mediated either by light-dependent photolyases or light-independent nucleotide excision repair (Pakker et al. 2000, van de Poll et al. 2002a).

In the planktonic Antarctic diatom *Chaetoceros dichaeta* L.A. Mangin, no CPD accumulation was detected in high light-acclimated cells (Buma et al. 2006). For polar benthic microalgae, only one study has addressed UV effects on DNA damage (Wulff et al. 2008a). In this study, only minimal CPD accumulation was observed under unnaturally high intensities of UVB, and the effect was completely reversed after recovery under photoreactivating radiation (Wulff et al. 2008a).

Besides UV-induced DNA photoproducts, damage to protein molecules represents the second major direct effect of UVB in algae. Typical target proteins are those associated with the plasmalemma or involved in photosynthesis, such as the D1 protein of photosystem II and the enzyme Rubisco in the Calvin cycle (Bischof et al. 2000). The native three-dimensional structure of proteins is a prerequisite for any specific function, and hence any structural damage can be reflected in loss of cell vitality because these molecules have multiple roles as enzymes, hormones and structural components. However, because proteins typically occur as numerous copies inside the algal cell, any UV-induced damage is less severe than DNA damage. By contrast, degradation and replacement of damaged protein molecules require energy, which might otherwise support more essential processes, such as DNA repair.

The replacement of any UV-damaged protein by *de novo* biosynthesis contributes to counteracting UV damage. The molecular mechanisms are well studied in Cyanobacteria and might be similar in polar micro- and macroalgae. Exposure to moderate doses of UVB results in an increased turnover rate of the D1 and D2 reaction centre subunits of photosystem II, thus rapidly replacing damaged proteins by newly synthesised polypeptides (Campbell et al. 1998, Máté et al. 1998). A key step in this repair process is the UVB-induced differential transcription of reaction centre protein encoding genes.

Other biomolecules, such as photosynthetic pigments, can also be destroyed under UVR, with the phycobilins (main pigments of Rhodophyta) being the most sensitive, and carotenoids generally being less affected than chlorophylls. In contrast, lipids, which are major compounds of all biological membranes, cannot absorb UVR, but are easily peroxidised through UV-induced reactive oxygen species (ROS, see below) (Bischof et al. 2006).

Induction of reactive oxygen species (ROS) and antioxidative strategies

Polar micro- and macroalgae perform oxygenic photosynthesis using water as an electron donor thereby releasing molecular oxygen, which can be accumulated and easily chemically converted to potentially damaging ROS. Photooxidative stress, including UVB, stimulates various cellular processes leading to the production of superoxide radicals (O_2^-), as well as singlet-oxygen ($^1O_2$) and hydroxyl radicals (OH^-). The sources and production sites of ROS are mainly related to photosynthetic activities, such as pseudocyclic photophosphorylation and the Mehler reaction, which stimulate the accumulation of hydrogen peroxide (Elstner 1990, Asada 1994).

Besides these internal processes, formation of ROS might also take place in the environment by UV-induced photoactivation of dissolved organic matter, photochemical degradation and liberation of excited electrons that initiate the reduction of molecular oxygen resulting in superoxide anion radicals (Cooper and Zika 1983). A second reduction step of superoxide radicals followed by protonation yields hydrogen peroxide, which is a powerful oxidant because of its relatively long half-life that allows long distance diffusion (Asada 1994). There are indications that UVB in Antarctic summer can stimulate the input of hydrogen peroxide from photochemical reactions and atmospheric wet deposition into shallow waters increasing its concentration from typical 20–300 nM up to 5000 nM (Abele et al. 1999).

ROS can cause extensive damage to DNA, proteins, and other biological molecules and structures (Bischof et al. 2006). Polar micro- and macroalgae have evolved a complex defence system against ROS including nonenzymatic antioxidants, such as ascorbate (vitamin C), tocopherol (vitamin E), carotenoids and reduced glutathione. However, these low-molecular weight antioxidants are chemically consumed and hence not considered the most efficient detoxifying agents. In contrast, antioxidant enzymes can efficiently counteract all UV-induced ROS and include superoxide dismutase, catalase, glutathione peroxidase and the enzymes involved in ascorbate-glutathione cycle like ascorbate peroxidase, mono-dehydroascorbate reductase, dehydroascorbate reductase and glutathione reductase (Schriek 2000, Aguilera et al. 2002b, Dummermuth et al. 2003). Whereas Antarctic diatoms have high enzymatic antioxidant activities even at low temperatures due to low activation energies (Schriek 2000), Arctic macroalgae are capable of upregulating some of their antioxidative enzymes under UVR, thereby allowing fast acclimation to changes in environmental radiation conditions (Aguilera et al. 2002b). In addition, some polar seaweeds might even synthesise and accumulate specific antioxidants, such as bromophenolic compounds (Dummermuth et al. 2003).

Physiological processes and acclimation

The effects of UVB exposure on biological systems are manifold, and any damage or disturbance of biomolecules will strongly affect physiological processes, such as photosynthesis, growth and reproduction.

Photosynthesis Photosynthesis is probably the most intensively studied process under UVR stress in polar algae. Owing to biochemical complexity of this process, numerous sites can be affected by the UV-waveband. These include, e.g., inhibition of energy transfer within the photosystem II reaction centre, the water splitting complex, or the light-harvesting complex, as well as pigment changes. In addition, key enzymes of the Calvin cycle, such as Rubisco, are typical molecular targets. The response of polar micro- and macroalgae towards UVB exposure is determined by the interplay of genetically fixed adaptation and physiological acclimation (Bischof et al. 2006).

Many micro- and macroalgae from the intertidal and supralittoral zone, such as the chlorophyte *Urospora penicilliformis* and *Prasiola crispa*, and the phaeophycean *Fucus distichus* Linnaeus, are only mildly affected by UVR, indicating a broad photophysiological tolerance (Hanelt et al. 1997, Holzinger et al. 2006, Roleda et al. 2009, 2010a). In particular, their ability for dynamic photoinhibition, a photoprotective mechanism by which excessive radiation energy absorbed is dissipated harmlessly as heat (Krause and Weis 1991), contributes to this radiation stress tolerance. Dynamic photoinhibition is characterised by a pronounced decrease in photosynthetic activity under high photon fluence rates especially at noon, followed by a fast and full recovery during subsequent exposure to low radiation conditions, e.g., in late afternoon, as documented for Arctic seaweeds in the field (Hanelt et al. 1997). In most polar seaweed species studied so far, UVB contributes significantly to some degree to photoinhibition of photosynthesis, whereas recovery of photosynthetic efficiency, regardless of life history phase, is always lower in deep-water species compared with species from shallow waters. However, the underlying photoprotective mechanisms involved need to be elucidated.

In contrast to intertidal seaweeds, all deep-water and understory algae from the sublittoral recover only partly and slowly, or even bleach, indicating chronic photoinhibition (Karsten et al. 2001). Whereas dynamic photoinhibition is a reversible physiological mechanism, chronic photoinhibition is reflected by photodamage to components of the photosynthetic apparatus, such as the D1 protein (Bischof et al. 2006) and requires *de novo* synthesis of proteins. Nevertheless, at least some macroalgal species are able to acclimate their photosynthetic performance to UVR, as documented in sporophytes of Arctic Laminariales collected at different depths (Bischof et al. 1998, 1999). The acclimation response included a reduction in the degree of photoinhibition, an effect that might be explained either by the activation of the anti-oxidative capacity, increased activity of repair and recovery mechanisms counteracting the inhibitory effects or by the formation of UV-screening compounds.

The sensitivity of photosynthesis in reproductive cells of Antarctic macroalgae to PAR and UVR is related to the observed vertical zonation pattern of the adult plants (Zacher et al. 2007a, Roleda et al. 2007a, 2008b, 2010b). This response was also reported in the early life stages of Arctic macroalgae (Roleda et al. 2006d). It seems that low light adaptation of photosynthesis is a typical feature of reproductive cells of macroalgae (Roleda et al. 2006a, d, Wiencke et al. 2007, Zacher et al. 2007a), which can be explained by differences in chlorophyll antenna size and number of chloroplasts present in reproductive cells compared with multicellular macroscopic stages. *In situ* solar radiation can, therefore, exert a significant effect on survival of spores in the field (Wiencke et al. 2006). Consequently, survival of propagules will be dependent on their immediate settlement on substrata at depths in between rock crevices, under boulders or under algal canopies where the prevailing low-light microenvironment is suitable for their germination.

The few studies of UVR effects on photosynthesis in benthic diatoms from polar waters indicate only temporary, low level photoinhibition facilitated by dynamic recovery and efficient DNA damage repair (Wulff et al. 2008a, b). In benthic microalgae, photosynthetic performance measured as optimum quantum yield of photosystem II (F_v/F_m) seems to be affected by UVA and UVB, but observed effects are mostly transient and disappear after 1 or 2 days of treatment (Wulff et al. 2008a, b, c).

The ability to cope with UVR after a period of darkness was tested in an Antarctic benthic diatom assemblage where the diatoms were kept in total darkness for 15 and 64 days, respectively, and then exposed to relatively high intensities of UVR (Wulff et al. 2008c). The diatom assemblage was dominated by the large species *Gyrosigma fasciola* (Ehrenberg) J.W. Griffith *et* Henfrey and *Pleurosigma obscurum* W. Smith, and these cells were able to resume photosynthetic activity after 64 days in darkness and coped with relatively high intensities of UVR compared with their natural habitat. However, although the cells surviving darkness could cope with the applied UVR, not all individual cells in the population survived the long dark period.

Growth Because the photosynthetic process in many species of polar micro- and macroalgae can acclimate to variable radiation stress, long-term effects of increasing UVR should be measured in growth and reproduction rather than photosynthesis alone. Growth and reproduction are typically related to ecological success (fitness) of individual algal species under given environmental conditions. In contrast to all other cellular processes, such as photosynthesis, respiration, anabolic pathways, etc., growth integrates well the impacts of all positive and negative abiotic factors and, thus, represents the most important physiological key parameter to describe the performance of algae in their polar ecosystem.

A simple growth model: $G = P - R - L$, proposed by Carr et al. (1997), follows a growth differentiation balance, where growth increment over time (G) is a function of the rate of biomass production through gross photosynthesis (P) and loss due to respiration (R) and tissue loss or decay (L). Under radiation stress (high PAR + UVR), photoinhibition of photosynthesis decreases potential carbon accretion (P) into plant dry matter (G) (Long et al. 1994). Dark respiration (R) represents the energy used to synthesise new biomass (growth respiration) and to maintain metabolic activity (maintenance respiration). Exposure to UVR causes molecular and cellular damage which might further increase loss due to respiration (R) by diverting more photosynthates to repair and/or protection (e.g., production of UV-sunscreen compounds), all of which results in decreasing growth.

Similar to photosynthetic performance, growth sensitivity of polar seaweeds to UVR is also strongly correlated to the vertical distribution (Aguilera et al. 1999, Michler et al. 2002, Roleda et al. 2007a). Moreover, growth rates of juvenile life stages of macroalgae are more susceptible to UVR exposure compared with adult plants, which is well documented for young kelp sporophytes (Roleda et al. 2005, 2006b). Because many growth studies on seaweeds were performed for only short periods on intact young plants or on discs excised from adult macroalgal thalli, UV-induced morphological deformation and tissue damage might easily have been overlooked (Aguilera et al. 1999). However, pronounced tissue necrosis and loss of parts of the thalli was reported in the UV-sensitive Arctic *Laminaria solidungula* J. Agardh after 1 week of daily exposure to 18 h UVR (Michler et al. 2002), indicating that such changes have a strong effect on growth performance.

In contrast to polar seaweeds and planktonic microalgae, studies of UVR effects on growth of benthic diatoms from Antarctic habitats are rare (Wulff et al. 2008a,b) and for the Arctic region virtually lacking. On the Antarctic Peninsula, Zacher et al. (2007c) studied UVA and UVB effects in a 15-week field experiment on an intertidal hard bottom platform. The growth of the microalgal assemblage dominated by the diatom genera *Cocconeis* and *Navicula* was unaffected by both UVA and UVB radiation. In a similar subtidal experiment, Campana et al. (2008) reached the same conclusion, the subtidal assemblage being dominated by the diatom species *Fragilaria striatula* Lyngbye, *Navicula* cf. *perminuta* Grunow and *Navicula* cf. *hanseni* Möller. In laboratory studies, growth of Antarctic benthic diatoms also seems to be unaffected by UVR. There are basically two strategies by which these microalgae respond to enhanced UVR: acclimation or avoidance. The first process includes repair and protection mechanisms, and, as recently shown for Antarctic benthic diatoms, they have only minimal DNA damage under UVB treatment and probably an elevated and efficient repair capacity (Wulff et al. 2008a). The second process includes migration and vertical movement (see below). In addition, because growth was always unaffected, these authors concluded that UVR does not seem to be a threat to benthic marine Antarctic diatoms.

Photoprotective mechanisms

Avoidance In contrast to sessile seaweeds, benthic diatoms can physically move away from harmful UVB, which represents one of the most effective avoidance strategies (Wulff et al. 2008a). However, it always requires an ability to detect this waveband. Furthermore, all algae face a dilemma with regard to solar radiation. Whereas sufficient PAR is an essential prerequisite for photosynthesis, too much UVR can be harmful. Migration and vertical movement in sediments in combination with self-shading definitely seems to be an efficient way for motile diatoms to receive the optimum light requirements and avoid long-term exposure to high UVR, thereby saving energy for photoprotective acclimation.

In contrast to motile benthic diatoms, most macroalgae must cope with prevailing radiation conditions. If these algae grow deep in the water column, they are never exposed to UVR (Karsten et al. 2001). In contrast, in the intertidal zone (eulittoral) or in sheltered coastal lagoons, macroalgae must cope with high radiation. Here, green algae of the genera *Ulva*, *Acrosiphonia* or *Urospora* often form mat- or turf-like assemblages, which exhibit steep gradients of UVB, but also of physiological responses. Whereas top layers are exposed to high surface irradiance and hence often bleach, bottom layers permanently experience very low radiation conditions or even remain in darkness (Bischof et al. 2002b, 2006). Another strategy for small macroalgae to avoid high radiation in shallower water is to grow under the protecting canopy of larger macroalgal taxa, such as kelps. Indeed, recruitment of juvenile kelps under the canopy of adult plants is considered to be an adaptive behaviour, which effectively protects these early life stages from radiation stress, therefore minimising ecological cost for protection and hence enabling allocation of more photosynthate for growth (Herms and Mattson 1992).

Another mechanism that might be involved in UVR protection and acclimation is the establishment of a physical barrier that shields the photosynthetic apparatus against damaging radiation (Karentz 1994). For example, increasing thallus thickness was observed to minimise UVB-induced DNA damage as outer cell layers can shade inner cells and present a longer pathlength for UVR absorption (Franklin and Forster 1997, Roleda et al. 2007a). Thallus translucence or opacity can also influence reflection, attenuation, scattering, absorption or transmittance of UV radiation (Caldwell et al. 1983). Among kelps, optically dark pigmented juvenile sporophytes show strong absorbance in the UV waveband (Roleda et al. 2005, 2006b), characteristic of the UV-absorbing phlorotannins accumulated within the outer cortical layer of the thalli of Laminariales (Lüder and Clayton 2004, Shibata et al. 2004). In *Hormosira banksii* (Turner) Decaisne (Fucales), a brown layer consisting of oxidised phenolic compounds released from the peripheral cells forms a protective lamina for the photosynthetic tissue beneath (Schoenwaelder 2002). UVR can, therefore, be attenuated by cellular UV-absorbing compounds and cell walls of the epidermal tissue, effectively reducing UV fluence before it reaches important physiological targets. UV-absorbing compounds located in the epidermis are reported to protect Arctic vascular plants from UVB radiation (Nybakken et al. 2004). More recently, Peschek et al. (2010) investigated the UVR screening capacity of macrothalli from marine chlorophytes, and reported highly efficient UVA and UVB screening for various members of the Cladophorales. Their UVB-screening potential varied between 40% and 85% of incoming UVB radiation. Although, the nature and localization of the underlying UV-absorbing compounds/structures are still unknown, they seem to be localized in the cell wall. One chemical candidate could be sporopollenins, which occur in the cell walls and zygospores of green microalgae, and are known as an almost undegradable macromolecule responsible for UV attenuation (Xiong et al. 1995). The search for flavonoid-like compounds at least in *Urospora penicilliformis* gave a negative result (Roleda et al. 2009b).

Photoprotective substances

One of the most important physiochemical acclimation mechanisms against UVR is the biosynthesis and accumulation of UV-sunscreen substances. Typically absorbing in the UVA and UVB range, these biomolecules have been invoked to function as passive shielding solutes by dissipating the absorbed short wavelength radiation energy in the form of harmless heat without generating photochemical reactions (Bandaranayake 1998). The most common photoprotective sunscreens in many polar micro- and macroalgae are the mycosporine-like amino acids (MAAs), a suite of chemically closely related, colourless, water-soluble, polar and, at cellular pH, uncharged or zwitterionic amino acid derivatives. MAAs exhibit a high molar absorptivity for UVA and UVB, and have been reported as photochemically stable molecules, which are prerequisites for their sunscreen function (Conde et al. 2000). Whereas MAAs have been mainly observed in numerous Rhodophyta (Karentz et al. 1991, Hoyer et al. 2001), Phaeophyceae and most Chlorophyta typically lack these compounds, except the green algal genus *Prasiola* which contains high concentrations of a unique MAA with an absorption maximum at 324 nm (Hoyer et al. 2001, Karsten et al. 2005). Most Phaeophyceae synthesise photoprotective phlorotannins under UV exposure (Pavia et al. 1997), these compounds will be reviewed in detail below. Whereas planktonic centric diatoms from Antarctica also accumulate MAAs under UVR (Buma et al. 2006), benthic pennate taxa are surprisingly poor in these compounds or even lack them (Wulff et al. 2008a).

Field and laboratory experiments in the Arctic with the intertidal red alga *Devaleraea ramentacea* (Linnaeus) Guiry showed a continuous decrease in photosynthetic performance under UVR with increasing collecting depths between 1 and 5 m. Total MAA concentration was also correlated with original sampling depth, i.e., shallow water isolates contained much higher amounts than algae from deeper waters, also indicating a strong correlation between MAA contents and degree of sensitivity of photosynthetic activity in *D. ramentacea* (Karsten et al. 1999). The sunscreen function of MAAs has been further inferred in other polar red algae from a decrease in concentration with increasing depth (Hoyer et al. 2001, 2003). Supra- and eulittoral Rhodophyta typically experience the strongest insolation, and consequently synthesise and accumulate highest MAA values, which are positively correlated with the natural UV doses (Karsten et al. 1998). In contrast, many red algal taxa growing in deep waters are biochemically incapable of producing MAAs (Hoyer et al. 2001, 2002; Karsten et al. 2001), which explains their strong sensitivity to high ambient solar radiation. These species do not experience UV exposure in nature, and, hence, there is no physiological need to synthesise and accumulate metabolically expensive nitrogen-containing MAAs. This in turn saves energy to better support other essential pathways, such as the biosynthesis of light-harvesting phycobilisomes to guarantee sufficient PAR absorption under the prevailing low-light conditions.

In the Antarctic red alga *Palmaria decipiens*, juveniles collected in the upper to mid sublittoral during winter contained low MAA concentrations, whereas mature plants collected in late spring and summer exhibited significantly higher values, indicating strong seasonal effects, which are related to the changing daylengths and radiation conditions (Post and Larkum 1993). Based on the MAA concentrations and the induction patterns after exposure to different defined radiation conditions, red algae can be physiologically classified into three categories (Hoyer et al. 2001): Type I – no MAAs at all, Type II – MAAs inducible in variable concentrations, and Type III – permanently high MAA values. Whereas Type I represents deep-water red algae of the lower sublittoral, Types II and III species grow from the supra- and eulittoral to the upper and mid-sublittoral zone. Experiments under defined radiation conditions demonstrated that the induction, biosynthesis and accumulation of MAAs are very flexible and species-specific processes. Whereas some polar Rhodophyta synthesise MAAs particularly under UVB, others do so under UVA or higher PAR (Hoyer et al. 2003). Although experimental evidence for a particular trigger mechanism is still missing, it is reasonable to assume that a signal transduction pathway is involved in the overall process leading to high MAA concentrations. Based on the different types of MAA induction patterns in polar red algae, the presence of various photoreceptors, most probably between the blue light and UVB range, must be considered (Kräbs et al. 2002).

The entire algal thallus does not respond uniformly to ambient UVR. Young apical or marginal zones, i.e., growing cells, synthesise and preferentially accumulate MAAs, leading to cross-sectional and longitudinal concentration gradients (Hoyer et al. 2001). Populations of *Devaleraea ramentacea* collected from very shallow water typically exhibit green apices and red basal parts. The more exposed green tips contain 5-fold higher MAA contents than the red bases (Karsten et al. 1999). Older tissue regions usually exhibit thicker cell walls and a leathery texture and are therefore optically opaque. The higher MAA concentrations in the most exposed thallus regions are essential to guarantee protection of the more delicate meristematic cells.

From an ecological standpoint, the ephemeral, tufty *Prasiola* is intriguing because of its ability to grow in polar regions subaerially on soil and rocks underneath or near seagull or penguin colonies (Holzinger et al. 2006), i.e., habitats rich in nitrogen-containing faeces. Considering a relation between MAA concentration and nitrogen availability in different species of the red alga *Porphyra* (Korbee et al. 2005), it seems that this nutrient might be a critical factor for the broad photophysiological tolerance of *Prasiola* under subaerial conditions (Holzinger et al. 2006). Seasonal studies on *Prasiola* in Antarctica indicate some variation in MAA concentrations but always with high minimum steady-state amounts (Jackson and Seppelt 1997). Consequently, it appears that the 324 nm MAA in *Prasiola* is expressed constitutively.

As well as functioning as natural UV-protective compounds, some MAAs, such as mycosporine-glycine, also have moderate antioxidant activity (Dunlap and Yamamoto 1995). In addition, the presumed biochemical precursor of MAAs, 4-deoxygadusol, exhibits strong antioxidant activity (Dunlap et al. 1998). The photophysicochemical properties of MAAs guarantee both a high UV-protective effectiveness in combination with antioxidant capabilities.

Phlorotannins are exclusively produced by brown algae and include hundreds of phenolic compounds (126–650 kDa) composed of the monomer phloroglucinol (Amsler and Fairhead 2005). The primary biological function of phlorotannins is the strengthening of cell walls, and one of the several suggested secondary functions is the protection of algae against harmful UVB radiation (Ragan and Glombitza 1986, Swanson and Druehl 2002). This role is based on the absorbance of phlorotannins in the UV region, with peaks at approximately 195 nm and 270 nm, whereas the absorbance shoulders extend from <200 nm to 400 nm (Ragan and Glombitza 1986, Pavia et al. 1997).

Experimental evidence showed that intra- and/or extracellular phlorotannins within very dense kelp spore suspensions used as biofilters shielded underlying kelp spores from UVB thereby allowing germination (Roleda et al. 2006c, Swanson and Druehl

2002). Such UVB protection of kelp spores by exuded phlorotannins might also occur in dense kelp beds (Swanson and Druehl 2002), although the sunscreen capability of phlorotannins diminishes with decreasing concentrations (Roleda et al. 2006c). Thus, different UVR responses of spore germination of the Arctic *Alaria esculenta* and *Laminaria digitata* at different dates within 2 weeks (Wiencke et al. 2006) are probably attributed to changing phlorotannin contents in the water column.

Whether UVR exposure can induce the formation and accumulation of phlorotannins in Phaeophyceae is still an open question because of contradictory results in the literature. Light-microscopic analysis of zoospores of the Arctic kelps *Alaria esculenta*, *Saccorhiza dermatodea* and *Laminaria digitata* indicated that UVB treatment led to an enlargement of phlorotannin-containing physodes, an aggregation of several small physodes to form larger ones, as well as to an exudation of physodes into the surrounding seawater (Wiencke et al. 2004, 2007, Roleda et al. 2006c). These observations were, however, not confirmed by an ultrastructural analysis of UVB-exposed zoospores of the UVR sensitive, cold-temperate *Laminaria hyperborea* (Steinhoff et al. 2008).

Moreover, zoospore suspensions of *Alaria esculenta*, *Laminaria digitata* and *Saccharina latissima*, juveniles and embryos of *Fucus gardneri* P.C. Silva, and sporophytes of *Desmarestia anceps* Montagne and *D. menziesii* J. Agardh did not show any concentration increase after UVB treatment (Henry and van Alstyne 2004, Fairhead et al. 2006, Müller et al. 2009). An accumulation of phlorotannin contents after treatment with UVB was also not detected in the seawater surrounding blades of *Macrocystis integrifolia* and zoospores of *A. esculenta*, *L. digitata* and *S. latissima* (Swanson and Druehl 2002, Müller et al. 2009). Otherwise, an induction of phlorotannins by UVR was observed in the blades of the giant kelp *M. integrifolia* (Swanson and Druehl 2002) and in sporophytic tissues of *Ascophyllum nodosum* (Linnaeus) Le Jolis, *S. latissima* and *Nereocystis luetkeana* (K. Mertens) Postels *et* Ruprecht (Pavia et al. 1997, Pavia and Brock 2000, Swanson and Fox 2007).

Generally, the often observed discrepancies between studies on UVR-sunscreen function of phlorotannins might be attributed to low replicate numbers, different exposure times and irradiation intensities, analytical methods and/or selected species/life stages studied, although ontogenetic, intra- and interspecific phlorotannin levels are known to be highly variable (Amsler and Fairhead 2005). In particular, the standard measurement of hundreds of compounds pooled as phlorotannins could lead to false negative or positive results, whereas the analysis of single phlorotannin compounds after UVB treatment might help to better understand the UV protective role of phlorotannins. Therefore, future studies should focus more attention on the simultaneous analysis of UVR responses of different algal species in the field, of their intra- and extracellular phlorotannin levels and compositions, and of their absorption properties to substantiate the role of phlorotannins as UV protectants in brown algae from polar environments.

Interactive effects of UVR with other abiotic and biotic factors

In nature, plants are subject to the influence of a multitude of abiotic and biotic factors (Alexieva et al. 2003). Stress factors do not operate individually or independently, so that spatial or temporal variations and interactions between and co-variation of stresses are the norm in the natural environment (Jones et al. 1993, Vinebrooke et al. 2004). Stressors act synergistically when their combined effect on biological components is larger than the responses to each single stressor, and antagonistically when the impact is smaller (Folt et al. 1999). It is also possible that plants exposed to a single stress agent are capable of increasing their resistance to subsequent unfavourable impacts (cross-adaptation; Alexieva et al. 2003).

In polar regions, several environmental factors have changed and are still changing in parallel, such as ozone depletion and the concomitant rise in UVB, as well as the increase in greenhouse gases (e.g., CO_2) and the resulting global warming and acidification of the oceans. Consequently, any increase in UVB does not have an independent effect on marine benthic micro- and macroalgae because changes in the other environmental factors have to be considered as well (Chapter 2; Zacher et al. 2009b).

Bothwell et al. (1994) reported that benthic diatoms benefit from UVR because their grazers were more susceptible to applied radiation. Also, in the field, grazers had a larger impact than UVR, although no negative effect of UVR on the grazers was apparent (Zacher et al. 2007c).

Hoffman et al. (2003) supported the hypothesis that temperature mitigates the net biological effect of UVR on macroalgae and *vice versa*. Correspondingly, the photosynthetic performances of sporophytes of the Antarctic green alga *Enteromorpha bulbosa* (Suhr) Montagne, the subantarctic *Ulva clathrata* (Roth) C. Agardh, the Arctic kelp *Alaria esculenta*, and the Arctic red algae *Palmaria palmata*, *Coccotylus truncatus* (Pallas) M.J. Wynne *et* J.N. Heine and *Phycodrys rubens* (Linnaeus) Batters were less impaired by UVB at higher temperatures compared with lower temperatures (van de Poll et al. 2002b, Rautenberger and Bischof 2006, Fredersdorf et al. 2009). In the latter three red algae, the recovery of photosynthesis after UVB stress was likewise improved at higher temperatures, although the influence of UVB acted predominantly on relative growth rate and CPD accumulation as a marker of DNA damage (van de Poll et al. 2002b). In contrast, the germination of zoospores and photosynthetic performance of sporophytes of the UVB-tolerant *A. esculenta* were mainly determined by temperature when thalli were exposed to both UVR and stressful temperature (Müller et al. 2008, Fredersdorf et al. 2009). Germination of highly UVB-vulnerable zoospores of the Arctic kelps *Saccharina latissima* and *Laminaria digitata* were more UVB affected at unfavourable lower and higher temperatures rather than at the ambient temperature of 7 °C (Müller et al. 2008). Germination of zoospores of the Helgoland ecotype of *L. digitata* was, however, less harmed by UVB at temperatures higher than 2 °C (Müller et al. 2008). In contrast, the sensitivity of germlings of three *Fucus* species from Helgoland to UVB radiation was enhanced with increasing temperature (Altamirano et al. 2003). Thus, the temperature dependence of UVR effects on seaweeds varies highly within and among species and genera.

There are only few ecophysiological studies on interactive effects of UVR and/or temperature with salinity in polar seaweeds, although significant amounts of freshwater regularly flow into polar waters due to rivers and melting of glaciers and snow (Karsten et al. 2003, Fredersdorf et al. 2009). These studies also reveal life stage-, intra- and interspecies-specific responses to interactive stress. In a study by Karsten et al. (2003), interactive effects of UVR and salinity on two shallow water Arctic red macroalgae, *Devaleraea ramentacea* and *Palmaria palmata*, were investigated. Whereas *D. ramentacea* exhibited euryhaline features and acclimated well to the

UVR applied, *P. palmata* can be characterised as a stenohaline plant because of its high mortality already under mild hyposaline conditions. In addition, the latter species had a limited ability to acclimate to changing PAR/UV radiation pointing to a relatively low physiological plasticity.

At the assemblage level, species-specific UVB responses of algae and distinct food preferences of grazers result in complex changes (Zacher et al. 2007b, Zacher and Campana 2008). For instance, the preference of snails for feeding on UVB-sensitive green algal recruits favoured the recruitment of leathery red algal recruits in the Antarctic intertidal phytobenthos (Zacher et al. 2007b). Although UVR effects on diversity and species composition of macroalgal assemblages were less pronounced, the biomass and recruit density of the assemblage was over the long-term more substantially affected by grazing (Zacher et al. 2007b, Zacher and Campana 2008). The effects of UVB and temperature on the recruitment and succession of the green alga *Ulva intestinalis* Linnaeus from Nova Scotia (Canada) were likewise mitigated by consumers, whereas nutrient supplements enhanced grazing pressure (Lotze et al. 2002). High levels of nutrients also abated detrimental UVB effects on a diatom-dominated, soft bottom assemblage off the Swedish west coast (Wulff et al. 2000). In Antarctic waters, which are characterised by generally high nutrient levels (Korb and Gerard 2000), a diatom-dominated, hard bottom assemblage was almost unaffected by UVB, although grazing structured the assemblage almost entirely (Zacher et al. 2007c). These few studies on interactive effects on micro- and macrophytobenthos of polar latitudes indicate that grazing pressure might have a generally stronger impact on the assemblage than UVB, whereas nutrients might influence the impact of grazing on algae. Comparable observations were made in the Arctic microbial food web, where negative UVB effects on the growth of nano- and picoautotrophs as well as heterotrophic nanoflagellates were much less substantial than mortality imposed by grazing (Wickham and Carstens 1998).

Nevertheless, there are numerous fundamental and still unanswered questions about the tolerance of benthic micro- and macroalgae to combined stress, especially at the respective ratio of PAR/UVR. One unsolved problem is related to the comparability of experiments undertaken in the field and in the laboratory because natural solar radiation (waveband ratios, intensities) is very difficult to artificially mimic. Consequently, PAR/UVR ratios in both approaches can be quite different and hence in the laboratory often so unnatural that the resulting data have to be interpreted with caution. Clearly, many more ecologically relevant studies on interactive effects, especially on the most sensitive developmental stages and on the algal assemblages are required for predictions of the effects of global change on polar marine ecosystems.

Ecological consequences

To determine ecological effects resulting from increased UVR is a difficult task. Probably, effects on the community or even ecosystem level had already taken place before the official discovery of the ozone hole over Antarctica in the 1970s. From field experiments in 10 different coastal regions ranging from tropical to polar waters, Wahl et al. (2004) concluded that any observed UV effect disappeared during species succession over 2–3 months. The Antarctic assemblage studied was dominated by benthic diatoms (Wahl et al. 2004). Although the few existing studies on benthic microalgae point to a capacity to acclimate to ambient and increasing UVR, much more ecological research is needed to identify the underlying mechanisms and interactions, and particularly to include several trophic levels under natural conditions. Considering only one trophic level could lead to misinterpretations of possible UV effects, e.g., the observed negative UV effects on grazers might be interpreted as a positive UV effect on benthic diatoms (Bothwell et al. 1994). Also, Xue et al. (2005) and Bancroft et al. (2007) stress the importance of including several trophic levels, as well as multiple stressors because UVB effects on one trophic level can lead to instabilities of a whole ecosystem favouring a few hardy species.

For macroalgae, colonisation stages and spores seem to be more sensitive to UVR compared with respective adult thalli. Consequently, in polar environments influenced by ice and scouring (frequent colonisation events) UVR could be an important controlling factor. In addition, a changed zonation pattern has been suggested (e.g., Bischof et al. 2006) with possible cascade effects at the assemblage level, including epiphytes and grazers. From temperate regions, Lotze et al. (2002) reported that although UVR could be a structuring force during early succession (early colonising species and microscopic recruits), grazers could mediate this effect over time. Including several abiotic factors (UV, temperature and nutrients) led to complex interactions of ecological controls on macroalgal recruitment, and various synergistic effects of abiotic and biotic (e.g., grazers) factors were described (Lotze and Worm 2002).

Although the number of field studies on benthic algae in polar areas has increased over the recent years, still very few take the ecosystem approach (Chapter 14; Wahl et al. 2004, Zacher et al. 2007b, Campana et al. 2009). For future studies and to better understand ecosystem consequences, this is clearly one of the greatest challenges to elucidate, e.g., UV effects on standing stock biomass, benthic primary production, biodiversity, food quality, trophic webs and interactions, etc.

Conclusion

Although considerable progress has been achieved in recent years on UVR effects, our present knowledge about the physiological capabilities of benthic micro- and macroalgae from polar regions is still fragmentary. Owing to the extreme remoteness of the polar regions and the infrequency of scientific studies, many open questions still have to be addressed. In particular, more experimental evidence is needed from both field and controlled laboratory studies to precisely document susceptibility, tolerances, acclimation and adaptation with an emphasis on the underlying mechanisms, which can only be elucidated by modern genomic, proteomic and metabolomic approaches. In addition, and as already mentioned, more ecosystem approaches and studies on interactive effects are needed to depict a more holistic picture.

References

Abele, D., G.A. Ferreyra and I. Schloss. 1999. H_2O_2 accumulation from photochemical production and atmospheric wet deposition in Antarctic coastal and off-shore waters of Potter Cove, King George Island. *Antarc. Sci. 11*: 131–139.

Aguilera, J., U. Karsten, H. Lippert, B. Vögele, E. Philipp, D. Hanelt and C. Wiencke. 1999. Effects of solar radiation on growth, photosynthesis and respiration of marine macroalgae from the Arctic. *Mar. Ecol. Prog. Ser. 191*: 109–119.

Aguilera, J., K. Bischof, U. Karsten, D. Hanelt and C. Wiencke. 2002a. Seasonal variation in ecophysiological patterns in macroalgae from an Arctic fjord. II. Pigment accumulation and biochemical defence systems against high light stress. *Mar. Biol. 140*: 1087–1095.

Aguilera, J., A. Dummermuth, U. Karsten, R. Schriek and C. Wiencke. 2002b. Enzymatic defenses against photooxidative stress induced by ultraviolet radiation in Arctic marine macroalgae. *Polar Biol. 25*: 432–441.

Alexieva, V., S. Ivanov, I. Sergiev and E. Karanov. 2003. Interaction between stresses. *Bulg. J. Plant Physiol. (Special Issue) 2003*: 1–17.

Altamirano, M., A. Flores-Moya and F.L. Figueroa. 2003. Effects of UV radiation and temperature on growth of germlings of three species of *Fucus* (Phaeophyceae). *Aquat. Bot. 75*: 9–20.

Amsler, C.D. and V.A. Fairhead. 2005. Defensive and sensory chemical ecology of brown algae. *Adv. Bot. Res. 43*: 1–91.

Asada, K. 1994. Production and action of active oxygen species in photosynthetic tissues. *In*: (C.H. Foyer and P.M. Mullineaux, eds) *Causes of photooxidative stress and amelioration of defence systems in plants*. CRC Press, Boca Raton, FL. pp. 77–104.

Bancroft, B.A., N.J. Baker and A.R. Blaustein. 2007. Effects of UVB radiation on marine and freshwater organisms: a synthesis through meta-analysis. *Ecol. Lett. 10*: 332–345.

Bandaranayake, W.M. 1998. Mycosporines: are they nature's sunscreens? *Nat. Prod. Rep. 15*: 159–172.

Bischof, K., D. Hanelt, H. Tüg, U. Karsten, P.E.M. Brouwer and C. Wiencke. 1998. Acclimation of brown algal photosynthesis to ultraviolet radiation in Arctic coastal waters (Spitsbergen, Norway). *Polar Biol. 20*: 388–395.

Bischof, K., D. Hanelt and C. Wiencke. 1999. Acclimation of maximal quantum yield of photosynthesis in the brown alga *Alaria esculenta* under high light and UV radiation. *Plant Biol. 1*: 435–444.

Bischof, K., D. Hanelt and C. Wiencke. 2000. UV-effects on photosynthesis and related enzyme reactions of marine macroalgae. *Planta 211*: 555–562.

Bischof, K., D. Hanelt, J. Aguilera, U. Karsten, B. Vögele, T. Sawall and C. Wiencke. 2002a. Seasonal variation in ecophysiological patterns in macroalgae from an Arctic fjord. I. Sensitivity of photosynthesis to ultraviolet radiation. *Mar. Biol. 140*: 1097–1106.

Bischof, K., G. Kräbs, C. Wiencke and D. Hanelt. 2002b. Solar ultraviolet radiation affects the activity of ribulose-1,5-biphosphate carboxylase-oxygenase and the composition of photosynthetic and xanthophyll cycle pigments in the intertidal green alga *Ulva lactuc*a L. *Planta 215*: 502–509.

Bischof, K., I. Gómez, M. Molis, D. Hanelt, U. Karsten, U. Lüder, M.Y. Roleda, K. Zacher and C. Wiencke. 2006. Ultraviolet radiation shapes seaweed communities. *Rev. Environ. Sci. Biotechnol. 5*: 141–166.

Bothwell, M.L., D.M.J. Sherbot and C.M. Pollock. 1994. Ecosystem response to solar ultraviolet-B radiation: influence of trophic-level interactions. *Science 265*: 97–100.

Britt, A.B. 1996. DNA damage and repair in plants. *Annu. Rev. Plant Physiol. Plant Mol. Biol. 47*: 75–100.

Buma, A.G.J., H.J. Zemmelink, K. Sjollema and W.W.C. Gieskes. 1996. UVB radiation modifies protein and photosynthetic pigment content, volume and ultrastructure of marine diatoms. *Mar. Ecol. Prog. Ser. 142*: 47–54.

Buma, A.G.J., T. van Oijen, W. van de Poll, M.J.W. Veldhuis and W.W.C. Gieskes. 2000. The sensitivity of *Emiliana huxleyi* (Prymnesiophyceae) to ultraviolet-B radiation. *J. Phycol. 36*: 296–303.

Buma, A.G.J., S.W. Wright, R. van den Enden, W.H. van den Poll and A.T. Davidson. 2006. PAR acclimation and UVBR-induced DNA damage in Antarctic marine microalgae. *Mar. Ecol. Prog. Ser. 315*: 33–42.

Cahoon, L.B. 1999. The role of benthic microalgae in neritic ecosystems. *Oceanogr. Mar. Biol. Annu. Rev. 37*: 47–86.

Caldwell, M.M., R. Robberecht and S.D. Flint. 1983. Internal filters: prospects for UV-acclimation in higher plants. *Physiol. Plant. 58*: 445–450.

Campana, G., M.L. Quartino, A. Yousif and A. Wulff. 2008. Impacts of UV radiation and grazing on the structure of a subtidal benthic diatom assemblage in Antarctica. *Ber. Polarforsch. Meeresforsch. 571*: 302–310.

Campana, G.L., K. Zacher, A. Fricke, M. Molis, A. Wulff, M. L. Quartino and C. Wiencke. 2009. Drivers of colonization and succession in polar benthic macro- and microalgal communities. *Bot. Mar. 52*: 655–668.

Campbell, D., M.J. Eriksson, G.Öquist, P. Gustafsson and A.K. Clarke. 1998. The cyanobacterium *Synechococcus* resists UV-B by exchanging photosystem II reaction-center D1 proteins. *Proc. Natl. Acad. Sci. USA 95*: 364–369.

Carr, G.M., H.C. Duthie and W.D. Taylor. 1997. Models of aquatic plant productivity: a review of the factors that influence growth. *Aquat. Bot. 59*: 195–215.

Conde, F.R., M.S. Churio and C.M. Previtali. 2000. The photo-protector mechanism of mycosporine-like amino acids. Excited-state properties and photostability of porphyra-334 in aqueous solution. *J. Photochem. Photobiol. 56B*: 139–144.

Cooper, J.C. and R.G. Zika. 1983. Photochemical formation of hydrogen peroxide in surface and ground waters exposed to sunlight. *Science 220*: 711–712.

De Brouwer, J.F.C., K. Wolfstein, G.K. Ruddy, T.E.R. Jones and L.J. Stal. 2005. Biogenic stabilization of intertidal sediments: the importance of extracellular polymeric substances produced by benthic diatoms. *Microb. Ecol. 49*: 501–512.

Duffy, J.E. and M.E. Hay. 2000. Strong impacts of grazing amphipods on the organization of a benthic community. *Ecol. Monogr. 70*: 237–263.

Dummermuth, A.L., U. Karsten, K.M. Fisch, G.M. König and C. Wiencke. 2003. Responses of marine macroalgae to hydrogen-peroxide stress. *J. Exp. Mar. Biol. Ecol. 289*: 103–121.

Dunlap, W.C. and Y. Yamamoto. 1995. Small-molecule antioxidants in marine organisms: antioxidant activity of mycosporine-glycine. *Comp. Biochem. Physiol. B 112*: 105–114.

Dunlap, W.C., B.E. Chalker, W.M. Bandaranayake and J.J. Wu Won. 1998. Nature's sunscreen from the Great Barrier Reef, Australia. *Int. J. Cosmet. Sci. 20*: 41–51.

Elstner, E.F. 1990. *Der Sauerstoff.* BI Wissenschaftlicher Verlag, Mannheim. pp. 529.

Fairhead, V.A., D.A. Amsler and J.B. McClintock. 2006. Lack of defence or phlorotannin induction by UV radiation or mesograzers in *Desmarestia anceps* and *D. menziesii* (Phaeophyceae). *J. Phycol. 42*: 1174–1183.

Folt, C.L., C.Y. Chen, M.V. Moore and J. Burnaford. 1999. Synergism and antagonism among multiple stressors. *Limnol. Oceanogr. 44*: 864–877.

Franklin, L.A. and R.M. Forster. 1997. The changing irradiance environment: consequences for marine macrophyte physiology, productivity and ecology. *Eur. J. Phycol. 32*: 207–232.

Frederick, J.E., H.E. Snell and E.K. Haywood. 1989. Solar ultraviolet radiation at the Earth's surface. *Photochem. Photobiol. 50*: 443–450.

Fredersdorf, J., R. Müller, S. Becker, C. Wiencke and K. Bischof. 2009. Interactive effects of radiation, temperature and salinity on different life history stages of Arctic kelp *Alaria esculenta* (Phaeophyceae). *Oecologia 160*: 483–492.

Glud, R.N., J. Woelfel, U. Karsten, M. Kühl and S. Rysgaard. 2009. Benthic microalgal production in the Arctic: applied methods and status of the current database. *Bot. Mar. 52*: 559–571.

Gribble, G.W. 2003. The diversity of naturally produced organohalogens. *Chemosphere 52*: 289–297.

Hanelt, D. 2004. The radiation, temperature and salinity regime in Kongsfjorden. *Ber. Polarforsch. Meeresforsch. 492*: 14–25.

Hanelt, D., C. Wiencke and W. Nultsch. 1997. Influence of UV radiation on photosynthesis of Arctic macroalgae in the field. *J. Photochem. Photobiol. B: Biol. 38*: 40–47.

Hannack, K., P. Kestler, O. Sicken and W. Westheide. 1998. The influence of UV radiation on number and ultrastructure of the endosymbiotic dinoflagellates in the sea anemone *Cereus pedunculatus* (Anthozoa: Actiniaria). *Helgol. Mar. Res. 51*: 487–502.

Henry, B.E. and K.L. van Alstyne. 2004. Effects of UV radiation on growth and phlorotannins in *Fucus gardneri* (Phaeophyceae) juveniles and embryos. *J. Phycol. 40*: 527–533.

Herms, D.A. and W.J. Mattson. 1992. The dilemma of plants: to grow or defend. *Q. Rev. Biol. 67*: 283–335.

Hessen, D.O., E. van Donk and T. Andersen. 1995. Growth responses, P-uptake and loss of flagellae in *Chlamydomonas reinhardtii*. *J. Plankton Res. 17*: 17–27.

Hoffman, J.R., L.J. Hansen and T. Klinger. 2003. Interactions between UV radiation and temperature limit inferences from single-factor experiments. *J. Phycol. 39*: 268–272.

Holzinger, A. and C. Lütz. 2006. Algae and UV radiation: effects on ultrastructure and related metabolic functions. *Micron 37*: 190–207.

Holzinger, A., C. Lütz, U. Karsten and C. Wiencke. 2004. The effect of ultraviolet radiation on ultrastructure and photosynthesis in the red macroalgae *Palmaria palmata* and *Odonthalia dentata* from Arctic waters. *Plant Biol. 6*: 568–577.

Holzinger, A., U. Karsten, C. Lütz and C. Wiencke. 2006. Ultrastructure and photosynthesis in the supralittoral green macroalga *Prasiola crispa* from Spitsbergen (Norway) under UV exposure. *Phycologia 45*: 168–177.

Holzinger, A., M.Y. Roleda and C. Lütz. 2009. The vegetative arctic freshwater green alga *Zygnema* is insensitive to experimental UV exposure. *Micron 40*: 831–838.

Hoyer, K., U. Karsten, T. Sawall and C. Wiencke. 2001. Photoprotective substances in Antarctic macroalgae and their variation with respect to depth distribution, different tissues and developmental stages. *Mar. Ecol. Prog. Ser. 211*: 117–129.

Hoyer, K., U. Karsten and C. Wiencke. 2002. Induction of sunscreen compounds in Antarctic macroalgae by different radiation conditions. *Mar. Biol. 141*: 619–627.

Hoyer, K., U. Karsten and C. Wiencke. 2003. Inventory of UV-absorbing mycosporine-like amino acids in polar macroalgae and factors controlling their concentration. *In*: (A.H.L. Huiskes, W.W.C. Gieskes, J. Rozema, R.M.L. Schorno, S.M. Van der Vies SM and W.J. Wolff, eds) *Proceedings of the VIII SCAR Meeting*. Backhuys Publishers, Leiden. pp. 56–62.

Huovinen, P.S., A.O.J. Oikari, M.R. Soimasuo and G.N. Cherr. 2000. Impact of UV radiation on the early development of the giant kelp (*Macrocystis pyrifera*) gametophytes. *Photochem. Photobiol. 72*: 308–313.

Jackson, A.E. and R.D. Seppelt. 1997. Physiological adaptations to freezing and UV radiation exposure in *Prasiola crispa*, an Antarctic terrestrial alga. *In*: (B. Battaglia, J. Valencia and D.W.H. Walton, eds) *Antarctic communities: species, structure, and survival.* Cambridge University Press, Cambridge. pp. 226–233.

Jones, H.G., T.J. Flowers and M.B. Jones. 1993. *Plants under stress. Biochemistry, physiology and ecology and their application to plant improvement.* Cambridge University Press, Cambridge. pp. 268.

Karentz, D. 1994. Ultraviolet tolerance mechanisms in Antarctic marine organisms. *Antarct. Res. Ser. 62*: 93–110.

Karentz, D., J.E. Cleaver and D.L. Mitchell. 1991. Cell survival characteristics and molecular responses of Antarctic phytoplankton to ultraviolet-B radiation. *J. Phycol. 27*: 326–341.

Karsten, U., T. Sawall, D. Hanelt, K. Bischof, F.L. Figueroa, A. Flores-Moya and C. Wiencke. 1998. An inventory of UV-absorbing mycosporine-like amino acids in macroalgae from polar to warm-temperate regions. *Bot. Mar. 41*: 443–453.

Karsten, U., K. Bischof, D. Hanelt, H. Tüg and C. Wiencke. 1999. The effect of UV-radiation on photosynthesis and UV-absorbing substances in the endemic Arctic macroalga *Devaleraea ramentacea* (Rhodophyta). *Physiol. Plant. 105*: 58–66.

Karsten, U., K. Bischof and C. Wiencke. 2001. Photosynthetic performance of Arctic macroalgae after transplantation from deep to shallow waters followed by exposure to natural solar radiation. *Oecologia 127*: 11–20.

Karsten, U., A. Dummermuth, K. Hoyer and C. Wiencke. 2003. Interactive effects of ultraviolet radiation and salinity on the ecophysiology of two Arctic red algae from shallow waters. *Polar Biol. 26*: 249–258.

Karsten, U., T. Friedl, R. Schumann, K. Hoyer and S. Lembcke. 2005. Mycosporine-like amino acids (MAAs) and phylogenies in green algae: *Prasiola* and its relatives from the Trebouxiophyceae (Chlorophyta). *J. Phycol. 41*: 557–566.

Korb, R.E. and V.A. Gerard. 2000. Nitrogen assimilation characteristics of polar seaweeds from differing nutrient environments. *Mar. Ecol. Prog. Ser. 198*: 83–92.

Korbee, N., P. Huovinen, F.L. Figueroa, J. Aguilera and U. Karsten. 2005. Availability of ammonium influences photosynthesis and the accumulation of MAAs in two *Porphyra* species (Bangiales, Rhodophyta). *Mar. Biol. 146*: 645–654.

Kräbs, G., K. Bischof, D. Hanelt, U. Karsten and C. Wiencke. 2002. Wavelength-dependent induction of UV-absorbing mycosporine-like amino acids in the red alga *Chondrus crispus*. *J. Exp. Mar. Biol. Ecol. 268*: 69–82.

Krause, G.H. and E. Weis. 1991. Chlorophyll fluorescence and photosynthesis, the basics. *Annu. Rev. Plant Physiol. Plant Mol. Biol. 42*: 313–349.

Laturnus, F. 2001. Marine macroalgae in polar regions as sources for volatile organohalogens. *Environ. Sci. Pollut. Res. 8*: 103–108.

Laturnus, F., K.F. Haselmann, T. Borch and C. Grøn. 2002. Terrestrial natural sources of trichloromethane (chloroform, $CHCl_3$) – an overview. *Biogeochemistry 60*: 121–139.

Li, Q., M. Osada, M. Kashihara, K. Hirohashi and A. Kijima. 2000. Effects of ultraviolet radiation on genetical inactivation and morphological structure of sperm of the Japanese scallop, *Patinopecten yessoensis*. *Aquaculture 186*: 233–242.

Lois, R. and B.B.N. Buchanan. 1994. Severe sensitivity to ultraviolet radiation in an *Arabidopsis* mutant deficient in flavonoid accumulation: II. Mechanisms of UV-resistance in *Arabidopsis*. *Planta 194*: 504–509.

Long, A. and R.E. Michod. 1995. Origin of sex for error repair I. Sex, diploidy, and haploidy. *Theor. Popul. Biol. 47*: 18–55.

Long, S.P., S. Humphries and P.G. Falkowski. 1994. Photoinhibition of photosynthesis in nature. *Annu. Rev. Plant Physiol. Plant Mol. Biol. 45*: 633–662.

Lotze, H.K. and B. Worm. 2002. Complex interactions of climatic and ecological controls on macroalgal recruitment. *Limnol. Oceanogr. 47*: 1734–1741.

Lotze, H.K., B. Worm, M. Molis and M. Wahl. 2002. Effects of UV radiation and consumers on recruitment and succession of a marine macrobenthic community. *Mar. Ecol. Prog. Ser. 243*: 57–66.

Lüder, U.H. and M.N. Clayton. 2004. Induction of phlorotannins in the brown algae *Ecklonia radiata* (Laminariales, Phaeophyta) in response to simulated herbivory – the first microscopic study. *Planta 218*: 928–937.

Lütz, C., H.K. Seidlitz and U. Meindl. 1997. Physiological and structural changes in the chloroplast of the green alga *Micrasterias denticulata* induced by UV-B simulation. *Plant Ecol. 128*: 54–64.

Makarov, M.V. and G.M. Voskoboinikov. 2001. The influence of ultraviolet-B radiation on spore release and growth of the kelp *Laminaria saccharina*. *Bot. Mar. 44*: 89–94.

Máté, Z., L. Sass, M. Szekeres, I. Vass and F. Nagy. 1998. UVB induced differential transcription of psbA genes encoding the D1 protein of photosystem II in the cyanobacterium *Synechocystis* 6803. *J. Biol. Chem. 273*: 17439–17444.

McKenzie, R.L., L.O. Björn, A. Bais and M. Ilyas. 2003. Changes in biologically-active ultraviolet radiation reaching the Earth's surface. *Photochem. Photobiol. Sci. 2*: 5–15.

McLean, R.J. and F. Pessoney. 1971. Formation and resistance of akinetes of *Zygnema*. *In*: (B.C. Parkaer and R.M.Brown Jr., eds) *Contributions in Phycology*. Allen Press, Lawrence, Kansas, USA. pp. 145–152.

Meindl, U. and C. Lütz. 1996. Effects of UV irradiation on cell development and ultrastructure of the green alga *Micrasterias*. *J. Photochem. Photobiol. 36*: 285–292.

Michler, T., J. Aguilera, D. Hanelt, K. Bischof and C. Wiencke. 2002. Long-term effects of ultraviolet radiation on growth and photosynthetic performance of polar and cold-temperate macroalgae. *Mar. Biol. 140*: 1117–1127.

Müller, R., C. Wiencke and K. Bischof. 2008. Interactive effects of UV radiation and temperature on microstages of Laminariales (Phaeophyceae) from the Arctic and North Sea. *Clim. Res.* 37: 203–213.

Müller, R., C. Wiencke, K. Bischof and B. Krock. 2009. Zoospores of three Arctic Laminariales under different UV radiation and temperature conditions: exceptional spectral absorbance properties and lack of phlorotannin induction. *Photochem. Photobiol. 85*: 970–977.

Nybakken, L., W. Bilger, U. Johanson, L.O. Björn, M. Zielke and B. Solheim. 2004. Epidermal UV-screening in vascular plants from Svalbard (Norwegian Arctic) *Polar Biol. 27*: 383–390.

Pakker, H., R.S.T. Martins, P. Boelen, A.G.J. Buma, O. Nikaido and A.M. Breeman. 2000. Effects of temperature on the photoreactivation of ultraviolet-B-induced DNA damage in *Palmaria palmata* (Rhodophyta). *J. Phycol. 36*: 334–341.

Pavia, H. and E. Brock. 2000. Extrinsic factors influencing phlorotannin production in the brown alga *Ascophyllum nodosum*. *Mar. Ecol. Prog. Ser. 193*: 285–294.

Pavia, H., G. Cervin, A. Lindgren and P.Åberg. 1997. Effects of UV-B radiation and simulated herbivory on phlorotannins in the brown alga *Ascophyllum nodosum*. *Mar. Ecol. Prog. Ser. 157*: 139–146.

Pescheck, F., K. Bischof and W. Bilger. 2010. Screening of ultraviolet-A and ultraviolet-B radiation in marine green macroalgae (Chlorophyta). *J. Phycol. 46*: 444–455.

Poppe, F., D. Hanelt and C. Wiencke. 2002. Changes in ultrastructure, photosynthetic activity and pigments in the Antarctic red alga *Palmaria decipiens* during acclimation to UV radiation. *Bot. Mar. 45*: 253–261.

Poppe, F., R.A.M. Schmidt, D. Hanelt and C. Wiencke. 2003. Effects of UV radiation on the ultrastructure of several red algae. *Phycol. Res. 51*: 11–19.

Post, A. and A.W. Larkum. 1993. UV-absorbing pigments, photosynthesis and UV exposure in Antarctica: comparison of terrestrial and marine algae. *Aquat. Bot. 45*: 231–243.

Ragan, M.A. and K.W. Glombitza. 1986. Phlorotannins, brown algal polyphenols. *Prog. Phycol. Res. 4*: 129–241.

Raper, J.R. and A.S. Flexer. 1970. The road to diploidy with emphasis on a detour. *Symp. Soc. Gen. Microb. 20*: 401–432.

Rautenberger, R. and K. Bischof. 2006. Impact of temperature on UV-susceptibility of two *Ulva* (Chlorophyta) species from Antarctic and Subantarctic regions. *Polar Biol. 29*: 988–996.

Remias, D., U. Lütz-Meindl and C. Lütz. 2005. Photosynthesis, pigments and ultrastructure of the alpine snow alga *Chlamydomonas nivalis*. *Eur. J. Phycol. 40*: 259–268.

Richter, A., S. Wuttke and K. Zacher. 2008. Two years of *in situ* UV measurements at the Dallmann Laboratory/Jubany Station. *Ber. Polarforsch. Meeresforsch. 571*: 12–19.

Risgaard-Petersen, N., S. Rysgaard, L.P. Nielsen and N.P. Revsbech. 1994. Diurnal variation of dentrification and nitrification in sediments colonized by benthic microphytes. *Limnol. Oceanogr. 39*: 573–579.

Roleda, M.Y., W.H. van de Poll, D. Hanelt and C. Wiencke. 2004. PAR and UVBR effects on photosynthesis, viability, growth and DNA in different life stages of two coexisting Gigartinales: implications for recruitment and zonation pattern. *Mar. Ecol. Prog. Ser. 281*: 37–50.

Roleda, M.Y., D. Hanelt and C. Wiencke. 2005. Growth kinetics related to physiological parameters in young *Saccorhiza dermatodea* and *Alaria esculenta* sporophytes exposed to UV radiation. *Polar Biol. 28*: 539–549.

Roleda, M.Y., C. Wiencke and U.H. Lüder. 2006a. Impact of ultraviolet radiation on cell structure, UV-absorbing compounds, photosynthesis, DNA damage and germination in zoospores of Arctic *Saccorhiza dermatodea*. *J. Exp. Bot. 57*: 3847–4856.

Roleda, M.Y., C. Wiencke and D. Hanelt. 2006b. Thallus morphology and optical characteristics affect growth and DNA damage by UV radiation in juvenile Arctic *Laminaria* sporophytes. *Planta 223*: 407–417.

Roleda, M.Y., M.N. Clayton and C. Wiencke. 2006c. Screening capacity of UV-absorbing compounds in spores of Arctic Laminariales. *J. Exp. Mar. Biol. Ecol. 338*: 123–133.

Roleda, M.Y., D. Hanelt and C. Wiencke. 2006d. Exposure to ultraviolet radiation delays photosynthetic recovery in Arctic kelp zoospores. *Photosynth. Res. 88*: 311–322.

Roleda, M.Y., C. Wiencke, D. Hanelt and K. Bischof. 2007a. Sensitivity of the early life stages of macroalgae from the Northern Hemisphere to ultraviolet radiation. *Photochem. Photobiol. 83*: 851–862.

Roleda, M.Y., K. Zacher, A. Wulff, D. Hanelt and C. Wiencke. 2007b. Photosynthetic performance, DNA damage and repair in gametes of the endemic Antarctic brown alga *Ascoseira mirabilis* exposed to ultraviolet radiation. *Austral Ecol. 32*: 917–926.

Roleda, M.Y., K. Zacher, A. Wulff, D. Hanelt and C. Wiencke. 2008a. Susceptibility of spores of different ploidy levels from Antarctic *Gigartina skottsbergii* (Gigartina, Rhodophyta) to ultraviolet radiation. *Phycologia 47*: 361–370.

Roleda, M.Y., K. Zacher, G.L. Campana, A. Wulff, D. Hanelt, M.L. Quartino and C. Wiencke. 2008b. Photosynthetic performance and impact of ultraviolet radiation on the reproductive cells of Antarctic macroalgae. *Ber. Polarforsch. Meeresforsch. 571*: 254–262.

Roleda, M.Y., G. Campana, C. Wiencke, D. Hanelt, M.L. Quartino and A. Wulff. 2009. Sensitivity of Antarctic *Urospora penicilliformis* (Ulotrichales, Chlorophyta) to ultraviolet radiation is life stage dependent. *J. Phycol. 45:* 600–609.

Roleda, M.Y., U. Lütz-Meindl, C. Wiencke and C. Lütz. 2010a. Physiological, biochemical, and ultrastructural responses of the green macroalga *Urospora penicilliformis* from Arctic Spitsbergen to UV radiation. *Protoplasma 243*: 105–116.

Roleda, M.Y., U.H. Lüder and C. Wiencke. 2010b. UV-susceptibility of zoospores of the brown macroalga *Laminaria digitata* from Spitsbergen. *Polar Biol. 33*: 577–588.

Schoenwaelder, M.E.A. 2002. Physode distribution and the effect of 'thallus sunburn' in *Homosira banksii* (Fucales, Phaeophyceae). *Bot. Mar. 45*: 262–266.

Schoenwaelder, M.E.A., C. Wiencke, M.N. Clayton and K.W. Glombitza. 2003. The effect of elevated UV radiation on *Fucus* spp. (Fucales, Phaeophyta) zygote and embryo development. *Plant Biol. 5*: 366–377.

Schriek, R. 2000. Effects of light and temperature on the enzymatic antioxidative defense systems in the Antarctic ice diatom *Entomoneis kufferathii* Manguin. *Ber. Polarforsch. Meeresforsch. 349*: 1–130.

Shibata, T., S. Kawaguchi, Y. Hama, M. Inagaki, K. Yamaguchi and T. Nakamura. 2004. Local and chemical distribution of phlorotannins in brown algae. *J. Appl. Phycol. 16*: 291–296.

Steinhoff, F.S., C. Wiencke, R. Müller and K. Bischof. 2008. Effects of ultraviolet radiation and temperature on the ultrastructure of zoospores of the brown macroalga *Laminaria hyperborea. Plant Biol. 10*: 388–397.

Swanson, A.K. and L.D. Druehl. 2000. Differential meiospore size and tolerance of ultraviolet light stress within and among kelp species along a depth gradient. *Mar. Biol. 136*: 657–664.

Swanson, A.K. and L.D. Druehl. 2002. Induction, exudation and the photoprotective role of kelp phlorotannins. *Aquat. Bot. 73*: 241–253.

Swanson, A.K. and C.H. Fox. 2007. Altered kelp (Laminariales) phlorotannins and growth under elevated carbon dioxide and ultraviolet-B treatments can influence associated intertidal food webs. *Glob. Chang. Biol. 13*: 1696–1709.

Teixido, N., J. Garrabou, J. Gutt and W.E. Arntz. 2007. Iceberg disturbance and successional spatial patterns: the case of the shelf Antarctic benthic communities. *Ecosystems 10*: 142–157.

van de Poll, W.H., D. Hanelt, K. Hoyer, A.G.J. Buma and A.M. Breeman. 2002a. Ultraviolet-B-induced cyclobutane-pyrimidine dimer formation and repair in Arctic marine macrophytes. *Photochem. Photobiol. 76*: 493–501.

van de Poll, W.H., A. Eggert, A.G.J. Buma and A.M. Breeman. 2002b. Temperature dependence of UV radiation effects in Arctic and temperate isolates of three red macrophytes. *Eur. J. Phycol. 37*: 59–68.

Vinebrooke, R.D., K.L. Cottingham, J. Norberg, M. Scheffer, S.I. Dodson, S.C. Maberly and U. Sommer. 2004. Impacts of multiple stressors on biodiversity and ecosystem functioning: the role of species co-tolerance. *Oikos 104*: 451–457.

Wahl, M., M. Molis, A. Davis, S. Dobretsov, S.T. Durr, J. Johansson, T. Kinley, D. Kirugara, M. Langer, H.K. Lotze, M. Thiel, J.C. Thomason, B. Worm and D.Z. Ben-Yosef. 2004. UV effects that come and go: a global comparison of marine benthic community level impacts. *Glob. Chang. Biol. 10*: 1962–1972.

Wessel, S., S. Aoki, P. Winkler, R. Weller, A. Herber, H. Gernandt and O. Schrems. 1998. Tropospheric ozone depletion in polar regions. A comparison of observations in the Arctic and Antarctic. *Tellus B 50*: 34–50.

Whitehead, R.F., S.J. de Mora and S. Demers. 2000. Enhanced UV radiation – a new problem for the marine environment. *In*: (S. de Mora, S. Demers and M. Vernet, eds) *The effects of UV radiation in the marine environment.* Cambridge University Press, Cambridge. pp. 1–34.

Wickham, S. and M. Carstens. 1998. Effects of ultraviolet-B radiation on two Arctic microbial food webs. *Aquat. Microb. Ecol. 16*: 163–171.

Wiencke, C., M.N. Clayton and M.E.A. Schoenwaelder. 2004. Sensitivity and acclimation to UV radiation of zoospores from five species of Laminariales from the Arctic. *Mar. Biol. 145*: 31–39.

Wiencke, C., M.Y. Roleda, A. Gruber, M.N. Clayton and K. Bischof. 2006. Susceptibility of zoospores to UV radiation determines upper depth distribution limit of Arctic kelps: evidence through field experiments. *J. Ecol. 94*: 455–463.

Wiencke, C., U.H. Lüder and M.Y. Roleda. 2007. Impact of ultraviolet radiation on physiology and development of zoospores of the brown alga *Alaria esculenta* from Spitzbergen. *Physiol. Plant. 130*: 601–612.

Woelfel, J., R. Schumann, P. Leopold, C. Wiencke and U. Karsten. 2009. Microphytobenthic biomass along gradients of physical conditions in Arctic Kongsfjorden, Svalbard. *Bot. Mar. 52:* 573–583.

Woelfel, J., R. Schumann, F. Peine, A. Flohr, A. Kruss, J. Tegowski, P. Blondel, C. Wiencke and U. Karsten. 2010. Microphytobenthos of Arctic Kongsfjorden (Svalbard, Norway) – Biomass and potential primary production along the shore line. *Polar Biol. 33*: 1239–1253.

Wulff, A., C. Nilsson, K. Sundbäck, S.A. Wängberg and S. Odmark. 1999. UV radiation effects on microbenthos – a four month field experiment. *Aquat. Microb. Ecol. 19*: 269–278.

Wulff, A., S.A. Wängberg, K. Sundbäck, C. Nilsson and G.J.C. Underwood. 2000. Effects of UVB radiation on a marine microphytobenthic community growing on a sand-substratum under different nutrient conditions. *Limnol. Oceanogr. 45*: 1144–1152.

Wulff, A., M.Y. Roleda, K. Zacher and C. Wiencke. 2008a. UV radiation effects on pigments, photosynthetic efficiency and DNA of an Antarctic marine benthic diatom community. *Aquat. Biol. 3*: 167–177.

Wulff, A., K. Zacher, D. Hanelt, A. Al-Handal and C. Wiencke. 2008b. UV radiation – a threat to Antarctic benthic marine diatoms? *Antarc. Sci. 20*: 13–20.

Wulff, A., M.Y. Roleda, K. Zacher and C. Wiencke. 2008c. Exposure to sudden light burst after prolonged darkness – a case study on benthic diatoms in Antarctica. *Diatom Res. 23*: 519–532.

Xiong, F., F. Lederer, J. Lukavsky and L. Nedbal. 1995. On the potential roles of sporopollenin and protein synthesis in the protection of freshwater algae against elevated ultraviolet-B radiation. *In:* (P. Mathis, P., ed.) *Photosynthesis: From Light to Biosphere.* Kluwer Academic Publishers, Dordrecht, the Netherlands. pp. 359–362.

Xue, L.G., Y. Zhang, T.G. Zhang, L.Z. An and X.L. Wang. 2005. Effects of enhanced ultraviolet-B radiation on algae and cyanobacteria. *Crit. Rev. Microb. 31*: 79–89.

Zacher, K. and G. Campana. 2008. UV radiation and consumer effects of an intertidal and subtidal macroalgal assemblage: a comparative study. *Ber. Polarforsch. Meeresforsch. 571*: 287–294.

Zacher, K., M.Y. Roleda, D. Hanelt and C. Wiencke. 2007a. UV effects on photosynthesis and DNA in propagules of three Antarctic seaweeds (*Adenocystis utricularis*, *Monostroma hariotii* and *Porphyra endiviifolium*). *Planta 225*: 1505–1516.

Zacher, K., A. Wulff, M. Molis, D. Hanelt and C. Wiencke. 2007b. Ultraviolet radiation and consumer effects on a field-grown intertidal macroalgal assemblage in Antarctica. *Glob. Chang. Biol. 13*: 1201–1215.

Zacher, K., D. Hanelt, C. Wiencke and A. Wulff. 2007c. Grazing and UV radiation effects on an Antarctic intertidal microalgal assemblage: a long-term field study. *Polar Biol. 30*: 1203–1212.

Zacher, K., M.Y. Roleda, A. Wulff, D. Hanelt and C. Wiencke. 2009a. Responses of Antarctic *Iridaea cordata* (Rhodophyta) tetraspores exposed to ultraviolet radiation. *Phycol. Res. 57*: 186–193.

Zacher, K., R. Rautenberger, D. Hanelt, A. Wulff and C. Wiencke. 2009b. The abiotic environment of polar marine benthic algae. *Bot. Mar. 52*: 483–490.

14 Drivers of colonization and succession in polar benthic macro- and microalgal communities

Gabriela Laura Campana, Katharina Zacher, Anna Fricke, Markus Molis, Angela Wulff, María Liliana Quartino and Christian Wiencke

Introduction

The ecology of benthic communities in polar regions was little studied until quite recently. Available knowledge has focused on the description of spatial and temporal patterns of the geographic distribution and vertical zonation of seaweeds and sessile invertebrates. Profound ecological understanding of the underlying processes and drivers was missing. This applies especially to the microphytobenthos, which has been barely investigated. Although information on the ecology of polar benthic algal communities is available (Chapter 3; Klöser et al. 1994, 1996, Quartino et al. 2001, Irving et al. 2005, Konar and Iken 2005, Wulff et al. 2009), it is a remarkable fact that the vast majority of investigations have focused on the macrozoobenthos. For these animal assemblages, the impact of ice scour on community structure and diversity has been the focus of most attention (Gutt 2001, Piepenburg 2005). Finally, most studies on primary producers have been of an observational nature, correlating patterns of diversity to the potential drivers (Cunningham and McMinn 2004, Quartino et al. 2005, Johnston et al. 2007). Long-term succession studies have allowed characterization of hard bottom macrobenthic assemblages in the Arctic (Beuchel and Gulliksen 2008 and references therein) and in Antarctica (Bowden et al. 2006, Barnes and Conlan 2007). These studies show that the succession of nearshore communities is either slow and continual or pulsed with inter-annual mass events of colonization (Barnes and Conlan 2007).

The colonization process of hard bottom polar benthic communities has been studied by clearing natural substrata and placement of settlement panels, followed by direct observation, analyses of underwater photographs or transfer of colonized substrata to the laboratory (Barnes and Conlan 2007, Beuchel and Gulliksen 2008 and references therein). Only recently, has primary succession of hard bottom polar benthic algae been studied in manipulative field experiments in relation to potential drivers of the community structure and diversity during colonization (Table 14.1). These studies have focused on (i) sudden changes in environmental conditions (Fricke et al. 2008), (ii) ultraviolet radiation (UVR, 280–400 nm) (Wahl et al. 2004), (iii) consumers (Konar 2007), and (iv) the combined effects of UVR and grazing (Zacher et al. 2007a, b, Campana et al. 2008a, b, Zacher and Campana 2008).

In the present chapter, we summarize studies addressing the effects of potential drivers of benthic algal primary succession in inter- and subtidal rocky shores of the

Table 14.1 Composition of algal communities that developed during colonization experiments in polar regions.

Location and habitat	Substratum	Manipulated Factors	Depth (m)	Duration	Taxa in developing communities[1])	Reference
Norway, (Arctic, 1), subtidal	Ceramic	UVR	0.04	3 months	**Bacillariophyta** **Heterokontophyta** [*Ectocarpus siliculosus* (Dillwyn) Lyngbye, *Chorda filum* (Linnaeus) J.B. Lamouroux] Chlorophyta Sessile invertebrates	Wahl et al. (2004)
Spitsbergen, (Arctic, 2), subtidal	Ceramic	Community age, transplantation	0.5	1 month	**Bacillariophyta** Chlorophyta	Fricke et al. (2008)
				2 months	**Bacillariophyta** **Chlorophyta** [*Ulothrix implexa* (Kützing) Kützing, *Urospora* sp.]	
			8	12 months	**Bacillariophyta** **Chlorophyta** [*Ulothrix implexa*] **Heterokontophyta** [*Pylaiella littoralis* (Linnaeus) Kjellman, *Pylaiella varia* Kjellman, *Ectocarpus siliculosus* (Dillwyn) Lyngbye] Rhodophyta Sessile invertebrates	
				21 months	**Bacillariophyta** **Chlorophyta** [*Ulva*, *Monostroma*] Heterokontophyta Rhodophyta Sessile invertebrates	
Stefansson Sound, (Arctic, 3), subtidal	Natural	Clearance, Grazing	6–7	0–2 years	No sessile organisms	Konar (2007)
				3 years	**Rhodophyta** [Corallinaceae] Sessile invertebrates	
Spitsbergen, (Arctic, 4), subtidal	Natural	Clearance	15	0–4 years	**Rhodophyta** [Corallinaceae] Motile invertebrates	Beuchel and Gulliksen (2008)
				4–14 years	**Rhodophyta** [Corallinaceae] Sessile invertebrates	

				14–24 years	**Rhodophyta** [Corallinaceae] **Heterokontophyta** [*Desmarestia*] Motile invertebrates Sessile invertebrates	
King George Island/Isla 25 de Mayo, (Antarctica, 5), intertidal	Ceramic	UVR, grazing	0–2	3.5 months	**Bacillariophyta** [*Navicula perminuta* Grunow, *Cocconeis costata var. antarctica* Manguin, *Cocconeis pinnata* Gregory *ex* Greville] **Chlorophyta** [*Monostroma hariotii* Gain] Rhodophyta (sporelings) Heterokontophyta	Zacher et al. (2007a, b)
(Antarctica, 6), subtidal	Ceramic	UVR	0.04	1.5 months	**Bacillariophyta**	Wahl et al. (2004)
King George Island/Isla 25 de Mayo, (Antarctica, 7), subtidal	Ceramic	UVR, grazing	2	3 months	**Bacillariophyta** [*Fragilaria striatula* Lyngbye, *Navicula* cf. *perminuta* Grunow, *Navicula* cf. *hanseni* Möller] **Chlorophyta** [*Urospora penicilliformis* (Roth) Areschoug, *Ulothrix*] Rhodophyta (sporelings)	Campana et al. (2008a,b)
King George Island/Isla 25 de Mayo, (Antarctica), (5) intertidal, (7) subtidal	Ceramic	UVR, grazing	0–2	2 months	**Bacillariophyta** **Chlorophyta** [*Monostroma hariotii*] Rhodophyta (sporelings)	Zacher and Campana (2008)
			2	2 months	**Bacillariophyta** **Chlorophyta** [*Urospora penicilliformis*] Rhodophyta (sporelings)	

1 Dominant algal groups are in bold font, dominant algal species are in square brackets. (Arctic, 1) 68° N, 13° E; (Arctic, 2) Ny-Ålesund harbor, 78°55′ N, 11°56′ E; (Arctic, 3) Boulder Patch, 70°20′ N, 147°40′ W; (Arctic, 4) Kvadehuken, 78°58.6′ N, 11°30.1′ E; (Antarctica, 5) Potter Cove, 62°14′ S, 58°41′ W; (Antarctica, 6) 68° N, 13° E; (Antarctica, 7) Potter Cove, 62°14′ S, 58°40′ W.
UVR, ultraviolet radiation.

Arctic and Antarctic. The main abiotic and biotic factors affecting community structure during colonization are discussed and we highlight gaps in information and propose future research perspectives.

Abiotic and biotic drivers of polar benthic algal succession

Abiotic factors

In general, colonization and succession of polar benthic primary producers are strongly influenced by abiotic parameters (see also chapter 2; Zacher et al. 2009). Benthic polar organisms are exposed to a variety of physical disturbance factors, such as ice scour and water turbulence, produced by tides and wave action (Barnes and Conlan 2007). Furthermore, other abiotic factors, such as the light regime and salinity, undergo marked seasonal changes (Chapter 2; Zacher et al. 2009) affecting functioning of benthic biota. Strongest environmental changes occur in the intertidal zone and, in fact, most polar macroalgae grow predominantly in the subtidal where changes are less pronounced (Wiencke et al. 2007). Almost no benthic microalgae were found on the rocky or soft bottom intertidal of Potter Cove, South Shetland Islands, Antarctica (Klöser 1998, A. Wulff personal observation), indicating that the environmental conditions rarely permit colonization by these organisms. Furthermore, in a comparative successional field study performed in Potter Cove, the algal biomass was much higher in the subtidal than in the intertidal (Zacher and Campana 2008), which is also a usual pattern in temperate rocky coasts of the Atlantic and Pacific (Lüning 1990). Other environmental factors, such as the types of substrata and the coastal profile, further influence the development of polar benthic algal communities in the subtidal zone (Barnes 1995, Klöser et al. 1996). In particular, algae have been shown to dominate upward-facing substrata, whereas animal groups dominate vertical or downward-facing substrata (Barnes 1995, Barnes et al. 1996).

Ice effects Ice has a significant impact on polar and subpolar benthic communities, but the relationships between corresponding physical and biological processes are not yet well understood (Gutt 2001). There are many types of ice impact on the benthic biota: fast ice can form an ice foot along the shoreline in winter. An ice foot can fundamentally affect intertidal communities by limiting the colonization of this habitat to periods when ice is absent (Barnes and Conlan 2007). Furthermore, the timing of fast ice release and movement in early spring can affect germination of overwintering macroalgal propagules (Kim 2001).

Anchor ice impacts the benthos by ripping free hard substrata (Bond et al. 1992, Barnes and Conlan 2007) or trapping organisms (Dayton 1979, 1989) that can be translocated over several kilometers. Generally, it is very rare in the Antarctic Peninsula and Scotia Arc region and occurs only sporadically in the Arctic (Gutt 2001, Smale and Barnes 2008). Anchor ice can be a determining factor for the vertical distribution of subtidal seaweeds in shallow coastal areas at high latitudes in Antarctica (Miller and Pearse 1991, Amsler et al. 1999). *Iridaea cordata* (Turner) Bory was shown to survive encasement by anchor ice and only this macroalgal species and a dense mat of benthic diatoms occupied the shallower subtidal in McMurdo Sound (Miller and Pearse 1991). The other macroalgae were restricted to deeper sites that were not heavily affected by anchor ice formation (Miller and Pearse 1991). Anchor ice has also been suggested as the cause of large mats of macroalgal drift spotted in McMurdo Sound (Amsler et al. 1999). Furthermore, anchor ice and frazil ice can lift benthic diatoms, which can be incorporated in ice algal communities (Reimnitz et al. 1987, Garrison et al. 1989).

Ice scouring can be a strong driver of the distribution and abundance of seaweeds, affecting zonation and diversity patterns (reviewed by Gutt 2001, Kim 2001). However, the impact of ice scour on benthic algal succession has been little reported in the literature (Bolton 1983, Keats et al. 1985, McCook and Chapman 1997). Studies in the subarctic region have revealed that annual algae quickly colonize the open space provided by ice-mediated removal of canopy species, leading to increased diversity (Keats et al. 1985).

Sea ice cover duration and thickness determine patterns of colonization of benthic algae. In East Antarctica, extensive stands of canopy-forming macroalgae were found at sites where sea ice break-up occurred during spring, but were absent at sites that retained sea ice cover into the summer (or even permanently) (Johnston et al. 2007). Thinner layers of sea ice cover allow deeper light penetration and can determine a downward shift in macroalgal colonization (when compared with sites with thicker or permanent ice cover) (Miller and Pearse 1991). However, shallow subtidal sites, which provide better irradiance levels, can be severely affected by both anchor ice and ice scouring and support very low macroalgal richness (Miller and Pearse 1991).

Hydrodynamics Shallow-water polar biota can be mechanically abraded by wave action either directly or indirectly through movement of boulders (Barnes and Conlan 2007). On temperate shores, strong wave action can even mobilize small rocks that might impact and damage intertidal organisms (Shanks and Wright 1986). A species-specific tolerance of turbulence can determine macroalgal community structure in the subtidal zone (Klöser et al. 1996). Furthermore, tides and wind-generated currents have been proposed as vectors of Antarctic benthic diatoms from intertidal to subtidal waters (Ahn et al. 1997). Thus, wind-driven resuspension of soft-sediment microphytobenthos might play a key role in benthic diatom dispersal and further settlement on both soft and hard substrata.

Light Polar regions are characterized by short periods of favorable light conditions and long periods of darkness due to the polar night and sea ice cover in winter and spring (Chapter 2; Kirst and Wiencke 1995, Zacher et al. 2009). The phenology of polar seaweeds is tuned to these strong seasonal changes (Chapter 9; Wiencke et al. 2007, 2009) and their juvenile stages are especially low-light adapted in most cases, although intertidal species are also able to tolerate high light conditions (Chapter 10; Wiencke et al. 1993, Hanelt et al. 1994, Gómez et al. 2009). Similarly, Antarctic benthic diatoms are mostly low-light adapted but are still capable of acclimation to excessive radiation, including UVR (Longhi et al. 2003, Wulff et al. 2008a). In addition, low-light adapted Arctic benthic microalgae photoacclimate at higher irradiance (not UVR) (Kühl et al. 2001). Physiological adaptation to low-light conditions seems to be correlated with vertical distribution patterns (Chapter 10; Longhi et al. 2003, Gómez et al. 2009). These patterns are further differentially influenced by the environmental gradients of disturbance, such as wave exposure, turbulence and ice scouring (Klöser et al. 1994, 1996, Quartino et al. 2001), all of which have decreasing impact with increasing depth (Laudien et al. 2007).

Salinity During the polar summer, glacial meltwater may influence the physical dynamics of the water column by increasing turbidity and reducing salinity (Chapter 2; Zacher et al. 2009), thus playing a critical role for the benthic biota (Klöser et al. 1993). Consequently, adult macroalgae from supra- and eulittoral zones in Antarctica are euryhaline (Wiencke et al. 2007). Little is known of the effects of low salinity on the survival of propagules or young germlings. First results from studies on Antarctic intertidal macroalgal spores demonstrated strong species-specific reactions. Whereas some species have no negative response to low salinities (20 psu), others did (K. Zacher unpublished data). In this context, it should be noted that meltwater can affect the marine ecosystem up to 100 km offshore (Dierssen et al. 2002).

Sedimentation In addition to salinity reduction, the inflow of meltwater during summer transports a high load of sediment and detritus, which results in an increasing turbidity in the water column (Wiencke et al. 2007). Thus, benthic algae are exposed to a reduction of underwater irradiance (including UVR at 280–400 nm) at sites that experience a strong land runoff influence (Chapter 2; Zacher et al. 2009).

In the proximity of glaciers, sedimentation rates can strongly influence the structure of benthic communities (Smale and Barnes 2008). Arctic glacial fjords have a declining gradient in sediment loading from the glacial front to the fjord mouth (Syvitski et al. 1989, Svendsen et al. 2002). Sites exposed to higher glacial sedimentation rates can support only pioneer faunal species, whereas a lower impact of sedimentation allows the development of more complex communities, which include benthic algae (Syvitski et al. 1989). High sedimentation can even cause smothering of benthic fauna (Smale and Barnes 2008). In Antarctica, sites experiencing prolonged sea ice cover, reduced currents and proximity to glaciers have organic sediment accumulations interspersed with only filamentous green algae (Johnston et al. 2007). Among microalgal assemblages, evidence from non-polar environments shows that epipelic benthic diatoms can migrate upwards as sediment accumulates, thereby restoring the oxygenation of the sediment surface (Wulff et al. 1997). Sedimentation and ice-rafted pebbles can also alter hard bottom habitats by creating pockets of overlying sediment where infaunal organisms are found (Hop et al. 2002).

Nutrients Similar to other regions (Lotze et al. 2001), development and recruitment of seaweeds is controlled by nutrient availability. In the Antarctic, seawater nutrient concentrations are not limiting for macroalgal growth throughout the year (Chapter 2; Drew and Hastings 1992, Zacher et al. 2009). In contrast, nutrient concentrations (particularly inorganic nitrogen) are relatively high during winter in the Arctic (Chapter 2; Zacher et al. 2009), and low in the summer season in Spitsbergen (Aguilera et al. 2002) and the Alaskan and Canadian High Arctic (Chapman and Lindley 1980, Dunton and Schell 1986). A common life strategy in polar seaweeds is the capacity to grow under very low-light conditions during late winter/spring, or even under sea ice cover (Wiencke et al. 2007). This is of particular importance for Arctic seaweeds, which can therefore utilize the high nutrient concentration in winter (Chapman and Lindley 1980, Dunton and Schell 1986) before the onset of phytoplankton bloom in spring and summer (Hop et al. 2002).

Fluctuations in the regional climate patterns can affect benthic succession, as a strong increase in brown algal canopy was observed in the Arctic, probably due to increased nutrient concentrations and also improved light conditions (Beuchel and Gulliksen 2008).

Biotic factors

Grazing The structure of polar algal assemblages can be strongly affected by biological interactions, including grazing (Sivertsen 1997, Kim 2001) and competition (Konar and Iken 2005). Herbivores have been identified as important structuring forces in marine benthic communities (Duffy and Hay 2000, Jenkins et al. 2005), as their grazing activities can (i) generate open space for colonizers, (ii) increase spatial heterogeneity (Sommer 2000), and (iii) mediate the outcome of competition among seaweeds (Worm et al. 2000). Although benthic micro- and macroalgae are very important carbon sources in polar waters as a direct food source for common benthic organisms (Brêthes et al. 1994, Iken et al. 1997, Iken 1999), there is little information on the role of herbivores in structuring these communities. Intense grazing on kelps by sea urchins can result in barren grounds that can later be recolonized by opportunistic species (Sivertsen 1997). In Antarctica, limpets have been shown to influence the distribution of intertidal algae at low shore levels by consuming benthic diatoms and early life stages of macroalgae (Kim 2001). Furthermore, mesograzer pressure probably plays a key structuring role in benthic algal communities by suppressing the presence of filamentous epiphytes in the subtidal (Peters 2003). Few filamentous species were found as epiphytes of macroalgae (Wiencke and Clayton 2002) and, in contrast, there was a high occurrence of filamentous endophytes (Peters 2003), most of which were shown to be highly palatable to amphipods (Chapter 5; Amsler et al. 2009a, b).

The success and survival of polar seaweeds can be critically affected by the existence of physical and chemical defenses against grazing (Chapter 5; Wiencke et al. 2007, Amsler et al. 2009b). Antigrazing defenses in Arctic seaweeds seem to be generally related to the toughness and morphology of the thallus (Wessels et al. 2006). In contrast, chemical defenses are widely distributed in dominant overstory brown algae and fleshy red macroalgae from Antarctica (Chapter 5; Amsler et al. 1998, 2005, 2009b, Fairhead et al. 2005). Chemically defended seaweeds were shown to mediate community level interactions that might have positive effects on macroalgal recruitment in both polar regions. The effect of sea urchin grazing on kelp beds can be mediated by the presence of a belt of the acid-defended brown seaweed *Desmarestia viridis* (O.F. Müller) J.V. Lamouroux (Molis et al. 2009). As a consequence, the presence of *D. viridis* indirectly protects kelp patches, which might serve as source areas during recolonization (Molis et al. 2009). In Antarctica, sea urchins can help sustain reproductive populations of *Iridaea cordata* and *Phyllophora antarctica* A. Gepp *et* E.S. Gepp (Amsler et al. 1999), which are the only two fleshy red algae that occur in the study site (McMurdo Sound) (Miller and Pearse 1991). Both macroalgae are chemically defended against grazing by the sympatric sea urchin *Sterechinus neumayeri* Meissner (Amsler et al. 1998). The sea urchins hold a cover of these drifting macroalgae that confers physical protection from predators, and consequently fertile thallus fragments are retained in the photic zone, potentially extending the algal distribution (Amsler et al. 1999).

The effects of grazers on developing algal communities in polar regions are variable. Motile grazers (crustaceans, molluscs and sea urchins) are the first species to appear on recently cleared areas in the Arctic, suggesting that they feed on the early colonizers

(Beuchel and Gulliksen 2008). However, a field study revealed a minor role of grazers on the initial recruitment of macroalgal species (Konar 2007). In contrast, grazing was a strong structuring force on early successional community stages (during the first 2 months) in inter- and subtidal sites in Antarctica (Zacher and Campana 2008). Here, grazing was proposed as a mechanism that controls the development of hard bottom communities dominated by sessile invertebrates and coralline algae over longer colonization times (Bowden et al. 2006). During early succession, grazers reduce algal biomass (Zacher et al. 2007b, Campana et al. 2008a) and shape seaweed species composition by favoring the colonization of firmly attached red algal recruits over less grazer-resistant macroalgal recruits and diatoms (Zacher et al. 2007b, Zacher and Campana 2008).

Diatom-dominated assemblages can be strongly controlled by grazers either by direct consumption or even indirectly by cell dislodgement (Nicotri 1977, Hillebrand et al. 2000). The effects of grazers might be seasonal, as shown by Sundbäck et al. (1996). In their investigation, meiofaunal grazing controlled microalgal biomass during spring and autumn but not during summer. As shown in the studies by Zacher et al. (2007a) and Campana et al. (2008b) conducted off King George Island/Isla 25 de Mayo, Antarctica, consumers can cause major biomass reductions of diatom assemblages during colonization, introduce a shift in species composition and alter the physiognomy of microbenthic communities (as defined by Steinman 1996). The absence of grazers favored the abundance of *Cocconeis* species (Zacher et al. 2007a) and also of colony-forming diatom species in a subtidal location (Campana et al. 2008b). The differential susceptibility of algae towards grazing (due to size, morphology, accessibility for herbivores and attachment strength) determined the shift in species composition and corroborates results from temperate regions (Nicotri 1977, Hillebrand et al. 2000). Furthermore, grazers caused changes in community diversity, which was either enhanced due to increased spatial heterogeneity (Zacher et al. 2007b) or negatively affected, measured as a reduction in the community evenness (Campana et al. 2008b). In general, the chemical ecology of microalgae is mostly unknown in Antarctic waters (Avila et al. 2007), and in particular, the presence of chemical defenses in polar benthic marine diatoms awaits elucidation. However, evidence from temperate regions revealed that polyunsaturated aldehydes produced by benthic diatoms had no negative effects on benthic copepods (Taylor et al. 2007).

Competition Whereas space and light are key resources for benthic algal communities in polar marine environments (Konar and Iken 2005, Wiencke et al. 2007), competitive interactions have been shown to control algal zonation in some polar regions (Klöser et al. 1996).

During early succession, when new space is available, diatoms, seaweed propagules and green filamentous macroalgae are early space-occupiers at shallow depths in the Arctic and Antarctic (Kim 2001, Fricke et al. 2008, Zacher and Campana 2008, see Table 14.1). Diatoms appear as superior competitors for space and might inhibit the settlement of red macroalgal propagules (Zacher and Campana 2008), as shown in other studies (Huang and Boney 1985). During these early successional stages, grazing may alter the outcome of competition among seaweeds. For instance, detrimental grazing on fast-growing and superior competitors, such as diatoms, favors the settlement of small, firmly attaching propagules of *Palmaria decipiens* (Reinsch) R.W. Ricker in Antarctica (Zacher and Campana 2008). At later stages of succession (12 and 21 months) in Arctic communities, a long-lasting colonization of macroalgal-dominated assemblages revealed a low rate of competitive displacement, as only few species became relatively more abundant (Fricke et al. 2008).

Although space is a limiting resource in a macroalgal-dominated community in the Arctic, very little of cleared natural substratum was occupied after 4 years (Konar 2007). This suggests a relatively slow pace in recolonization and indicates the vulnerability of polar ecosystems. A combination of frequent physical disturbance and slow macroalgal recruitment in the absence of vegetative regrowth has been suggested as the major reasons for the very low rate of recolonization (Konar 2007).

Effects of global change on polar benthic algal succession

The Atlantic sector of the Arctic Ocean and the Antarctic Peninsula are the regions most strongly affected by global change, namely, increased ultraviolet-B radiation (UVBR, 280–315 nm) (Whitehead et al. 2000), climate warming and ice retreat (Chapter 2; 12; Barnes and Conlan 2007, Clarke et al. 2007, Müller et al. 2009, Zacher et al. 2009). These environmental changes are expected to have strong influences on benthic communities, particularly in coastal zones (Smale and Barnes 2008).

Ultraviolet radiation UVR is known to be a stress factor that can negatively affect polar benthic primary producers and especially early life stages of macroalgae, which are more susceptible to UVR than their adult stages (Chapter 13; Roleda et al. 2007, Karsten et al. 2009). Thus, particularly during early succession, species-specific or life stage-specific algal susceptibility to UV stress are expected to alter colonization patterns and, consequently, the structure of the hard bottom benthic systems (Lotze et al. 2002, Molis and Wahl 2004, Wahl et al. 2004, Zacher et al. 2007b, Zacher and Campana 2008).

In the Arctic, UVR significantly affected the diversity and structure of field-grown macroalgal communities that developed on unglazed ceramic tiles (Wahl et al. 2004, A. Fricke unpublished data). In Norway, a UVBR-mediated reduction of canopy species reduced the further settlement of UV-sensitive species (Wahl et al. 2004). Also on Spitsbergen, ultraviolet-A radiation (UVAR, 315–400 nm) in particular significantly affected the structure of the seaweed assemblages, depending on the exposure time and the successional stage (A. Fricke unpublished data). UVAR, which is characterized by naturally high fluxes (Cullen and Neale 1994), had a negative impact on diversity and biomass of benthic assemblages in non-polar regions (Wahl et al. 2004). Field studies performed in Antarctica revealed a strong impact of UVR on newly developing benthic communities, altering the species composition of intertidal and subtidal seaweed communities (Zacher et al. 2007b, Zacher and Campana 2008, G.L. Campana unpublished data). In other environments, UVR had a structuring effect on seaweed assemblages: the effect was more pronounced during the first 2 months of the experiments and disappeared as succession proceeded (Lotze et al. 2002, Wahl et al. 2004, but see Dobretsov et al. 2005). In contrast, ambient UVR had no effect on community structure during succession in other temperate habitats, although transient detrimental effects were induced by treatments using UVB lamps that simulated future levels of enhanced UVBR (Molis and Wahl 2009).

Comparing micro- and macrobenthic assemblages, a high tolerance to UVR was evident in microalgal assemblages as no biomass reduction or species shift was observed under UVR in Antarctic assemblages (Chapter 13; Wahl et al. 2004, Zacher et al. 2007a, Campana et al. 2008b, Karsten et al. 2009) and even positive effects were found for Arctic assemblages (A. Fricke unpublished data). Laboratory experiments showed that even elevated levels of UVBR caused no effects on community structure parameters (biomass and species composition) of soft bottom Antarctic diatoms (Wulff and Zacher 2008, Wulff et al. 2008b, c) and only transient effects on functional variables (photosynthetic efficiency and cell number). Overall, the high resistance of benthic diatoms to UVR is comparable to studies performed in non-polar regions (Peletier et al. 1996, Wulff et al. 2000). As protective mechanisms, such as vertical migration (Underwood et al. 1999, Waring et al. 2007), or production of UV-absorbing mycosporine-like amino acids (MAAs) (Hernando and Ferreyra 2005) were absent in the hard bottom polar assemblages studied (Campana et al. 2008b), the nature of their high UV-resistance awaits elucidation. Moreover, resistant diatom mats can attenuate UVR (Vinebrooke and Leavitt 1999) with a possible positive consequence for the recruitment of sensitive early life stages of macroalgae (Roleda et al. 2007).

Interactive effects between UVR and ecological drivers of succession, such as grazing, are strong in both freshwater (Bothwell et al. 1994) and marine environments (Lotze et al. 2002). In the Antarctic intertidal zone, the presence of grazers counteracted the negative effects of UVR on macrobenthic algae, and, overall, grazing effects on biomass were stronger in the UV-shielded subtidal communities (Zacher et al. 2007b, Campana et al. 2008a, G.L. Campana unpublished data). A possible explanation for these UVR-grazer interactions might be the production of UV-protecting substances, such as MAAs (Hoyer et al. 2001) or phlorotannins (Swanson and Fox 2007), which may alter algal chemical composition and therefore palatability (Pavia et al. 1997). Furthermore, UVR can have strong detrimental effects on the grazers themselves (Obermüller et al. 2007). For instance, in short-term experiments, grazing rates of the Antarctic amphipod *Gondogeneia antarctica* Chevreux were negatively affected by UVR (G.L. Campana unpublished data).

Elevated temperatures Owing to global warming, littoral areas are expected to experience a wider range of temperatures, including higher maximum values (Barnes and Conlan 2007). This could have consequences for the structure of coastal benthic assemblages in polar regions (Chapter 12; Hop et al. 2002, Barnes and Conlan 2007, Müller et al. 2009). Physiological constraints and invasion by algae or higher trophic level organisms could alter community structure and succession patterns with unknown consequences.

Several North Atlantic polar to cold-temperate seaweeds are expected to expand their distribution into higher latitudes but retreat along the NE-Atlantic coastline by the end of this century (Chapter 12; Müller et al. 2009). For example, the Atlantic influence in the Svalbard region is expected to increase, leading to change in the structure of benthic communities (Hop et al. 2002). In this regard, it is very probable that the boreal macroalga *Laminaria hyperborea* (Gunnerus) Foslie might expand its northern distribution (Chapter 12; Hop et al. 2002, Müller et al. 2009). There are already recent records of this species from southern Spitsbergen (Isfjorden) (Peltikhina 2002, Olsen et al. 2004).

Warming can potentially exert an influence on the structure of benthic communities in Antarctica, as Antarctic macroalgae are rather stenothermal (Chapter 10; Wiencke and tom Dieck 1989, Gómez et al. 2009). Whereas the distribution of key macroalgal species along the Antarctic Peninsula will probably not change until the end of this century (Chapter 12; Müller et al. 2009), warmer waters could favor the establishment and spread of non-indigenous cold-temperate species (Chapter 12; Barnes and Conlan 2007, Müller et al. 2009). In fact, there are already reports on the appearance of non-indigenous macroalgae in Antarctica (Chapter 12; Müller et al. 2009). *Ulva intestinalis* Linnaeus was found growing in dense mats on intertidal rocks, after it was probably introduced by ships (Clayton et al. 1997). The presence of such species has been regarded as an indication of water temperature changes (Clayton et al. 1997). The cold-temperate brown alga *Petalonia fascia* (O.F. Müller) Kuntze was recorded for the first time on Deception Island (Antarctica) by Clayton et al. (1997). It has also been found a second time following cultivation of intertidal communities from King George Island/Isla 25 de Mayo (Zacher et al. 2007b). There has been a third record of this species after cultivation of thallus fragments of *Delesseria lancifolia* (Hooker f.) J. Agardh collected at Palmer Station pier, Anvers Island (Amsler et al. 2009a). According to Amsler and collaborators, the absence of previous records of *P. fascia* in the study site would support the hypothesis that this species has been recently introduced by ships. However, the authors also point out that *P. fascia* might have remained undetected due to i) its relatively cryptic, crustose microthallus or ii) its palatable blade-forming phase, which might be readily subject to mesograzing (Amsler et al. 2009a). Finally, the presence of *Blidingia minima* (Nägeli *ex* Kützing) Kylin, *Chordaria linearis* (Hooker and Harvey) Cotton and *Rhodymenia coccocarpa* (Montagne) M.J. Wynne cited as *R. subantarctica* R.W. Ricker (Wiencke and Clayton 2002) are also considered to be related to climate warming (Chapter 12; Müller et al. 2009).

Sea ice reduction Temporal and spatial reduction in sea ice cover is expected to alter the underwater radiation regime, with strong consequences for primary producers (Clarke et al. 2007). A higher underwater irradiance can increase the area available for colonization by benthic algae (Chapter 12; McClintock et al. 2008, Müller et al. 2009) and favor benthic production (Hop et al. 2002). The cover of brown algae varied strongly between years in relation to climatic fluctuations at Kvadehuken (Svalbard) (Beuchel et al. 2006). Between 1994 and 1996, warmer winters with less sea ice cover, earlier melting and increased winter precipitation were likely to increase light and nutrients (Beuchel et al. 2006, Beuchel and Gulliksen 2008). Concomitantly, there was a decline in sea anemones and an increase in macroalgal cover and sea urchin density in a long-term colonization study during this period (Beuchel and Gulliksen 2008) (Table 14.1). Following a cold winter with prolonged ice-cover during spring, a strong decline in brown algal cover was observed at the same site (F. Beuchel and B. Gulliksen personal communication, cited in Hop et al. 2002). Other Arctic areas are also expected to provide new areas for colonization by macroalgae (Chapter 12; Müller et al. 2009). As mentioned (see abiotic factors section), reduced sea ice formation – both spatially and temporally – could determine a higher complexity in the structure of benthic macroalgal assemblages in Antarctica (Johnston et al. 2007) and favor the colonization of deeper zones (Miller and Pearse 1991).

Simulated sea ice break-up was shown to be a strong driver of secondary succession in Arctic benthic communities. A transplantation experiment from 8 m to 0.5 m water depth (Fricke et al. 2008) showed major detrimental effects on macroalgal abundance and richness. These changes were related to the sudden shift in the underwater light conditions and salinity (due to meltwater inflow) (Fricke et al. 2008). The different acclimation potential of the macroalgal species and a species-specific capacity to recover from stress were regarded as the mechanisms that reshaped the composition of the communities (Fricke et al. 2008). Microphytobenthos seemed to be more tolerant than seaweeds to the near surface conditions in this experiment (Fricke et al. 2008). The abundance of diatoms was not affected, and it was suggested that a shift in species composition from high irradiance-sensitive species to more resistant diatom species occurred, explaining the lack of transplantation effects (Fricke et al. 2008). Indeed, in Antarctica, benthic microphytobenthos has marked successional shifts in assemblage species composition and also changes in the ability of individual cells to photoacclimate to the higher irradiances after sea ice cover break-up (McMinn et al. 2004). Finally, expected higher underwater irradiance was suggested to affect benthic algal-herbivore interactions, as the increased energy available for photosynthesis could intensify the investment in antigrazing defenses (McClintock et al. 2008).

Retreating glaciers Climate warming has been associated with retreating glaciers and increased ice melting in polar areas (Svendsen et al. 2002, Cook et al. 2005). This might generate open space for benthic colonizers but also alter properties of the water column (due to reduced transparency and salinity) and increase ice disturbance processes (Schloss et al. 2008, Smale and Barnes 2008), factors that affect algal colonization (see abiotic factors section). A reduction in underwater radiation due to increased sedimentation might alter macroalgal lower depth distribution limits, which can be predicted by their light demands for photosynthesis and growth (Gómez et al. 1997). Thus, a shift in the lower distribution limit of subtidal macroalgae to shallower depths can be expected due to a higher sediment impact in coastal waters. Additionally, increased ice disturbances in the proximities of glaciers are expected to affect benthic communities. The colonization of a deglaciated fjord in a subantarctic area revealed a directional succession, with increasing abundance and diversity from the terminus of the glacier to the fjord mouth, mainly driven by the decreasing impact of ice scouring (Pugh and Davenport 1997). Overall, information about benthic algal colonization on recently formed ice-free areas in polar regions is very much needed, as a significant change in benthic communities is expected due to increased levels of stress and disturbance (Smale and Barnes 2008).

The response of hard bottom benthic algal communities to UVR and simulated ice cover break-up has been tested at the benthic community level in polar regions. However, other factors related to global change, such as rising temperatures, salinity and sediment input change, still need to be assessed. Results from a temperate intertidal site showed that increased thermal stress led to suppression of microalgal primary production but had no measurable effect on grazing rates (Morelissen and Harley 2007). However, these authors pointed out that there could be a negative effect on grazer populations under more stressful conditions (Morelissen and Harley 2007). To the best of our knowledge, community level interactions between temperature and other important drivers of succession, such as grazing, are still lacking for polar regions.

Testing ecological models and hypotheses with polar communities Owing to the small number of manipulative studies using benthic polar communities, rigorous experimental testing of ecological theory and models is scarce. The most advanced attempt was made in observational studies from the boreal intertidal rocky shores of the Gulf of Saint Lawrence, Nova Scotia (Scrosati and Heaven 2007, 2008). Here, diversity measures were related to stress levels (wave exposure and ice scour) to test the environmental stress hypothesis. The results suggested that patterns in benthic community diversity follow the predictions, as the most species-rich communities were found at sites experiencing intermediate levels of stress. The authors point out, however, that manipulative experiments are necessary to validate suggested drivers of the observed pattern. The "intermediate disturbance hypothesis (IDH)" has been the focus of most attention in polar studies and it has been used to interpret effects of ice scour on community diversity (Gutt 2001, Piepenburg 2005). Whereas some studies report that the effect of scouring icebergs supported the IDH (Laudien et al. 2007), others did not (Smale 2007). To date, polar seaweed communities have not been included in disturbance studies, but owing to an expected increase in the impact of icebergs, at least in Antarctica, fast growing and competitively inferior seaweed species could prevail here in the future (Smale and Barnes 2008). Nevertheless, there seems to be confusion about what should be considered as 'disturbance'. Several authors define changes in temperature, UVR or pollution regimes as a disturbance (Barnes and Conlan 2007, Smale and Barnes 2008). However, these should be classified as stress because stress consists of conditions that restrict production, whereas disturbance is associated with the partial or complete destruction of biomass (Grime 1977). In this context, it is also surprising that biotic disturbances, such as grazing by sea urchins, have not been addressed so far.

Concluding remarks and future perspectives

Polar benthic algae are exposed to highly seasonal abiotic factors and frequent episodes of physical disturbance (mainly related to ice abrasion), and affected areas are repeatedly recolonized (Barnes and Conlan 2007). Early successional stages are controlled by a complex interaction of abiotic factors and biological interactions. Intertidal sites are subjected to stronger physical disturbance and stress (Chapter 2; Zacher et al. 2009) and, in both polar regions, macroalgae occur mostly in the subtidal (Wiencke et al. 2007). These factors show decreasing impact with increasing depth and can determine the structure and complexity of the algal assemblages (Chapter 3; Kim 2001, Quartino et al. 2005, Zacher and Campana 2008, Wulff et al. 2009).

Polar algal primary succession is mainly driven by rapid colonizers, such as benthic diatoms, which have inhibitory or facilitating effects on ephemeral or annual macroalgae in intertidal and subtidal sites (Fricke et al. 2008, Zacher and Campana 2008), whereas annual and perennial macroalgal colonization is generally slow (Bowden et al. 2006, Konar 2007, Zacher et al. 2007b, Fricke et al. 2008, G.L. Campana unpublished data). Grazing is perhaps the most important biological factor structuring polar benthic algal communities at early stages of succession. Depending on species-specific susceptibility to consumers, grazing can have positive, negative or neutral effects on macroalgal succession (Konar 2007, Zacher et al. 2007b, G.L. Campana unpublished data). Microalgal assemblages are strongly affected by grazers, which can control biomass and introduce a shift in species composition, thereby altering the physiognomy of the assemblages (Zacher et al. 2007a, Campana et al. 2008b).

In the Arctic, Beuchel and Gulliksen (2008) could not identify a "climax stage" in a recolonization study over a 24-year period, presumably due to the constant disturbances at the site and also due to climate-related fluctuations, which were unrelated to the local disturbance regimes. Similarly, Barnes (1996) states that the frequency and influence of ice-mediated disturbance probably prevents community development in the shallow sublittoral of Signy Island (Antarctica) and that the concept of "climax community", as observed in boreal shallow sublittorals, cannot be applied for faunal studies of short colonization time (2 years) at relatively shallow depth. Deeper (>15 m) sites not heavily affected by physical disturbance might allow the development into a climax community that is dominated by large brown overstory algae (Desmarestiales) (Chapter 3; Wiencke et al. 2007, Wulff et al. 2009) that seem to be constant over time (C.D. Amsler personal communication). However, there is a substantial lack of information on algal succession over long colonization periods in macroalgal-dominated habitats for both polar regions (Table 14.1) and especially for deep subtidal sites.

So far, naturally cleared boulders and artificial substrata have been employed to study community succession in polar regions (Table 14.1). Natural and artificial substrata have been widely used in other regions and are considered to be the most powerful method to study succession dynamics in controlled field experiments (Foster and Sousa 1985). Artificial substrata transferred to the laboratory allowed the quantification and identification of microalgae and early stages of macroalgal colonizers in polar regions (Zacher et al. 2007a,b, Campana et al. 2008a,b, Fricke et al. 2008, Zacher and Campana 2008). Developing communities on cleared boulders were studied by direct observation or photographic image analyses (Konar 2007, Beuchel and Gulliksen 2008) and allowed long-term non-destructive analyses of community succession. On the one hand, artificial substrata provide uniform settlement conditions (e.g., size and rugosity), allowing standardized replication (Foster and Sousa 1985, Stanwell-Smith and Barnes 1997) and easy handling by scuba divers. Clay tiles (unglazed terracotta tiles) can be a good substratum for microphytobenthic algae and the developing communities were shown to be very similar to those on natural substrata (Lane et al. 2003). On the other hand, procedural controls are needed, as artifactual effects can be introduced due to the substratum material, size and position on the seabed (Foster and Sousa 1985). In this regard, consideration should be given to preparing experimental substrata from natural rock (Konar 2007). Also, natural substrata cut from the rock can be sterilized and reinstalled in the field, in exactly the same place from which it was removed. However, weather and logistic constraints could restrict this approach in polar regions, particularly in subtidal habitats. Artificial panels thus offer advantages in polar regions as they can provide rapid deployment, recording of data or retrieval of samples (Stanwell-Smith and Barnes 1997) but designs should include procedural controls.

More information on polar algal succession is required to better understand the magnitude of abiotic and biotic drivers on the composition, diversity and ecosystem functions of benthic communities. In particular, long-term succession studies over several years, with regular surveys on microphytobenthos and macroalgal development, are very much needed in polar systems. In general, the effects of biological interactions on algal succession in polar regions are still poorly studied and should be one promising avenue for future research. To date, assumptions on disturbance effects have been deduced exclusively from correlation studies of patterns of recolonization (Gutt 2000). Future research on polar algal community ecology should greatly benefit from carefully designed manipulative experiments using natural substrata or settlement panels. By doing so, advanced ecological concepts, such as (i) the interactive effects of disturbance and productivity (Huston 1979, Kondoh 2001), (ii) the effects of variation in disturbance (Benedetti-Cecchi 2003), (iii) the Menge-Sutherland model (Menge and Sutherland 1987), or (iv) of keystone predation (Menge et al. 1994) could be tested in carefully designed field experiments.

It is also an important task to investigate the potential changes of benthic communities due to global change. Ambient UVR can shape species composition of polar seaweed communities during early succession and exert negative and persistent effects on later stages (after 1 and 2 years of colonization). Increased UVBR can affect the composition of seaweed communities and thereby alter trophic interactions with unknown consequences. The high UV resistance of microalgal assemblages can have positive implications for macroalgal recruitment by providing UV-free space for UV-sensitive propagules and germlings but also negative effects due to pre-emption of space. Expected climate change is likely to affect algal succession as (i) increased temperature might impose physiological constraints on stenothermal polar seaweeds and (ii) lead to altered patterns of predation and competition (Barnes and Conlan 2007), (iii) the reduction of sea ice cover might lead to strong shifts in the algal community structure, and finally (iv) retreating glaciers provide open space for colonization but alter nutrient supply, reduce salinity and increase the intensity and frequency of ice disturbance and sedimentation. The proposed studies would strongly advance our understanding of polar benthic algae, which inhabit one of the most dramatically changing environments.

References

Aguilera, J., K. Bischof, U. Karsten, D. Hanelt and C. Wiencke. 2002. Seasonal variation in ecophysiological patterns in macroalgae from an Arctic fjord. II. Pigment accumulation and biochemical defence systems against high light stress. *Mar. Biol. 140*: 1087–1095.

Ahn, I.Y., H. Chung, J.S. Kang and S.H. Kang. 1997. Diatom composition and biomass variability in nearshore waters of Maxwell Bay, Antarctica, during the 1992/1993 austral summer. *Polar Biol. 17*: 123–130.

Amsler, C.D., J.B. McClintock and B.J. Baker. 1998. Chemical defense against herbivory in the Antarctic marine macroalgae *Iridaea cordata* and *Phyllophora antarctica* (Rhodophyceae). *J. Phycol. 34*: 53–59.

Amsler, C.D., J.B. McClintock and B.J. Baker. 1999. An Antarctic feeding triangle: defensive interactions between macroalgae, sea urchins and sea anemones. *Mar. Ecol. Prog. Ser. 183*: 105–114.

Amsler, C.D., K. Iken, J.B. McClintock, M.O. Amsler, K.J. Peters, J.M. Hubbard, F.B. Furrow and B.J. Baker. 2005. Comprehensive evaluation of the palatability and chemical defenses of subtidal macroalgae from the Antarctic Peninsula. *Mar. Ecol. Prog. Ser. 294*: 141–159.

Amsler, C.D., M.O. Amsler, J.B. McClintock and B.J. Baker. 2009a. Filamentous algal endophytes in macrophytic Antarctic algae: prevalence in hosts and palatability to mesoherbivores. *Phycologia 48*: 324-334.

Amsler, C.D., K. Iken, J.B. McClintock and B.J. Baker. 2009b. Defenses of polar macroalgae against herbivores and biofoulers. *Bot. Mar. 52*: 535–545.

Avila, C., S. Taboada and L. Núñez-Pons. 2007. Antarctic marine chemical ecology: what is next? *Mar. Ecol. 29*: 1–71.

Barnes, D.K.A. 1995. Sublittoral epifaunal communities at Signy Island, Antarctica. II. Below the ice-foot zone. *Mar. Biol. 121*: 565–572.

Barnes, D.K.A. 1996. Low levels of colonisation in Antarctica: the role of bryozoans in early community development. *In*: (D.P. Gordon, A.M. Smith and J.A. Grant-Mackie, eds) *Bryozoans in space and time*. Proceedings of the 10th International Bryozoology Conference, Wellington, New Zealand. National Institute of Water and Atmospheric Research Ltd, Wellington. pp. 19–27.

Barnes, D.K.A. and K.E. Conlan. 2007. Disturbance, colonization and development of Antarctic benthic communities. *Phil. Trans. R. Soc. B 362*: 11–38.

Barnes, D.K.A., P. Rothery and A. Clarke. 1996. Colonisation and development in encrusting communities from the Antarctic intertidal and sublittoral. *J. Exp. Mar. Biol. Ecol. 196*: 251–265.

Benedetti-Cecchi, L. 2003. The importance of the variance around the mean effect size of ecological processes. *Ecology 84*: 2335–2346.

Beuchel, F. and B. Gulliksen. 2008. Temporal patterns of benthic community development in an Arctic fjord (Kongsfjorden, Svalbard): results of a 24-year manipulation study. *Polar Biol. 31*: 913–924.

Beuchel, F., B. Gulliksen and M.L. Carroll. 2006. Long-term patterns of rocky bottom macrobenthic community structure in an Arctic fjord (Kongsfjorden, Svalbard) in relation to climate variability (1980–2003). *J. Mar. Syst. 63*: 35–48.

Bolton, J.J. 1983. Effects of short-term ice scouring on a Newfoundland rocky shore community. *Astarte 12*: 39–43.

Bond, G., H. Heinrich, W. Broecker, L. Labeyrie, J. McManus, J. Andrews, S. Huon, R. Jantschik, S. Clasen, C. Simet, K. Tedesco, M. Klas, G. Bonani and S. Ivy. 1992. Evidence for massive discharges of icebergs into the North Atlantic ocean during the last glacial period. *Nature 360*: 245–249.

Bothwell, M.L., D.M.J. Sherbot and C.M. Pollock. 1994. Ecosystem response to solar ultraviolet-B radiation: influence of trophic-level interactions. *Science 265*: 97–100.

Bowden, D.A., A. Clarke, L.S. Peck and D.K.A. Barnes. 2006. Antarctic sessile marine benthos: colonisation and growth on artificial substrata over three years. *Mar. Ecol. Prog. Ser. 316*: 1–16.

Brêthes, J.C., G.A. Ferreyra and S. de la Vega. 1994. Distribution, growth and reproduction of the limpet *Nacella (Patinigera) concinna* (Strebel 1908) in relation to potential food availability, in Esperanza Bay (Antarctic Peninsula). *Polar Biol. 14*: 161–170.

Campana, G.L., F.R. Momo, M.L. Quartino and G.A. Ferreyra. 2008a. Effects of UV radiation and grazing on biomass and primary production of subtidal benthic algae in Antarctica. *Ber. Polarforsch. Meeresforsch. 571*: 278–286.

Campana, G.L., M.L. Quartino, A. Yousif and A. Wulff. 2008b. Impacts of UV radiation and grazing on the structure of a subtidal benthic diatom assemblage in Antarctica. *Ber. Polarforsch. Meeresforsch. 571*: 302–310.

Chapman, A.R.O. and J.E. Lindley. 1980. Seasonal growth of *Laminaria solidungula* in the Canadian High Arctic in relation to irradiance and dissolved nutrient concentration. *Mar. Biol. 57*: 1–5.

Clarke, A., E.J. Murphy, M.P. Meredith, J.C. King, L.S. Peck, D.K.A. Barnes and R.C. Smith. 2007. Climate change and the marine ecosystem of the western Antarctic Peninsula. *Phil. Trans. R. Soc. B 362*: 149–166.

Clayton, M.N., C. Wiencke and H. Klöser. 1997. New records of temperate and sub-Antarctic marine benthic macroalgae from Antarctica. *Polar Biol. 17*: 141–149.

Cook, A.J., A.J. Fox, D.G. Vaughan and J.G. Ferrigno. 2005. Retreating glacier fronts on the Antarctic Peninsula over the past half-century. *Science 308*: 541–544.

Cullen, J. and P. Neale. 1994. Ultraviolet radiation, ozone depletion and marine photosynthesis. *Photosynth. Res. 39*: 303–320.

Cunningham, L. and A. McMinn. 2004. The influence of natural environmental factors on benthic diatom communities from the Windmill Islands, Antarctica. *Phycologia 43*: 744–756.

Dayton, P.K. 1979. Observation of growth, dispersal and dynamics of some sponges in McMurdo Sound, Antarctica. *In*: (C. Levian and N. Bourny-Esnault, eds) *Conservation problems in Antarctica*. Centre National de la Recherche Scientifique, Paris. pp. 271–283.

Dayton, P.K. 1989. Interdecadal variation in an Antarctic sponge and its predators from oceanographic climate shifts. *Science 245*: 1484–1486.

Dierssen, H.M., R.C. Smith and M. Vernet. 2002. Glacial meltwater dynamics in coastal waters west of the Antarctic Peninsula. *Proc. Natl. Acad. Sci. USA 99*: 1790–1795.

Dobretsov, S.V., P.Y. Qian and M. Wahl. 2005. Effect of solar ultraviolet radiation on the formation of shallow, early successional biofouling communities in Hong Kong. *Mar. Ecol. Prog. Ser. 290*: 55–65.

Drew, E.A. and R.M. Hastings. 1992. A year-round ecophysiological study of *Himantothallus grandifolius* (Desmarestiales, Phaeophyta) at Signy Island, Antarctica. *Phycologia 31*: 262–277.

Duffy, J.E. and M.E. Hay. 2000. Strong impacts of grazing amphipods on the organization of a benthic community. *Ecol. Monogr. 70*: 237–263.

Dunton, K.H. and D.M. Schell. 1986. Seasonal carbon budget and growth of *Laminaria solidungula* in the Alaskan High Arctic. *Mar. Ecol. Prog. Ser. 31*: 57–66.

Fairhead, V.A., C.D. Amsler, J.B. McClintock and B.J. Baker. 2005. Within thallus variation in chemical and physical defences in two species of ecologically dominant brown macroalgae from the Antarctic Peninsula. *J. Exp. Mar. Biol. Ecol. 322*: 1–12.

Foster, M.S. and W.P. Sousa. 1985. Succession. *In*: (M.M. Littler and D.S. Littler, eds) *Handbook of phycological methods – ecological field methods: macroalgae*. Cambridge University Press, Cambridge. pp. 269–290.

Fricke, A., M. Molis, C. Wiencke, N. Valdivia and A.S. Chapman. 2008. Natural succession of macroalgal-dominated epibenthic assemblages at different water depths and after transplantation from deep to shallow water on Spitsbergen. *Polar Biol. 31*: 1191–1203.

Garrison, D.L., A.R. Close and E. Reimnitz. 1989. Algae concentrated by frazil ice: evidence from laboratory experiments and field measurements. *Antarct. Sci. 1*: 313–316.

Gómez, I., G. Weykam, H. Klöser and C. Wiencke. 1997. Photosynthetic light requirements, daily carbon balance and zonation of sublittoral macroalgae from King George Island (Antarctica). *Mar. Ecol. Progr. Ser. 148*: 281–293.

Gómez, I., A. Wulff, M.Y. Roleda, P. Huovinen, U. Karsten, M.L. Quartino, K. Dunton and C. Wiencke. 2009. Light and temperature demands of marine benthic microalgae and seaweeds in polar regions. *Bot. Mar. 52*: 593–608.

Grime, P.J. 1977. Evidence for the existence of three primary strategies in plants and its relevance to ecological and evolutionary theory. *Am. Nat. 111*: 1169–1194.

Gutt, J. 2000. Some driving forces structuring communities of the sublittoral Antarctic macrobenthos. *Antarct. Sci. 12*: 297–313.

Gutt, J. 2001. On the direct impact of ice on marine benthic communities, a review. *Polar Biol. 24*: 553–564.

Hanelt, D., I.M. Jaramillo, W. Nultsch, S. Senger and R. Westermeier. 1994. Photoinhibition as a regulative mechanism of photosynthesis in marine algae of Antarctica. *Ser. Cient. INACH 44*: 67–77.

Hernando, M.P. and G.A. Ferreyra. 2005. The effects of UV radiation on photosynthesis in an Antarctic diatom (*Thalassiosira* sp.): does vertical mixing matter? *J. Exp. Mar. Biol. Ecol. 325*: 35–45.

Hillebrand, H., B. Worm and H.K. Lotze. 2000. Marine microbenthic community structure regulated by nitrogen loading and grazing pressure. *Mar. Ecol. Prog. Ser. 204*: 27–38.

Hop, H., T. Pearson, E.N. Hegseth, K.M. Kovacs, C. Wiencke, S. Kwasniewski, K. Eiane, F. Mehlum, B. Gulliksen, M. Wlodarska-Kowalczuk, C. Lydersen, J.M. Weslawski, S. Cochrane, G.W. Gabrielsen, R.J.G. Leakey, O.J. Lønne, M. Zajaczkowski, S. Falk-Petersen, M. Kendall, S.Å. Wängberg, K. Bischof, A.Y. Voronkov, N.A. Kovaltchouk, J. Wiktor, M. Poltermann, G. di Prisco, C. Papucci and S. Gerland. 2002. The marine ecosystem of Kongsfjorden, Svalbard. *Polar Res. 21*: 167–208.

Hoyer, K., U. Karsten, T. Sawall and C. Wiencke. 2001. Photoprotective substances in Antarctic macroalgae and their variation with respect to depth distribution, different tissues and developmental stages. *Mar. Ecol. Prog. Ser. 211*: 117–129.

Huang, R. and A.D. Boney. 1985. Individual and combined interactions between littoral diatoms and sporelings of red algae. *J. Exp. Mar. Biol. Ecol. 85*: 101–111.

Huston, M. 1979. A general hypothesis of species diversity. *Am. Nat. 113*: 81–101.

Iken, K. 1999. Feeding ecology of the Antarctic herbivorous gastropod *Laevilacunaria antarctica* Martens. *J. Exp. Mar. Biol. Ecol. 236*: 133–148.

Iken, K., E.R. Barrera-Oro, M.L. Quartino, R.J. Casaux and T. Brey. 1997. Grazing by the Antarctic fish *Notothenia coriiceps*: evidence for selective feeding on macroalgae. *Antarct. Sci. 9*: 386–391.

Irving, A.D., S.D. Connell, E.L. Johnston, A.J. Pile and B.M. Gillanders. 2005. The response of encrusting coralline algae to canopy loss: an independent test of predictions on an Antarctic coast. *Mar. Biol. 147*: 1075–1083.

Jenkins, S.R., R.A. Coleman, P. Della Santina, S.J. Hawkins, M.T. Burrows and R.G. Hartnoll. 2005. Regional scale differences in the determinism of grazing effects in the rocky intertidal. *Mar. Ecol. Prog. Ser. 287*: 77–86.

Johnston, E.L., S.D. Connell, A.D. Irving, A.J. Pile and B.M. Gillanders. 2007. Antarctic patterns of shallow subtidal habitat and inhabitants in Wilke's Land. *Polar Biol. 30*: 781–788.

Karsten, U., A. Wulff, M.Y. Roleda, R. Müller, F.S. Steinhoff, J. Fredersdorf and C. Wiencke. 2009. Physiological responses of polar benthic algae to ultraviolet radiation. *Bot. Mar. 52*: 639–654.

Keats, D.W., G.R. South and D.H. Steele. 1985. Algal biomass and diversity in the upper subtidal at a pack-ice disturbed site in Eastern Newfoundland. *Mar. Ecol. Prog. Ser. 25*: 151–158.

Kim, D. 2001. Seasonality of marine algae and grazers of an Antarctic rocky intertidal, with emphasis on the role of the limpet *Nacella concinna* Strebel (Gastropoda: Patellidae). *Ber. Polarforsch. Meeresforsch. 397*: 1–136.

Kirst, G.O. and C. Wiencke. 1995. Ecophysiology of polar algae. *J. Phycol. 31*: 181–199.

Klöser, H. 1998. Habitats and distribution patterns of benthic diatoms in Potter Cove (King George Island) and its vicinity. *Ber. Polarforsch. 299*: 95–105.

Klöser, H., G.A. Ferreyra, I.R. Schloss, G. Mercuri, F. Laturnus and A. Curtosi. 1993. Seasonal variation of algal growth conditions in sheltered Antarctic bays: the example of Potter Cove (King George Island, South Shetlands). *J. Mar. Syst. 4*: 289–301.

Klöser, H., G. Mercuri, F. Laturnus, M.L. Quartino and C. Wiencke. 1994. On the competitive balance of macroalgae at Potter Cove (King George Island, South Shetlands). *Polar Biol. 14*: 11–16.

Klöser, H., M.L. Quartino and C. Wiencke. 1996. Distribution of macroalgae and macroalgal communities in gradients of physical conditions in Potter Cove, King George Island, Antarctica. *Hydrobiologia 333*: 1–17.

Konar, B. 2007. Recolonization of a high latitude hard-bottom nearshore community. *Polar Biol. 30*: 663–667.

Konar, B. and K. Iken. 2005. Competitive dominance among sessile marine organisms in a high Arctic boulder community. *Polar Biol. 29*: 61–64.

Kondoh, M. 2001. Unifying the relationships of species richness to productivity and disturbance. *Proc. R. Soc. Lond. B 268*: 269–271.

Kühl, M., R.N. Glud, J. Borum, R. Roberts and S. Rysgaard. 2001. Photosynthetic performance of surface-associated algae below sea ice as measured with a pulse-amplitude-modulated (PAM) fluorometer and O_2 microsensors. *Mar. Ecol. Prog. Ser. 223*: 1–14.

Lane, C.M., K.H. Taffs and J.L. Corfield. 2003. A comparison of diatom community structure on natural and artificial substrata. *Hydrobiologia 493*: 65–79.

Laudien, J., M. Herrmann and W.E. Arntz. 2007. Soft bottom species richness and diversity as a function of depth and iceberg scour in Arctic glacial Kongsfjorden (Svalbard). *Polar Biol. 30*: 1035–1046.

Longhi, M.L., I.R. Schloss and C. Wiencke. 2003. Effect of irradiance and temperature on photosynthesis and growth of two Antarctic benthic diatoms, *Gyrosigma subsalinum* and *Odontella litigiosa*. *Bot. Mar. 46*: 276–284.

Lotze, H.K., B. Worm and U. Sommer. 2001. Strong bottom-up and top-down control of early life stages of macroalgae. *Limnol. Oceanogr. 46*: 749–757.

Lotze, H.K., B. Worm, M. Molis and M. Wahl. 2002. Effects of UV radiation and consumers on recruitment and succession of a marine macrobenthic community. *Mar. Ecol. Prog. Ser. 243*: 57–66.

Lüning, K. 1990. *Seaweeds: their environment, biogeography and ecophysiology*. John Wiley, New York. pp. 527.

McClintock, J.B., H. Ducklow and W. Fraser. 2008. Ecological responses to climate change on the Antarctic Peninsula. *Am. Sci. 96*: 302–310.

McCook, L.J. and A.R.O. Chapman. 1997. Patterns and variations in natural succession following massive ice-scour of a rocky intertidal seashore. *J. Exp. Mar. Biol. Ecol. 214*: 121–147.

McMinn, A., J.W. Runcie and M. Riddle. 2004. Effect of seasonal sea ice breakout on the photosynthesis of benthic diatom mats at Casey, Antarctica. *J. Phycol. 40*: 62–69.

Menge, B.A. and J.P. Sutherland. 1987. Community regulation: variation in disturbance, competition, and predation in relation to environmental stress and recruitment. *Am. Nat. 130*: 730–757.

Menge, B.A., E.L. Berlow, C.A. Blanchette, S.A. Navarrete and S.B. Yamada. 1994. The keystone species concept: variation in interaction strength in a rocky intertidal habitat. *Ecol. Monogr. 64*: 250–286.

Miller, K.A. and J.S. Pearse. 1991. Ecological studies of seaweeds in McMurdo Sound, Antarctica. *Am. Zool. 31*: 35–48.

Molis, M. and M. Wahl. 2004. Transient effects of solar ultraviolet radiation on the diversity and structure of a field-grown epibenthic community at Lüderitz, Namibia. *J. Exp. Mar. Biol. Ecol. 302*: 51–62.

Molis, M. and M. Wahl. 2009. Comparison of the impacts of consumers, ambient UV, and future UVB irradiance on midlatitudinal macro-epibenthic assemblages. *Global Change Biol. 15*: 1833–1845.

Molis, M., H. Wessels, W. Hagen, U. Karsten and C. Wiencke. 2009. Do sulphuric acid and the brown alga *Desmarestia viridis* support community structure in Arctic kelp patches by altering grazing impact, distribution patterns, and behaviour of sea urchins? *Polar Biol. 32*: 71–82.

Morelissen, B. and C.D.G. Harley. 2007. The effects of temperature on producers, consumers, and plant-herbivore interactions in an intertidal community. *J. Exp. Mar. Biol. Ecol. 348*: 162–173.

Müller, R., T. Laepple, I. Bartsch and C. Wiencke. 2009. Impact of oceanic warming on the distribution of seaweeds in polar and cold-temperate waters. *Bot. Mar. 52*: 617–638.

Nicotri, M.E. 1977. Grazing effects of four marine intertidal herbivores on the microflora. *Ecology 58*: 1020–1032.

Obermüller, B., S. Puntarulo and D. Abele. 2007. UV-tolerance and instantaneous physiological stress responses of two Antarctic amphipod species *Gondogeneia antarctica* and *Djerboa furcipes* during exposure to UV radiation. *Mar. Environ. Res. 64*: 267–285.

Olsen, B.R., K. Sjøtun, T. Høsæter and O.J. Lønne. 2004. Population structure of *Laminaria digitata* (Hudson) J.V. Lamouroux from three different areas in the Northeast Atlantic ocean. *Abstract book: XVIIIth International Seaweed Symposium, Bergen, Norway*: 75.

Pavia, H., G. Cervin, A. Lindgren and P. Åberg. 1997. Effects of UV-B radiation and simulated herbivory on phlorotannins in the brown alga *Ascophyllum nodosum. Mar. Ecol. Prog. Ser. 157*: 139–146.

Peletier, H., W.W.C. Gieskes and A.G.J. Buma. 1996. Ultraviolet-B radiation resistance of benthic diatoms isolated from tidal flats in the Dutch Wadden Sea. *Mar. Ecol. Prog. Ser. 135*: 163–168.

Peltikhina, T.S. 2002. Distribution peculiarities and stocks of *Laminaria* algae in the area of Isfjord of the Spitsbergen Archipelago. *In*: (G.G. Matishov and G.A. Tarasov, eds) *The complex investigations of the Spitsbergen nature*. Murmansk Biological Institute, Murmansk. pp. 168–174 (in Russian with English abstract).

Peters, A.F. 2003. Molecular identification, distribution and taxonomy of brown algal endophytes, with emphasis on species from Antarctica. *In*: (A.R.O. Chapman, R.J. Anderson, V. Vreeland and I.R. Davison, eds) *Proceedings of the 17th International Seaweed Symposium*. Oxford University Press, New York. pp. 293–302.

Piepenburg, D. 2005. Recent research on Arctic benthos: common notions need to be revised. *Polar Biol. 28*: 733–755.

Pugh, P.J.A. and J. Davenport. 1997. Colonisation vs. disturbance: the effects of sustained ice-scouring on intertidal communities. *J. Exp. Mar. Biol. Ecol. 210*: 1–21.

Quartino, M.L., H. Klöser, I.R. Schloss and C. Wiencke. 2001. Biomass and associations of benthic marine macroalgae from the inner Potter Cove (King George Island, Antarctica) related to depth and substrate. *Polar Biol. 24*: 349–355.

Quartino, M.L., H.E. Zaixso and A.L. Boraso de Zaixso. 2005. Biological and environmental characterization of marine macroalgal assemblages in Potter Cove, South Shetland Islands, Antarctica. *Bot. Mar. 48*: 187–197.

Reimnitz, E., E.W. Kempema and P.W. Barnes. 1987. Anchor ice, seabed freezing, and sediment dynamics in shallow Arctic seas. *J. Geophys. Res. C 92*: 14671–14678.

Roleda, M.Y., C. Wiencke, D. Hanelt and K. Bischof. 2007. Sensitivity of the early life stages of macroalgae from the Northern Hemisphere to ultraviolet radiation. *Photochem. Photobiol. 83*: 851–862.

Schloss, I.R., G.A. Ferreyra, O. González, A. Atencio, V.L. Fuentes, G. Tosonotto, G. Mercuri, R. Sahade, M. Tatián and D. Abele. 2008. Long-term hydrographic conditions and climate trends in Potter Cove. *Ber. Polarforsch. Meeresforsch. 571*: 382–389.

Scrosati, R. and C. Heaven. 2007. Spatial trends in community richness, diversity, and evenness across rocky intertidal environmental stress gradients in eastern Canada. *Mar. Ecol. Prog. Ser. 342*: 1–14.

Scrosati, R. and C. Heaven. 2008. Trends in abundance of rocky intertidal seaweeds and filter feeders across gradients of elevation, wave exposure, and ice scour in eastern Canada. *Hydrobiologia 603*: 1–14.

Shanks, A.L. and W.G. Wright. 1986. Adding teeth to wave action: the destructive effects of wave-borne rocks on intertidal organisms. *Oecologia 69*: 420–428.

Sivertsen, K. 1997. Geographic and environmental factors affecting the distribution of kelp beds and barren grounds and changes in biota associated with kelp reduction at sites along the Norwegian coast. *Can. J. Fish. Aquat. Sci. 54*: 2872–2887.

Smale, D.A. 2007. Ice disturbance intensity structures benthic communities in nearshore Antarctic waters. *Mar. Ecol. Prog. Ser. 349*: 89–102.

Smale, D.A. and D.K.A. Barnes. 2008. Likely responses of the Antarctic benthos to climate-related changes in physical disturbance during the 21st century, based primarily on evidence from the West Antarctic Peninsula region. *Ecography 31*: 289–305.

Sommer, U. 2000. Benthic microalgal diversity enhanced by spatial heterogeneity of grazing. *Oecologia 122*: 284–287.

Stanwell-Smith, D. and D.K.A. Barnes. 1997. Benthic community development in Antarctica: recruitment and growth on settlement panels at Signy Island. *J. Exp. Mar. Biol. Ecol. 212*: 61–79.

Steinman, A.D. 1996. Effects of grazers on freshwater benthic algae. *In*: (R.J. Stevenson, M.L. Bothwell and R.L. Lowe, eds) *Algal ecology – freshwater benthic ecosystems*. Academic Press, San Diego, CA. pp. 341–373.

Sundbäck, K., P. Nilsson, C. Nilsson and B. Jönsson. 1996. Balance between autotrophic and heterotrophic components and processes in microbenthic communities of sandy sediments: a field study. *Estuar. Coast. Mar. Sci. 43*: 689–706.

Svendsen, H., A. Beszczynska-Møeller, J.O. Hagen, B. Lefauconnier, V. Tverberg, S. Gerland, J.B. Ørbæk, K. Bischof, C. Papucci, M. Zajaczkowski, R. Azzolini, O. Bruland, C. Wiencke, J.-G. Winther and W. Dallmann. 2002. The physical environment of Kongsfjorden-Krossfjorden, an Arctic fjord system in Svalbard. *Polar Res. 21*: 133–166.

Swanson, A.K. and C.H. Fox. 2007. Altered kelp (Laminariales) phlorotannins and growth under elevated carbon dioxide and ultraviolet-B treatments can influence associated intertidal food webs. *Global Change Biol. 13*: 1696–1709.

Syvitski, J.P.M., G.E. Farrow, R.J.A. Atkinson, P.G. Moore and J.T. Andrews. 1989. Baffin Island Fjord macrobenthos: bottom communities and environmental significance. *Arctic 42*: 132–247.

Taylor, R.L., G.S. Caldwell, H.J. Dunstan and M.G. Bentley. 2007. Short-term impacts of polyunsaturated aldehyde-producing diatoms on the harpacticoid copepod, *Tisbe holothuriae*. *J. Exp. Mar. Biol. Ecol. 341*: 60–69.

Underwood, G.J.C., C. Nilsson, K. Sundbäck and A. Wulff. 1999. Short-term effects of UVB radiation on chlorophyll fluorescence, biomass, pigments, and carbohydrate fractions in a benthic diatom mat. *J. Phycol. 35*: 656–666.

Vinebrooke, R.D. and P.R. Leavitt. 1999. Differential responses of littoral communities to ultraviolet radiation on an Alpine lake. *Ecology 80*: 223–237.

Wahl, M., M. Molis, A. Davis, S. Dobretsov, S. Dürr, J. Johansson, J. Kinley, D. Kirugara, M. Langer, H.K. Lotze, M. Thiel, J.C. Thomason, B. Worm and D. Zeevi Ben-Yosef. 2004. UV effects that come and go: a global comparison of marine benthic community level impacts. *Global Change Biol. 10*: 1962–1972.

Waring, J., N.R. Baker and G.J.C. Underwood. 2007. Responses of estuarine intertidal microphytobenthic algal assemblages to enhanced ultraviolet B radiation. *Global Change Biol. 13*: 1398–1413.

Wessels, H., W. Hagen, M. Molis, C. Wiencke and U. Karsten. 2006. Intra- and interspecific differences in palatability of Arctic macroalgae from Kongsfjorden (Spitsbergen) for two benthic sympatric invertebrates. *J. Exp. Mar. Biol. Ecol. 329*: 20–33.

Whitehead, R.F., S.J. de Mora and S. Demers. 2000. Enhanced UV radiation – a new problem for the marine environment. *In*: (S. de Mora, S. Demers and M. Vernet, eds) *The effects of UV radiation in the marine environment*. Cambridge University Press, Cambridge. pp. 1–34.

Wiencke, C. and I. tom Dieck. 1989. Temperature requirements for growth and temperature tolerance of macroalgae endemic to the Antarctic region. *Mar. Ecol. Prog. Ser. 54*: 189–197.

Wiencke, C. and M.N. Clayton. 2002. *Antarctic seaweeds. Synopses of the Antarctic benthos*, 9. A.R.G. Gantner Verlag K.G., Ruggell, Lichtenstein. pp. 239.

Wiencke, C., J. Rahmel, U. Karsten, G. Weykam and G.O. Kirst. 1993. Photosynthesis of marine macroalgae from Antarctica: light and temperature requirements. *Bot. Acta. 106*: 78–87.

Wiencke, C., M.N. Clayton, I. Gómez, K. Iken, U.H. Lüder, C.D. Amsler, U. Karsten, D. Hanelt, K. Bischof and K. Dunton. 2007. Life strategy, ecophysiology and ecology of seaweeds in polar waters. *Rev. Environ. Sci. Biotechnol. 6*: 95–126.

Wiencke, C., I. Gómez and K. Dunton. 2009. Phenology and seasonal physiological performance of polar seaweeds. *Bot. Mar. 52*: 585–592.

Worm, B., H.K. Lotze and U. Sommer. 2000. Coastal food web structure, carbon storage, and nitrogen retention regulated by consumer pressure and nutrient loading. *Limnol. Oceanogr. 45*: 339–349.

Wulff, A. and K. Zacher. 2008. Short-term effects of UV radiation on benthic diatoms. *Ber. Polarforsch. Meeresforsch. 571*: 295–301.

Wulff, A., K. Sundbäck, C. Nilsson, L. Carlson and B. Jönsson. 1997. Effect of sediment load on the microbenthic community of a shallow-water sandy sediment. *Estuaries 20*: 547–558.

Wulff, A., S.Å. Wängberg, K. Sundbäck, C. Nilsson and G.J.C. Underwood. 2000. Effects of UVB radiation on a marine microphytobenthic community growing on a sand-substratum under different nutrient conditions. *Limnol. Oceanogr. 45*: 1144–1152.

Wulff, A., M.Y. Roleda, K. Zacher and C. Wiencke. 2008a. Exposure to sudden light burst after prolonged darkness – a case study on benthic diatoms in Antarctica. *Diatom Res. 23*: 519–532.

Wulff, A., M.Y. Roleda, K. Zacher and C. Wiencke. 2008b. UV radiation effects on pigments, photosynthetic efficiency and DNA of an Antarctic marine benthic diatom community. *Aquat. Biol. 3*: 167–177.

Wulff, A., K. Zacher, D. Hanelt, A. Al-Handal and C. Wiencke. 2008c. UV radiation – a threat to Antarctic benthic marine diatoms? *Antarct. Sci. 20*: 13–20.

Wulff, A., K. Iken, M.L. Quartino, A. Al-Handal, C. Wiencke and M.N. Clayton. 2009. Biodiversity, biogeography and zonation of marine benthic micro- and macroalgae in the Arctic and Antarctic. *Bot. Mar. 52*: 491–507.

Zacher, K. and G.L. Campana. 2008. UV and grazing effects on an intertidal and subtidal algal assemblage: a comparative study. *Ber. Polarforsch. Meeresforsch. 571*: 287–294.

Zacher, K., D. Hanelt, C. Wiencke and A. Wulff. 2007a. Grazing and UV radiation effects on an Antarctic intertidal microalgal assemblage: a long-term field study. *Polar Biol. 30*: 1203–1212.

Zacher, K., A. Wulff, M. Molis, D. Hanelt and C. Wiencke. 2007b. Ultraviolet radiation and consumer effects on a field-grown intertidal macroalgal assemblage in Antarctica. *Global Change Biol. 13*: 1201–1215.

Zacher, K., R. Rautenberger, A. Wulff, D. Hanelt and C. Wiencke. 2009. The abiotic environment of polar marine benthic algae. *Bot. Mar. 52*: 483–490.

15 Conclusion and outlook

Future perspectives on the investigation of polar benthic algae

Christian Wiencke and Margaret N. Clayton

The main aim of this book and the special issue of *Botanica Marina* was to summarise progress achieved during the last few decades in the field of marine benthic algal biology for both polar regions and to give a brief overview of current investigations. In this way, gaps in our current state of knowledge were identified in the various papers and are placed here in a broader context as a basis for future research.

With respect to biodiversity and biogeography, there are clear gaps in our knowledge, especially for benthic microalgae from both polar regions and for red macroalgae from the Antarctic, as pointed out by Wulff et al. (2009; Chapter 3) and Hommersand et al. (2009; Chapter 4), respectively. The biodiversity and ecology of polar benthic algae must be studied more intensively in order to understand their evolutionary history and their physiological performance under present environmental conditions. In view of global climate changes, their diversity and ecology must be documented before the changes are too great!

Knowledge on the Antarctic red algal flora, the systematic group with the greatest number of seaweed species endemic to the Antarctic region, needs to be expanded by studies from additional localities in the Antarctic and sub-Antarctic regions. Although considerable progress has been made, as documented by Hommersand et al. (2009; Chapter 4), many more data are necessary to get a full overview over this important component of the Antarctic marine algal flora. Hypotheses on taxonomic and biogeographical relationships of the Southern Ocean red algae presented by this research group (op. cit.) must be substantiated by modern molecular analyses. This will allow us to understand the evolution of this algal group over geological times characterized by strong plate tectonic events and climate changes.

While considerable progress has been made on understanding the defenses of Antarctic seaweeds against herbivory, much less is known on macroalgal-herbivore relationships in the Arctic. Hence, many more studies are needed on Arctic communities, as pointed out by Amsler et al. (2009; Chapter 5). In Antarctic seaweed communities, the roles of macroalgal secondary metabolites in chemical defence and functioning of these systems deserve further studies not only with respect to herbivory, but also to biofouling. Phlorotannins from Antarctic brown algae clearly have the potential to deter herbivores, as shown by Iken et al. (2009; Chapter 6). There are, however, conflicting results, which could certainly be explained if we knew more about the molecular structure of phlorotannins, an open field for future research. With this knowledge it should be possible to study the effect of these compounds on grazers in a better way, focusing not only on the deterrent taste of phlorotannins but also on their effects on the digestive system of consumers.

As the database for benthic microalgal production in polar regions is so small – as is the case for biodiversity of this algal group – further studies are urgently needed in this field of research as pointed out by Glud et al. (2009; Chapter 7). The chapter by Woelfel et al. (2009; Chapter 8) is a baseline study on microphytobenthic biomass in a glacial fjord in the Arctic. Similar studies are needed in other regions on soft and hard bottom and as well on shallow offshore areas both in the Arctic and the Antarctic. Such investigations should include not only biomass studies but in particular also non-invasive as well as traditional approaches to benthic microalgal activity under present and changing environmental conditions. The rapidly decreasing sea ice cover in the Arctic will certainly lead to better light conditions for benthic primary producers. On the other hand, inflow of turbid meltwater from glaciers and thawing permafrost will counteract the expected improvement of light availability in coastal waters. The result of these processes is impossible to predict, but could be best monitored in long-term observatory-based measurements of benthic primary production in coastal areas. These measurements should include continuous recordings of the environmental variables in the intertidal and different water depths. In particular, data on the radiation conditions (photosynthetically active and UV radiation), salinity, water turbidity, sedimentation rates and nutrient levels are urgently needed, as becomes apparent also in the chapter by Zacher et al. (2009; Chapter 2).

The seasonal development of polar seaweeds is strongly synchronized with the seasonally changing light conditions in the polar regions (Wiencke et al. 2009; chapter 9). Several of the underlying physiological mechanisms have been studied already. Open questions range from the perception of daylength, the most probable signal controlling phenology in kelps and kelp-like algae, processes related to storage and remobilisation of photosynthetic products, translocation of these compounds to the meristems and to other morpho-functional processes such as light-independent carbon fixation. In this context, more (especially field) data are necessary on the physiological performance of polar seaweeds in winter. Without a thorough understanding of these physiological processes we will not understand the ecological performance of polar seaweeds today and under conditions of global change.

The same arguments apply also to the general ecophysiology of benthic macro- and microalgae (Chapter 10, 13; Gómez et al. 2009, Karsten et al. 2009). In particular, it is necessary to characterize the thresholds of tolerance to light, UV radiation, temperature, freezing and other stresses in order to understand the ecology of individual species. More experimental evidence is needed to precisely document the underlying mechanisms of acclimation and adaptation. These can only be elucidated by use of modern genomic, proteomic and metabolomic approaches. Additionally, more studies are needed examining the interactive effects of various environmental factors, as pointed out by Becker et al. (2009; Chapter 11). Such studies will deliver important new insights into the physiological and ecological performance of algae, which cannot be obtained through unifactorial experiments.

This knowledge will also allow us to predict the impact of global changes on marine benthic algae. The chapter by Müller et al. (2009; Chapter 12) sets a good baseline for understanding how global warming affects the geographic distribution of seaweeds. We must, however, recognize that the database for many species, especially for seaweeds from the Southern Ocean is still meagre, as is the case for benthic microalgae from both polar regions. Global warming and enhanced UV radiation through stratospheric ozone depletion will certainly affect the composition of polar seaweed communities during early succession and also at later stages (Campana et al. 2009; Chapter 14). Ecological studies are urgently needed to document how benthic communities are affected by increasing water temperatures, which will certainly lead to altered patterns of predation and competition, and by changes in the radiation regime through either enhanced UV radiation or changing light availability due to increasing turbidity or decreasing sea ice cover. Retreating glaciers provide new space for colonization but reduced salinity affects nutrient concentrations and increase in ice disturbance and sedimentation have unknown effects on the benthic communities.

Polar regions are not only parts of the world most strongly affected by global warming and enhanced UV radiation, they are also likely highly susceptible to threats from ocean acidification as a consequence of human use of fossil fuels. As fossil fuel CO_2 enters the ocean surface, it reacts with seawater to form bicarbonate and protons, thereby consuming carbonate ions. The result of this process is an increase in CO_2 and bicarbonate concentrations and a decrease in seawater pH and carbon ion concentration (Feely et al. 2004). As the solubility of gases is higher in colder waters, polar regions will be most strongly affected. The effects of ocean acidification on marine benthic algae and their communities today are in general poorly understood. Elevated CO_2 levels may enhance primary production, although species with a carbon-concentrating mechanism (Giordano et al. 2005) may be much less affected. Calcification of calcareous algae will be reduced (Leclerc et al. 2000). There is only one publication available addressing this question in polar waters (McClintock et al. 2009). Therefore, we urgently need to know the responses of marine benthic algae to elevated CO_2 levels and ocean acidification and also more about the underlying mechanisms and about possible acclimation processes on the individual and community level. Furthermore, as outlined above: we need to study the interactive effects with other environmental variables in order to predict the consequences for coastal marine ecosystems in the polar regions.

Overall, although much has been achieved in the last few decades, there are many open questions. A great deal of work is pure basic research. In this respect it would be most interesting to show how marine benthic algae have evolved and adapted to their physical, chemical and biotic environments in coastal ecosystems of the Arctic and Antarctic. A second important focus should be the response of species, communities and coastal ecosystems in both polar regions to global climate changes. Coastal regions, especially in the Arctic and on the Western Antarctic Peninsula, are currently undergoing the greatest and most rapid climate-mediated change of any region on Earth (Zacher et al. 2009; Chapter 2), prompting the urgency of more intensive studies.

References

Amsler, C.D., K. Iken, J.B. McClintock and B.J. Baker. 2009. Defenses of polar macroalgae against herbivores and biofoulers. *Bot. Mar. 52*: 535–545.

Becker, S., B. Walter and K. Bischof. 2009. Freezing tolerance and photosynthetic performance of polar seaweeds at low temperatures. *Bot. Mar. 52*: 609–616.

Campana, G.L., K. Zacher, A. Fricke, M. Molis, A. Wulff, M.L. Quartino and C. Wiencke. 2009. Drivers of colonization and succession in polar benthic macro- and microalgal communities. *Bot. Mar. 52*: 655–667.

Feely, R.A., C.L. Sabine, K. Lee, W. Berelson, J. Kleypas, V.J. Fabry and F.J. Millero. 2004. Impact of anthropogenic CO_2 on the $CaCO_3$ system in the oceans. *Science 305*: 362–366.

Giordano, M., J. Beardall and J.A. Raven. 2005. CO_2 concentrating mechanisms in algae: mechanisms, environmental modulation, and evolution. *Annu. Rev. Plant Biol. 56*: 99–131.

Glud, R.N., J. Woelfel, U. Karsten, M. Kühl and S. Rysgaard. 2009. Benthic microalgal production in the Arctic: applied methods and status of the current database. *Bot. Mar. 52*: 559–571.

Gómez, I., A. Wulff, M.Y. Roleda, P. Huovinen, U. Karsten, M.L. Quartino, K. Dunton and C. Wiencke. 2009. Light and temperature demands of marine benthic microalgae and seaweeds in polar regions. *Bot. Mar. 52*: 593–608.

Hommersand, M.H., R.L. Moe, C.D. Amsler and S. Fredericq. 2009. Notes on the systematics and biogeographical relationships of Antarctic and sub-Antarctic Rhodophyta with descriptions of four new genera and five new species. *Bot. Mar. 52*: 509–534.

Iken, K., C.D. Amsler, M.O. Amsler, J.B. MacClintock and B.J. Baker. 2009. Field studies on deterrent properties of phlorotannins in Antarctic brown algae. *Bot. Mar. 52*: 547–557.

Karsten, U., A. Wulff, M.Y. Roleda, R. Müller, F.S. Steinhoff, J. Fredersdorf and C. Wiencke. 2009. Physiological responses of polar benthic algae to ultraviolet radiation. *Bot. Mar. 52*: 639–654.

Leclerc, N., J.P. Gattuso and J. Jaubert. 2000. CO_2 partial pressure controls the calcification rate of a coral community. *Global Change Biol. 6*: 329–334.

McClintock, J.B., R.A. Angus, M.R. Mcdonald, C.D. Amsler, S.A. Catledge and Y.K. Vohra. 2009. Rapid dissolution of shells of weakly calcified Antarctic benthic macroorganisms indicates high vulnerability to ocean acidification. *Antarct. Sci. 21*: 449–456.

Müller, R., T. Laepple, I. Bartsch and C. Wiencke. 2009. Impact of oceanic warming on the distribution of seaweeds in polar and cold-temperate waters. *Bot. Mar. 52*: 617–638.

Wiencke, C., I. Gómez and K. Dunton. 2009. Phenology and seasonal physiological performance of polar seaweeds. *Bot. Mar. 52*: 585–592.

Woelfel, J., R. Schumann, P. Leopold, C. Wiencke and U. Karsten. 2009. Microphytobenthic biomass along gradients of physical conditions in Arctic Kongsfjorden, Svalbard. *Bot. Mar. 52*: 573–583.

Wulff, A., K. Iken, M.L. Quartino, A. Al-Handal, C. Wiencke and M.N. Clayton. 2009. Biodiversity, biogeography and zonation of marine benthic micro- and macroalgae in the Arctic and Antarctic. *Bot. Mar. 52*: 491–507.

Zacher, K., R. Rautenberger, D. Hanelt, A. Wulff and C. Wiencke. 2009. The abiotic environment of polar benthic algae. *Bot. Mar. 52*: 483–490.

Acknowledgements

Chapter 4:

We are indebted to S.-M. Lin, Paul W. Gabrielson and D. Wilson Freshwater for unpublished sequences cited here and to Fran Hommersand for her help in curating the Antarctic collections at NCU. M.H. Hommersand thanks NSF–PEET award DEB-0328491 for support. Recent collections in the Amsler personal herbarium used in this study were made possible by National Science Foundation awards OPP-9814538, OPP-0125181, and OPP-0442769 to C.D. Amsler and J.B. McClintock. We thank Craig Aumack, Maggie Amsler, and Katrin Iken who contributed significantly to these collections. Descriptions were translated into Latin by Mark Garland. Special thanks go to Susan Whitfield for her help in preparing the polar projection map in Figure 4.6.

Chapter 5:

We are very grateful to our past Antarctic field teams and to the employees and subcontractors of Raytheon Polar Services Company and Antarctic Support Associates for performing or assisting with much of the research reviewed in this chapter. We also thank M. Amsler, C. Aumack, P. Bucolo, C. Wiencke, and two anonymous reviewers for constructive comments on earlier versions of the text and thank the National Science Foundation for supporting much of the work under awards OPP-9526610, OPP-9530735, OPP-9814538, OPP-9901076, OPP-0125152, OPP-0125181, OPP-0442769, and/or OPP-0442857.

Chapter 6:

We would like to express our special thanks to J. Hubbard, K. Peters, B. Furrow, A. Mahon and C. Petrie for their valuable assistance during the field work. We are grateful to the employees and subcontractors of Raytheon Polar Services Company and Antarctic Support Associates for logistical support while in Antarctica. Two anonymous reviewers provided constructive comments that improved the chapter. This work was conducted under National Science Foundation awards to CDA and JBM (OPP-9814538) and to BJB (OPP-9901076).

Chapter 7:

We sincerely acknowledge Olga Kimmins for help translating the Russian literature that was made available to us. The work was supported by the Danish Natural Science Research Council, National Environmental Research Council (UK – NE/F012691/1 & NE/F020406/1), The commission for scientific Reserach in Greenland KVUG (GCRC6507) and the Deutsche Forschungsgemeinschaft for financial support (UK899/12-1/2/3). JW and UK also thank the crew at the AWIPEV-base at Ny-Ålesund (Svalbard), and the German dive team (under supervision of Max Schwanitz) for assistance in the field. We thank Dr Phil Hwang for the calculation on periods of open-water in the coastal Arctic and two anonymous reviewers for constructive criticism improving the chapter.

Chapter 8:

This work has been performed at the Ny-Ålesund International Arctic Environmental Research and Monitoring Facility and under the agreement on scientific cooperation between the Alfred Wegener Institute for Polar and Marine Research (AWI), Germany and the University of Rostock, Germany. The authors thank the crew at the AWIPEV-base at Ny-Ålesund and especially the German dive team (supervision Max Schwanitz) for assistance in the field, collecting the samples and further support in 2006. We thank Solvig Görs for intensive field assistance, sample preparation and measurements in the lab. Special thanks to Juliane Buß (University of Rostock, Germany) for determination of POC as well as to Anita Flohr and Nadine Dolata for cell counting. We are also grateful to Birgit Sadkowiak (Leibniz Institute of Baltic Sea Research, Rostock, Germany) for the silicate measurements. The sediment tubes were made by Peter Kumm in the Institute of Chemistry workshop at the University of Rostock, thank you very much. Thanks to Ruth Müller and Anna Fricke (AWI) for providing PAR data, as well as to Florian Peine (University of Rostock) for creating the maps. We gratefully acknowledge financial support by the German Research Council (DFG, KA899/12-1/2/3) and the AWI.

Chapter 9:

Many data described in this chapter would not have been produced without the generous financial support of the Deutsche Forschungsgemeinschaft (to CW), the German Academic Exchange Service (to IG), and the National Science Foundation (to KHD). This help is greatly acknowledged here. Finally, CW thanks the AWI Center for Scientific Diving, especially Max Schwanitz, for collecting seaweeds and for support during underwater works. The Instituto Antártico Chileno kindly supported two expeditions to Base Frei, King George Island. Parts of the study were performed at the International Arctic Environmental Research and Monitoring Facility in Ny-Ålesund, Svalbard and at the Dallmann Laboratory, annex to Jubany Station on King George Island, under the agreement on scientific cooperation between the Alfred Wegener Institute and the Dirección Nacional del Antártico, Buenos Aires.

Chapter 10:

This chapter was written in the framework of grants from the following institutions: German Research Foundation (DFG), German Academic Exchange Service (DAAD), the EU-ARCFAC program at Ny-Ålesund, the Chilean Fund for Science and Technology (FONDECYT 1080171- 1060503), Intituto Antartico Chileno (INACH T-20-09), the US National Science Foundation and the Minerals Management Service Alaska Environmental Studies Program.

Chapter 11:

Funding by the Deutsche Forschungsgemeinschaft (Bi 772/7-1,2,3) is gratefully acknowledged. We are also grateful to Max Schwanitz and the AWI diving team for logistic support and sample supply. Part of the study has been performed at the Ny Ålesund International Arctic Environmental Research and Monitoring Facility, and part has been performed upon agreement of cooperation between the Dirección Nacional del Antárctico and the Alfred Wegener Institute.

Chapter 12:

We are extremely grateful to colleagues W. Adey, A.R.O. Chapman, K. Gunnarsson, P.M. Pedersen, R. Rauja, and J. Rico, who all provided unpublished data about distribution of seaweeds in the Atlantic Sector, to J. Bartsch for final touches to the text and to two anonymous reviewers and A.R.O. Chapman who made valuable suggestions for improving the text. We also thank M.Y. Roleda and F.S. Steinhoff for providing unpublished data about temperature demands of *Laminaria solidungula* meiospores and A. Preisler for unpublished information about sporogenesis of *Saccharina latissima*. We acknowledge the modelling group WCRP's Working Group on Coupled Modelling (WGCM) for making available the WCRP CMIP3 multi-model dataset. Support of this dataset is provided by the Office of Science, U.S. Department of Energy. The present study was partially prepared at the Biodiversity and Climate Research Centre (BiK-F), Frankfurt a.M. and supported by the research funding programme "LOEWE – Landes-Offensive zur Entwicklung Wissenschaftlich-Ökonomischer Exzellenz" of the Hessian Ministry of Higher Education, Research, and the Arts.

Chapter 13:

The authors thank the German Research Foundation (DFG), the Bremen International Graduate School for Marine Sciences (GLOMAR) and the Helmholtz Association for financial support through several projects over the past decades, as well as the Ny-Ålesund International Research and Monitoring Facility for their support, and the AWI diving team for providing all types of field samples.

Chapter 14:

A significant part of the work presented here was performed at Dallmann Laboratory, annex to Jubany Station, within the frame of the scientific collaboration existing between Instituto Antártico Argentino (IAA)/Dirección Nacional del Antártico and Alfred Wegener Institute for Polar and Marine Research. Another part of the work was performed at the Ny-Ålesund International Research and Monitoring facility, Spitsbergen (Svalbard). Grants from the Argentinean National Council of Scientific and Technological Research (CONICET) and IAA to GC, Swedish Institute (SI) to GC and KZ, and the German Research Foundation (DFG) to KZ are gratefully acknowledged.

Index

Acanthococcus antarcticus 86
Acrosiphonia 197, 283
– *arcta* 28, 29, 31, 35, 36, 184, 200
– *sonderi* 28, 29
– spp. 27, 37, 199
Actinocyclus curvatulus 24
Adelaide Island 32
Adenocystis utricularis 28, 31, 36, 122–128, 182, 184, 186, 199, 205
Aeodes 89
Agarum clathratum 28, 30, 206, 261
Ahnfeltia plicata 27, 28, 30, 59
Ahnfeltiopsis 89
Alaria esculenta 27, 28, 30, 104, 198, 200, 205, 206, 260, 286
Alaria sp. 30
Amphidinium glabrum
amphiequatorial 36
Amphora
– cf. *coffeaeformis* 24
– *coffeaeformis* 24
– *laevis* var. *laevissima* 24
– sp. 199
anchor ice 14, 37, 302, 303
Antarctic Bottom Water 90, 91
Antarctic Circumpolar Current (ACC) 9, 35, 89, 90, 91, 92
Antarctic Ice Shelf 257
Antarctic Peninsula 10–12, 15, 16, 30, 31, 40, 53–93, 102, 121, 181, 182, 208, 209, 238, 241, 250–252, 255, 257, 259, 282, 302, 307, 309, 323
Antarctic Polar Front (Antarctic Convergence) 23, 67, 92, 93, 219
Antarcticophyllum
– *aequabile* 57
– *subantarcticum* 58
Antarcticothamnion polysporum 65
Antarctocolax lambii 73
antifouling 2, 102, 112, 123, 127, 132, 133
antifreeze proteins 195, 207
antimicrobial 112, 121, 123, 128, 133
Antithamnion antarcticum 64
Anvers Island 30, 55-87, 122, 239, 250–252, 309
Arctic Ocean 9, 16, 34, 56, 58, 104, 141, 155–157, 182, 211, 238, 257, 307
artificial substrata 312
Ascophyllum nodosum 27, 28, 286
ascorbate 279
Ascoseira mirabilis 28, 31, 108, 122–129, 183, 186, 188, 190, 196–199, 204, 278
attenuation coefficient 13, 15, 16, 155, 169, 172, 174, 200, 283
Auckland Islands 53, 92
Austropugetia crassa 75–77, 80, 82
avoidance 106, 282

Baffin Bay 29, 238, 239, 241, 256
Balleny Islands 53
Ballia 56, 59, 89
– *callitricha* 28, 31, 36, 56, 57, 59, 89, 204
– *sertularioides* 57
Bangia atropurpurea 54, 273, 274
Bangiadulcis atropurpurea 28, 29
barren ground 232, 263, 305
base pair distances 62, 79, 58, 86, 88, 89, 91–93
Bear Island 27
Beaufort Sea 30, 37, 38, 174, 183, 185, 186, 189, 200, 208, 210
Benguela Current 92
Benguela upwelling system 92
Berkeleya rutilans 24
biofilm 163, 272
biofouling 2, 101, 107, 112, 113, 321
biogeographic province 53, 89, 92
Biremis ambigua 24
Blidingia minima 258, 309
Bowiesia pulchra 60
Brittany 60, 243, 250, 257, 259, 262
bromophenolic compounds 280
Byssus purpurea 55

^{14}C approach 144, 145
Callithamnion
– *membranaceum* 56
– *sertularioides* 57
– *simile* 64
Callophyllis 81
– *atrosanguinea* 81, 82
– *cristata* 200
– *tenera* 81
– *variegata* 82
Callymenia antarctica 62
Campbell Island 92

Campbell Plateau 91
Canadian Arctic 10, 30, 35, 242, 243
canopy 15, 27, 40, 260, 262, 272, 283, 303–304, 307
Cape Hallett 32
Cape Horn 64, 65, 70, 86, 252
catalase 279
β-carboxylation 187
Cenozoic 89
Ceramium involutum 64
Chaetoceros dichaeta 278
Chaetomorpha 30
– *linum* 28, 30
– *melagonium* 27, 28, 30
Chlamydomonas
– *nivalis* 277
– *reinhardtii* 277
chloroplast 201, 203, 273–277, 281
Chondrus crispus 246, 249, 254, 257, 258
Chorda
– *filum* 300
– sp. 30
Chordaria
– *flagelliformis* 27–29, 37, 200
– *linearis* 246, 251, 252, 255, 258, 259, 309
Choreocolax delesseriae 72
chronic photoinhibition 280
Chukchi Sea 30, 174, 210
circannual rhythm 11, 184, 189, 206
circum-equatorial ocean circulation 87
Cladophora
– *repens* 124, 130
– *rupestris* 27, 28
Clathromorphum
– *lemoineanum* 58
– *obtectulum* 28, 32, 57
climate model simulation 238
climax 262, 312
Cocconeis 40, 282, 306
– *costata* var. *antarctica* 301
– *pinnata* 301
– *scutellum* 24
– sp. 306
Coccotylus truncatus 28, 30 287
Codium fragile ssp. *fragile* 263
Colaconema membranaceum 56
cold shock proteins 195, 207
cold-water history 9, 34, 35, 207, 232
competition 157, 263, 305, 306, 313, 323
Conferva
– *atropurpurea* 54
– *carnea* 54
Corallina 197, 200, 263
cretaceous 87, 89
Crozet 53, 88, 92, 251
crustal blocks 89
Curdiea racovitzae 28, 31, 62, 108, 109
cyclobutane dimers (CPDs) 277, 278, 287
Cyclotella sp. 276
Cylindrotheca closterium 24, 175, 202, 276
Cystoclonium obtusangulum 86
Cystoseira 258
– *baccata* 258
– sp. 258
Cystosphaera jacquinotii 28, 31, 32, 35, 103, 107, 122–124, 126–133

D1 protein 221, 224, 230, 231, 278, 280
D1/D2 reaction centre 279, 280
Dasyptilon pellucidum 65
Davis Station 34
de novo synthesis of proteins 230, 279, 280
decomposition processes 209
deep-water passage 90
Delesseria 89
– *antarctica* 70
– *carnosa* 69
– *lancifolia* 15, 70, 309
– *ligulata* 72
– *quercifolia* 71
– *salicifolia* 55, 70, 183
– *stephanocarpa* 70
Delisea pulchra 60, 103
depth distribution 12, 31, 39, 122, 184, 197, 200, 243, 244, 260, 278, 310
Desmarestia 31, 301
– *aculeata* 27–29, 200
– *anceps* 14, 15, 28, 31, 32, 39, 103–107, 110–112, 122, 126-132, 182, 196, 199, 200, 208–210, 250, 286
– *antarctica* 28, 31, 32, 103, 122–124, 128–130, 182, 196, 198, 208, 246, 250, 255, 257
– *chordalis* 28, 32
– *confervoides* 36
– *menziesii* 28, 31, 32, 103–107, 110–112, 122–125, 128–132, 182, 186, 200, 205, 208–210, 286
– *muelleri* 244, 252, 256
– sp. 30, 32, 111, 251
– spp. 37, 55, 111, 210
– *viridis* 27, 28, 36, 104, 106, 189, 305
Devaleraea ramentacea 27–30, 35, 204, 244, 246, 253, 256, 259, 284, 285, 287

Dictyosiphon foeniculaceus 28, 30
diffusive boundary layer (DBL) 146, 150
Dilsea integra 28, 30, 35
dimethylsulphoniopropionate (DMSP) 103, 207
Disko Island 29, 242
distributional change 237, 242, 244, 257, 259-261
DNA damage 4, 39, 273, 277–279, 281–283, 287
DNA photoproducts 278
DNA repair 278, 279
Drake Passage 88, 90
Dumontia
– *contorta* 27, 28, 196, 199
– *incrassata* 27
Durvillaea antarctica 92, 93
dynamic photoinhibition 230, 231, 280

East Gondwana 87, 90, 91
ecological models 311
ecotype 244–248, 251, 252, 259, 287
Ectocarpus siliculosus 300
eddy correlation 149, 157
El Niño Southern Oscillation (ENSO) 261
Elachista fucicola 27, 28
Emiliania huxleyi 277
endemism 2, 10, 16, 30, 34–36, 92
endophyte 40, 104, 105, 305
energy flux 209
Enteromorpha bulbosa 28, 31, 36, 37, 184, 199, 287
Entomoneis
– *gigantea* 24
– – var. *septentrionalis* 24
kjellmanii 24
– *paludosa* 24
– – var. *borealis* 24
– – var. *hyperborea* 24
environmental factors 2, 9, 196, 211, 212, 287, 302, 322
environmental stress hypothesis 311
Eocene 90, 91
Erythrotrichia carnea 54
eurythermal 206, 222
excision repair 278
extracellular polymeric substances 163, 207, 272

Falklandiella harveyi 65
Falklands 239, 251, 252, 257
Faroes 244
filamentous algae 104, 105
food web 40, 122, 142, 164, 272, 288
Fragilaria 33, 211
– sp. 206
– *striatula* 24, 40, 282 301
Franz Joseph Land 243
freezing 3, 4, 11, 12, 207, 221–223, 227–232, 242, 260, 322
Fucus 221, 272
– *cartilagineus* 61
– *cordatus* 85
– *distichus* 4, 11–12, 27–30, 39, 200, 221–224, 227–232, 280
– *evanescens* 27–29, 232
– *gardneri* 286
– *plicatus* 59
– *serratus* 231
– *spiralis* 232
– spp. 37
– *vesiculosus* 27, 28, 133

Gaina mollis 73, 79, 83, 93
Galapagos 63, 260
gametogenesis 182, 202, 205, 206, 243, 247–249, 252, 260
Gammarellus homari 40, 106, 109
gastropod 40, 104, 106, 131
Geminocarpus geminatus 28, 36, 182
genetic buffering hypothesis 278
geographic distribution 4, 34, 35, 182, 183, 189, 206, 211, 237–263, 299, 322
Georgiella confluens 28, 31, 55, 64, 65, 204
germination 4, 201, 202, 206, 211, 243, 246, 247, 273, 281, 285–287, 302
Gigartina 55, 85, 197
– *circumcincta* 85
– *orbicularis*62
– *skottsbergii* 28, 85, 108, 184, 197, 278
– spp. 196
Glaphyrymenia pustulosa 79
global change 15, 237, 307, 310, 313, 322
global warming 4, 16, 189, 212, 238, 241, 242, 251, 261–263, 287, 308, 322, 323
glutathione 279, 280
Gondogeneia antarctica 40, 105–111, 124–127, 131, 308
Gracilaria
– *nigrescens* 86
– *obtusangula* 86
– *prolifera* 63

grazing 39, 40, 107, 121, 122, 130–132, 176, 288, 299–301, 305–312
Greenland 10, 11, 14, 16, 29, 33, 38, 142, 155, 164, 198, 199, 209–211, 238, 239, 243–246, 249, 256, 258
Griffithsia antarctica 65
gross photosynthesis 143, 146, 149, 150, 281
gross primary production 144–151, 155
Gymnogongrus 89
– *antarcticus* 84, 183, 204
– *turqueti* 28, 31, 60, 64, 84, 183
– *turquetii* 84
Gyrosigma 33
– *fasciola* 25, 281
– *subsalinum* var. *antarcticum* 203
– *tenuissimum* 25
– – var. *hyperborea* 25

Halopteris obovata 105
Halosiphon tomentosus 205
Halymenia variegata 82
hard bottom 243, 250, 272, 282, 288, 299, 304, 306–308, 310
Heard Island 251
Helgoland 202, 237, 244, 246, 287
herbivory 103, 107–113, 321
Hiatella arctica 40
high light 10, 39, 203, 221, 230–232, 278, 303
Hildenbrandia 55
– *lecannellieri* 55
– sp. 32
Himantothallus grandifolius 14, 28, 103, 122, 182, 200, 250
Hormosira banksii 283
Hornsund 27
Hudson Bay 238, 243
Hyas araneus 16
Hydrolapathum stephanocarpum 70
Hydrolithon subantarcticum 58, 59
Hymenena venosa 73
Hymenocladiopsis 63
– *crustigena* 63, 183
– *prolifera* 63
– sp. 63

ice disturbance 310, 313, 323
ice foot 12, 302
ice scour 12, 14, 30, 38, 92, 221, 238, 244, 249, 251, 271, 299, 302, 303, 310, 311
induced defense model (IDM) 111
interactive effects 4, 224, 231, 286–289, 308, 313, 322
intermediate disturbance hypothesis (IDH) 311
intertidal algae 11, 27, 30, 305
invasion 4, 16, 258, 261, 308
Iridaea 89
– *cordata* 28, 85, 103, 184, 206, 255, 302
– *laminarioides* 106
– *mawsonii* 86
– *papillosa* 85

Kallymenia antarctica 62
Kerguelen 53, 57, 60-65, 82, 87, 250
King George Island 10–14, 30, 40, 53, 104, 200, 203, 206, 208, 209, 222, 237, 239, 252, 271, 306, 309
Kola Peninsula 153, 256
Kongsfjorden 3, 10–14, 27, 33, 163–177, 201, 202, 206, 207, 210, 211, 222, 223, 242
Kvaleya epilaeve 261

Labrador Sea 240
Laevilacunaria antarctica 40, 104, 106, 111, 131
Laingia hookeri 69
Lambia antarctica 28, 35, 37
Laminaria 183
– *digitata* 27, 183, 200, 205, 258, 286
– *farlowii* 260
– *hyperborea* 27, 183, 202, 245, 257, 286, 308
– *ochroleuca* 258
– *saccharina* 27,277
– *solidungula* 11, 27, 183, 196, 208, 242, 282
– sp. 208
Leniea lubrica 75–78
Leptophytum
– *coulmanicum* 28, 32, 58
– *foecundum* 197, 199
– *laeve* 58, 59, 261
Lessonia vadosa 246, 252, 253, 256
lethal boundary 245, 249
lethal limit 243, 244, 249, 257, 259
Licmophora gracilis 25
life cycle (life history) 36, 182–184, 195, 201, 202, 205, 206, 211, 231, 238, 242, 243, 249, 259, 263, 272, 273, 276, 280
light climate 11, 17, 174, 176

light compensation point 38, 151, 185, 197, 200, 208
light transmittance 10, 13
light-independent carbon fixation 187, 322
limpets 40, 58, 78, 103, 305
Lithophyllum
– *aequabile* 57, 58
– *decipiens* f. *subantarcticum* 59
– *discoideum* f. *aequabile* 57
– *discoideum* f. *aequabilis* 57
– *discoideum* f. *wandelica* 57
– *discoideum* f. *wandelicum* 57
– *subantarcticum* 58
Lithothamnion
– *antarcticum* 59
– *coulmanicum*58
– *foecundum* 58
– *granuliferum* 58
– *kerguelenum* f. *obtectula* 57
– *kerguelenum* f. *obtectulum* 57
– *manginii* 59
– sp. 31
– *tophiforme* 261
Lola 35
long distance dispersal 2, 87, 89, 90
low light 10–13, 38, 39, 110, 176, 184–187, 195, 200–203, 208, 211, 222, 223, 231, 281, 284, 303, 304
Lützow-Holm Bay 32

Macquarie Island 55, 60, 63, 82, 92, 250
macroalgal assemblage 30–32, 36–37, 40, 272, 288, 309
Macrocystis 208, 260
– *integrifolia* 286
– *pyrifera* 260, 262, 277
macrograzer 130, 131, 272
macroherbivores 102, 103, 107, 108, 113
Marie Byrd Land 88, 89
marine transgression 90, 91
Marion Island 53
Mazzaella laminarioides 106
McDonald Island 53
McMurdo Sound 32, 53, 59, 85, 86, 103, 112, 127, 128, 174, 198, 199, 210, 302, 303, 305
mean base pair distance 91
Mehler reaction 221, 279
Meiodiscus concrescens 56, 57, 93
Melobesia
– *patena* 59
– *verrucata* var. *Antarctica* 59
meltwater 10–13, 38, 181, 304, 310, 322
membrane fluidity 195, 221, 230
Membranipora membranacea 263
Membranoptera alata 66
mesograzer 40, 107, 111, 130, 131, 305
mesoherbivore 2, 102, 104, 107, 111, 113
mesozoic 9, 35, 87
metabolic carbon balance 186, 200, 208, 209
Micrasterias denticulata 277
microalgal assemblage 196, 202, 210, 282, 304, 308, 312
microcontinent 90
microelectrode 146, 149
microherbivore 102
microplate movements 88
Microrhinus carnosus 69
microscopic stage 3, 36, 201, 205, 273
microtubule 277
migration 88, 151, 175, 237, 258, 259, 272, 282, 308
minimum light requirement 154, 197, 202
mitochondria 273, 274, 276
mollusc 176, 237, 305
Monostroma 300
– *endiviaefolium* 54
– *endiviifolium* 54
– *hariotii* 28, 31, 37, 199, 205, 301
Murman coast 27
Mya truncata 40
mycosporine like amino acids (MAAs) 4, 273, 284, 285, 308
Myriogramme 71, 72, 196
– *livida* 71, 72
– *manginii* 59, 71, 198, 210, 244, 251, 255, 257
mangini 71
– *smithii* 72, 73

Nacella concinna 40
Navicula
– *cancellata* 25
– cf. *hanseni* 282, 301
– cf. *perminuta* 40, 282, 301
– *directa* 25
– – var. *directa* 25
– *glaciei* 25, 207
– *perminuta* 40, 282, 301
– *spicula* 25
– *transitans* 25
– – var. *derasa* 25
– – var. *transitans* 25

Neodilsea integra 35
Neotethys Ocean 87
Nereocystis luetkeana 201, 286
Nereoginkgo 78, 81
– *adiantifolia* 74, 78
– *populifolia* 74–77
net photosynthesis 146, 151, 157
net primary production 145, 149, 151, 152, 208
Neuroglossum 72
– *binderianum* 72
– *delesseriae* 72
– *ligulatum* 28, 31, 72
Newfoundland 182, 239, 242, 257, 266
Nitophyllum 71, 196
– *lividum* 71
– *mangini* 71
– *manginii* 59, 71, 198, 251
– *smithii* 72
Nitzschia
– *angularis* 25
– *aurariae* 202
– cf. *aurariae* 175
– *closterium* 276
– *cylindrus* 25
– *distans* 25
– – var. *erratica* 25
– *hybrida* 25
non-indigenous cold-temperate species 309
Notophycus 35
– *fimbriatus* 87
Nototanais antarcticus 40
Notothenia coriiceps 40, 102, 106, 121, 124, 130
Nova Scotia 243, 257, 263, 288, 311
Novaya Zemlya 153, 243, 257
nucleolus 276
nutrient concentration 12, 38, 164, 169, 170, 176, 181, 187, 304, 305, 323

ocean acidification 323
oceanic warming 237, 241, 261, 263
Odontaster validus 102, 106, 107, 124–126, 131
Odontella aurita 26
– – var. *aurita* 26
– – var. *obtusa* 26
Odonthalia dentata 27, 28, 30, 273, 274, 276
Oligocene 90, 91
optimal defense theory 110
optimum quantum yield (F_v/F_m) 202, 223–229, 281
oxidative stress 4, 224, 230, 231, 276, 279

Pachymenia 62
– *orbicularis* 28, 36, 62, 84, 108
– *saxicola* 63
– sp. 62
pack-ice 245
pairwise base distances 88, 90
Palmaria 56, 196, 224
– *callophylloides* 57, 93
– *decipiens* 4, 28, 56, 105, 124, 183, 197, 221, 250, 273, 306
– *mollis* 93
– *palmata* 27, 28, 273, 274, 287
Pantoneura 70, 86
– *baerii* 29, 35
– *fabriciana* 66
– *plocamioides* 29, 31, 66, 69, 86
Paraglossum 70, 89
Paralia sulcata 26
Patagonia 252, 257
pCO_2 concentration 90
perennial 4, 12, 27, 31, 37, 80, 182, 183, 187, 189, 242, 244, 245, 249, 250–252, 260, 262, 311
periphyton 24, 26
Petalonia fascia 258, 309
Petrocelis polygyna 29, 35
Peyssonnelia harveyana 60
Phaeurus antarcticus 29, 31, 112, 122–130, 182, 183
phenolic compounds 280, 283, 285
phlorotannin content 111, 122, 286
phlorotannin level 122, 131, 286
photodamage 280
photoperiod 181, 237, 242, 243, 250, 259
photoprotective mechanisms 12, 205, 280, 282
photoprotective substances 283
photoreactivation 278
photosynthesis/respiration ratio 204
photosynthetic capacity (P_{max}) 149, 151, 185, 186, 190, 200, 204
photosynthetic efficiency (α) 38, 185, 195, 196, 200, 211, 276, 280, 308
photosynthetic pigments 11, 187, 231, 279
photosynthetically active radiation (PAR) 10, 271
phototaxis 201
phycobilisome 273, 274, 284

Phycodrys 89
– *antarctica* 70, 71
– *austrogeorgica* 15, 71, 273, 274
– *quercifolia* 71
– *rubens* 27, 29, 30, 200, 206, 287
Phyllophora
– *ahnfeltioides* 59, 84, 183, 204
– *antarctica* 29, 32, 85, 103, 113, 198, 305
– *appendiculata* 84
– *coccocarpa* 63
– *cuneifolia* 63
phylogenetic 2, 35, 88, 211
Phymatolithon
– *foecundum* 29, 58, 197, 199
– *lenormandii* 58
– spp. 38
– *tenue* 29
physodes 276, 278, 286
Picconiella plumosa 29, 31, 73
Pinnularia quadratarea 25
Plagiotropis longa 26
Plakarthrium punctatissimum 131
plastoglobuli 275, 276
plate tectonics 2, 87, 89
Platysiphon verticillatus 29, 35
Pleurosigma obscurum 281
Pliocene 90
Plocamium 55, 61, 89
– *cartilagineum* 29–32, 61, 104 ,106, 204
– *coccineum* var. *secundatum* 61
– *hookeri* 61
– *secundatum* 61
Plumariopsis
– *eatonii* 65
– *peninsularis* 55, 65
polar day 201, 276
polar night 10, 13, 181, 303
polychaete 40, 176
Polycoelia laciniata 79
Polysiphonia
– *abscissa* 73
– *stricta* 29
Porphyra 285
– *endiviifolia* 29, 31, 35, 37, 54, 199, 205
– *endiviifolium* 54
– *plocamiestris* 55, 84
Powell Basin 88
Prasiola crispa 29, 31, 32, 39, 276, 280
predator 101, 104, 106, 130, 272, 305
Prince Edward Island 53
productivity of polar seaweeds 208
Prostebbingia gracilis 105, 111
Pseudogomphonema kamtschaticum 26
Pseudolithophyllum aequabile 57
pseudoperennial 12, 183, 189, 244, 250, 251
Pseudophycodrys
– *phyllophora* 69
– *pulcherrima* 69
Pteronia plumosa 73
Pterothamnion
– *antarcticum* 64
– *simile* 64
Pterygophora californica 260
Ptilonia magellanica 60
Ptilota
– *confluens* 64
– *eatonii* 65
– *gunneri* 27, 29, 104, 106, 200
– *harveyi* 65
– *serrata* 261
Pugetia
– *delicatissima* 81
– *fragilissima* 81
Punctaria glacialis 29, 30, 35
Pylaiella
– *littoralis* 27, 29, 37, 182, 300
– *varia* 300

Q_{10} value 204

rbcL sequence 2, 60–72, 79, 81, 88
reactive oxygen species (ROS) 221, 230, 231, 279
recolonization 92, 305, 307, 312, 313
refuge population 242, 260
reproduction boundary 245, 249
reproduction limit 243, 244, 259
resource allocation model 109
retreating glaciers 5, 310, 313, 323
Rhabdonema arcuatum 26
Rhodochorton
– *concrescens* 56
– *purpureum* 55, 93
Rhodokrambe laingioides 66–68
Rhodomela
– *confervoides* 29, 30
– *tenuissima* 29, 30
Rhodophyllis 89
Rhodymenia 56
– *coccocarpa* 63, 258, 309
– *cuneifolia* 63
– *decipiens*

– *montagneana* 61
– *subantarctica* 63, 309
– *variegata* var. *atrosanguinea* 81
Ross Island 250, 256
Ross Sea 14, 32, 37, 90, 103, 251, 257
rubisco 207, 278, 280

Saccharina latissima 27, 29, 183, 196, 244, 254-259, 276, 286
Saccorhiza dermatodea 27, 29, 200, 243, 256, 277, 286
Sarcodia montagneana 61, 62
Sarcothalia
– *circumcincta* 85
– *papillosa* 85
Scagelia americana 29, 30
Scotia Sea 90
Scott Island 53
sea ice break-up 13, 181, 183, 188, 303, 310
sea ice cover 10, 13, 35, 155, 157, 181, 256, 302, 304, 309, 310, 313, 322, 323
sea surface temperature (SST) 237, 253–256, 261
sea urchin grazing 305
sea urchins 40, 102, 262, 305, 311
seafloor spreading 87
season anticipator 3, 182–184, 189, 231
season responder 182, 184, 187, 189
seasonal temperature range 239, 240
seasonality 2, 9, 170, 181, 182, 184, 188, 200, 208
Secchi depth 168, 169, 176
secondary succession 310
sediment/water interface 272
sedimentation 5, 14, 164, 201, 204, 304, 310, 313, 322, 323
Severnaya Zemlya 243
shade adaptation 187, 195, 197, 201, 202
short-day conditions 182, 245, 259
Silicularia rosea 40
soft bottom 3, 14, 33, 37, 203, 211, 255, 288, 302, 308
South Atlantic Ocean 87, 90
South Georgia 53–94, 239, 240, 250–252, 258
South Orkney Islands 10, 54, 57, 70, 78, 81, 87, 174
South Shetland Islands 10, 53–88, 222, 238, 252, 302
South Tasman Rise 91
Spain 239, 245, 257, 259
Spathulospora 57
species pair 91
species richness 27, 31, 32
Sphacelaria
arctica 29, 30
– *callitricha* 57
– *plumose* 29, 30, 200
Spitsbergen 9–13, 163, 165, 181, 182, 198, 199, 206, 207, 211, 222, 223, 242–246, 249, 256–258, 273, 276, 277, 304, 307, 308
Spongomorpha spp. 27
sporogenesis 243–249, 259, 327
standing crop 208
Stefansson Sound 30, 243, 300
stenothermal 206, 222, 232, 309, 313
Sterechinus neumayeri 103, 305
Stictyosiphon tortilis 29, 30
stratospheric ozone depletion 4, 111, 271, 323
Strongylocentrotus droebachiensis 40, 104, 106, 109
subduction 260, 261
substratum characteristics 14
superoxide dismutase 279
supralittoral 11, 31, 280
Svalbard 23, 27, 29, 33, 35, 104, 106, 142, 153, 163–177, 210, 221, 232, 208, 309
Synarthrophyton patena 57, 59
Synedropsis 211

Tabularia
– *investiens* 26
– *tabulata* 26
Tasmanian gateway 91
temperate floras 90
temperature ecotype 244, 245, 252, 259
Terra Nova Bay 32
Tethyan Seaway 87
Thalassiosira nordenskioeldii 276
Thamnophora magellanica 60
thermocline 261
thylakoids 273–276
tidal range 11
Tierra del Fuego 36, 55, 58, 85, 251, 252, 257
tocopherol 279
Tonicella marmorea 40
Trachyneis
– *aspera* 26, 199, 202
– – var. *aspera* 26
transition region 257

trans-West Antarctic passageways 90
Trematocarpus
– *acicularis* 62
– *antarcticus* 29, 31, 61, 89, 183, 204
Triceratium arcticum 26
trumpet cells 187
trumpet hyphae 188
Turnerella pennyi 29, 35, 261

Ulothrix 12, 301
– *flacca* 29
– *implexa* 300
– spp. 27, 31
ultrastructural changes 272, 276
ultraviolet radiation (UVR) 107, 271, 299
Ulva
– *clathrata* 287
– *intestinalis* 16, 258, 288, 309
Ulvaria obscura 199
underwater irradiance 304, 309
Urococcus foslieanus 29
Urospora
– *penicilliformis* 27, 31, 37, 200, 275–280, 283, 301
– sp. 300

Varimenia macropustulosa 63, 75–77, 82, 108
vertical migration 308
vertical zonation 12, 27, 272, 281, 299
Vestfold Hills 32, 55, 65, 73, 84
vicariance events 87
Victoria Land 32, 85, 250

water transparency 10, 13, 169, 176, 200, 202
Weddell Sea 88, 90, 91, 251, 256
West Spitsbergen Current 9
West Wind Drift 53, 89, 92, 93
whole core incubation 145

Yoldiella 176

Zelandia 91

Printed and bound by CPI Group (UK) Ltd, Croydon, CR0 4YY

Printed and bound by CPI Group (UK) Ltd, Croydon, CR0 4YY
13/09/2026
14960854-0003